COSMETICS ADDITIVES

COSMETICS ADDITIVES

An Industrial Guide

by

Ernest W. Flick

np

NOYES PUBLICATIONS

Park Ridge, New Jersey, U.S.A.

Library of Congress Catalog Card Number: 90-46775
ISBN: 0-8155-1255-4 :
Printed in the United States

Published in the United States of America by
Noyes Publications
Mill Road, Park Ridge, New Jersey 07656

10 9 8 7 6 5 4 3 2 1

Library of Congress Cataloging-in-Publication Data

Flick, Ernest W.
Cosmetics additives : an industrial guide / by Ernest W. Flick.
p. cm.
Includes index.
ISBN 0-8155-1255-4 :
1. Cosmetics--Additives. I. Title
TP983.F56 1990
668'.55--dc20 90-46775
CIP

To Another Flick Generation

Alain Amsellen (1948–1983) and Patricia A. Horny

and

Ann Elise and Eric Taylor

and

Gary Taylor and Pamela T. Davis

and

Paul, Ceri, Donald, Julie Taylor

and

Allyn Rey

Preface

This book describes approximately 4,000 cosmetics additives currently available for industrial use. It has been compiled from information received from 84 manufacturers and distributors of these materials.

The cosmetics and personal care products industry has a $60 billion market worldwide, and future growth is expected both for "prestige" products and the mass market. In the U.S. the market breaks down roughly into the following major segments: hair care—20%, fragrances—20%, skin care—14%, makeup—17%, deodorants and skin and body lotions—10%, and oral care products—10%. Areas of expected growth include products targeted toward ethnic markets and working women.

Applications for cosmetics additives cover a wide product range. For example: hairsprays, sunscreens, skin cleansers, depilatories, lip gloss, mouthwash, massage and body oils, nail treatments, shaving products, baby products, eye care products, bath products, moisturizers, acne products, dandruff treatments, toothpaste, astringents, to name but a few.

Generic types of raw materials which are used as cosmetics additives include: emulsifiers, emollients, preservatives, binders, stabilizers, wetting agents, dispersants, foaming agents, pearlizers, gelling/stiffening agents, surfactants, and viscosity builders. Examples of chemicals which may be employed as additives are: polyalkylene glycols, alkyl ethoxy sulfates, fatty acid esters, fatty alcohols, sulfosuccinates, tertiary amine derivatives, alkyl phenols, glycerides, acrylate copolymers, polysorbates, lanolin derivatives, celluloses, salt, vitamins, "natural products," "botanicals," "exotic oils," mineral oils, vegetable oils and aloe vera.

The data included represent selections from manufacturers' descriptions, made at no cost to, nor influence from, the makers or distributors of the materials.

Only the most recent information has been included. It is believed that all of the products listed here are currently available, which will be of interest to readers concerned with product discontinuances.

The book lists the following product information, as available, in the manufacturer's own words:

(1) Company name and product category,

(2) Trade name and product number,

(3) Product description including properties and applications, as presented by the supplier.

Products are presented by company, and the companies are listed alphabetically. The table of contents is organized in such a way as to serve as a subject index to the book. Also included is a Trade Name Index, for easy and rapid location of products by the reader. In addition, another section, which will be useful, contains the Suppliers' Addresses. It can be found immediately following the Product Information section.

My fullest appreciation is expressed to the companies and organizations which supplied the data included in the book.

September 1990

Ernest W. Flick

NOTICE

Contents and Subject Index

PRODUCT INFORMATION 1

Aceto Corp. 2

Aroma/Flavor/Food Chemicals 2

Akzo Chemie America 3

ELFACOS ST9 and ELFACOS ST37 3

ELFACOS C26 5

ELFACOS E200 6

ELFACOS GT282 7

Akzo International Salt Co. 8

Sodium Chloride for Household and Personal Products 8

Albright & Wilson 9

Product Index 9

Lauryl Sulfates 9

EMPICOL

Lauryl Ethoxy Sulfates 9

Formulated Concentrates 10

Fatty Acid Monoalkylolamides 10

EMPILAN

Fatty Acid Dialkylolamides 10

Fatty Acid Esters 11

Tertiary Amine Derivatives 11

EMPIGEN

Imidazoline Amphoterics 12

Fatty Alcohols 12

LAUREX

Fatty Alcohol Ethoxylates 13

Mono Alkyl-Sulfo-Succinates 13

Miscellaneous 14

EMPIWAX, EMPICRYL, BRIPHOS, EMPIPHOS

Alcolac 15
ALCONATE Sulfosuccinates and Sulfosuccinamates 15
CYCLOCHEM Fatty Acid Esters, Fatty Alcohols, Nonionic Wax Blends. 17
Pearlizing Agents. 17
CYCLOSHEEN
Emulsifiers, Viscosity Builders 17
Emollients, Viscosity Builders 18
Lubricants, Polishes 18
CYCLOMIDE Alkanolamides. 19
1:1 Monoethanolamides, 1:1 Diethanolamides Superamides, 2:1 Diethanolamides, 1:1 Monoisopropanolamides, Amidoamines
CYCLOMOX Amine Oxides 22
CYCLORYL Formulated Shampoo and Detergent Concentrates. 23
CYCLOTERIC Amphoterics 25
Amido-Betaines, Betaines, Glycinate, Imidazoline, Propionates, Sultaine
CYCLOTON Cationic Surfactants. 27
DERMALCARE Skin Care Products 29
Cosmetic Esters, Specialty Emollient Blends and Emulsifiers, Formulated Skin Cleansers
SIPEX Alkyl and Alkyl Ether Sulfates 31
SIPON Lauryl Sulfates and Lauryl Ether Sulfates 33
SIPONATE Alkyl Aryl Sulfonates and Alpha Olefin Sulfonates 36
SIPONIC Alkoxylated Nonionic Surfactants. 38
Cetyl/Stearyl Alcohol Ethoxylates, Octylphenol Ethoxylates, Alcohol Ethoxylates, Tertiary Thio Ethoxylates, Tridecyl Alcohol Ethoxylates, Linear Alcohol Ethoxylates, Nonylphenol Ethoxylates
SIPOTHIX Specialty Polymers. 43
Allied Signal Inc. 44
A–C Polyethylenes, A–C Copolymers, and ACUMIST Micronized Polyethylene Powders. 44
Homopolymers, Oxidized Homopolymer, High Density Oxidized Homopolymers, Copolymers, Micronized Polyethylene Waxes ACUMIST, POLYMIST
Amerchol Corp. 52
Lanolin Derivatives 52
ACETULAN, AMERCHOL, AMERLATE, MODULAN, OHLAN, POLYLAN, SOLULAN
Chemical Specialties. 56
AMEROXOL, AMERSCREEN, GLUCAM, GLUCAMATE, GLUCATE
Aqualon Co. 58
AQUALON Cellulose Gum 58
KLUCEL Hydroxypropylcellulose 59
NATROSOL Plus CS, Grade 330 60
NATROSOL 250 Hydroxyethylcellulose 61

Avatar Corp. 62
CRYOSEPT I and CRYOSEPT II Preservatives 62
BASF Corp. 63
Products for the Cosmetics Industry 63
Polymers 63
LUVISKOL, ULTRAHOLD, LUVISET
Quaternary Compounds 64
LUVIQUAT
Emulsifiers. 64
CREMOPHOR, PLURONIC
Fat and Oil Components. 64
LUVITOL
Solubilizers 65
EO/PO Block Copolymers. 65
Sunscreen Agents 66
UVINUL
Active Agents 67
Thickeners 67
KATIORAN, LUTROL
Polyethylene Glycols 67
Vitamins 68
Solvents. 69
PALATINOL
Neutralizing Agents 69
QUADROL
Various 69
TRILON
Colorants–Dyes and Pigments 70
SICOMET
Bio-Botanica, Inc. 76
Aloe Vera Gel 76
Bath Complexes 77
BBC MINERAL COMPLEX, BBC MOISTURE TROL, BBC RELAXING COMPLEX
Botanicals in Cosmetics 78
Shampoo Herbal Complexes 94
BIO-CHELATED NEUTRAL HENNA PLUS, BIO-CHELATED NEUTRAL HENNA EXTRACT, BIO-CHELATED RED HENNA EXTRACT, BIO-DANDRA PLEX
Hair Complexes. 96
BIO-CHELATED NUTRA PLANT
Herbal Skin Care Complexes 97
BIO-CHELATED DERMA-PLEX, BIO-CHELATED SAUNA DERM
Brooks Industries Inc. 98
Active Ingredients 98
Essential Fatty Acids 98
EFA-PLEX
Tissue Respiratory Factors–Live Yeast Cell Derivative 98

BIODYNES, DERMASOME
"Bioengineered" Protein Mineral Complexes 98
BIOMIN CINQUE, BIOMIN ACQUACINQUE
"Bioengineered" Protein Mineral Complexes 99
Super-Oxide Dismutase-Protein Bound 99
BIOCELL
Substantive Humectants 99
QUAMECTANT
Nucleic Acid Extract–Containing RNA and DNA 99
BIOPLEX
Cosmetic Preservatives 99
MIKROKILL
Emulsifiers 100
BROOKSWAX, RELAXER CONCENTRATE
Phosphate Esters 100
BROPHOS
Lanolin and Lanolin Derivatives 101
Pure Lanolins and Lanolin Oil 101
IVARLAN
Lanolin Absorption Bases 101
IVARBASE
Lanolin Alcohol Derivatives 101
Lanolin Ethoxylates 102
Proteins, Amino Acids and Derivatives 103
Collagen Derived 103
HYDROCOLL, SOLU-COLL, QUAT-COLL, ETHA-COLL, OLEO-COLL, FOAM-COLL, COLLA-MOIST, SOLU-COLL, COLLAMINO
Elastin Derived 105
SOLU-LASTIN
Keratin Derived 105
HYDROKERATIN, ETHA-KERATIN, OLEO-KERATIN, KERAMINO
Silk Derived 106
FIBRO-SILK, SOLUSILK, ETHA-SILK
Soyabean Derived 107
SOLU-SOY, FOAM-SOY, ETHA-SOY, OLEO-SOY
Vegetable Derived 107
SOLU-VEG
Single Cell Organism Derived 107
HYDROCELL
Whole Milk Derived Proteins 108
HYDROMILK, MILKAMINO
Serum Proteins 108
SERUMPRO
Marine Proteins 108
SOLU-MAR
Glycoproteins and Mucoproteins (Mucins) 108
GLYPROSOL, MUSOL

Whole Dermal Tissue Extract 109
BIOPOL

Capital City Products Co. 110
Emulsifiers and Surfactants 110
CAPMUL Mono- and Diglycerides, CAPMUL Sorbitan Esters, CAPMUL Ethoxylated Esters, CAPROL Polyglycerol Esters, ACCOMEEN Ethoxylated Fatty Amines, ACCOMID Surfactants, ACCONON Surfactants
Specialty Fats—Fractionated Non-Lauric Hard Butter 120
ARKOPOL
Specialty Fats—Standard Lauric Hard Butter 121
Specialty Oils and Bases 124
CAPTEX Oils 124
HYDROKOTE Specialty Bases 125
STEROTEX Powdered Lubricants 128
Vegetable Oil Products 129
Vegetable Oils, Specialty Oils, Flaked Stearines, Lauric Oils
CAPITAL Fatty Acids 131

CasChem Inc. 134
Emollient Esters 134
NATURECHEM Non-Comedogenic Fatty Acid Esters 134
Emollients, Emulsifiers
WICKENOL and WAXENOL Emollient Esters 136
Emollients, Emollients Blended Esters, Solid Emollients
Specialty Products 142
Emulsifiers—SURFACTOL Non-Ionic Surfactants 143
SOLARSCREEN O Sunscreen 143
Emulsifiers—SURFACTOL Cationic Surfactants 144
Waxes—CASTORWAX Hydrogenated Castor Oils 144
Cosmetic Grade Castor Oils 145

Costec, Inc. 146
COJOBA Jojoba Oil 146
COLLAGEN Proteins for Skin Care 147
SOLLAGEN
Skin Lipids 147
LIPITEIN, ELASTEIN
COSEPT Parabens 148
COSEPT Preservatives 149
NUOSEPT
COVERA 154
COVERA—Dry 155
Proteins for Hair Care 156
PEPTEIN

Croda Inc. 157
Absorption Bases 157
CREMBA
Alkanolamides (INCROMIDES) 158
Amine Amides (INCROMINES) 162

Amine Oxides (INCROMINE OXIDES) . . . 164
Amino Acids . . . 168
AMINO GLUTEN, CROSILK, CROTEIN
Betaines–Amphoterics (INCRONAMS) . . . 170
Cationic Polymers (CRODACELS) . . . 172
Cationic Salts (INCROMATES) . . . 173
Emollient Esters (CRODAMOLS) . . . 176
CRILL, CRODALAN
Ethoxylated (Nonionic) Emulsifiers . . . 179
CETOMACROGOL, PROCETYL, VOLPO
Fatty Acids . . . 182
CRODACID, SYNCROWAX, SKLIRO
Fatty Alcohols (CRODACOLS) . . . 183
NOVOL
Humectants (INCROMECTANTS) . . . 185
Lanolin . . . 186
SUPER CORONA, SUPERFINE
Lanolin Derivatives . . . 187
ACYLAN, CRODALAN, FLUILAN, HARTOLAN, SUPER HARTOLAN, LANEXOL, LANPOLAMIDE, POLYCHOL, SOLAN, SUPER SOLAN, SUPER SOLANGEL, SUPER STEROL
Nonionics (INCROCAS and INCROPOL) . . . 191
Pearling Agents (CRODAPEARLS) . . . 193
Phosphate Ester Emulsifiers (CRODAFOS) . . . 194
Propoxylates Couplers and Emollients . . . 196
PROCETYL, PROMYRISTYL, PROSTEARYL, PROVOL
Proteins . . . 197
CLEARCOL, COLLASOL, CRODYNE, CROLASTIN, CROMOIST, CRONECTIN, CROSILK, CROTEIN, ELASTOSOL, HYDROSOY, KERASOL, RETICUSOL
Protein Derivatives . . . 202
AMINOFOAM, CROQUAT
Quaternary Cosmetic Conditioners (INCROQUATS) . . . 205
Self-Emulsifying Waxes . . . 209
COSMOWAX, POLAWAX
Sugar Esters (CRODESTAS) . . . 211
Sulfosuccinates (INCROSULS) . . . 213
SUPER REFINED Oils . . . 215
CROSSENTIAL
Synthetic Waxes (SYNCROWAXES) . . . 219
Cyprus Industrial Minerals . . . 221
CYPRUS Talc Products for Cosmetic Applications . . . 221
SUPREME, ALTALC, ACT, BRILLIANTE, SUPRAFINO, ULTRAFINO, TOPNOTE, VERTAL, REGAL, CASTLE, AURA, SUPRA
Desert King Corp. . . . 223
FOAMATION 50 Natural Foaming Agent . . . 223
ESTRAN-Series Cosmetic Grade Jojoba Oils . . . 224

Dow Chemical U.S.A. 226
DOWICIL 200 Preservative 226
Dow Polyglycols 227
Polyethylene Glycols E-Series, Methoxypolyethylene Glycols MPEG, Polypropylene Glycols L-Series, Polypropylene Glycols P-Series, Polyglycol Copolymers
METHOCEL Cellulose Ethers 232
Dow Corning Corp. 233
Silicones 233
Amodimethicone (and) Tallowtrimononium Chloride (and) Nonoxynol-10. 233
Cyclomethicones 233
Cyclomethicone (and) Dimethicone Copolyol 233
Cyclomethicone (and) Dimethiconol 234
Dimethicone 234
Dimethicone (and) Trimethylsiloxysilicate 234
Dimethicone Copolyol 235
Phenyl Trimethicone 235
Methicone 235
Stearoxytrimethylsilane (and) Stearyl Alcohol 235
Trimethylsilylamodimethicone 235
Trimethylsilylamodimethicone (and) Octoxynol-40 (and) Isolaureth-6 (and) Propylene Glycol 235
DOW CORNING
Emulsions and Compounds 243
Amino-Functional Silicones, Antifoam Compounds
Dragoco, Inc. 244
DRAGOCO EXTRAPONES 244
DRAGOSANTOL 246
DuPont 247
Chemicals and Pigments 247
ALBONE 35, 50, 70 CG Hydrogen Peroxide, AVITEX Surface Active Agents/Softeners/Static Control Agents, DUPONOL Surface Active Agents, OXONE Monopersulfate Compound
Eastman Chemical Products, Inc. 250
EASTMAN Distilled Monoglycerides 250
MYVEROL Distilled Monoglycerides 251
MYVEROL Distilled Propylene Glycol Monoester 252
MYVEROL Succinylated Monoglyceride 253
MYVAPLEX Concentrated Glyceryl Monostearates 253
MYVACET Distilled Acetylated Monoglycerides 254
MYVATEX Emulsifiers 255
MIGHTY SOFT, DO CONTROL, TEXTURE LITE, LIQUID LITE, MONOSET
MYVATEM Dispersing Agents 258
Products for Cosmetics and Personal Care 259
Alcohols 259
Glycols, Oxo Alcohols

Antioxidants 262
TENOX
Chemical Intermediates 263
Emulsifiers 264
Concentrated Glyceryl Monostearate, Dispersing Agents, Distilled Acetylated Monoglycerides, Distilled Monoglycerides, Distilled Propylene Glycol Monoester, Emulsifier Blends
Succinylated Monoglycerides 266
Plasticizers 266
KODAFLEX
Polyethylene Waxes 267
EPOLENE
Polymers 267
Water-Dispersible Polymers 267
Solvents 268
Esters, Glycol Ethers
EKTASOLVE
Glycol Ether Esters, Ketones
Specialty Chemicals 270
TECQUINOL
Vitamin E (Natural Source) 271
ECC America Inc. 272
BENTOLITE Bentonites 272
CLAYTONE XL Bentonite 273
FLOWTONE Organic Rheological Additives 274
GELWHITE Montmorillonite 275
MINERAL COLLOID Montmorillonite 276
Finetex, Inc. 277
FINQUAT CT Highly Cationic Conditioner 277
FINSOLV TN/FINSOLV SB Ester Emollients 278
SURFINE Carboxylated Surfactants 279
TAURANOL Sodium Cocoyl Isethionate 280
TAURANOL Sulfoalkyl Amido Surfactants 281
Taurate/Carboxylate Gels 282
Florida Food Products, Inc. 283
Aloe Vera Gel 283
ALOE-CON
Freeman Industries, Inc. 286
Products for the Cosmetics Industry 286
Products of United Guardian Inc. 286
AQUATHIX, ALCOLITE, DERMAFORM, KARAJEL, LUBRAJEL, LUBRASLIDE, LUBRAQUAT CONC., POLYJEL HV, SHEERSKIN, SKIN LITE, SUPERTI, ULTRATI, UNITWIX
Products of Paninkret Chemicals & Pharmaceuticals GmbH and Dr. F.K. Marcus GmbH 288
Natural Colors, Pectin and Derivatives, Proteins and Derivatives

ZEIN
Proteins
KERAMOIS, SILKPRO,
Exotic Plant Oils, Rice Derivatives, Sodium Bromate, Vitamins, Custom Designed Potencies, Other Specialties
SILKALL
H.B. Fuller . . . 290
Specialty Waxes . . . 290
LUNACERA, LUNACERIN, COSMERIN
GAF Chemicals Corp. . . . 297
Copolymer Series . . . 297
GAFFIX VC-713 Cationic Fixative Polymer . . . 298
POLECTRON 430 Opacifier . . . 298
GAFQUAT Series Quaternary Copolymers . . . 299
GAFQUAT HS-100 Resin . . . 300
GANTREZ SP-215 Resin . . . 300
GANEX Alkylated Polyvinylpyrrolidone . . . 301
GANEX Series Copolymers . . . 302
GANTREZ ES Anionic Copolymers . . . 303
PVP/VA Series Copolymers . . . 304
GE Silicones . . . 305
AF-72 Silicone Antifoam Emulsion . . . 305
AF-75 Silicone Antifoam Emulsion . . . 305
AF-9020 Silicone Antifoam Emulsion . . . 306
SF-18(350)/SF-96/VISCASIL Fluids . . . 307
SF-96 (5) Silicone Fluid . . . 308
SE-30 Silicone Gum . . . 308
SF-1188 Silicone Fluid . . . 308
SF-1173/SF-1202/SF-1204 Silicone Fluids . . . 309
SF-1214 Silicone Fluid . . . 310
SF-1228 Silicone Fluid . . . 310
SF-1236 Silicone Fluid . . . 311
SS-4230 Silicone Fluid . . . 311
SS-4267 Silicone Fluid . . . 311
SM-2059 Silicone Emulsion . . . 312
SM-2101 Silicone Emulsion . . . 312
Givaudan . . . 313
PARSOL MCX . . . 313
Goldschmidt Chemical Corp. . . . 314
ABIL Silicones . . . 314
ABIL QUAT
ABIL Silicone Oils . . . 315
Methyl Silicone Oils, Phenylmethyl Silicone Oils, Volatile Silicone Oils
ABIL Silicone Surfactants . . . 316
Polyether Polysiloxane Copolymers, W/O Emulsifiers
ABIL Special Products . . . 317
Cosmetic Oils/Waxes, Quaternary Compounds, Silicone Betaine

ABIL Waxes 318
Additional Products 319
LACTIL, TEGO, TEGAMINE, ANTIL
Betaines/Amine Oxides 320
TEGAMINE
Esters and Emulsifiers 321
TEGIN, TEGACID, PROTEGIN
O/W-Emulsifiers for Acid, Alkali or Salt Stable Emulsions 322
TEGO-CARE, TEGINACID, EMULGATOR
Special Products 324
Aminoxide 324
AMINOXID
Conditioning Agent 324
Pearlizing Agents 324
Thickening Agent 325
ANTIL
Cosmetic Oils/Wax Esters 325
PUR-OBA
Humectants 325
LACTIL
Surfactants: Amido Alkyl Betaines 326
TAGAT Polyoxyethylene Glycerol Fatty Acid Esters 327
TEGIN Edible Acid Esters of Monodiglycerides 330
1,2-Propylene Glycol Ester, Citric Acid Ester
TEGIN Glycerol Fatty Acid Partial Esters 331
Palmitic Stearic Acid Esters, Oleic Acid Ester, Isostearic Acid Ester
TEGIN Glycol Fatty Acid Esters 333
Ethylene Glycol Esters
W/O Emulsifiers 334
PROTEGIN
B.F. Goodrich 335
CARBOPOL Water Soluble Resins 335
W.R. Grace & Co. 337
SYLOID Silicas 337
SYLOX Silicas 338
Haarman & Reimer 339
Perfumery Components 339
Hexcel Chemical Products 348
Quaternary Ammonium Salts 348
SUMQUAT, ANTI-STAT, CETATS, BRETOL, BROMAT
Hoechst Celanese Corp. 351
Cosmetics Specialties 351
Detergent Bases 351
Alkyl Ether Sulfates
GENAPOL
Amide Ether Sulfates, Secondary Alkane Sulfonate
HOSTAPUR

Isethionate
HOSTAPON
Sarcosinates
MEDIALAN
Amine Oxide
GENAMINOX
Methyl Taurides
Cosmetic Thickeners 353
Polymers
HOSTACERIN
Cellulose Ethers
TYLOSE
Emulsifiers for Creams and Lotions. 354
Oil-In-Water Emulsifiers
HOSTAPHAT, GENAMIN
Water-In-Oil Emulsifiers
HOE S 2721
Quaternaries, Pearlescing Agents
OCTOPIROX Antidandruff Agent 356
Products for Cosmetics. 357
Surfactants. 366
TYLOSE Water-Soluble Cellulose Ethers 370
Hydroxyethylcellulose, Methylhydroxyethylcellulose, Sodium Carboxymethylcellulose
Geo. A. Hormel & Co. 375
Collagen Family of Proteins. 375
PEPTEIN CAA Collagen Amino Acids. 376
PEPTEIN 2000. 377
Huls 378
Raw Materials for Cosmetic Products 378
Anionic Surfactants. 378
Fatty Alcohol Ether Sulfates
MARLINAT, SERDET
Fatty Alcohol Sulfates, Sulfonates
MARLON, MARLOPON
Fatty Alcohol Carboxylates, Sulfosuccinates
SERVOXYL
Phosphoric Acid Esters
MARLOPHOR
Ampholytes 379
SERVO AMFOLYT
Nonionic Surfactants 380
Fatty Alcohol Polyglycol Ethers
MARLIPAL
Fatty Alcohol Polyalkylene Glycol Ethers
MARLOX
Superfatting Agents and Thickeners 380
Fatty Acid Polyglycol Esters
MARLOSOL

Superfatting Agents and Foam Stabilizers 381
Fatty Acid Alkanolamides
MARLAMID
Fatty Acid Monoethanolamide Polyglycol Ethers
DIONIL
Silk Luster Agents/Pearlescent Agents 381
Silk Luster Components, Pearlescent Components
Emulsifiers. 382
MARLOWET
Polyethylene Glycols 383
Solvents. 383
Propellants 383
DRIVOSOL
Fragrances, Fragrance Precursors 384
AROVA
Humko Chemical Division. 385
HYSTRENE Behenic and Arachidic Acids 385
HYSTRENE Tallow Type Acids. 386
HYSTRENE/INDUSTRENE Stearic and Palmitic Acids 387
HYSTRENE Lauric and Myristic Acids 389
KEMESTER: Specialty Esters 391
KEMESTER: Homosalate. 394
KEMSTRENE: Glycerine 395
Quaternary Ammonium Chlorides 397
KEMAMINE
Tertiary Amines 399
ICI Specialty Chemicals 401
ATLAS Sorbitol Products 401
SORBO
Mannitol USP. 403
Nonionic Surfactants for Cosmetics and Pharmaceuticals 404
ARLACEL, ARLAMOL, ARLATONE, BRIJ, MYRJ, SPAN, TWEEN
Inolex Chemical Co.. 413
Biocides. 413
BRONOPOL, MYACIDE, LEXGARD
Concentrates and Blends. 414
LEXATE
Emollients 415
LEXOL
Emulsifiers. 418
LEXEMUL, LEXQUAT
Proteins. 421
Hydrolysates 421
LEXEIN, POLYPEPTIDE
Specialties 422
Condensates. 422
MAYPON, SUPERPRO

Specialties 425
Surfactants. 426
LEXAINE, LEXAMINE
Jojoba Growers & Processors Inc. 428
Jojoba Products 428
HYDROBA, JOJOBEADS, TRANSJOJOBA, SONORA, JOJOBUTTER
Lipo Chemicals Inc. 431
LIPO Glyceryl, Ethylene Glycol, and Propylene Glycol Esters 431
LIPOCOL Polyoxyethylene Ethers 432
LIPONATE Emollient Esters. 435
LIPONIC Humectants 437
LIPAMIDE
LIPOLAN: Lanolin Derivatives 437
LIPOPEG Polyoxyethylene Glycol Esters. 438
LIPOSORB Polyoxyethylene Sorbitan Esters 440
LIPOSORB Sorbitan Esters 441
LIPOVOL MOS Series 442
LIPOVOL Natural Oils 443
LIPOWAX Emulsifying Waxes 445
LIPO Natural Abrasives 445
Lipo Miscellaneous Products 446
LIPOBEE, ORGASOL, UNICIDE, UNIPERTAN, UNIPHEN, SAFESTER, UNIPABOL, UNITRIENOL, UVATONE
Lonza Inc. 447
AMPHOTERGE K/AMPHOTERGE K-2 447
Dr. Madis Laboratories, Inc. 448
Cosmetic Ingredients 448
VERAGEL Aloe Vera Gel. 449
VERAGEL Lipoid: Aloe Vera Leaf Extract 451
McIntyre Group Ltd. 453
MACKAM/Amphoterics 453
MACKAM/Betaines 455
MACKAMIDE/Alkanolamides 458
MACKAMIDE/Modified 461
MACKANATE/Sulfosuccinates 463
MACKANATE/Monoesters, MACKANATE/Diesters
MACKADET/Blends 467
MACKINE/Amidoamines 468
MACKALENE/Cationic Conditioners 470
MACKAMINE/Amine Oxides 471
MACKERNIUM/Quaternary Ammonium Compounds 473
MACKESTER/Organic Esters 474
MACKPRO/Quaternized Proteins 475
MACKSTAT DM/DMDM Hydantoin 476
MACKOL/Fatty Alcohols 476
MACKPEARL/Liquid: Pearling Agents 476
Mearl Corp. 477

BIJU TX and BIJU ULTRA TX Concentrates 477
Frost (Pearlescent) Qualities 477
MEARLMAID, FLAMENCO VELVET
Metallic and Iridescent Qualities 478
FLAMENCO, CLOISONNÉ
CLOISONNÉ Color Pigments 479
NU-ANTIQUE
DUOCHROME Iridescent Colors 481
FLAMENCO Pearl and Color Pigments 483
GEMTONE Color Pigments 484
MEARLMAID Natural Pearl Essence 486
Standard MEARLMAID Pastes for Lotions and Makeups, Standard MEARLMAID Pastes for Nail Enamels
TIMICA Luster Pigments 488
White Pearlescents 488
SILKWHITE, PEARLWHITE, SPARKLE
Metallic and Earth Colors 489
VELVET Color Pigments 490
M. Michel and Co., Inc. 491
CACHALOT Fatty Alcohols 491
MICHEL XO-150 Series Guerbet-Reaction Branched-Chain Alcohols 492
Miranol Inc. 493
MIRANOL Amphoteric Anionic Salts 493
MIRANOL Amphoteric Surfactants (Carboxylates) 496
MIRANOL Amphoteric Surfactants (Sulfonates) 503
MIRAPOL Cationic Polymers 504
MIRATAINE Aminopropionates 506
MIRATAINE Betaines and Sultaines 507
Miscellaneous Products 510
MIRANATE, CEDAPAL, CEDEPON
Mona Industries, Inc. 513
MONAMATES 513
MONAMINES and MONAMIDS 516
1:2 FA-Diethanolamides, 1:1 FA-Diethanolamides, 1:1 FA-Monoethanolamides, 1:1 FA-Monoisopropanolamides
MONAQUAT 531
MONATERIC 535
MONATERIC Surfactants 537
Group I: Imidazoline Derived, Group II: Betaines, Group III: Blends, Group IV: Proprietary Compounds
MONAWET Surfactants 545
Miscellaneous 549
AVAMID
Napp Chemicals Inc. 550
Pharmaceutical and Cosmetic Products 550
Preservatives, Generics, Sulfonamides, Steroids, Hormones, Intermediates/Miscellaneous

Penreco 552
Petrolatums 552
White Petrolatum USP 552
PENRECO ULTIMA, PENRECO SUPER, PENRECO SNOW, PENRECO REGENT, PENRECO LILY, PENRECO CREAM
Petrolatum (USP) 553
PENRECO ROYAL, PENRECO BLOND, PENRECO AMBER
Technical Petrolatum 553
PENRECO RED
Ointment Based White Petrolatum, USP 554
Mineral Jelly 555
White Mineral Oils 556
Mineral Oil USP 556
DRAKEOL
Light Mineral Oil NF 557
Technical Mineral Oil 559
PENETECK, PAROL
Petrolite Specialty Polymers Group 561
Microcrystalline Waxes and Synthetic Polymers 561
VICTORY, BE SQUARE, SILTEX, VYBAR, UNILIN, UNITHOX, PETRONAUBA
Pfizer Chemical Division 563
Pfizer Citric Acid 563
Pfizer Fragrance and Flavor Modifiers in Pharmaceutical and Cosmetic Applications 564
VELTOL
Pfizer Products for the Pharmaceutical and Cosmetic Industries 565
Acidulants, Antibiotics, Buffering Agents, Mineral Products, Preservatives, Vitamins, Other Pharmaceutical Products
Sorbitol in Pharmaceuticals and Cosmetics 566
Pilot Chemical Co. 567
CALFOAM Alcohol and Ether Sulfates 567
Pokonobe Industries, Inc. 568
Unrefined, Refined, and USP/NF Specialty Vegetable Oils for the Food, Cosmetic, and Pharmaceutical Industries 568
Polyesther Corp. 569
Product List 569
Fatty Acids, Fractionated Fatty Acids, Specialty Fatty Acids, Fatty Alcohols, Glycerine, Fatty Acid Esters, Miscellaneous Oils, Polysynlane, Pure Silk Powder
PPG/Mazer Chemicals 570
AVANEL S-150: Sodium C12-15 Pareth-15 Sulfonate 570
JORDAPON CI Sodium Cocoyl Isethionate 571
MASIL 556: Phenyl Trimethicone 573
MASILWAX 135 Silicone Wax 574
M-QUAT Hair Conditioners 575
Procter & Gamble 576
Fatty Alcohols 576

Glycerine 579
SUPEROL, STAR

Protameen Chemicals Inc. 580
Amine Condensate 1:2 FA-Diethanolamides–Monoethanolamides . . . 580
PROTAMIDE
Amphoterics 581
PROTACHEM, PROTERIC
Hydrolyzed Proteins 581
PROCHEM
Vitamins 581
Ethoxylated Aliphatic Amines 582
PROTOX, PROTAQUAT
Fatty Esters and Glyceryl Esters 584
PROTACHEM
Lanolin and Lanolin Derivatives 586
PROTOLAN
Polyethylene Glycol Esters 587
PROTAMATE
Polyethylene Sorbitan Esters 590
PROTASORB
Polyoxyethylene Castor Oil Derivatives 591
Polyoxyethylene Ethers 592
PROCOL
Sorbitan Esters 595
Preservatives 595
PROTASTAT
Superamides 1:1 FA-Diethanolamides–Monoethanolamides 596
Alkyl Phenol Ethoxylates 597
Polyoxyethylene Phosphate Esters 597
PROTAPHOS
Methyl Taurate Esters 597
PROTAPON

Quantum Chemical Corp. 598
EMERY CP/USP Glycerines 598
EMERY Short-Chain and Coconut Fatty Acids 599
Even-Carbon Acids, Odd-Carbon Acids, Coconut Fatty Acids
QUANTUM Unsaturated Fatty Acids 604
Monounsaturated Acids 604
EMERSOL
Polyunsaturated Acids 604

Quest International 607
Fragrance Ingredients 607
AGRUDOR, AMBERLYN, AMBER-SEC, AMBRASOL, AMBRETTILYN, ANTHER, APPLINAL, AQUANAL, ARBOROMA, AURANTION, AVALONE, BANGALOL, BEAUVERTATE, BOURGEONAL, CALYXOL, CERVOLIDE, CHANDILYN, CHRYSANTHAL, CISTULATE, CITRATHAL, CITROFURAN, CIVET, CUIRONAL, DECUMAL, DEWBERRY, DUPICAL, EFETAL, ELINTAAL, ELINTAAL FORTE, EMPETAL,

EPITONE, EQUINOL, EVERNIA, FELVINONE, FENYRANE, FIORIVERT, FLORANE, FLOROCYCLENE, FRESCILE, FRUITLISE, FRUTONILE, GARDAMIDE, GARDOCYCLENE, GYRANE, HEPTAVERT, HERBOXANE, INONYL ACETATE and INONYL ACETATE EXTRA, ISOJASMONE, ISOLONGIFOLANONE, JACINTHFLOR, JASILYN, JASMACYCLENE, JASMATONE, JASMIN, JASMOPYRANE, JASMOPYRANE FORTE, JESSATE, LEATHER BASE, LIGANTRAAL, LIGUSTRAL, LIGUVERT, LIME BASE, LIMETTAL, LIXETONE, MACEAL, MANZANATE, MEFRANAL, MELONIS, MEVANTRAAL, MOSS BASE, MOUSSE DE MER, MUSK, NEOBERGAMATE, NEOBERGAMATE FORTE, NEOBERGAMATE SPECIAL, NEOLAVANDATE, ORTHOLATE, PELARGENE, PETIOLE, PIVAROSE, PRENYL ACETATE, RHUBACITRIL, RHUBAFURAN, ROSANIA, SAUTANE, SINOCITRIL, TANGENIL, TETRAHYDROCONVALOL, TOP ROSE, TRASEOLIDE, TUBEROSE, VERDALIA, VERDILYN, VERDINAL, VERDORACINE, VERTELON, VIGOROSE, VIOTRIL, YLANG

Rewo 632
REWOTERIC Amphoteric Surfactants 632
Glycine Derivatives, Salt-Free Amphoteric Tensides, Anionic Mixtures, Betaine, Sulfobetaines

Rheox, Inc. 635
BENTONE Organophilic Clay Rheological Additives 635
BENTONE GEL Rheological Additives 636
Hydrophilic Clay Rheological Additives 637
MACALOID
THIXCIN R Rheological Additive 638

Rhone Poulenc 639
ALIPAL CO HF Series Anionics 639
EMULPHOR EL, ON Series Nonionics 640
GAFAC Series Anionics 641
GAFAMIDE CDD 518 Superamide 642
IGEPAL CO, CA, DM Series Nonionics 643
IGEPON A, T Series Anionics 644

Roche Chemical Division 645
Biotin 645
Citric Acid, USP, FCC 646
Panthenol (Provitamin B5) 647
Vitamin A and Vitamin A with Vitamin D3 Products Available for Topical Formulations 649
AQUAPALM
Vitamin A 650
Vitamin C (as Ascorbyl Palmitate) 651
Vitamin E 652

Rohm and Haas 653
KATHON CG Microbiocide 653

Sandoz Chemicals Corp. . . . 654
CARTARETIN F-4 and F-23 Water Soluble Cationic Polymers . . . 654
SANDOPAN TFL Conc. . . . 654
SANDOPAN Carboxylated Surfactants . . . 655
VELSAN . . . 656
Scher Chemicals, Inc. . . . 657
Amido-Amines . . . 657
SCHERCODINE
Amido-Amine Salts . . . 659
KATEMUL
Amine Oxides . . . 660
SCHERCAMOX
Betaine Amphoterics . . . 661
SCHERCOTAINE
1-1 Fatty Acid Diethanolamides . . . 664
SCHERCOMID
Imidazolinium Amphoterics . . . 667
SCHERCOTERIC
Monoalkanolamides . . . 670
Monohydric Alcohol Di/Tri Esters . . . 672
SCHERCEMOL
Monohydric Alcohol Esters . . . 674
SCHEROBA
Polyhydric Alcohol Esters . . . 680
Sunscreens . . . 681
DIPSAL
Specialty Quats (Mono) . . . 682
SCHERCOQUAT
Specialty Quats (Bis) . . . 686
Sulfosuccinates . . . 687
SCHERCOPOL
Schweizerhall, Inc. . . . 689
Cosmetic Raw Materials . . . 689
Shell Chemical Co. . . . 690
NEODOL Alcohols . . . 690
NEODOL Ethoxylates . . . 692
Sherex Chemical Co. . . . 701
Personal Care Surfactants and Specialty Products . . . 701
Nonionics: Ethoxylated Monodiglycerides . . . 701
VARONIC
Nonionics: Ethoxylated Alcohols . . . 702
AROSURF
Nonionics: Ethoxylated Amines . . . 703
Alkanolamides: Diethanolamides . . . 707
VARAMIDE
Amine Oxides . . . 708
VAROX

Concentrates 709
VARIFOAM
Quaternaries: Monoalkyl Trimethyl Quaternaries. 709
VARISOFT
Quaternaries: Alkyl Dimethyl Benzyl Quaternaries 710
Quaternaries: Dialkyl Dimethyl Quaternaries 711
ADOGEN
Methyl Dialkoxy Alkyl Quaternary. 711
VARIQUAT
Sulfosuccinates 712
VARSULF
Amphoterics: Glycinates. 713
VARION
Amphoterics: Betaines 713
Amphoterics: Sulfobetaines. 714
Fatty Acid Esters 715
STARFOL
Fatty Acids 717
HYDROFOL
Glycerine. 717
Fatty Alcohols 718
ADOL
Sonneborn Division 719
MULTIWAX Microcrystalline Waxes 719
SONNEBORN Petrolatums 721
PROTOPET, ALBA, PERFECTA, FONOLINE, SONO JELL, TECH PET
Stepan Co. 725
Cationics 725
Methosulfate 725
STEPANQUAT
Stearyl Dimethyl Benzyl Ammonium Chloride 725
AMMONYX
Cetyl Trimethyl Ammonium Chloride 726
Oleyl Dimethyl Benzyl Ammonium Chloride 726
Anionics 726
Alkyl Sulfates. 726
STEPANOL
Alcohol Ether Sulfates 728
STEOL
Alpha Olefin Sulfonate. 729
BIO-TERGE
Linear Alkylbenzene Sulfonate 729
BIO-SOFT
Sarcosinate. 729
MAPROSYL
Sulfoacetates 729
LATHANOL

Sulfosuccinates 730
ANIONYX, STEPAN-MILD
Blends 730
Nonionics/Amphoterics 731
Alkanolamides 731
NINOL
Ethoxylated Alkanolamides 732
AMIDOX
Amine Oxides 732
Betaines 733
AMPHOSOL
Polymeric Resins 733
Hair Care Fixatives 733
STEPANHOLD
Esters 734
Emollients 734
NEOBEE
Base 736
WECOBEE
Emulsifiers, Opacifiers 736
Pearlescent Agents, Auxiliary Emulsifiers 738
Emulsifiers, Viscosity Builders 740
DREWMULSE Mono and Diglycerides 741
KESSCO Alcohol Esters 742
KESSCO Glycerol Esters 746
KESSCO Glycol Esters 750
KESSCO Polyethylene Glycol Esters 753
KESSCO Specialty Products 765
Synthetic Waxes 765
KESSCOWAX, KESSCOLIN
NEOBEE Triglycerides and Propylene Glycol Derivatives 768
Coconut Oils 768
WECOBEE 769
Sutton Laboratories, Inc. 770
GERMABEN II Preservative System 770
GERMABEN II-E Preservative 771
GERMALL II Preservative 772
GERMALL 115 Preservative System 773
SUTTOCIDE A Sodium Hydroxymethylglycinate 774
Allantoin 774
Tic Gums, Inc. 775
Classical Gum Arabic (Acacia) TIC PRETESTED 1 FCC Powder 775
TIC PRETESTED Colloid 710H Powder 776
TIC PRETESTED Tragacanth "C" FCC Powder 777
TRI-K Industries, Inc. 778
Product List by C.T.F.A. Nomenclature 778
ALCLOXA, ALDIOXA, ARIANOR, TRI-K E.F.A., FIBRONEX, COLLNECTIN, PLASMANECTIN, PRONECTIN, TRI-DERM,

DETAINE, KELATE, TRISTAT, TRISEPT, TRI-TEIN, TRI-LASTIN, TRI-COL, TRI-QUAT, NEWSULFUR, TRILANE, SUNVEIL, TAKANAL
Union Carbide Chemicals and Plastics Co., Inc. 781
CARBOWAX Polyethylene Glycols 781
R.T. Vanderbilt Co., Inc. 786
Bactericides/Fungicides 786
VANCIDE
Dispersing Agents 787
DARVAN
Anionic Surfactants 787
VANSEAL
Emulsion Stabilizer/Suspending Agent 788
VEEGUM
Xanthan Gum Thickener/Suspending Agent 788
RHODIGEL
Lipoamino Acids and Their Salts 789
LIPACIDE, LIPOPROTEOL
RHODIGEL Xanthan Gum 790
VEEGUM 791
Van Dyk 792
CERAPHYL and CERASYNT 792
CERASYNT and EMULSYNT 805
EMULSYNT and ESCALOL 806
ESCALOL and FOAMOLE 807
FOAMOLE 808
BI-LITE Pearlescent Pigments 809
CHROMA-LITE Pearlescent Pigments 813
LUSTRA-PEARL Pigments 821
PEARL-GLO Pearlescent Pigments 827
SPECTRA-PEARL Pigments 832
Waitaki International Biosciences 837
The Cosmetic Industry 837
Neutral Lipids 843
Witco 844
Surfactants for Cosmetics and Toiletries 844
EMCOL, EMPHOS, WITCAMIDE, WITCOLATE, WITCONATE, WITCONOL
Zschimmer & Schwarz 848
Basic Material for Cosmetic Industry 848
Fatty Alcohol Sulfates 848
SULFETAL
Fatty Alcohol Ether Sulfates 849
ZETESOL
Sulfosuccinic Acid Semi-Esters 851
SETACIN
Fatty Acid Alkylolamides and Superfatting Agents 853
OXYPON, PURTON

Fatty Alcohol Polyglycol Ether 856
OXETAL
Emulsifiers. 857
MULSIFAN
Amphoteric Surfactants 861
AMPHOTENSID
Other Special Products 862

SUPPLIERS' ADDRESSES 867

TRADE NAME INDEX 873

PRODUCT INFORMATION

ACETO CORP.: Aroma/Flavor/Food Chemicals:

New Additions:
Aldehyde C-8 (Octyl Aldehyde)
Aldehyde C-10 (N Decanal)
Aldehyde C-12 (Lauric)
Amyl Cinnamic Aldehyde
Hexyl Cinnamic Aldehyde

Acetophenone PG (99%)
Amyl Acetate FCC
Amyl Butyrate FCC
Amyl Salicylate FCC
Benzaldehyde NF
Benzoic Acid Technical
Benzyl Acetate PQ FCC
Benzyl Butyral Lactate FCC
p-Tertiary Butyl Cyclohexanol
o-Tertiary Butyl Cyclohexyl Acetate
p-Tertiary Butyl Cyclohexyl Acetate
BHA Food Grade
BHT Food Grade
Cinnamaldehyde FCC
Ethyl Aceto Acetate FCC
Ethyl Butyrate FCC
Ethyl Lactate FCC
Ethyl Salicylate FCC
Iso Amyl Acetate FCC
Iso Amyl Alcohol 98% (Natural)
Iso Amyl Butyrate FCC
Iso Amyl Salicylate Extra FCC
Isobutyric Acid FCC
d-Limonene
Melilotin (Benzodihydro Pyrone)
Methyl Salicylate NF FCC
Phenyl Ethyl Acetate FCC
Phenyl Propyl Alcohol FCC
Propionic Acid FCC
Propyl Gallate Food Grade
Sodium Benzoate NF Powder
Tannic Acid Alsok
Triacetin
Trimethyl Cyclohexanol

AKZO CHEMIE AMERICA: ELFACOS ST9 and ELFACOS ST37:

Chemical and Physical Properties:

Trade Name:

ELFACOS ST9:
CTFA adopted name: PEG-45/Dodecyl Glycol Copolymer

ELFACOS ST37:
CTFA adopted name: PEG-22/Dodecyl Glycol Copolymer

Typical data:
ELFACOS ST9:
average molecular weight (GPC): approx. 4,000
solidification point: 25C (paste to solid)
active content: 99% min.
colour Gardner: 5 max.
density: approx. 0,94 (30C)
flash point (Pensky Martens): approx. 170C

ELFACOS ST37:
average molecular weight (GPC): approx. 2,300
solidification point: 10C
active content: 99% min.
colour Gardner: 5 max.
density: approx. 0,95 (25C)
flash point (Pensky Martens): approx. 175C

Appearance:
- liquid of medium viscosity (ELFACOS ST37)
- pasty solid with low melting point (ELFACOS ST9)
- little odour
- low volatility
- chemical resistance to weak acids, bases and oxidizing influences

Solubility:
Due to their chemical structures, the ELFACOS ST products in certain ranges of concentrations and in aqueous media build colloidal dispersed systems with a partly gel-like character. Solubility of the ELFACOS ST products in some of the substances used for cosmetics:

Soluble:
peanut oil
olive oil
isopropylmyristate
2-octyldodecanoate
ethyl alcohol
halogenated hydrocarbons
soya oil
paraffin oil
capryl/caprylic acid triglyceride
vaseline (heat)
ether

Dispersable:
propylene glycol
glycerol
sorbitol

AKZO CHEMIE AMERICA: ELFACOS ST9 and ELFACOS ST37(Continued):

Suggested Use:
- cosmetic emulsions
- decorative cosmetics
- oleophilic bath formulations
- hair products

ELFACOS ST9 and ST37 are used as:
- storage stabilizers for W/O-emulsions
- emollients for emulsions and decorative cosmetics

ELFACOS ST9 and ST37 have remarkable properties due to:
- their high capacity to adsorb water and oil
- their optimal effects with low dosage.

Fields of applications:
- Creams and lotions.
- Decorative cosmetics.

Applicational properties:

Compared with conventional ointments and creams which contain mainly wool grease products as emulsifying agents, the emulsification and storage stability is considerably improved by the use of a suitable combination of primary emulsifiers and ELFACOS ST.

In general the emulsions prepared with ELFACOS ST are very compatible with the usual emulsifiers and additives. Even the effectiveness of cationics is not or only slightly impaired by ELFACOS ST.

The choice of suitable emulsifiers and stabilisers and the ratio of these is critical for the formulation of stable W/O-emulsions. In general the stabilities of emulsions are improved by ELFACOS ST.

ELFACOS ST products assist the emulsifier as a binding agent for oil and water. ELFACOS ST builds a frame work in which both oil and water are lodged and anchored.

ELFACOS ST has an excellent adsorbing capacity for water. Therefore emulsions with a high water content can be prepared with ELFACOS ST. Such emulsions, despite having a high water content do not tend to dry out at the surface.

The liquid form of ELFACOS ST37 does not only ease manufacture but also has a favourable influence on the consistency and spreading properties of the emulsions.

Due to its solid consistency ELFACOS ST9 is especially suitable for the preparation of decorative cosmetics.

AKZO CHEMIE AMERICA: ELFACOS C26:

ELFACOS C26:
CTFA adopted name: Hydroxyoctacosanyl Hydroxystearate

Typical Data:
Acid value: 5-10
Saponification value (1,5 hrs): 75-90
Melting point: approx. 80C
Iodine value: max. 15
Hydroxyl value: approx. 120
pH (5% in IPA/H20:1/1): 5-6
Colour Gardner (ASTM-D-1544-68): max. 7
Bulk Density kg/m3: approx. 450

Appearance:
- pellets
- cream coloured
- little odour
- exceptional stability against hydrolysis
- chemical resistance to oxidizing influences

Solubility:
ELFACOS C26 is compatible with other synthetic and natural oils and waxes.

Soluble when melted in:
peanut oil
soya oil
olive oil
paraffin oil
isopropyl myristate
castor oil
vaseline
candelilla wax
carnauba wax
micro wax
sorbitan esters
halogenated hydrocarbons

ELFACOS C26 is less dispersable in strong polar substances e.g. glycerol, propylene glycol and sorbitol.

Suggested Use:
ELFACOS C26 is mainly used as:
- consistency regulating agent for W/O-emulsions with emulsions with emulsion-stabilising properties.
- waxy substance for decorative cosmetics.

Moreover ELFACOS C26 can be applied in:
- O/W-emulsions to improve the "body" of the emulsion.
- liquid make-up improving the "body".
- pharmaceutical preparations (emulsions and ointments)

ELFACOS C26 has remarkable properties due to its:
- constant composition.
- optimum effect as a consistency regulator in emulsions and decorative cosmetics.

AKZO CHEMIE AMERICA: ELFACOS E200:

ELFACOS E200:
CTFA adopted name: Methoxy PEG-22/Dodecyl Glycol Copolymer

Typical Data:
Average mol. wt. (GPC): approx. 1,800
Solidification point (DAB7): approx. 25C
Active matter: 98% min.
water: 70/30): approx. 6-7
pH (5% in methanol/
Colour Gardner: 5 max.
Density (30C): approx. 0,96 g/cm3
Flashpoint (Pensky Martens): approx. 167C

Appearance:
This non-ionic high molecular product is solid at room temperature. It has to be slightly heated up to 30C and mixed thoroughly before use.
ELFACOS E200 has a weak odour and is stable against weak acids and bases and oxidizing influences.

Solubility:
Soluble:
In various substances, used in cosmetics, sometimes at elevated temperatures:

peanut oil	vaseline
soya oil	carnauba wax
olive oil	microcrystalline wax
castor oil	sorbitan ester
paraffin oil	ethanol

Dispersable:

glycerine	sorbitol
	propylene glycol

Suggested Use:
To emulsify water in oil in all cosmetic and pharmaceutical formulae, ELFACOS E200 does an effective job especially in those cases where a relatively high percentage active matter (like sodium chloride, urea, propylene glycol and salicylic acid) containing emulsions are needed.

Applicational properties:
- nonionic W/O-emulsifier.
- lipophilic solubiliser.
- compatible with all other cosmetic ingredients.
- stable against electrolytes, mild acids and bases.
- good dermatological and toxicological properties.
- in combination with hydrophilic emulsifiers as an O/W complex emulsifier.
- increasing the water binding capacity of ointments.
- for difficult soluble active substances in lipophilic bases.
- for insoluble active substance in lipophilic bases as a wetting agent.

AKZO CHEMIE AMERICA: ELFACOS GT282:

ELFACOS GT282 is a nonionic polyalkylene glycol with an average molecular weight of about 3000.

The C.T.F.A. adopted name is: Talloweth-60 Myristyl Glycol.

Main applicational properties of ELFACOS GT are:

- Effective thickener for aqueous solutions of anionic, nonionic, cationic and ampohoteric surfactants.
- Because of its chemical structure ELFACOS GT 282 is inherently hydrolysis resistant.
- The thickening effect is practically insensitive to the presence of electrolytes or the pH of the system.
- The type of surfactant mainly determines the thickening characteristics. In addition, the viscosity is a function of temperature, pH and presence of electrolytes.

The application for ELFACOS GT 282 in all kinds of shampoo- and bath preparations is extensively described. Further suggested applications are:

- Hydrogen peroxide solutions for hair colouring and bleaching
- Emulsifier for cosmetic creams and lotions of the oil in water type

ELFACOS GT 282 S is a waxy solid.
ELFACOS GT 282 L is a pumpable solution of ELFACOS GT 282 S in water.

ELFACOS GT 282 S:

Average molecular weight: 3000
Active content %: >99
Water %: < 0.1
Melting point C: approx. 48
pH value 5% in MeOH/H20 70:30: 6-7
Appearance: off-white pellets
Flash point (Pensky Martin) C: 168
Specific Gravity @ 25C: 1.070
Bulk density kg/m3: approx. 550

ELFACOS GT 282 L:

Average molecular weight: 3000
Active content %: 15
Water %: 85
pH value: 5-6
Viscosity 20C (Brookfield LTV) mPa-s: approx. 5000
Appearance: clear to slightly opaque solution
Refractive Index n 20D: 1.3540
Specific gravity 25C: 1.018

ELFACOS GT 282 L contains preservative: 0.05% KATHON CG

AKZO INTERNATIONAL SALT CO.: Sodium Chloride for Household and Personal Products:

Alberger Grades

Coarse:
Sodium Chloride (NaCl) %: 99.95
Screening-USS Mesh: 10-50
Cheese:
Sodium Chloride (NaCl) %: 99.95
Screening-USS Mesh: 20-50
Topping:
Sodium Chloride (NaCl) %: 99.95
Screening-USS Mesh: 20-70
Flake:
Sodium Chloride (NaCl) %: 99.95
Screening-USS Mesh: 30-70
Butter:
Sodium Chloride (NaCl) %: 99.95
Screening-USS Mesh: 40-70
Fine Flake:
Sodium Chloride (NaCl) %: 99.95
Screening-USS Mesh: 50-100
Fine Flour:
Sodium Chloride (NaCl) %: 99.95
Screening-USS Mesh: 80-200
Mich Med.:
Sodium Chloride (NaCl) %: 99.95
Screening-USS Mesh: 20-70

Microsized:
66:
Sodium Chloride (NaCl) %: 99.95
Screening-USS Mesh: 200-325
86:
Sodium Chloride (NaCl) %: 99.89
Screening-USS Mesh: 325
95:
Sodium Chloride (NaCl) %: 99.89
Screening-USS Mesh: 325

Granulated:
Compressed Pretzel:
Sodium Chloride (NaCl) %: 99.95
Screening-USS Mesh: 16-30
CMF Pan Run:
Sodium Chloride (NaCl) %: 99.95
Screening-USS Mesh: 20-70
Table:
Sodium Chloride (NaCl) %: 99.89
Screening-USS Mesh: 40-70
Common Pan Run:
Sodium Chloride (NaCl) %: 99.89
Screening-USS Mesh: 20-70

ALBRIGHT & WILSON: Product Index:

Lauryl Sulfates:

Sodium lauryl sulphate
powder
EMPICOL LX, LZ, LZP, 0045, 0185, 0193

needles
EMPICOL LMV/T, LXZ, LZV, LZV/E, LZV/D
LM45, LX28, LXS95, LZ34

Ammonium lauryl sulfate
liquid/paste
EMPICOL AL30/T, AL70

Monoethanolamine lauryl sulphate
liquid
EMPICOL LQ33/T, LQ70

Magnesium lauryl sulphate
liquid
EMPICOL ML26

Triethanolamine lauryl sulphate
liquid
EMPICOL TL40/T

Lauryl Ethoxy Sulphates:

Sodium lauryl ethoxy sulphate
liquid
EMPICOL ESA, ESA70, ESB3, ESB3D, MD, ESB50, ESB70, ESC3, ESC70

Ammonium lauryl ethoxy sulphate
liquid
EMPICOL EAA, EAA70, EAB/T, EAB70, EAC70

Magnesium lauryl ethoxy sulphate
liquid
EMPICOL EGB

Sodium/magnesium lauryl ethoxy sulphate
liquid
EMPICOL BSD

ALBRIGHT & WILSON: Product Index(Continued):

Formulated Concentrates:
Built triethanolamine lauryl sulphate
liquid
EMPICOL TLP/T

Pearly shampoo concentrate
pearly liquid
EMPICOL XC35

Fatty Acid Monoalkylolamides:
Coconut monoethanolamide
flake
EMPILAN CME

Lauric monoethanolamide
flake
EMPILAN LME

Coconut isopropanolamide
flake
EMPILAN CIS

Lauric isopropanolamine
flake
EMPILAN LIS

Fatty Acid Di-Alkylolamides:
Coconut diethanolamide
liquid
EMPILAN CDE, CDE/X, 2502

Coconut diethanolmide 1:2
liquid
EMPILAN CDX

Lauric diethanolamide
waxy solid
EMPILAN LDE

Lauric diethanolamide 1:2
waxy solid
EMPILAN LDX

Oleic diethanolamide 1:2
liquid/paste
EMPILAN 2121

Linoleic diethanolamide
liquid/paste
EMPILAN 2125
Lauric/myristic diethanolamide
waxy solid
EMPILAN 2624

ALBRIGHT & WILSON: Product Index (Continued):

Fatty Acid Dialkylolamides (Continued):

Coconut monoethanolamide ethoxylate
paste
EMPILAN MAA, LP2
liquid
EMPILAN LP10

Fatty Acid Esters:

Glycerol monostearate, self-emulsifying
powder
EMPILAN GMS SE

Glycerol monostearate, non self-emulsifying
powder
EMPILAN GMS NSE

Ethylene glycol monostearate
flake
EMPILAN EGMS

Tertiary Amine Derivatives:

Lauryl dimethyl betaine
liquid
EMPIGEN BB

Coco amido propyl dimethyl betaine
liquid
EMPIGEN BS

Alkyl amido propyl dimethyl betaine
liquid
EMPIGEN BT

Lauryl dimethyl amine oxide
liquid
EMPIGEN OB

Myristyl dimethyl amine oxide
liquid
EMPIGEN OH

Lauryl ethoxy dimethyl amine oxide
liquid
EMPIGEN OY

ALBRIGHT & WILSON: Product Index(Continued):

Tertiary Amine Derivatives(Continued):

Alkyl amido propyl dimethyl amine oxide
liquid
EMPIGEN OS/A

Benzalkonium chloride
liquid
EMPIGEN BAC, BAC90 (paste)

Cetylstearyl trimethyl ammonium methosulfate
liquid
EMPIGEN CM

Alkyl amido propyl trimethyl ammonium chloride
liquid
EMPIGEN CSC

Hydrogenated tallow dimethyl benzyl ammonium chloride
paste
EMPIGEN BCM75

Imidazoline Amphoterics:

Coconut imidazoline amphoteric
liquid
EMPIGEN CDR10, CDR30

Imidazoline amphoteric/sodium lauryl ethoxy sulphate blend
liquid
EMPIGEN XDR121, XDR123

Imidazoline amphoteric/sodium lauryl sulphate blend
liquid
EMPIGEN XDR302

Fatty Alcohols:

Narrow cut lauryl alcohol
liquid/solid
LAUREX NC

Medium cut lauryl alcohol
liquid/solid
LAUREX LI

Broad cut lauryl alcohol
liquid/solid
LAUREX LM

Cetyl stearyl alcohol
flake
LAUREX CS

ALBRIGHT & WILSON: Product Index(Continued):

Fatty alcohol ethoxylates:

Lauryl ethoxylate (2EO)
liquid
EMPILAN KB2

Lauryl ethoxylate (3EO)
liquid
EMPILAN KB3

Lauryl ethoxylate (12EO)
solid
EMPILAN KB12

Synthetic alcohol ethoxylate (5EO)
liquid/paste
EMPILAN KA5

Synthetic alcohol ethoxylate (8EO)
waxy solid
EMPILAN KA8

Synthetic alcohol ethoxylate (10EO)
waxy solid
EMPILAN KA10

Cetyl stearyl ethoxylate (20EO)
flake
EMPILAN KM20

Cetyl stearyl ethoxylate (50EO)
flake
EMPILAN KM50

Mono Alkyl-Sulpho-Succinates:

Disodium lauric monoethanolamide sulphosuccinate
liquid/solid
EMPICOL SCC

Disodium lauryl ethoxy sulphosuccinate
liquid/solid
EMPICOL SDD

Disodium undecylinic monoethanclamide sulphosuccinate
liquid/solid
EMPICOL SEE

ALBRIGHT & WILSON: Product Index(Continued):

Mono Alkyl Sulpho-succinates(Continued):

Disodium cocomonoethanolamide ethoxy sulphosuccinate
liquid/solid
EMPICOL SGG

Disodium alkyl ethoxy sulphosuccinate
liquid/solid
EMPICOL SFF

Disodium lauryl sulphosuccinate
liquid/solid
EMPICOL SLL, SLL/P (powder)

Disodium alkyl sulphosuccinate
liquid/solid
EMPICOL STT

Miscellaneous:

Self emulsifying wax
waxy flake
EMPIWAX SK

Pearling agent
pearly liquid
EMPICOL 0627

Dispersing agent
liquid/paste
EMPICRYL APD/B

Alkyl ethoxy phosphate ester
liquid
EMPICOL 0216
liquid
BRIPHOS 03D

Sodium tripolyphosphate
powder
EMPIPHOS STP

ALCOLAC: ALCONATE Sulfosuccinates and Sulfosuccinamates:

ALCONATE surfactants are a family of specialty sulfosuccinates and sulfosuccinamates that exhibit novel performance properties in a wide range of formulated cosmetic and industrial products. Recognized as functional irritation mollifying agents, ALCONATE surfactants are ideal additives for the preparation of ultra-mild skin care, shampoo, and related personal care products. Select ALCONATE surfactants also function as emulsifiers, foaming agents, dispersants and penetrating agents for numerous industrial and latex applications.

Sulfosuccinates:

ALCONATE 2CH:
Dicyclohexyl Sodium Sulfosuccinate
% Activity: 43
Form: Paste
Emulsifier in emulsion polymerization or post-add in latex compounding.

ALCONATE CPA:
Disodium Cocamido MIPA-Sulfosuccinate
% Activity: 40
Form: Liquid
Effective anti-irritation agent for mild shampoos, liquid handsoaps, bath and shower products.

ALCONATE SBDO:
Dioctyl Sodium Sulfosuccinate
% Activity: 70
Form: Liquid
Outstanding wetting and surface active agent. Dispersant and penetrating agent for industrial and mining applications.

ALCONATE SBF-12:
Disodium Lauryl Sulfosuccinate
% Activity: 40
Form: Paste
Irritation mollifying agent for high foam shampoos, toilet soaps and personal care products.

ALCONATE SBFA-30:
Disodium Monolaureth Sulfosuccinate
% Activity: 30
Form: Liquid
Exceptionally mild, good flash foam. Used in bath gels and baby care products.

ALCOLAC INC.: ALCONATE Sulfosuccinates and Sulfosuccinamates (Continued):

Sulfosuccinates(Continued):

ALCONATE SBG-280:
Disodium Oleamido MEA-Sulfosuccinate
% Activity: 30
Form: Liquid
Excellent thickener and conditioner. Imparts sheen to hair and soft feel to skin.

ALCONATE SBL-203:
Disodium Lauramido MEA-Sulfosuccinate
% Activity: 40
Form: Liquid
Improves flash foam of anionic systems. Produces brittle, tack-free residue for carpet shampoos.

ALCONATE SBN-862:
Disodium Nonoxynol-10 Sulfosuccinate
% Activity: 32
Form: Liquid
Emulsifier and dispersant for emulsion polymerization of vinyl acetate and acrylates.

ALCONATE SBR-3:
Disodium Ricinoleamido MEA-Sulfosuccinate
% Activity: 39
Form: Liquid
Counter irritant for anionic bases. Skin protectant.

ALCONATE SBU-185:
Disodium Undecylenamido MEA-Sulfosuccinate
% Activity: 45
Form: Liquid
Demonstrates anti-microbial activity for dandruff removal shampoos and medicated treatments.

Sulfosuccinamate:

ALCONATE SBTA-269:
Disodium Alkyl (C18) Sulfosuccinamate
% Activity: 36
Form: Liquid
Generates copious, stable foam in latex systems.

ALCOLAC: CYCLOCHEM Fatty Acid Esters, Fatty Alcohols, Nonionic Wax Blends:

CYCLOCHEM fatty esters, alcohols, and emulsifying waxes exhibit unique performance and aesthetic properties in specialty cosmetic, hair care, and industrial applications. Their many uses include emollients, emulsifiers, thickeners, stabilizers, pearlizers, dispersants, and lubricants.

Pearlizing Agents:

CYCLOCHEM EGDS:
Glycol Distearate
Form: Flakes
Melting Range C: 60-67
Opacifier/Pearlizing agent in personal care and detergent systems.

CYCLOCHEM EGMS:
Glycol Stearate
Form: Flakes
Melting Range C: 57-62
CYCLOCHEM SEG:
Glycol Stearate
Form: Flakes
Melting Range C: 55-60
Pearlizing agent in shampoos, liquid handsoaps, and liquid detergent products. Also serves as emulsion stabilizer and viscosity builder in these systems.

CYCLOSHEEN 202:
Glycol Stearate, Emulsifiers
Form: Lotion/Paste
Pearl concentrate for cold blend formulations. Also contains viscosity building, foam boosting, and conditioning agents.

Emulsifiers, Viscosity Builders:

CYCLOCHEM NI:
Emulsifying Wax
Form: Flakes
Melting Range C: 48-51
CYCLOCHEM POL:
Emulsifying Wax
Form: Solid
Melting Range C: 48-52
Nonionic emulsifying waxes used in creams, lotions, and ointments. Stable in caustic and thioglycolate bases. Excellent viscosity building agents.

ALCOLAC: CYCLOCHEM Fatty Acid Esters, Fatty Alcohols, Nonionic Wax Blends(Continued):

CYCLOCHEM PEG 200, 300, 400, 600, 6000 Esters:
Mono & Diester Laurates, Stearates & Oleates
Form: Various
Coupling and spreading agents for bath oils and personal care products. Pigment dispersing aids and wetting agents. Useful in metal lubricants formulations. Viscosity builders for lotions, shampoos, and creme rinses.

Emollients, Viscosity Builders:

Cetyl Alcohol, NF:
Cetyl Alcohol
Form: Flake
Melting Range C: 45-50
Cetyl-Stearyl Alcohol:
Cetearyl Alcohol
Form: Flake
Melting Range C: 52-56
Cetostearyl Alcohol, NF:
Cetearyl Alcohol
Form: Flake
Melting Range C: 43-53
Stearyl Alcohol, NF:
Stearyl Alcohol
Form: Flake
Melting Range C: 55-60
Emollients, viscosity builders, and opacifiers in creams, lotions, makeup bases, and creme rinse conditioners. Lubricants in industrial applications.

Lubricants, Polishes:

CYCLOCHEM GMO:
Glyceryl Mono Oleate
Form: Liquid
Emulsifier and anti-stat for plastic industries. Lubricant in metal-working industry.

CYCLOCHEM PETS:
Pentaerythritol Tetrastearate
Form: Flake/Bead
Melting Range C: 59-63
External lubricant and anti-stat for plastic extrusion, metal-working and mold releases.

ALCOLAC: CYCLOMIDE Alkanolamides:

Alcolac offers an extensive line of CYCLOMIDE alkanolamides for cosmetic and industrial applications. These versatile compounds serve as performance and viscosity builders for shampoos, skin cleaners, toiletries, detergents, dishwashes, and general purpose cleaners. CYCLOMIDE amides also are extremely useful in specialty applications such as industrial lubricants, anticorrosion agents, dispersants, couplers, and emulsifiers.

1:1 Monoethanolamides:

CYCLOMIDE C212:
Cocamide MEA
% Amide: 95
Form: Flake
Versatile foam booster, stabilizer and viscosity building agent for shampoo and personal care systems. Used in liquid and powdered cosmetics and detergents.

CYCLOMIDE L203:
Lauramide MEA
% Amide: 95
Form: Flake
High purity amide used in the formulation of premium quality shampoos and high foaming cosmetic products. Especially useful in powdered foam bath applications.

CYCLOMIDE S280:
Stearamide MEA
% Amide: 95
Form: Flake
Superb viscosity building and pearlizing agent for liquid and paste shampoos, toiletries, and conditioner systems. Excellent, high melting lubricant for industrial applications.

1:1 Diethanolamides Superamides:

CYCLOMIDE DC212/S:
Cocamide DEA
% Amide: 85
Form: Liquid
Economical foam booster and viscosity builder. Used in shampoos, bubble baths, liquid handsoaps, dishwashes, and household cleaners.

ALCOLAC: CYCLOMIDE Alkanolamides(Continued):

1:1 Diethanolamides Superamides(Continued):

CYCLOMIDE DC212/SE:
Cocamide DEA
% Amide: 95
Form: Liquid
CYCLOMIDE KD:
Cocamide DEA
% Amide: 95
Form: Liquid
High performance cosmetic grade amides. Display exceptional viscosity building properties in high foaming shampoos and personal care products.

CYCLOMIDE DL203/S:
Lauramide DEA
% Amide: 95
Form: Solid
CYCLOMIDE DL207/S:
Lauramide DEA (Lauric/Myristic)
% Amide: 95
Form: Solid
Outstanding foam boosting, stabilizing properties. Greatly enhances viscosity and performance properties in handsoaps and related cosmetic preparation.

CYCLOMIDE LE:
Lauramide DEA
% Amide: 95
Form: Liquid
Unique liquid Lauramide DEA. Exhibits the same excellent performance properties displayed by conventional, solid Lauramide DEA. Requires no heat.

CYCLOMIDE DIN295/S:
Linoleamide DEA
% Amide: 85
Form: Liquid
CYCLOMIDE DO280/S:
Oleamide DEA
% Amide: 85
Form: Liquid
CYCLOMIDE DS280/S:
Stearamide DEA
% Amide: 95
Form: Solid
Specialty cosmetic grade amides that exhibit exceptional viscosity building properties in formulated products. Excellent thickeners for gel and economy shampoos, handsoaps, and bath preparations. Contribute conditioning properties to hair care and skin cleaner formulations.

ALCOLAC: CYCLOMIDE Alkanolamides(Continued):

2:1 Diethanolamides:

CYCLOMIDE DC212:
Cocamide DEA
Form: Liquid
Industrial grade amide with high free amine content. Used in high pH cleaners, degreasers, and floor strippers. Good foam boosters and stabilizers.
CYCLOMIDE DC212/M:
Modified Cocamide DEA
Form: Liquid
Modified industrial grade amide. Functional ingredient in light and heavy duty detergents, degreasers, and all-purpose cleaners.

CYCLOMIDE DO280:
Oleamide DEA
Form: Liquid
CYCLOMIDE DIN295:
Linoleamide DEA
Form: Liquid
CYCLOMIDE DS280:
Stearamide DEA
Form: Solid
CYCLOMIDE RODEA:
Ricinoleamide DEA
Form: Liquid
Specialty fatty amides that demonstrate excellent viscosity building and grease cutting properties in formulated cleaner systems. Also excellent lubricants for metal treatment applications.

1:1 Monoisopropanolamides:

CYCLOMIDE LIPA:
Lauramide MIPA
% Amide: 95
Form: Flake
Solid amide used in powdered dishwash and bath & toiletry formulations. Produces brittle, dry residue for carpet shampoo applications.

Amidoamines:

CYCLOMIDE CODI:
Cocamidopropyl Dimethylamine
% Amide: 98
Form: Liquid
CYCLOMIDE SODI:
Stearamidopropyl Dimethylamine
% Amide: 98
Form: Solid
Conditioning and anti-stat agents in low pH creme rinse conditioners. Excellent emulsifiers and viscosity building additives for cationic emulsions.

ALCOLAC: CYCLOMOX Amine Oxides:

CYCLOMOX amine oxides are polar nonionic surface active agents that provide performance boosting and conditioning properties in a variety of personal care and industrial applications.

CYCLOMOX C:
Cocamine Oxide
% Activity: 30
Form: Liquid
CYCLOMOX CO:
Cocamidopropylamine Oxide
% Activity: 30
Form: Liquid
CYCLOMOX L:
Lauramine Oxide
% Activity: 30
Form: Liquid
CYCLOMOX LO:
Lauramidopropylamine Oxide
% Activity: 30
Form: Liquid
CYCLOMOX SO:
Stearamidopropylamine Oxide
% Activity: 30
Form: Paste
Features: Foam boosters, stabilizers, and viscosity builders for anionic based systems. Contribute excellent conditioning properties to formulated shampoo products. Used in shampoos, liquid handsoaps, bubble baths, liquid dishwashes, bathroom scours, and light duty cleaners.

ALCOLAC: CYCLORYL Formulated Shampoo and Detergent Concentrates:

CYCLORYL concentrates are professionally formulated blends that have been specially developed to produce optimized shampoo, personal care, and industrial cleaner products. With simple water dilution, these concentrates readily produce economical finished systems that demonstrate exceptional performance properties.

Personal Care & Specialty Concentrates:

CYCLORYL ALC:
PEG 80 Sorbitan Laurate, Sodium Trideceth Sulfate, PEG 150 Distearate, et al
% Concentration: 62
Features: Economical baby shampoo concentrate specially formulated to produce viscous, non-irritating shampoos and skin cleanser products.

CYCLORYL ANL:
Sodium C14-16 Olefin Sulfonate, Sodium Laureth Sulfate, Lauramide DEA
% Concentration: 50
Optimized high foaming base used in the formulation of quality shampoos, handsoaps, and bath and toiletry preparations.

CYCLORYL CBS:
PEG 80 Sorbitan Laurate, Sodium Trideceth Sulfate, Lauroamphocarboxyglycinate, et al
% Concentration: 40
Premium baby shampoo concentrate used in the preparation of high performance, non-irritating, baby care formulations.

CYCLORYL CN:
Sodium Laureth Sulfate, Builders
% Concentration: 64
High activity concentrate for preparation of economy shampoos, bubble baths, and skin cleanser products.

CYCLORYL EW:
Ammonium Lauryl Sulfate, Lauramide DEA
% Concentration: 40
Versatile premium quality concentrate used in the preparation of generic matches for commercial hair care, bath gels, bubble and handsoap systems.

ALCOLAC: CYCLORYL Formulated Shampoo and Detergent Concentrates (Continued):

CYCLORYL GSC:
Cocoamphodiacetate, Sodium Laureth Sulfate
% Concentration: 30
Mild bath gel/shampoo concentrate developed for use in high viscosity, low irritation personal care preparations.

CYCLORYL M1:
Sodium Lauryl Sulfate, Stearamide MEA, Glycol Stearate, Cocamide MEA
% Concentration: 35
Completely formulated pearlescent base developed for the preparation of pearl shampoo, bubble bath, handsoap, and pet products. Requires no heat.

CYCLORYL NWC:
Sodium Laureth Sulfate, Cocamide DEA, TEA-Lauryl Sulfate
% Concentration: 62
Multi-functional, concentrated blend custom formulated to be used in the preparation of both economy and high performance cosmetic and pet care products.

CYCLORYL XL-M:
DEA Lauryl Sulfate and DEA Cocaminopropinate
% Concentration: 36
Rich lathering base used in the formulation of mild shampoo, bubble bath, and skin cleanser products, generates an elegant, luxurious foam.

ALCOLAC: CYCLOTERIC Amphoterics:

Alcolac offers a full line of gentle, non-irritating, high-foaming amphoteric surfactants. CYCLOTERICS help improve performance and lower production costs in a wide variety of shampoo, conditioning, skin care, and industrial products. Effective over wide pH ranges, they are compatible with anionics, cationics, and nonionics.

Amido-Betaines:

CYLOTERIC BET-C 30:
Cocamidopropyl Betaine (Cosmetic Grade)
% Activity: 30
CYCLOTERIC BET-W:
Cocamidopropyl Betaine (Technical Grade)
% Activity: 30
CYCLOTERIC BET-CB:
Cocamidopropyl Betaine (Cosmetic, Glycerine Free)
% Activity: 30

Foam boosters, foaming agents, thickeners, and conditioning agents. Used as performance boosters in shampoo, cosmetic and industrial applications. Effective irritation mollifying agents for use in baby shampoo and ultra-mild personal care products.

CYCLOTERIC BET-O 30:
Oleamidopropyl Betaine
% Activity: 30

Unique hair conditioning ingredient for detangling, conditioning shampoos. Excellent viscosity building properties for bath gels and gel shampoos.

CYCLOTERIC BET-OD 40:
Capric/Caprylic Amido Betaine
% Activity: 40

Low foam wetting agent for pressure sprays, mechanical washes, detergent scours and low foam cleaners.

Betaines:

CYCLOTERIC BET-C 41:
Coco Betaine
% Activity: 41
CYCLOTERIC BET-L 31:
Lauryl Betaine
% Activity: 31

High foaming conditioning agent used in shampoos, foam baths, and liquid handsoaps.

ALCOLAC: CYCLOTERIC Amphoterics(Continued):

Betaines(Continued):

CYCLOTERIC BET-OB 50:
Oleyl Betaine
% Activity: 50
Viscosity building/gelling agent used in bath gel and shampoo applications. Imparts soft, elegant feel to skin.

Glycinate:

CYCLOTERIC BET-T2 40:
Dihydroxyethyl Tallow Glycinate
% Activity: 40
Acid stable, viscosity building agent for industrial applications. Excellent conditioning ingredient in premium quality shampoos.

Imidazoline:

CYCLOTERIC 1398:
Cocoamphodiacetate
% Activity: 40
Low eye sting surfactant used in the preparation of baby shampoo, baby bath, and ultra-mild personal care products.

Propionates:

CYCLOTERIC CAPA:
Cocaminopropionic Acid
% Activity: 40
CYCLOTERIC SLIP:
Sodium Lauriminodipropiaonate
% Activity: 30
High foaming, protein substantive conditioning agents used in shampoos, skin cleansers, and foam baths. Also, excellent surfactants for hard surface cleaners, car washes, and industrial foamers.

Sultaine:

CYCLOTERIC BET-CS:
Cocamidopropyl Hydroxy Sultaine
% Activity: 40
Exceptional foaming agent and performance booster for high foaming shampoos and related cosmetic products. Effective irritation ameliorating ingredient for baby care formulations.

ALCOLAC: CYCLOTON Cationic Surfactants:

CYCLOTON cationic surfactants are protein and fiber substantive quaternary ammonium compounds that serve as the active ingredients in creme rinse conditioners, fabric softeners, cationic emulsions, and anti-static treatments. Alcolac offers a broad line of these concentrated and formulated bases to meet essentially all application requirements.

CYLCLOTON M242C/29:
Cetrimonium Chloride
% Activity: 29
Form: Liquid
Cold water dispersible quaternary surfactant. Forms transparent dispersions in water. Versatile conditioning agent in all types of hair conditioner and hair treatment applications.

CYLCOTON D261C/70:
Ditallowalkonium Chloride
% Activity: 70
Form: Soft Paste
CYCLOTON D261C/75:
Ditallowalkonium Chloride
% Activity: 75
Form: Firm Paste
Concentrated base for preparation of premium fabric softeners, anti-static treatments, and hair conditioner products. CTFA adopted name Quaternium-18.

CYLCLOTON 7LUF:
Olealkonium Chloride
% Activity: 50
Form: Liquid
Pumpable conditioner base for formulation of high viscosity, creme rinse systems.

CYCLOTON M270C/18:
Stearalkonium Chloride
% Activity: 25
Form: Soft Paste
CYLCOTON M270C/85:
Stearalkonium Chloride
% Activity: 85
Form: Flake
Widely used cationic base for preparation of opaque and pearlescent creme rinse conditioners. Provide excellent anti-stat, comb-out, and detangling properties in formulated hair care products.

ALCOLAC: CYCLOTON Cationic Surfactants(Continued):

CYCLOTON SCS:
Stearalkonium Chloride, Builders
Form: Flake
Economical formulated hair conditioner base. Forms attractive, viscous, creme rinse products that display excellent conditioning properties.

CYCLOTON D256B/99:
Cetyl Ethyl Dimethyl Ammonium Bromide
% Activity: 99
Form: Powder
CYCLOTON M242B/99:
Cetrimonium Bromide
% Activity: 99
Form: Powder
CYCLOTON M214B/99:
Myrtrimonium Bromide
% Activity: 99
Form: Powder
Unique high purity cationic surfactants. Contributes exotic elegant feel to formulated hair conditioner products. Used in the formulation of light hair conditioners, hair conditioner sprays, mousses, and anti-static sprays.

CYCLOTON CT100:
Cetrimonium Bromide, Builders
Form: Flake
Formulated creme rinse concentrate. Produces viscous, premium quality hair care products that exhibit unique conditioning properties.

CYCLOTON 75C:
Ditallow Based Methylsulfate Quaternary
% Activity: 75
Form: Liquid
Economy softener concentrate for household and institutional applications. Used as textile softener, anti-stat, and in paper and pulp processing.

ALCOLAC: DERMALCARE Skin Care Products:

DERMALCARE skin care products are a custom line of fatty emollients, emulsifiers, and formulated cleanser blends specially developed to protect and enhance the natural appearance and texture of healthy skin.

Cosmetic Esters:

DERMALCARE EGMS/SE:
Glycol Stearate, Self Emulsifying
Form: Flake
Range C: 56-62
DERMALCARE GMS:
Glyceryl Stearate
Form: Flake
Range C: 58-63
DERMALCARE GMS/SE:
Glyceryl Stearate, Self Emulsifying
Form: Flake
Range C: 58-63
DERMALCARE GMS-165:
Glyceryl Stearate, PEG 100 Stearate
Form: Flake
Range C: 53-57
DERMALCARE GTIS:
Triisostearin
Form: Liquid
DERMALCARE HL:
Hexyl Laurate
Form: Liquid
DERMALCARE LVL:
Lauryl Lactate
Form: Liquid
DERMALCARE MM/M:
Myristyl Myristate
Form: Flake
Range C: 37-39
DERMALCARE MST:
Myristyl Stearate
Form: Flake
Range C: 43-45
DERMALCARE PGMS:
Propylene Glycol Stearate
Form: Flake
Range C: 33-38
DERMALCARE SDG:
PEG 2 Stearate
Form: Solid
Range C: 43-47
DERMALCARE SS:
Stearyl Stearate
Form: Flake
Range: 53-55

ALCOLAC: DERMALCARE Skin Care Products(Continued):

Cosmetic Esters(Continued):

High purity cosmetic grade fatty esters used as emollients, moisturizers, lubricants, and conditioning agents in emulsified hair care and skin care systems. They impart a soft, elegant feel when formulated into creams, lotions, bath oils, anti-perspirants, sunscreens, medicated bases, ointments, and related personal care products. Viscosity building agents for cosmetic emulsions and gelling/stiffening agents for lipstick, makeup, and deodorant applications.

Specialty Emollient Blends & Emulsifiers:

DERMALCARE C-20:
Ceteth 20
Form: Flake
Range C: 47-52

DERMALCARE SPS:
Cetyl Esters
Form: Flake
Range C: 46-48
Cosmetic grade emulsifier and emulsion stabilizer for creams, lotions, anti-perspirants, creme rinse conditioners, and personal care products. Acid and alkali stable.

DERMALCARE 326A:
Synthetic Beeswax
Form: Flake
Range C: 51-54
Unique blend of natural fatty acids and alcohols specially developed to match the performance properties of beeswax.

Formulated Skin Cleansers:

DERMALCARE COSO:
Potassium Cocoate
Form: Liquid
Range C: 40
Economy handsoap base for bathroom dispensers.

DERMALCARE WHC:
Modified Amide
Form: Liquid
Range C: 100
Emulsifying base for preparation of waterless hand cleaners.

DERMALCARE 1673:
Sodium Laureth Sulfate, Builders
Form: Liquid
Range C: 35
High performance base for the preparation of ultra-mild face and skin cleanser formulations.

ALCOLAC: SIPEX Alkyl and Alkyl Ether Sulfates:

SIPEX alkyl and alkyl ether sulfates are a diverse group of anionic surfactants custom developed for specialty cosmetic, emulsion polymerization, and industrial uses. Excellent wetting agents, these compounds demonstrate unique performance properties in a wide variety of commercial applications.

SIPEX BOS:
Sodium 2-Ethylhexyl Sulfate
% Activity: 40
Form: Liquid
Low foaming anionic surfactant that displays excellent wetting properties at high pH ranges. Used in emulsion polymerization, industrial cleaner and agricultural washing and peeling applications.

SIPEX CAV:
Sodium Isodecyl Sulfate
% Activity: 40
Form: Liquid
Exhibits exceptional surfactant and wetting properties at extreme temperatures. Recommended applications include household and industrial cleaners, emulsion polymerization and textile treatments.

SIPEX EST-30:
Sodium Trideceth Sulfate
% Activity: 30
Form: Liquid

SIPEX EST-75:
Sodium Trideceth Sulfate
% Activity: 75
Form: Liquid
Economical high foaming surfactants used in the formulation of mild shampoos, cosmetics, and household cleaners. Also used in the emulsion polymerization of PVC and styrene/butadiene resins.

SIPEX ME 60:
Sodium Myreth Sulfate
% Activity: 58
Form: Liquid
Concentrated surfactant base for preparation of elegant shampoos and cosmetic products. Produces luxurious creamy lather in formulated personal care systems.

ALCOLAC: SIPEX Alkyl and Alkyl Ether Sulfates(Continued):

SIPEX NB 60:
Sodium Alkyl Ether Sulfate
% Activity: 60
Form: Liquid
Industrial grade foaming agent for de-dusting treatments, air drilling, wall board foaming, and brine water baths. Displays excellent electrolyte and heavy ion tolerance.

SIPEX OLS:
Sodium Octyl Sulfate
% Activity: 33
Form: Paste
Rapid wetting, low foaming surfactant designed for use in metal cleaners, dishwashing detergents, rinse aids, and emulsion polymerization applications. Especially useful in electrolyte baths for metal cleaning and as a textile mercerizing agent.

SIPEX TDS:
Sodium Tridecyl Sulfate
% Activity: 25
Form: Liquid
Branched chain alkyl sulfate developed for emulsion polymerization of PVC, styrene, and acrylic systems. Also an excellent wetting agent for general detergent applications.

SIPEX 280:
Ammonium Nonoxynol-4 Sulfate
% Activity: 58
Form: Liquid
Versatile high foaming surfactant with excellent wetting, dispersing, and emulsifying properties. Used in shampoo, skin cleansers, light duty cleaners, and emulsion polymerization applications.

ALCOLAC: SIPON Lauryl Sulfates and Lauryl Ether Sulfates:

SIPON lauryl sulfates and lauryl ether sulfates are universally accepted standards of the industry for shampoo, cosmetic, industrial, and emulsion polymerization applications.

Lauryl Sulfates:

SIPON L-22:
Ammonium Lauryl Sulfate
% Activity: 28
High foaming base for the preparation of low pH shampoos, bubble baths and cosmetic products.

SIPON L-22HV:
Ammonium Lauryl Sulfate
% Activity: 28
High viscosity version of SIPON L-22. Used in the formulation of viscous shampoos, bath gels, and personal care products.

SIPON LCP:
Sodium Lauryl Sulfate
% Activity: 30
Low cloud point SLS. Specifically designed for emulsion polymerization application where it promotes good stability and particle uniformity.

SIPON LSB:
Sodium Lauryl Sulfate
% Activity: 29
Low salt, high foaming base used in the formulation of shampoos, bubble baths, hand cleaners, cosmetic emulsions, carpet shampoos, and detergent systems.

SIPON SB:
Sodium Lauryl Sulfate
% Activity: 29
High purity SLS recommended for emulsion polymerization, textile scouring, metal cleaning, and agricultural washing.

SIPON UB:
Sodium Lauryl Sulfate
% Activity: 30
Premium grade SLS for emulsion polymerization applications including vinyl and vinylidene chlorides, styrene and acrylic monomers. Choice surfactant for carboxylated SBR and acrylic froth applications.

ALCOLAC: SIPON Lauryl Sulfates and Lauryl Ether Sulfates (Continued):

Lauryl Sulfates(Continued):

SIPON LD:
DEA-Lauryl Sulfate
% Activity: 40

SIPON LT-6:
TEA-Lauryl Sulfate
% Activity: 40

Cosmetic grade base for the preparation of mild, high foaming shampoos, bubble baths, skin cleansers, and related cosmetic products.

SIPON LM:
Magnesium Lauryl Sulfate
% Activity: 27

Versatile foaming base used in both cosmetic and industrial applications. Excellent soil and grease suspending agent. Produces brittle, tack-free residue for carpet shampoos.

Lauryl Ether Sulfates:

SIPON EAY:
Ammonium Laureth (1) Sulfate
% Activity: 26

SIPON EA:
Ammonium Laureth (3) Sulfate
% Activity: 27

Multi-functional foaming bases used in the preparation of mild shampoos, bubble baths, skin cleansers and related cosmetic products.

SIPON MA 360:
Ammonium Laureth (3) Sulfate
% Activity: 58

Economical concentrated ALES base with excellent hard water tolerance. Commonly used in shampoo, bath, and personal care products. Also widely used in light duty cleansers, liquid dishwash, and fabric washes.

ALCOLAC: SIPON Lauryl Sulfates and Lauryl Ether Sulfates (Continued):

Lauryl Ether Sulfates(Continued):

SIPON ESY:
Sodium Laureth (1) Sulfate
% Activity: 25
SIPON ES 2:
Sodium Laureth (2) Sulfate
% Activity: 26
SIPON ES:
Sodium Laureth (3) Sulfate
% Activity: 27
Cosmetic grade SLES used in mild shampoos, bath products, skin cleansers, handsoaps, and high foaming toiletries. Excellent bases for preparation of high viscosity, formulated products.

SIPON NA 61:
Sodium Laureth (3) Sulfate
% Activity: 58
High activity, economical SLES concentrate. Universal foaming agent used in preparation of cost-effective shampoos, cosmetic products, light duty liquids, dishwash, and general purpose cleaners.

SIPON N70:
Sodium Laureth (2) Sulfate
% Activity: 70
Highly concentrated SLES slurry. Recommended for use in high foaming shampoos, bath gels, and cosmetic preparations.

SIPON ES-7:
Sodium Laureth (7) Sulfate
% Activity: 28
SIPON ES-12:
Sodium Laureth (12) Sulfate
% Activity: 60
Specialty SLES bases that are exceptionally mild to the skin and eyes. Suggested applications include light face cleansers, baby baths, and perfume solubilizers.

ALCOLAC: SIPONATE Alkyl Aryl Sulfonates and Alpha Olefin Sulfonates:

SIPONATE surfactants are widely used bases for the preparation of shampoos, cosmetic skin cleansers, and household detergents. These compounds are also excellent emulsifiers for the polymerization of vinyl, acrylic, styrene, and various monomer systems. Other applications include agricultural emulsifiers, metal cleaning and automotive specialties.

Alkyl Aryl Sulfonates:

SIPONATE ABSA:
Dodecyl Benzene Sulfonic Acid
% Activity: 98
Form: Liquid
Concentrated free acid that with neutralization forms high foaming surfactant used in economy dishwashes, all purpose cleaners, degreasers, floor strippers, and car washes.

SIPONATE DDB 40:
Sodium Dodecyl Benzene Sulfonate
% Activity: 40
Form: Liquid
High foaming surfactant base for the preparation of liquid dishwashes, general purpose cleaners, degreasers, and heavy duty detergents.

SIPONATE DDB 60T:
TEA-Dodecyl Benzene Sulfonate
% Activity: 60
Form: Liquid
Surfactant base for the preparation of high foaming liquid dishwashes, car washes, and light duty cleaners. Also used in shampoo, bubble bath, and hand cleaner formulations.

SIPONATE DS-4:
Sodium Dodecyl Benzene Sulfonate
% Activity: 23
Form: Liquid
Emulsifier widely use in the polymerization of styrene/butadiene, vinyl chloride and acrylic polymers. FDA approved for use in fruits and vegetable washes.

ALCOLAC: SIPONATE Alkyl Aryl Sulfonates and Alpha Olefin Sulfonates(Continued):

Alkyl Aryl Sulfonates(Continued):

SIPONATE LDS-10:
Sodium Dodecyl Benzene Sulfonate
% Activity: 98
Form: Flake
Concentrated high purity flake form of Sodium DDBSA. Used in the formulation of light duty and heavy duty detergents. Especially useful in powdered cleaners and scours. Linear molecule offers high biodegradability.

SIPONATE 330:
Amine Salt, Alkyl Benzene Sulfonate
% Activity: 90
Form: Liquid
Unique specialty surfactant developed to solubilize and disperse heavy greases and oils. Uses include automotive engine degreasers, metal cleaners, oil spill cleanups, and heavy duty emulsification. Excellent emulsifier for agricultural formulations.

Alpha Olefin Sulfonates:

SIPONATE A246L:
Sodium C14-16 Olefin Sulfonate
% Activity: 40
Form: Liquid
High foaming technical grade detergent and emulsifier for use in shampoos, bubble baths, liquid handsoaps, dishwashes, car washes, pet shampoos and general puropose cleaners.

SIPONATE A246LX:
Sodium C14-16 Olefin Sulfonate
% Activity: 40
Form: Liquid
Cosmetic grade surfactant base for preparation of high foaming shampoo, handsoap, and light duty detergent systems. Displays excellent stability over wide pH ranges.

SIPONATE 301-10F:
Sodium C14-16 Olefin Sulfonate
% Activity: 98
Form: Flake
SIPONATE 301-10P:
Sodium C14-16 Olefin Sulfonate
% Activity: 98
Form: Powder
Concentrated dry detergent. Especially useful for powdered shampoo, bubble bath, handsoap, fine fabric cleaners, and general purpose cleaners.

ALCOLAC: SIPONIC Alkoxylated Nonionic Surfactants:

SIPONIC alkoxylated fatty alcohols, octylphenols, nonylphenols, and mercaptans are a broad family of nonionic surfactants that have been specially derived for use in the formulation of a wide variety of cosmetic, household cleaner, and industrial applications. Superb wetting agents, emulsifiers, and emulsion stabilizers, these compounds serve as primary detergents and soil suspending agents in diverse heavy duty and light duty cleaner formulations. Specialty SIPONIC surfactants are also choice additives for the preparation of emulsified cosmetic and hair care systems.

SIPONIC E Series:
Cetyl/Stearyl Alcohol Ethoxylates:

SIPONIC E-2:
Ceteareth 4
Mole of Ethylene Oxide: 4
HLB: 8.0
% Activity: 100

SIPONIC E-3:
Ceteareth 6
Mole of Ethylene Oxide: 6
HLB: 10.1
% Activity: 100

SIPONIC E-5:
Ceteareth 10
Mole of Ethylene Oxide: 10
HLB: 12.4
% Activity: 100

SIPONIC E-7:
Ceteareth 15
Mole of Ethylene Oxide: 15
HLB: 14.3
% Activity: 100

SIPONIC E-10:
Ceteareth 20
Mole of Ethylene Oxide: 20
HLB: 15.3
% Activity: 100

SIPONIC E-15:
Ceteareth 30
Mole of Ethylene Oxide: 30
HLB: 16.9
% Activity: 100

Primary emulsifiers and emulsion stabilizers for cosmetic creams, lotions, creme rinse conditioners, depilatories, and hair relaxer treatments. Extremely stable over wide pH ranges and compatible with both anionic and nonionic surfactants and highly reactive oxidizing and reducing agents. Also used as leveling agents and dyeing assistants in textile applications and as emulsion polymerization surfactants, usually in combination with anionics.

ALCOLAC: SIPONIC Alkoxylated Nonionic Surfactants(Continued):

SIPONIC F Series:
Octylphenol Ethoxylates:

SIPONIC F-90:
Octoxynol 9
Mole of Ethylene Oxide: 9
HLB: 13.5
% Activity: 100

SIPONIC F-160:
Octoxynol 16
Mole of Ethylene Oxide: 16
HLB: 15.8
% Activity: 100

SIPONIC F-300:
Octoxynol 30
Mole of Ethylene Oxide: 30
HLB: 17.3
% Activity: 70

SIPONIC F-400:
Octoxynol 40
Mole of Ethylene Oxide: 40
HLB: 17.9
% Activity: 70

SIPONIC F-707:
Octoxynol 70
Mole of Ethylene Oxide: 70
HLB: 18.7
% Activity: 70

Stabilizers in emulsion polymerization of vinyl and acrylic latices. Wetting agents, dispersants, detergents and emulsifiers in industrial and household cleaner applications. Also used as emulsifiers for herbicides and insecticides.

SIPONIC Y-500-70:
Oleyl Alcohol Ethoxylate
Oleth 25
Mole of Ethylene Oxide: 25
HLB: 16.1
% Activity: 70

Emulsifier and stabilizer for cosmetic creams, lotions, and hair care treatments. Wetting agent and emulsifier for emulsion polymerization and industrial applications.

ALCOLAC: SIPONIC Alkoxylated Nonionic Surfactants(Continued):

SIPONIC L Series:
Alcohol Ethoxylates:

SIPONIC L-4:
Laureth 4
Mole of Ethylene Oxide: 4
HLB: 9.7
% Activity: 100

SIPONIC L-7-90:
Laureth 7
Mole of Ethylene Oxide: 7
HLB: 12.1
% Activity: 90

SIPONIC L-12:
Laureth 12
Mole of Ethylene Oxide: 12
HLB: 14.5
% Activity: 100

SIPONIC L-25:
Laureth 23
Mole of Ethylene Oxide: 23
HLB: 16.9
% Activity: 100

Coupling agents, solubilizers, and emulsion stabilizers for cosmetic and hair care systems. Used in combination with anionic surfactants for emulsion polymerization applications. Excellent ingredients for coning and textile spin finishes.

Tertiary Thio Ethoxylates:

SIPONIC 260:
Alkyl Mercaptan Ethoxylate
Mole of Ethylene Oxide: 6
HLB: 11.0
% Activity: 100

SIPONIC SK:
Alkyl Mercaptan Ethoxylate
Mole of Ethylene Oxide: 8
HLB: 12.7
% Activity: 100

SIPONIC 218:
Alkyl Mercaptan Ethoxylate
Mole of Ethylene Oxide: 10
HLB: 13.9
% Activity: 100

Exceptional wetting agents and surfactants. Functional over wide pH ranges, temperatures and water hardnesses. Outstanding detergents for metal cleaning and household and industrial cleaners/scours and degreasers. Excellent emulsifiers for herbicides and insecticides.

ALCOLAC: SIPONIC Alkoxylated Nonionic Surfactants(Continued):

SIPONIC TD Series:
Tridecyl Alcohol Ethoxylates:

SIPONIC TD-3:
Trideceth 3
Mole of Ethylene Oxide: 3
HLB: 7.9
% Activity: 100

SIPONIC TD-6:
Trideceth 6
Mole of Ethylene Oxide: 6
HLB: 11.4
% Activity: 100

SIPONIC TD-990:
Trideceth 9
Mole of Ethylene Oxide: 9
HLB: 13.8
% Activity: 90

SIPONIC TD-12:
Trideceth 12
Mole of Ethylene Oxide: 12
HLB: 14.5
% Activity: 100

Versatile emulsifiers and emulsion stabilizers for cosmetic systems. Excellent wetting properties. Also used as detergents for metal cleaners, textile scours, household cleaners, and industrial applications.

Linear Alcohol Ethoxylates:

SIPONIC 25-3:
POE (3) C12 C15
Mole of Ethylene Oxide: 3
HLB: 7.9
% Activity: 100

SIPONIC 25-7:
POE (7) C12 C15
Mole of Ethylene Oxide: 7
HLB: 12.2
% Activity: 100

SIPONIC 25-9:
POE (9) C12 C15
Mole of Ethylene Oxide: 9
HLB: 13.3
% Activity: 100

SIPONIC 91-6:
POE (6) C9 C11
Mole of Ethylene Oxide: 6
HLB: 12.5
% Activity: 100

Biodegradable detergents, wetting agents, and emulsifiers for household and industrial cleaners. Excellent couplers and solubilizing agents for perfumes and organic additives.

ALCOLAC: SIPONIC Alkoxylated Nonionic Surfactants(Continued):

SIPONIC NP Series:
Nonylphenol Ethoxylates:

SIPONIC NP-4:
Nonoxynol 4
Mole of Ethylene Oxide: 4
HLB: 8.9
% Activity: 100

SIPONIC NP-6:
Nonoxynol 6
Mole of Ethylene Oxide: 6
HLB: 10.9
% Activity: 100

SIPONIC NP-7:
Nonoxynol 7
Mole of Ethylene Oxide: 7
HLB: 11.7
% Activity: 100

SIPONIC NP-8:
Nonoxynol 8
Mole of Ethylene Oxide: 8
HLB: 12.3
% Activity: 100

SIPONIC NP-9:
Nonoxynol 9
Mole of Ethylene Oxide: 9
HLB: 12.9
% Activity: 100

SIPONIC NP-9.5:
Nonoxynol 9.5
Mole of Ethylene Oxide: 9.5
HLB: 13.2
% Activity: 100

SIPONIC NP-10:
Nonoxynol 10
Mole of Ethylene Oxide: 10
HLB: 13.6
% Activity: 100

SIPONIC NP-13:
Nonoxynol 13
Mole of Ethylene Oxide: 13
HLB: 14.4
% Activity: 100

ALCOLAC: SIPONIC Alkoxylated Nonionic Surfactants(Continued):

SIPONIC NP Series(Continued):
Nonylphenol Ethoxylates(Continued):

SIPONIC NP-15:
Nonoxynol 15
Mole of Ethylene Oxide: 15
HLB: 15.1
% Activity: 100

SIPONIC NP-40:
Nonoxynol 40
Mole of Ethylene Oxide: 40
HLB: 17.8
% Activity: 100

SIPONIC NP-407:
Nonoxynol 40
Mole of Ethylene Oxide: 40
HLB: 17.8
% Activity: 70

SIPONIC NP-707:
Nonoxynol 70
Mole of Ethylene Oxide: 70
HLB: 18.6
% Activity: 70

Primary emulsifiers for a wide variety of personal care, household and industrial formulations. Excellent detergents and wetting agents for heavy duty and light duty cleaner systems. Function as dispersants, solubilizers, couplers, and co-emulsifiers for specialty cosmetic and industrial applications. Also used as primary and auxiliary surfactants for emulsion polymerization.

ALCOLAC: SIPOTHIX Specialty Polymers:

SIPOTHIX specialty polymers are complex acrylate thickeners chemically designed to enhance the consistency and performance properties of detergent, cosmetic, and hair care systems. At neutral to alkaline pH ranges, these polymers efficiently build the viscosities of virtually all formulated products and are therefore well suited for numerous specialty applications.

SIPOTHIX 1941:
Acrylate/Steareth-20 Methylacrylate Copolymer
% Solids: 30
Form: Opaque Liquid

SIPOTHIX H-65:
Ethylacrylate/Methylacrylic Acid Copolymer
% Solids: 30
Form: Opaque Liquid

pH sensitive thickening agents designed to build the viscosities of liquid detergent, shampoo, and cosmetic preparations. Optimum thickening occurs at slightly alkaline pH ranges (7-9).

ALLIED SIGNAL INC.: A-C Polyethylenes, A-C Copolymers, and ACUMIST Micronized Polyethylene Powders:

A-C polyethylenes, copolymers and ACUMIST powders are polyethylene waxes which, because of their unique combination of physical properties and particle characteristics, are used as additives to improve performance and processing in industries such as:

- Adhesives
- Ceramics
- Coatings
- Cosmetics
- Inks
- Paper Coatings
- Plastics
- Powdered Metals
- Rubber

The A-C Performance Additives product line includes polyethylene homopolymers, oxidized homopolymers (both high and low density), ethylene-vinyl acetate copolymers and ethylene-acrylic acid copolymers. These products, depending on grade, are available in prilled, granular, atomized, ground and micronized forms.

Typical Properties/Particle Characteristics:
Homopolymers:

A-C 617A:
- Hardness dmm: 7.0
- Density g/cc: 0.91
- Acid Number mg KOH/g: Nil
- Mettler Drop Point C(F): 102(216)
- Typical Sieve Analysis (% Retained):
  - 40: 17.4
  - 60: 43.3
  - 80: 19.8

A-C 6A:
- Hardness dmm: 4.0
- Density g/cc: 0.92
- Acid Number mg KOH/g: Nil
- Mettler Drop Point C(F): 106(223)
- Typical Sieve Analysis (% Retained):
  - 40: 17.4
  - 60: 43.3
  - 80: 19.8

A-C 7A:
- Hardness dmm: 2.5
- Density g/cc: 0.92
- Acid Number mg KOH/g: Nil
- Mettler Drop Point C(F): 107(225)
- Typical Sieve Analysis (% Retained):
  - 40: 17.4
  - 60: 43.3
  - 80: 19.8

ALLIED SIGNAL INC.: A-C Polyethylenes, A-C Copolymers, and ACUMIST Micronized Polyethylene Powders(Continued):

Homopolymers(Continued):

A-C 8A:
Hardness dmm: 1.0
Density g/cc: 0.93
Acid Number mgKOH/g: Nil
Mettler Drop Point C(F): 116(241)
Typical Sieve Analysis (% Retained):
20: 0.4
40: 17.4
60: 43.3
80: 19.8

A-C 9A:
Hardness dmm: 0.5
Density g/cc: 0.94
Acid Number mgKOH/g: Nil
Mettler Drop Point C(F): 117(243)
Typical Sieve Analysis (% Retained):
20: 0.4
40: 17.4
60: 43.3
80: 19.8

A-C 9F:
Hardness dmm: 0.5
Density g/cc: 0.94
Acid Number mgKOH/g: Nil
Mettler Drop Point C(F): 117(243)
Typical Sieve Analysis (% Retained):
40: 0.2
60: 0.4
80: 2.9
100: 33.1
200: 50.7

Oxidized Homopolymer:

A-C 629A:
Hardness dmm: 5.5
Density g/cc: 0.93
Acid Number mgKOH/g: 15
Mettler Drop Point C(F): 102(216)
Typical Sieve Analysis (% Retained):
40: 5.9
60: 39.4
80: 24.6
100: 24.6

ALLIED SIGNAL INC: A-C Polyethylenes, A-C Copolymers, and ACUMIST Micronized Polyethylene Powders(Continued):

High Density Oxidized Homopolymers:

A-C 316A:
Hardness dmm: <0.5
Density g/cc: 0.98
Acid Number mgKOH/g: 16
Mettler Drop Point C(F): 140(284)
Typical Sieve Analysis (% Retained):
60: 14.2
80: 29.7
100: 16.3
200: 18.0
325: 13.2
>325: 8.6

A-C 316:
Hardness dmm: <0.5
Density g/cc: 0.98
Acid Number mgKOH/g: 16
Mettler Drop Point C(F): 140(284)
Typical Sieve Analysis (% Retained):
40: 56.1
60: 29.4
80: 5.4

A-C 392:
Hardness dmm: <0.5
Density g/cc: 0.99
Acid Number mgKOH/g: 30
Mettler Drop Point C(F): 138(280)
Typical Sieve Analysis (% Retained):
40: 46.0
60: 25.8
80: 8.5
100: 4.5

A-C 395:
Hardness dmm: <0.5
Density g/cc: 1.00
Acid Number mgKOH/g: 41
Mettler Drop Point C(F): 137(279)
Typical Sieve Analysis (% Retained):
20: 5.2
40: 43.2
60: 28.8
80: 9.4
100: 5.1

ALLIED SIGNAL INC.: A-C Polyethylenes, A-C Copolymers, and ACUMIST Micronized Polyethylene Powders(Continued):

High Density Oxidized Homopolymers(Continued):

395A:
- Hardness dmm: <0.5
- Density g/cc: 1.00
- Acid Number mgKOH/g: 41
- Mettler Drop Point C(F): 137(279)
- Typical Sieve Analysis (% Retained):
  - 60: 14.1
  - 80: 23.3
  - 100: 12.8
  - 200: 29.2
  - 325: 14.5

Copolymers:

Ethylene-Acrylic Acid A-C 540A:
- Hardness dmm: 2.0
- Density g/cc: 0.93
- Acid Number mgKOH/g: 40
- Mettler Drop Point C(F): 108(226)
- Typical Sieve Analysis (% Retained):
  - 40: 14.5
  - 60: 37.2
  - 80: 21.6
  - 100: 14.3
  - 200: 8.5

Ethylene-Vinyl Acetate A-C 400A:
- Hardness dmm: 9.5
- Density g/cc: 0.92
- % Vinyl Acetate: 13
- Mettler Drop Point C(F): 95(203)
- Typical Sieve Analysis (% Retained):
  - 40: 33.5
  - 60: 35.4
  - 80: 13.6
  - 100: 7.4
  - 200: 3.3

ALLIED SIGNAL INC.: A-C Polyethylenes, A-C Copolymers, and ACUMIST Micronized Polyethylene Powders(Continued):

Micronized Polyethylene Waxes ACUMIST:

A-12:
Hardness dmm: <0.5
Density g/cc: 0.99
Acid Number mg KOH/g: 30
Mettler Drop Point C(F): 136(277)
Particle Size Distribution in Microns: Mean Value: 12.9

A-18:
Hardness dmm: <0.5
Density g/cc: 0.99
Acid Number mgKOH/g: 30
Mettler Drop Point C(F): 136(277)
Particle Size Distribution in Microns: Mean Value: 15.4

B-6:
Hardness dmm: <0.5
Density g/cc: 0.96
Acid Number mgKOH/g: Nil
Mettler Drop Point C(F): 126(259)
Particle Size Distribution in Microns: Mean Value: 7.5

B-9:
Hardness dmm: <0.5
Density g/cc: 0.96
Acid Number mgKOH/g: Nil
Mettler Drop Point C(F): 126(259)
Particle Size Distribution in Microns: Mean Value: 9.3

B-12:
Hardness dmm: <0.5
Density g/cc: 0.96
Acid Number mgKOH/g: Nil
Mettler Drop Point C(F): 126(259)
Particle Size Distribution in Microns: Mean Value: 10.7

B-18:
Hardness dmm: <0.5
Density g/cc: 0.96
Acid Number mgKOH/g: Nil
Mettler Drop Point C(F): 126(259)
Particle Size Distribution in Microns: Mean Value: 20.6

ALLIED SIGNAL INC.: A-C Polyethylenes, A-C Copolymers, and ACUMIST Micronized Polyethylene Powders(Continued):

Micronized Polyethylene Waxes ACUMIST(Continued):

C-5:
Hardness dmm: 1.0
Density g/cc: 0.95
Acid Number mgKOH/g: Nil
Mettler Drop Point C(F): 121(250)
Particle Size Distribution in Microns: Mean Value: 6.8

C-9:
Hardness dmm: 1.0
Density g/cc: 0.95
Acid Number mgKOH/g: Nil
Mettler Drop Point C(F): 121(250)
Particle Size Distribution in Microns: Mean Value: 9.4

C-12:
Hardness dmm: 1.0
Density g/cc: 0.95
Acid Number mgKOH/g: Nil
Mettler Drop Point C(F): 121(250)
Particle Size Distribution in Microns: Mean Value: 12.2

C-18:
Hardness dmm: 1.0
Density g/cc: 0.95
Acid Number mgKOH/g: Nil
Mettler Drop Point C(F): 121(250)
Particle Size Distribution in Microns: Mean Value: 16.2

ALLIED SIGNAL INC.: A-C Polyethylenes and Copolymers For Personal Care Products: Typical Advantages:

Applications: Eye Shadow
Recommended A-C Polyethylene Grades: A-C 617, 617A, 6, 6A

Applications: Eye Mascara
Recommended A-C Polyethylene Grades: A-C 617, 617A, 6, 6A

Applications: Hair Pomade
Recommended A-C Polyethylene Grades: A-C 617, 617A, 6A, 9

Applications: Lipstick
Recommended A-C Polyethylene Grades: A-C 617, 617A, 9, 9A, 540, 540A, 400

Applications: Lip Gloss
Recommended A-C Polyethylene Grades: A-C 617, 617A, 9, 9A, 540, 540A, 400

Applications: Stick Products
Recommended A-C Polyethylene Grades: A-C 617, 617A, 9, 9A, 540, 540A, 400

Applications: Medicated Gels
Recommended A-C Polyethylene Grades: A-C 617, 617A, 9, 9A

Applications: Medicated Ointments
Recommended A-C Polyethylene Grades: A-C 617, 617A, 9, 9A

Special Properties & Advantages:
- Gel is thixotropic, easy to apply.
- Gel is not sensitive to temperature changes.
- Gel is water resistant.
- Gel is good suspension medium.
- Gel is inert medium for medication.
- Gel deposits an emollient film that soothes and protects.
- A-C polyethylene gels are low in cost.
- A-C polyethylenes are inert and safe.

Applications: Deodorants & Antiperspirants
Recommended A-C Polyethylene Grades: A-C 400, POLYMIST B-6, B-12
Products are inert and safe as gellant or dispersant.

ALLIED SIGNAL INC.: A-C Polyethylenes and Copolymers for Personal Care Products: Typical Advantages(Continued):

Applications: General Purpose Creams Both W/O and O/W
Recommended A-C Polyethylene Grades: A-C 617, 617A, 540, 540A

Applications: Cleansing Cream
Recommended A-C Polyethylene Grades: A-C 617, 617A, 540, 540A

Applications: Protective Hand Cream
Recommended A-C Polyethylene Grades: A-C 617, 617A, 540, 540A

Special Properties & Advantages:
Emollient film soothes and protects.
Good dispersibility and stability.
Low cost, inert and safe.
Film protects against chemicals.

Applications: Acne Scrub Cream
Recommended A-C Polyethylene Grades: A-C 9A, 9F
Inert, safe, non-irritating abrasive.

Applications: Hand Scrub Soap
Recommended A-C Polyethylene Grades: A-C 9A, 9F, 316A
Inert, safe, non-irritating abrasive.

Applications: Facial Scrub Cream
Recommended A-C Polyethylene Grades: A-C 617A, 9A, 9F
Inert, safe, non-irritating abrasive.

Applications: Sticky Gels
Recommended A-C Polyethylene Grades: A-C 400, 430, 540, 540A, 617, 617A
Good adhesion to most surfaces.

Applications: Pressed Powders, Eye Shadow, Etc.
Recommended A-C Polyethylene Grades: POLYMIST B-6, B-12
Micronized polyethylenes for use as flatting or texturizing agents.

AMERCHOL CORP.: Lanolin Derivatives:

ACETULAN:

Water-thin fraction of acetylated lanolin alcohols. Hydrophobic (oil-miscible) emollient with remarkable velvety feel. Anhydrous spreading agent, penetrant, co-solvent and plasticizer. Reduces "greasiness" of normally oily or greasy formulations.

CTFA Adopted Name: Acetylated Lanolin Alcohol

AMERCHOL L-101:

Multisterol extract containing highly surface active selected lanolin sterols and complex higher alcohols in their free forms only. Superb natural moisturizer, penetrant and emollient. Nonionic w/o emulsifier. Stabilizer for o/w emulsions and multiphase dispersions. Anhydrous, oil-miscible, low viscosity fluid.

CTFA Adopted Name: Mineral Oil and Lanolin Alcohol

AMERCHOL L-99:

Liquid absorption base containing broader spectrum of lanolin sterols and complex higher alcohols than in AMERCHOL L-101.

CTFA Adopted Name: Mineral Oil and Lanolin Alcohol

AMERCHOL L-500:

Multisterol extract containing a higher concentration of the surface active lanolin sterols and higher alcohols of AMERCHOL L-99.

CTFA Adopted Name: Lanolin Alcohol and Mineral Oil

AMERCHOL 400:

Relatively lipophilic fraction of lanolin alcohols, primarily triterpene and aliphatic in nature. Unique moisturizing emollient with lubricant and barrier properties.

CTFA Adopted Name: Lanolin Alcohol and Petrolatum

AMERCHOL RC:

Special concentrate of the lanolin alcohol fraction of AMERCHOL 400. Moisturizer and conditioner with characteristic nontacky feel.

CTFA Adopted Name: Lanolin Alcohol and Petrolatum

AMERCHOL CAB:

Multisterol extract containing select lanolin sterols and higher alcohols in their free forms only. Unusually emollient. Nonionic w/o emulsifier and stabilizer for o/w systems with special application to pharmaceutical vehicles.

CTFA Adopted Name: Petrolatum and Lanolin Alcohol

AMERCHOL CORP.: Lanolin Derivatives(Continued):

AMERCHOL C:
Highly emollient multisterol absorption base containing select fractions of surface active lanolin sterols and sterol esters.
CTFA Adopted Name: Petrolatum, Lanolin and Lanolin Alcohol

AMERCHOL H-9:
Special multisterol absorption base containing cholesterol esters as well as free sterols. Natural emollient. Nonionic w/o emulsifier. Absorbs unusually large amounts of water.
CTFA Adopted Name: Petrolatum, Lanolin and Lanolin Alcohol

AMERCHOL BL:
Enriched multisterol absorption base containing a high concentration of sterol esters as well as free sterols and higher alcohols. Nonionic w/o emulsifier.
CTFA Adopted Name: Lanolin, Mineral Oil and Lanolin Alcohol

AMERLATE P:
Isopropyl ester of the unique fatty acids occurring in lanolin. Remarkable buttery consistency with spreading, penetrating, lubricating and moisturizing properties. Distinctive satiny feel. Wetting agent for cosmetic pigments.
CTFA Adopted Name: Isopropyl Lanolate

AMERLATE W:
Isopropyl lanolate. More economical than AMERLATE P because of process simplification. Wetting and dispersing agent for pigmented products. Improves spread and ease of application of waxy products.
CTFA Adopted Name: Isopropyl Lanolate

AMERLATE LFA:
Lanolin fatty acids. Distinctive because they also contain branched chain and hydroxy acids. Carefully isolated to prevent loss of functional groups particularly the hydroxyls and side chains. Performance is significantly different from conventional fatty acids. Excellent pigment wetting and dispersing agent. Versatile o/w or w/o emulsifier for soap emulsions depending on the valence of the neutralizer. Stabilizes propellant dispersions in water-based aerosol foams.
CTFA Adopted Name: Lanolin Acid

AMERCHOL CORP.: Lanolin Derivatives(Continued):

AMERLATE WFA:

Lanolin fatty acids similar to AMERLATE LFA but somewhat more economical because of the manufacturing process.

CTFA Adopted Name: Lanolin Acid

MODULAN:

Acetylated USP lanolin (US 2,725,334 and foreign patents). Contains the constituents of lanolin modified by acetylation. This treatment of lanolin greatly improves oil solubility and sharply reduces the allergenicity of lanolin. Retains the conditioning and protective benefits of lanolin while eliminating undesirable features.

CTFA Adopted Name: Acetylated Lanolin

OHLAN:

Hydroxylated lanolin. Greatly increased structural hydroxyl content by chemical reaction for enhanced surface and interfacial activity. Superior nonionic w/o emulsifier. Stabilizer for o/w systems. Pigment wetting and dispersing agent. Adds tack, gloss, body and adhesion for makeup.

CTFA Adopted Name: Hydroxylated Lanolin

POLYLAN:

Polyunsaturated liquid ester of linoleic acid and selected lanolin alcohols. Combines beneficial effects of essential unsaturated fatty acids with the most active lanolin components.

CTFA Adopted Name: Lanolin Linoleate

SOLULAN 5:

Ethylene oxide (5 moles) ether of a select fraction of lanolin alcohols and sterols. Nontacky lubricant, emollient, moisturizer and conditioner. Nonionic w/o emulsifier, auxiliary emulsifier and stabilizer for o/w systems. Wetting and dispersing aid for cosmetic pigments, solvent for bromo acid dyes.

CTFA Adopted Name: Laneth-5

SOLULAN 16:

Ethoxylated (16 moles) lanolin alcohols. Nonionic solublizer, wetting agent and o/w emulsifier. Compatible with high concentrations of electrolytes. Excellent conditioning agent in shampoos; foam stabilizer.

CTFA Adopted Name: Laneth-16

SOLULAN 25:

Ethoxylated (25 moles) lanolin alcohols. Nonionic o/w emulsifier and solubilizer. Homolog of SOLULAN 16.

CTFA Adopted Name: Laneth-25

AMERCHOL CORP.: Lanolin Derivatives(Continued):

SOLULAN 75:
Highly purified water-soluble lanolin. Polyoxyethylene (75 moles) ether of whole lanolin. Economical. Not defatting. Does not interfere with foam.
CTFA Adopted Name: PEG-75 Lanolin

SOLULAN L-575:
50% aqueous solution of SOLULAN 75. Convenient fluid form of water-souble lanolin. Adds conditioning effects to those water-based systems in which lanolin is not soluble.
CTFA Adopted Name: PEG-75 Lanolin

SOLULAN 97:
Completely acetylated ethoxylated (9 moles) ester-ether lanolin derivative. Excellent conditioner for skin and hair. Plasticizer for many resins.
CTFA Adopted Name: Laneth-9 Acetate

SOLULAN 98:
Partially acetylated ethoxylated (10 moles) ester-ether lanolin derivative. Nonionic o/w emulsifier, solubilizer and pigment wetting agent.
CTFA Adopted Name: Laneth-10 Acetate

SOLULAN C-24:
Polyethoxylated (24 moles) cholesterol. Convenient, economical means for incorporating emollient cholesterol in clear aqueous and alcoholic preparations. Viscosity control agent for soap emulsions. Plasticizer for water-soluble resins.
CTFA Adopted Name: Choleth-24

SOLULAN PB-2:
Propoxylated (2 moles) lanolin alcohols. Wetting agent for cosmetic pigments. Naturally-derived hydrophobic emollient. Adds gloss, body and tack to a variety of product forms. Especially suitable for hair grooms and heavily pigmented makeup items.
CTFA Adopted Name: PPG-2 Lanolin Ether

SOLULAN PB-5:
Propoxylated (5 moles) lanolin alcohols. Pigment wetter and glosser. Water-resistant conditioner for skin and hair, effective at low concentrations in shampoos.
CTFA Adopted Name: PPG-5 Lanolin Ether

AMERCHOL CORP.: Lanolin Derivatives(Continued):

SOLULAN PB-10:
Propoxylated (10 moles) lanolin alcohols. Spreading agent, plasticizer, pigment wetting agent, glosser and hydrophobic emollient. Co-solvent for AMERSCREEN P u.v. absorber.
CTFA Adopted Name: PPG-10 Lanolin Ether

SOLULAN PB-20:
Propoxylated (20 moles) lanolin alcohols. Plasticizer, glosser, spreading agent and conditioner. Co-solvent for AMERSCREEN P u.v. absorber.
CTFA Adopted Name: PPG-20 Lanolin Ether

AMERCHOL CORP.: Chemical Specialties:

AMEROXOL OE-2:
Ethoxylated (2 moles) oleyl alcohol. Designated "fragrance grade" because of exceptionally low odor. Nonionic w/o emulsifier and spreading agent for bath oils, etc. Auxiliary emulsifier for o/w systems.
CTFA Adopted Name: Oleth-2

AMEROXOL OE-10:
Ethoxylated (10 moles) oleyl alcohol. Odorless solubilizer, especially for fragrance products. Nonionic o/w emulsifier.
CTFA Adopted Name: Oleth-10

AMEROXOL OE-20:
Ethoxylated (20 moles) oleyl alcohol. Practically odorless "fragrance grade" solubilizer.
CTFA Adopted Name: Oleth-20

AMERSCREEN P:
U.V. absorber. Propoxylated (2 moles) ethyl para-aminobenzoate. Highly active absorber in burning range. Prevents burning but permits tanning. Unusually stable to heat, oxidation and u.v. rays. Safety and efficacy data file available. Category I for cosmetic use.
CTFA Adopted Name: Ethyl dihydroxypropyl PABA

GLUCAM E-10:
Ethoxylated (10 moles) glucose derivative. Rich, uniform rubout and smooth after-feel. High film gloss. Effective humectant. Interesting alternative to glycerine, sorbitol, propylene glycol and other polyols.
CTFA Adopted Name: Methyl Gluceth-10

AMERCHOL CORP.: Chemical Specialties(Continued):

GLUCAM E-20:

Ethoxylated (20 moles) glucose derivative. Specially developed polyol for use on skin and hair. Rich emollient feel, adds gloss. Humectant and foam enricher. Most active of GLUCAM series.

CTFA Adopted Name: Methyl Gluceth-20

GLUCAM P-10:

Propoxylated (10 moles) glucose derivative. Water-soluble emollient with rich feel. Enhances film gloss. Increases wetness and improves texture of foams. Used in hydroalcoholic systems such as astringents and after-shaves.

CTFA Adopted Name: PPG-10 Methyl Glucose Ether

GLUCAM P-20:

Propoxylated (20 moles) glucose derivative. Unique water-soluble oil with broad range of miscibility in water, alcohol and most organic oils and solvents (except hydrocarbons). Distinctive emollient feel and high film gloss. Serves as perfume fixative.

CTFA Adopted Name: PPG-20 Methyl Glucose Ether

GLUCAMATE SSE-20:

Ethoxylated (20 moles) methyl glucose sesquistearate. An o/w emulsifier (HLB-15). Forms unique nonionic primary emulsifier system in combination with GLUCATE SS (q.v.). Active solubilizer for perfume oils.

CTFA Adopted Name: Methyl Gluceth-20 Sesquistearate

GLUCATE SS:

Methyl glucoside sesquistearate. A w/o emulsifier (HLB-6). Forms unique nonionic primary emulsifying system when used in combination with GLUCAMATE SSE-20. This emulsifying system provides unusual degree of viscosity recovery in o/w emulsions after shear stress.

CTFA Adopted Name: Methyl Glucose Sesquistearate

AQUALON CO.: AQUALON Cellulose Gum:

Purified Sodium Carboxymethylcellulose

AQUALON cellulose gum is the sodium salt of carboxymethylcellulose with a minimum purity of 99.5%. It is an anionic water-soluble polymer, supplied as a powder of white to light cream color, and is available in a number of grades and viscosity types for use in foods, pharmaceuticals, cosmetics, water-based coatings, and many other products.

This polymer provides a variety of functions for aqueous systems, including thickening, rheology control, binding, and oil- and grease-resistance for films. It is also a protective colloid and suspending agent. Cellulose gum is available in a variety of particle sizes; this facilitates its handling in commercial operations.

Typical Properties:

Sodium carboxymethylcellulose content (dry basis, as shipped), % min: 99.5
Moisture content (as shipped), %, max: 8.0
Bulk density, g/ml: 0.75
Specific gravity of film: 1.59
Specific gravity of 2% solution at 25C: 1.0068
pH of 2% solution: 7.5
Solution bulking value, gal/lb: 0.0652
Biological oxygen demand (BOD), ppm:
7H type: 11,000
7L type: 17,300

Grades and Types: Viscosity Range, cps:

Viscosity Type:

7H4, 9H4: 1% by Weight in Water: 2,500-6,000
7HF: 1% by Weight in Water: 1,500-3,000
7H3S: 1% by Weight in Water: 1,000-2,800

9M31, 12M31: 2% by Weight in Water: 800-3,100
7M, 9M8, 12M8P: 2% by Weight in Water: 400-800
7M2: 2% by Weight in Water: 100-200
7L: 2% by Weight in Water: 25-50

7L2: 4% by Weight in Water: 50-200
7L1: 4% by Weight in Water: 25-50

AQUALON CO.: KLUCEL Hydroxypropylcellulose:

KLUCEL hydroxypropylcellulose is a nonionic water-soluble cellulose ether with a versatile combination of properties. It combines organic solvent solubility, thermoplasticity, and surface activity with the thickening and stabilizing properties of other water-soluble cellulose polymers.

Outstanding Characteristics:

KLUCEL is soluble in many polar organic solvents and in water below 38C, but is insoluble in water above 45C. KLUCEL is highly surface-active, with low surface and interfacial tensions of solutions. It has a wide range of compatibility with latexes and with synthetic and natural colloids. KLUCEL is available in a wide range of viscosities of varying concentrations in both water and anhydrous ethanol. It is thermoplastic and can be injection molded and extruded. In films and coatings, KLUCEL is heat-sealable, extremely flexible without plasticizers, and nontacky at high humidity. The Food and Drug Administration has cleared KLUCEL for use as an intentional food additive.

CAS Number: 9004-64-2
CAS Name: Cellulose, 2-hydroxypropyl ether

Grades and Types:

KLUCEL is produced in two grades: standard and premium. The standard grades are for industrial uses; the premium grades are for foods, pharmaceuticals, and cosmetics. All viscosity types are available in each grade.

Typical Properties:

- Appearance: off-white, tasteless powder
- Particle size:
  - Through 30 mesh, % min: 85
  - Through 20 mesh, % min: 99
- Ash content (calculated as Na2SO4), % max: 0.5
- Ash content, premium grades (F), calculated as Na2SO4, % max (excluding silica): 0.2
- Moisture content (as packed), %, max: 5.0
- Bulk density, g/ml: 0.5 (varies with type)
- Softening temperature, C: 100-150
- Burnout temperature in N2 or O2, C: 450-500
- Solubility
  - in water: clear, smooth solutions below 38C
  - in organic solvents: Dissolves easily in many polar organic solvents to give clear, smooth solutions

AQUALON CO.: NATROSOL Plus CS, Grade 330: Pharmaceutical and Cosmetics Grade Modified Hydroxyethylcellulose:

NATROSOL Plus CS, Grade 330, modified hydroxyethylcellulose is an associative cellulosic thickener designed to provide thickening and viscosity stability in both aqueous and surfactant systems. This polymer has potential use in a wide variety of applications, including pharmaceutical and cosmetic products. NATROSOL Plus CS can function as a thickener, binder, stabilizer, film-former, or suspending agent.

In Shampoo Systems:

NATROSOL Plus CS has excellent compatibility with anionic, nonionic, and amphoteric surfactants. NATROSOL Plus CS is soluble at all temperatures, and heat stabilities of shampoos thickened with this polymer are expected to be excellent. The hydrophobic groups of this polymer will associate with the surfactant hydrophobes.

Chemical and Physical Properties:

NATROSOL Plus CS is a nonionic water-soluble polymer that contains both hydroxyethyl and long-chain alkyl groups. The unique combination of hydrophobic side chains and a hydrophilic polymer backbone differentiates this new polymer from traditional cellulosics. The hydrophobes interact through intramolecular and intermolecular associations, increasing the viscosity and changing the rheology of NATROSOL Plus CS solutions. NATROSOL Plus CS is known to associate with the hydrophobic groups of surfactants, latexes, and pigments.

Solutions of NATROSOL Plus CS form clear films upon drying. Thanks to its nonionic nature, NATROSOL Plus CS blends well with natural polymers and pigments, and is compatible with a wide range of additives.

Typical Properties:

Form as supplied: white to off-white powder
Moisture content (as packed), %: 0-5.0
Ash content (calculated as Na_2SO_4), %, max: 4
Bulk density, g/ml: 0.75
Bulking value, gal/lb: 0.10
Brookfield viscosity, 1%, cps: 150-500
Solution appearance: clear
Hydration time, minutes, pH 7.2: 6-20
Surface tension, dynes/cm: 60-65

AQUALON CO.: NATROSOL 250 Hydroxyethylcellulose:

A Nonionic Water-Soluble Cellulose Ether

NATROSOL 250 hydroxyethylcellulose, a nonionic water-soluble polymer, is a white, free-flowing granular powder insoluble in organic solvents, yet it is easily dispersed in cold or hot water to give solutions of varying viscosities. Chemically, it is cellulose that has been etherified with hydroxyethyl groups to give the desired properties.

NATROSOL 250 is used as a thickener, protective colloid, binder, stabilizer, and suspending agent, particularly in applications where a nonionic material is desired. The HH type is the most efficient nonionic thickener available. All viscosity types are obtainable in easy-dispersing form, designated R (e.g., 250HHR). These R materials have been treated to delay hydration of the particle and thus prevent lumping as the dry powder is added to water. A B grade of NATROSOL (e.g. 250HBR) with improved resistance to biological degradation is also available.

Typical Properties: Solution Viscosity(Brookfield at 25C, cps):

Viscosity Type:

HH:1%: 3,400-5,000
H4:1%: 2,600-3,300
H: 1%: 1,500-2,500
MH:1%: 800-1,500

M: 2%: 4,500-6,500
K: 2%: 1,500-2,500
G: 2%: 150- 400
E: 2%: 25- 105

J: 5%: 150- 400
L: 5%: 75- 150

All viscosity types have the following typical properties:
Ash (calculated as Na2SO4), % max: 5.5
pH (2% solution): 6.0-8.5
Color: white to light tan
Moisture (as packed), %, max: 5
Particle size: Passing U.S. No. 40 mesh, %, min: 90

AVATAR CORP.: CRYOSEPT I and CRYOSEPT II Preservatives:

CRYSOSEPT I:

CRYOSEPT I is a preservative complex that allows the cosmetic formulator to incorporate methyl and propyl paraben at common usage levels via an easy to handle liquid. Incorporating CRYOSEPT I at a level of 1% introduces the parabens without the application of additional process heat. Clear at room temperature, and when used at 1% or below, CRYOSEPT I is completely soluble in most personal care formulations.

Specifications:
- Appearance, 25C: Clear liquid, free from foreign matter.
- Color, 25C: Water white to very pale copper.
- Odor: Characteristic, mild not offensive.
- pH, 25C: 6.00-7.00
- Total Solids, %: 28.0-32.0
- Specific Gravity, 25C: 1.050-1.090

Ingredient Declaration:
Propylene Glycol, Methyl Paraben, Propyl Paraben

CRYOSEPT II:

CRYOSEPT II is a liquid preservative complex that allows the cosmetic formulator to incorporate broad spectrum anti-microbial protection without the application of additional process heat or the attendent risk of agglomeration. CRYOSEPT II, incorporated at a level of 1% introduces imidizolidinyl urea, methyl paraben, and propyl paraben at conventional concentrations in one straightforward step. Clear at room temperature, and when used at 1% or below, CRYOSEPT is completely soluble in most personal care formulations.

Specifications:
- Appearance, 25C: Clear liquid, free from foreign matter.
- Color, 25C: Water white to very pale yellow.
- Odor: Characteristic, mild not offensive.
- pH, 25C: 6.50-7.00
- Total Solids, %: 38.5-42.5
- Specific Gravity, 25C: 1.140-1.180

Ingredient Declaration:
Propylene Glycol, Imidazolidinyl Urea, Methyl Paraben, Propyl Paraben

BASF CORP.: Products for the Cosmetics Industry:

Polymers:

CTFA Name: PVP:

LUVISKOL K12 Powder
LUVISKOL K12 Solution approx. 50%
LUVISKOL K17 Powder
LUVISKOL K17 Solution approx. 50%
LUVISKOL K30 Powder
LUVISKOL K30 Solution approx. 30%
LUVISKOL K60 Solution approx. 45%
LUVISKOL K80 Powder
LUVISKOL K80 Solution approx. 20%
LUVISKOL K90 Powder
LUVISKOL K90 Solution approx. 20%

CTFA Name: PVP/VA Copolymer:

LUVISKOL VA73E
LUVISKOL VA73I
LUVISKOL VA73W
LUVISKOL VA64 Powder
LUVISKOL VA64E
LUVISKOL VA64I
LUVISKOL VA64W
LUVISKOL VA55E
LUVISKOL VA55I
LUVISKOL VA37E
LUVISKOL VA37I
LUVISKOL VA37PM
LUVISKOL VA28E
LUVISKOL VA28I

CTFA Name: PVP/VA/Vinyl Propionate Copolymer:

LUVISKOL VAP343E
LUVISKOL VA343I
LUVISKOL VA343PM

CTFA Name: Acrylate/Acrylamide Copolymer:

ULTRAHOLD 8

CTFA Name: Vinyl Acetate/Crotonic Acid Copolymer:

LUVISET CA66:

CTFA Name: Vinyl Acetate/Crotonic Acid/Vinyl Propionate Copolymer:

LUVISET CAP:

CTFA Name: Rosin Acrylate:

Hairspray Additive S

BASF CORP.: Products for the Cosmetics Industry(Continued):

Quaternary Compounds:

CTFA Name: Polyquaternium-16:

LUVIQUAT FC370
LUVIQUAT FC550
LUVIQUAT FC905
LUVIQUAT HM552

CTFA Name: Hydroxyethyl Cetyldimonium Phosphate:

LUVIQUAT MONO CP:

Emulsifiers:

CREMOPHOR A6:
CTFA: Ceteareth-6 (and) Stearyl Alcohol

CREMOPHOR A11:
CTFA: Ceteareth-11

CREMOPHOR A25:
CTFA: Ceteareth-25

CREMOPHOR S9:
CTFA: PEG-9 Stearate

CREMOPHOR WO7:
CTFA: PEG-7 Hydrogenated Castor Oil

PLURONIC L72:
CTFA: Poloxamer 212

PLURONIC L121:
CTFA: Poloxamer 401

Fat and oil components:

LUVITOL EHO:
Cetearyl Octanoate

LUVITOL HP:
Hydrogenated Polysiobutene

BASF CORP.: Products for the Cosmetics Industry(Continued):

Solubilizers:

CREMOPHOR RH40:
CTFA: PEG-40 Hydrogenated Castor Oil

CREMOPHOR RH410:
CTFA: PEG-40 Hydrogenated Castor Oil (and) Water

CREMOPHOR RH455:
CTFA: PEG-40 Hydrogenated Castor Oil (and) Propylene Glycol (and) Water

CREMOPHOR RH60:
CTFA: PEG-60 Hydrogenated Castor Oil

CREMOPHOR EL:
CTFA: PEG-35 Castor Oil

CREMOPHOR NP10:
CTFA: Nonoxynol-10

CREMOPHOR NP14:
CTFA: Nonoxynol-14

PLURONIC F68:
CTFA: Poloxamer 188

EO/PO Block Copolymers:

PLURONIC L35:
CTFA: Poloxamer 105

PLURONIC L64:
CTFA: Poloxamer 184

PLURONIC P65:
CTFA: Poloxamer 185

PLURONIC F68:
CTFA: Poloxamer 188

PLURONIC L72:
CTFA: Poloxamer 212

PLURONIC F87:
CTFA: Poloxamer 237

PLURONIC P103:
CTFA: Poloxamer 333

BASF CORP.: Products for the Cosmetics Industry(Continued):

EO/PO Block Copolymers(Continued):

PLURONIC F108:
CTFA: Poloxamer 338

PLURONIC L121:
CTFA: Poloxamer 401

LUTROL F127:
CTFA: Poloxamer 407

TETRONIC 304:
CTFA: Poloxamine 304

Sunscreen Agents:

UVINUL P25:
(Previously LUSANTAN 25)
CTFA: PEG-25 PABA

UVINUL M40:
CTFA: Benzophenone-3

UVINUL MS40:
CTFA: Benzophenone-4

UVINUL D49:
CTFA: Benzophenone-6

UVINUL DS49:
CTFA: Benzophenone-9

UVINUL D50:
CTFA: Benzophenone-2

UVINUL 400:
CTFA: Benzophenone-1

UVINUL 490:
CTFA: Benzophenone-11

UVINUL N35:
CTFA: Etocrylene

UVINUL N539:
CTFA: Octocrylene

BASF CORP.: Products for the Cosmetics Industry(Continued):

Active Agents:

(±)-alpha-Bisabolol rac.:
(-)-alpha-Bisabolol nat.:
CTFA: Bisabolol

D-Panthenol USP:
CTFA: Panthenol

D-Panthenol 50P:
CTFA: Panthenol (and) Propylene Glycol

Vitamin E Nicotinate C:
CTFA: Tocopheryl Nicotinate

Thickeners:

KATIORAN AF:
CTFA: Stearamide DEA (and) Ceteareth-25

LUTROL FC127:
CTFA: Poloxamer 407

Polyethylene Glycols:

LUTROL E300:
CTFA: PEG 6

LUTROL E400:
CTFA: PEG 8

LUTROL E1500:
CTFA: PEG 32

LUTROL E4000:
CTFA: PEG 75

LUTROL E6000:
CTFA: PEG 150

BASF CORP.: Products for the Cosmetics Industry(Continued):

Vitamins:

Vitamin A acetate:
CTFA: Retinyl Acetate

Vitamin A palmitate:
CTFA: Retinyl Palmitate

Vitamin A propionate:
CTFA: Retinyl Propionate

Vitamin A/D mixtures:
CTFA: Retinyl Palmitate (and) Ergocalciferol

Vitamin B1 hydrochloride:
CTFA: Thiamine HCL

Vitamin B1 mononitrate:
CTFA: Thiamine Nitrate

Vitamin B2:
CTFA: Riboflavin

Vitamin B5:
CTFA: Calcium Pantothenate

Vitamin B6:
CTFA: Pyridoxine HCL

Vitamin C:
CTFA: Ascorbic Acid

Sodium ascorbate:
CTFA: Sodium Ascorbate

Vitamin E:
CTFA: Tocopherol

Vitamin E acetate:
CTFA: Tocopheryl Acetate

Vitamin E Nicotinate C:
CTFA: Tocopheryl Nicotinate

BASF CORP.: Products for the Cosmetics Industry(Continued):

Solvents:

1,2-Propanediol USP:
CTFA: Propylene Glycol

n-Propanol USP:
CTFA: Propyl Alcohol

Solvent PM:
CTFA: Methoxypropanol

PALATINOL M:
CTFA: Dimethyl Phthalate

Neutralizing agents:

Triethanolamine Pure C:
CTFA: Triethanolamine

QUADROL:
CTFA: Tetrahydroxypropyl Ethylenediamine

Various:

PALATINOL A:
CTFA: Diethyl Phthalate

PALATINOL M:
CTFA: Dimethyl Phthalate

TRILON BS:
CTFA: EDTA

TRILON B Powder:
CTFA: Tetrasodium EDTA

TRILON B Liquid:
CTFA: Tetrasodium EDTA

TRILON BD:
CTFA: Disodium EDTA

BASF CORP.: Products for the Cosmetics Industry(Continued):

Colorants:

1. Dyes:

Beta-Carotene Cryst.:
CTFA: Carotene

Beta-Carotene 10A:
CTFA: Vegetable oil (and) Carotene

Beta-Carotene 20A:
CTFA: Vegetable oil (and) Carotene

SICOMET Uranin 75:
Colour Index: 45350
FDA Listed As: D&C Yellow No. 8
CTFA: Acid Yellow 73 Sodium Salt

SICOMET Quinoline 70E104:
Colour Index: 47005
CTFA: Acid Yellow 3

SICOMET Tartrazine 85E102:
Colour Index: 19140
FDA Listed As: FD&C Yellow No. 5
CTFA: Acid Yellow 23

SICOMET Yellow F11920:
Colour Index: 11920

SICOMET Yellow Z2787:
Colour Index: 13015 + 15510

SICOMET Sunset Yellow 85E110:
Colour Index: 15985
FDA Listed as: FD&C Yellow No. 6
CTFA: Sunset Yellow

SICOMET Red F12150:
Colour Index: 12150

SICOMET COCHINEAL Red 70E124:
Colour Index: 16255

SICOMET ERYTHROSINE 85E127:
Colour Index: 45430
FDA Listed as: FD&C Red No. 3
CTFA: Acid Red 51

BASF CORP.: Products for the Cosmetics Industry(Continued):

Colorants(Continued):

1. Dyes(Continued):

SICOMET AZORUBINE 85E122:
Colour Index: 14720

SICOMET AMARANTH 85E123:
Colour Index: 16185

SICOMET Red S45410:
Colour Index: 45410
FDA Listed as: D&C Red No. 28
CTFA: Acid Red 92

SICOMET Red B45170:
Colour Index: 45170
FDA Listed as: D&C Red No. 19
CTFA: Basic Violet 10

SICOMET Blue S42735:
Colour Index: 42735

SICOMET INDIGOTINE 85E132:
Colour Index: 73015
FDA Listed as: FD&C Blue No. 2

SICOMET Blue S74180:
Colour Index: 74180

SICOMET Blue S42045:
Colour Index: 42045

SICOMET Patent Blue 80E131:
Colour Index: 42051

SICOMET Blue S42090:
Colour Index: 42090
FDA Listed as: FD&C Blue No. 1
CTFA Name: Acid Blue 9

SICOMET Green Z6120:
Colour Index: 19140 + 42051

SICOMET Green Z2755:
Colour Index: 47005 + 42051

SICOMET Green S10020:
Colour Index: 10020

BASF CORP.: Products for the Cosmetic Industry(Continued):

Colorants(Continued):

1. Dyes(Continued):

SICOMET Green Z2901:
Colour Index: 19140 + 61570 + 15985

SICOMET Green S61570:
Colour Index: 61570
FDA Listed as: D&C Green No. 5
CTFA Name: Acid Green 25

SICOMET Brown S20170:
Colour Index: 20170

SICOMET Brilliant Black 80E151:
Colour Index: 28440

2. Pigments:

SICOMET QUINOLINE Yellow Lake E104:
Colour Index: 47005:1

SICOMET Yellow P11680:
Colour Index: 11680
CTFA: Pigment Yellow 1

SICOMET Yellow D11680:
Colour Index: 11680

SICOMET TARTRAZINE Lake E102:
Colour Index: 19140:1
FDA Listed as: FD&C Yellow No. 5 Aluminum Lake

SICOMET Yellow 10:
Colour Index: 77492
FDA Listed as: Iron Oxides
CTFA: Iron Oxides

SICOMET Sunset Yellow Lake E110:
Colour Index: 15985:1
FDA Listed as: FD&C Yellow No. 6 Aluminum Lake

SICOMET Orange P12075:
Colour Index: 12075
FDA Listed as: D&C Orange No. 17
CTFA: Pigment Orange 5

BASF CORP.: Products for the Cosmetics Industry(Continued):

Colorants(Continued):

2. Pigments(Continued):

SICOMET Red 30:
Colour Index: 77491
FDA Listed as: Iron Oxides
CTFA: Iron Oxides

SICOMET COCHINEAL Red Lake E124:
Colour Index: 16255:1

SICOMET ERYTHROSINE Lake E127:
Colour Index: 45430:1
FDA Listed as: FD&C Red No. 3 Aluminum Lake

SICOMET AZORUBINE Lake E122:
Colour Index: 14720:1

SICOMET AMARANTH Lake E123:
Colour Index: 16185:1

SICOMET Red P12085:
Colour Index: 12085
FDA Listed as: D&C Red No. 36
CTFA: Pigment Red 4

SICOMET Red P12490:
Colour Index: 12490

SICOMET Red P15630 Ca:
Colour Index: 15630:2

SICOMET Red P15850 Ca:
Colour Index: 15850:1
FDA Listed as: D&C Red No. 7
CTFA: Pigment Red 57:1

SICOMET Red P15880 Ca:
Colour Index: 15880:1
FDA Listed as: D&C Red No. 34
CTFA: Pigment Red 63:1

SICOMET Violet P77742:
Colour Index: 77742
FDA Listed as: Manganese Violet
CTFA: Manganese Violet

BASF CORP.: Products for the Cosmetics Industry(Continued):

Colorants(Continued):

2. Pigments(Continued):

SICOMET Violet P77007G:
Colour Index: 77007
FDA Listed as: Ultramarines
CTFA: Ultramarine Violet

SICOMET Blue P77007F:
Colour Index: 77007
FDA Listed as: Ultramarines
CTFA: Ultramarine Blue

SICOMET Blue P77007E:
Colour Index: 77007
FDA Listed as: Ultramarines
CTFA: Ultramarine Blue

SICOMET Blue P74160:
Colour Index: 74160
CTFA: Pigment Blue 15

SICOMET Blue D74160:
Colour Index: 74160

SICOMET Blue P77510:
Colour Index: 77510
FDA Listed as: Ferric Ferrocyanide
CTFA: Ferric Ferrocyanide

SICOMET INDIGOTINE Lake E132:
Colour Index: 73015:1
FDA Listed as: FD&C Blue No. 2 Aluminum Lake

SICOMET Green P74260:
Colour Index: 74260
CTFA: Pigment Green 7

SICOMET Green F74260:
Colour Index: 74260

SICOMET Green Lake:
Colour Index: 47005 + 73015

SICOMET Green P77288:
Colour Index: 77288
FDA Listed as: Chrom Oxide Greens
CTFA Name: Chromium Oxide Greens

BASF CORP.: Products for the Cosmetics Industry(Continued):

Colorants(Continued):

2. Pigments(Continued):

SICOMET Brown ZD2691:
Colour Index: 77491 + 11680 + 10006

SICOMET Brown 70:
Colour Index: 77491 + 77492 + 77499
FDA Listed as: Iron Oxides
CTFA: Iron Oxides

SICOMET Brown D77491:
Colour Index: 77491

SICOMET Brown Lake:
Colour Index: 15985 + 14720 + 28440

SICOMET Brown 75:
Colour Index: 77491 + 77499
FDA Listed as: Iron Oxides
CTFA: Iron Oxides

SICOMET Black 80:
Colour Index: 77499
FDA Listed as: Iron Oxides
CTFA: Iron Oxides

SICOMET Black 85:
Colour Index: 77499
FDA Listed as: Iron Oxides
CTFA Name: Iron Oxides

BIO-BOTANICA, INC.: Aloe Vera Gel (CTFA designation):

Correct botanical name: Aloe Barbadensis Miller
or
Aloe Vera Linne

Bio-Botanica presents:

Aloe Vera Gel 1:2 (Equivalent to Aloe Vera Gel 1:1 liq.)

Aloe Vera Gel 1:10 concentrate

Aloe Vera Gel 1:40 concentrate

Aloe Vera Gel 1:50 concentrate

Aloe Vera Gel 1:100 concentrate

Aloe Phytogel 1:199 Spray Dried powder

Aloe Vera Oil Bases

Aloe Vera Linne is a perennial herb that grows mainly in tropical and sub tropical areas.

It originates from Africa, spread out to the Mediterranean, East and West Indies, and the Americas. It is cultivated for its clear inner GEL.

This Aloe Vera Gel or mucilage has the appearance of colorless gelatin with hairlike connective veins. This gel has been used since ancient times by natives of many tropical countries for sunburns, cuts, bruises and a multitude of other ailments.

All concentrates are available as food grade or cosmetic grade, the difference being the preservative system used.

BIO-BOTANICA, INC.: Bath Complexes:

Code:

3157:
Chemical/Trade Name: BBC MINERAL COMPLEX (Bio-Botanica)
CTFA Adopted Name: Alfalfa Extract (and) Nettle Extract (and) Parsley Extract (and) Borage Extract (and) Horsetail Extract (and) Bladderwrack Extract (and) Red Raspberry Extract (and) Clover Blossom Extract
Approximate Use Levels: MINERAL COMPLEX is a highly concentrated botanical extract which has been formulated to mineralize the bath water. Bio-Botanica has chosen those botanicals which are reputedly used for this purpose. MINERAL COMPLEX can also be used in creams, lotions, etc. wherever a high botanical source of minerals would be required, usually incorporated at 0.5-5% in aqueous phase.

3158:
Chemical/Trade Name: BBC MOISTURE TROL (Bio-Botanica)
CTFA Adopted Name: Slippery Elm Bark (and) Comfrey Extract (and) Horsetail Extract (and) Oat Extract (and) Althea Extract (and) Orange Flower Water (and) Borage Extract (and) Chamomile Extract (and) Lemongrass Extract (and) Rose Extract (and) Peppermint Extract
Approximate Use Levels: Moisture complex is a highly concentrated botanical extract formulated with those herbs which are noted for their hydrating qualities. This complex can be added to bath blends, creams, lotions, wherever a moisturizing formula would be advantageous in creams, lotions, ointments. MOISTURE TROL is highly mucilaginous and water soluble, usually incorporated at 1-5% in aqueous phase.

3159:
Chemical/Trade Name: BBC RELAXING COMPLEX (Bio-Botanica)
CTFA Adopted Name: Lavender Extract (and) Chamomile Extract (and) Passion Flower Extract (and) Peppermint Extract (and) Horsetail Extract (and) Rose Extract
Approximate Use Levels: Relaxing complex is a highly concentrated botanical extract formulated with those herbs which are noted for their tranquilizing, sedating properties, that lend themselves beautifully when incorporated in a relaxing bath blend, usually incorporated at 1-3% in aqueous phase.

BIO-BOTANICA, INC.: Botanicals in Cosmetics:

Code:

3400:
Strength: F.E.
3000:
Strength: 1:1 PG
3200:
Strength: 5:1 PG
Common Name: Agrimony
Botanical Name: (Agrimonia eupatoria)
CTFA Name: (Agrimony Extract)
Folkloric Use: Its action is a mild astringent, diuretic and a mild tonic.

3403:
Strength: F.E.
3003:
1:1 PG
3203:
5:1 PG
Common Name: Alfalfa
Botanical Name: (Medicago sativa)
CTFA Name: (Alfalfa Extract)
Folkloric Use: Used in facial steams, as a mild exfoliant (contains protease) in oils, creams, lotions, hair rinses, bath gels, oils, etc. Alfalfa is also high in minerals.

3401A:
Strength: F.E.
Common Name: Aloe
Botanical Name: (Aloe Vera)
Folkloric Use: Sunburn, burns, creams, lotions, hair treatments, cuts, scratches, irritated skin, insect bites, etc.

3401:
Strength: F.E.
3001:
Strength: 1:1 PG
3201:
Strength: 5:1 PG
Common Name: Arnica
Botanical Name: (Arnica montana)
CTFA Name: (Arnica)
Folkloric Use: Various uses as a stimulant to increase blood circulation including application to unbroken skin that has been bruised (black and blue marks). It also combines well with various other herbs to stimulate hair growth, but it is only to be applied to unbroken skin.

BIO-BOTANICA, INC.: Botanicals in Cosmetics(Continued):

Code:

3406:
Strength: F.E.
3007:
Strength: 1:1 PG
3207:
Strength: 5:1 PG
Common Name: Balm of Gilead
Botanical Name: (Populus candicans)
Folkloric Use: Aromatic antiseptic, resinous substance, has been used on blemishes, sores, bruises, cuts, various skin diseases.

3407:
Strength: F.E.
3008:
Strength: 1:1 PG
3208:
Strength: 5:1 PG
Common Name: Barberry
Botanical Name: (Barberis vulgaris)
CTFA Name: (Berberis)
(Berberis Extract)
Folkloric Use: Used as a stimulant in shampoos and hair rinses. Contains a yellow coloring matter (Berberine).

3408:
Strength: F.E.
3009:
Strength: 1:1 PG
3209:
Strength: 5:1 PG
Common Name: Bayberry
Botanical Name: (Myrica cerifera)
CTFA Name: (Bayberry Wax)
Folkloric Use: Astringent, circulatory, stimulant has been used for indolent ulcers. Contains Myricinic Acid (an acrid astringent resin) Tannic and garlic acid, a red coloring substance plus a minute amount of volatile oil. Used in shampoos and hair rinses.

BIO-BOTANICA, INC.: Botanicals in Cosmetics(Continued):

Code:

3411:
Strength: F.E.
3012:
Strength: 1:1 PG
3212:
Strength: 5:1 PG
Common Name: Black Walnut
Botanical Name: (Juglans Nigra)
CTFA Name: (Black Walnut Extract)
Folkloric Use: Contains Juglone, a coloring matter. 5-Hydroxy-1, 4 naphtha quinone, a yellow crystalline substance. Walnut extract is an old fashioned hair dye employed alone or in combination with Henna. Walnut is also used externally for its antiseptic properties in many kinds of skin diseases. It is astringent, anthelmintic and detergent.

3412:
Strength: F.E.
3013:
Strength: 1:1 PG
3213:
Strength: 5:1 PG
Common Name: Bladderwrack
Botanical Name: (Fucus vesiculosis)
CTFA Name: (Bladderwrack Extract)
Folkloric Use: Fucus contains a gelatinous substance, algin, mannitol, B-carotene, Zeaxanthin, traces of iodine, bromine and various other minerals. It has been used in obesity, rheumatism, myxoedema, cellutitis, slimming activity, as a wash for psoriasis, massage for cellulite, sprains and bruises.

3414:
Strength: F.E.
3015:
Strength: 1:1 PG
3215:
Strength: 5:1 PG
Common Name: Blueberry Leaf
Botanical Name: (Vaccinium myrtillus)
Folkloric Use: Astringent, used in sores, wounds, and ulcers.

BIO-BOTANICA, INC.: Botanicals in Cosmetics(Continued):

Code:

3421:
Strength: F.E.
3022:
Strength: 1:1 PG
3222:
Strength: 5:1 PG
Common Name: Burdock
Botanical Name: (Arctium lappa)
CTFA Name: (Burdock Root Extract)
Folkloric Use: Lappa contains inulin, flavinoids. It also contains an antibacterial substance (J. Am. Chem, Soc. 67 948). Used as a poultice for boils and abscesses, eczema, psoriasis, scaly skin, cutaneous eruptions, cystitis, gout, anorexia nervosa, can be used in facial steams. Has been combined with Comfrey, Licorice, Fennel to restore skin tone and smoothness. For use in hair conditioners, shampoos, creams, lotions, gels and baths.

3424:
Strength: F.E.
3025:
Strength: 1:1 PG
3225:
Strength: 5:1 PG
Common Name: Capsicum
Botanical Name: (Capsicum annum)
CTFA Name: (Capsicum Extract, Capsicum, Capsicum Oleo-Resin)
Folkloric Use: Cayenne Pepper contains a crystalline pungent principal capsaicin. Red coloring matter and a fatty oil spasmolytic, carminative counter-irritant, antiseptic, rubefacient used for Neuralgia, rheumatic pains in combination with tincture of myrrh is very antiseptic. Used in hair tonics to stimulate follicole along with Nettles, Jaborandi, Colocynth, etc.

3426:
Strength: F.E.
3028:
Strength: 1:1 PG
3227:
Strength: 5:1 PG
Common Name: Cat-nip
Botanical Name: (Nepeta cataria)
Folkloric Use: Catnip has been used for reduced swellings, puffy swollen eyes in combination with Gobenadora and wintergreen. It is useful for dandruff and various scalp disorders.

BIO-BOTANICA, INC.: Botanicals in Cosmetics(Continued):

Code:

3428:
Strength: F.E.
3031:
Strength: 1:1 PG
3230:
Strength: 5:1 PG
Common Name: Chamomile Flowers
Botanical Name: (Matricaria chamomilla L)
CTFA Name: (Chamomile Extract)
(Chamomile Oil)
Folkloric Use: Chamomile is one of the most widely used herbs in the Kingdom of Botanicals. It is Carminative tonic, anti-inflammatory, healing, chamomile contains (Azulene). Used in facial steams to reduce puffiness and cleanse the pores and impurities. Used in oils to remove aches and pains. Chamomile will add highlights to light hair and is compatible with neutral henna, calendula, mullein in shampoo rinses, lotions, creams, etc.

3431:
Strength: F.E.
3033:
Strength: 1:1 PG
3233:
Strength: 5:1 PG
Common Name: Cherry Bark, Wild
Botanical Name: Prunus virgini
CTFA Name: (Wild Cherry)
(Wild Cherry Extract)
Folkloric Use: Soothing, astringent, sedating used in hair conditioning, as a rinse for ease of combing, will add body to hair.

3432:
Strength: F.E.
3034:
Strength: 1:1 PG
3234:
Strength: 5:1 PG
Common Name: Chickweed Herb
Botanical Name: (Stellaria media)
Folkloric Use: Antipruritic, Vulnerary, emollient. Has been used as an ointment in eczema, psoriasis, indolent ulcer also as a poultice for carbuncles or abcesses. Stellaria combines well in ointments with althea or slippery elm bark, comfrey can be used in creams, lotions, baths, facial steams. For inflamed eyes, swelling, redness of face, erysipelas.

BIO-BOTANICA, INC.: Botanicals in Cosmetics(Continued):

Code:

3434:
Strength: F.E.
3036:
Strength: 1:1 PG
3236:
Strength: 5:1 PG
Common Name: Colts Foot Lvs.
Botanical Name: (Tussilago farfara)
CTFA Name: (Coltsfoot Leaf Extract)
Folkloric Use: Emollient, has been used as a poultice for welts and swelling.

3435:
Strength: F.E.
3037:
Strength: 1:1 PG
3237:
Strength: 5:1 PG
Common Name: Comfrey Leaf
Botanical Name: (Symphytum officinale L)
CTFA Name: (Comfrey Extract)

3436:
Strength: F.E.
3038:
Strength: 1:1 PG
3238:
Strength: 5:1 PG
Common Name: Comfrey Root
Botanical Name: (Symphytum officinale L)
CTFA Name: (Comfrey Extract)
Folkloric Use: Comfrey is one of the most useful Botanicals that can be employed in cosmetics. Comfrey contains Allantoin C6H6 N4O3 a very powerful cell proliferant. Allantoin has been used in supperating wounds, resistant ulcers to stimulate growth of healthy tissue. Comfrey is also astringent, antihaemorrhagic and demulcent. It has been used in chronic varicose ulcers on wounds, fractures, even for hernias (10-15% extract in ointment base). Comfrey can be used in creams, lotions, ointments, salves, douches, hair rinses, shampoos, massage and body oils. Comfrey combines well with ulmus fulva, fillipendula, althea, aloe, coneflower, plus other botanicals too numerous to mention. Comfrey has a wide and varied reputation. Range 2-15%.

BIO-BOTANICA, INC.: Botanicals in Cosmetics(Continued):

3440:
Strength: F.E.
3041:
Strength: 1:1 PG
3242:
Strength: 5:1 PG
Common Name: Corn Silk
Botanical Name: (Zea Mays)
Folkloric Use: Corn Silk has been employed in face masks and packs. It is soothing and emollient.

3445:
Strength: F.E.
3045:
Strength: 1:1 PG
3247:
Strength: 5:1 PG
Common Name: Dandelion Lvs.
Botanical Name: (Taraxacum officinale)
CTFA Name: (Dandelion Extract)
3446:
Strength: F.E.
3046:
Strength: 1:1 PG
3248:
Strength: 5:1 PG
Common Name: Dandelion Rt.
Botanical Name: (Taraxacum officinale)
CTFA Name: (Dandelion Root)
Folkloric Use: Dandelion is useful in facial steams, packs, various skin complaints, eczema. Dandelion contains the following nutrients per ounce. Vitamin A-7000 units, Vitamin B, C and G.

3449:
Strength: F.E.
3048A:
Strength: 1:1 PG
3252:
Strength: 5:1 PG
Common Name: Echinacea
Botanical Name: (Cone flower)
(Echinacia Augustifolia et al)
CTFA Name: (Coneflower Extract)
Folkloric Use: Cone flower is noted to be antiseptic, healing for skin conditions or carbuncles, boils, wounds, ulcers, burns, bed sores, bites, stings or poisonous insects. Echinacea contains inulin, sucrose, betaine, two isomeric 2-methyltetra decadienes, echinacein. (Neoherculin aasanshool) echinacoside (a caffeic acid glycoside), resins, fatty acids. It has been formerly used in medicine to increase resistance to infection, Merck Index 8th Edition.

BIO-BOTANICA, INC.: Botanicals in Cosmetics(Continued):

<u>Code:</u>

3450:
Strength: F.E.
3049:
Strength: 1:1 PG
3253:
Strength: 5:1 PG
Common Name: Elder Flowers
Botanical Name: (Sambucus canadensis)
CTFA Name: (Sambucus)
(Sambucus Extract)
Folkloric Use: Herbalists often call the Elder tree the cosmetic tree as almost all parts will aid in complexion beauty such as softening the skin, removing freckles, preserving and softening the skin. It will reduce swellings, tumors, etc., soreness, inflammations, joint stiffness.

3451:
Strength: F.E.
3050:
Strength: 1:1 PG
3255:
Strength: 5:1 PG
Common Name: Elecampane
Botanical Name: (Inula helenium)
Folkloric Use: Has been used as a wash for various skin disorders, as a bactericide, itching skin, wounds and rashes. Elecampane contains 40% inulin, 1-2% of alantol, an oil containing alantolactone, alantoic acid, alantol and camphor, mucilage and inulenin. Merck Index 7th Edition.

3453:
Strength: F.E.
3052:
Strength: 1:1 PG
3257:
5:1 PG
Common Name: Eyebright
Botanical Name: (Euphrasia officinalis)
CTFA Name: (Euphrasia, Euphrasia Extract)
Folkloric Use: Eyebright is reportedly used as a wash for sore eyes and is astringent, anti-inflammatory. Herbalists also use it locally in conjunctivitis. Usually combined with Golden Seal Root and Witch Hazel. Euphrasia contains glycosides including aucubin, a volatile oil.

BIO-BOTANICA, INC.: Botanicals in Cosmetics(Continued):

Code:

3455:
Strength: F.E.
3054:
Strength: 1:1 PG
3259:
Strength: 5:1 PG
Common Name: Fennel Seed
Botanical Name: (Foeniculum vulgare)
CTFA Name: (Fennel, Fennel Extract)
Folkloric Use: Used in facial steams to sooth the skin in eyewashes. Combines well with Eyebright.

3458:
Strength: F.E.
3057:
Strength: 1:1 PG
3262:
Strength: 5:1 PG
Common Name: Garlic
Botanical Name: (Allium sativum)
CTFA Name: (Garlic Extract)
Folkloric Use: Garlic is a source of organic sulphur. Applied on swellings, sores, pimples and acne.

3459:
Strength: F.E.
3058:
Strength: 1:1 PG
3263:
Strength: 5:1 PG
Common Name: Gentian
Botanical Name: (Gentiani lutea)
CTFA Name: (Gentian Extract)
Folkloric Use: Gentian is tonic, astringent, cleansing.

3442:
Strength: F.E.
3042:
Strength: 1:1 PG
3244:
Strength: 5:1 PG
Common Name: Geranium (wild)
Botanical Name: (Gernaium maculatum)
CTFA Name: Geranium Extract
Folkloric Use: Astringent, tonic, styptic useful in sore mouth, indolent ulcers. Can be used in douches, as a deodorant wash, in baths, for oily skin in creams, lotions, hair rinse, conditioners, shampoo also good for swollen aching feet.

BIO-BOTANICA, INC.: Botanicals in Cosmetics(Continued):

Code:

3461:
Strength: F.E.
3060:
Strength: 1:1 PG
3265:
Strength: 5:1 PG
Common Name: Ginseng
Botanical Name: (Panax quinquefolia)
CTFA Name: (Ginseng Extract)
Folkloric Use: Ginseng contains, saponins or panaxosides. It is said to be an excellent skin conditioner in masks, packs, moisturizing lotions, night creams, etc.

3463:
Strength: F.E.
3063:
Strength: 1:1 PG
3268:
Strength: 5:1 PG
Common Name: Golden Seal Rt.
Botanical Name: (Hydrastis canadensis)
Folkloric Use: Golden Seal has a wide and varied reputation, hydrastis contains the alkaloids, hydrastine, berberine and canadine. Berberine is the yellow coloring substance. Golden Seal Root has been employed in eczema, pruritis, otorrhoea conjunctivitis. Golden Seal is astringent, makes an excellent yellow hair dye. Can be combined with chamomile, marigold, henna to brighten blond hair. Can be employed in cases of acne, dandruff and mouth washes.

3144:
Strength: F.E.
Common Name: Henna (neutral)
Botanical Name: (Lawsonia alba or inermis)
CTFA Name: Henna Extract
Folkloric Use: Neutral Henna is an excellent conditioning agent. Will add body and highlight to hair. It will not impart any color to the hair. Can be used in shampoos, hair rinses, conditioners.

BIO-BOTANICA, INC.: Botanicals in Cosmetics(Continued):

3466:
Strength: F.E.
3067:
Strength: 1:1 PG
3271:
Strength: 5:1 PG
Common Name: Hops
Botanical Name: (Humulus lupulus)
CTFA Name: (Hops, Hops Extract, Hops Oil)
Folkloric Use: Hops flowers contain volatile oil consisting mainly of humulene, bitter principles and tannins. Hops is sedative, hypnotic, bactericidal. Has been used on crural ulcers. Will give body to the hair. Useful in treatment of dandruff, ringworm, sores, tetters and discolorations, in combination with chamomile will reduce swellings and alleviate pain and itching.

3468:
Strength: F.E.
3069:
Strength: 1:1 PG
3273:
Strength: 5:1 PG
Common Name: Horsetail Grass
Botanical Name: (Equisetum arvense)
CTFA Name: (Horsetail Extract)
Folkloric Use: Horsetail is an excellent source of soluble vegetal silica. It is diuretic, astringent, healing and combines well with comfrey for skin disorders. Putrid wounds, gangrenous ulcers external bleeding. Will add strength and sheen to hair.

3471:
Strength: F.E.
3072:
Strength: 1:1 PG
3276:
Strength: 5:1 PG
Common Name: Hyssop
Botanical Name: (Hyssopus officinallis)
Folkloric Use: Hyssop can be employed in baths as a diaphoretic in facial steams for cleansing the pores. Has been used to aid healing of wounds and ulcers.

BIO-BOTANICA, INC.: Botanicals in Cosmetics(Continued):

Code:

3472:
Strength: F.E.
3073:
Strength: 1:1 PG
3278:
Strength: 5:1 PG
Common Name: Ivy
Botanical Name: (Glechoma hederacea)
CTFA Name: (Ivy, Ivy Extract)
Folkloric Use: Useful for control of cellulite decongesting tonic.

3473:
Strength: F.E.
3074:
Strength: 1:1 PG
3279:
Strength: 5:1 PG
Common Name: Jaborandi
Botanical Name: (Pilocarpus jabcrandi)
Folkloric Use: Has a reputation as a stimulant to hair growth. It contains pilocarpine. Can be used in shampoos, hair rinses, conditioning hair tonics, lotions to stimulate the follicle. Combines well with colo-cynth, tinct capsicum, mullein as a hair tonic.

3479:
Strength: F.E.
3080:
Strength: 1:1 PG
3287:
Strength: 5:1 PG
Common Name: Lemongrass
Botanical Name: (Andropogon citratum)
CTFA Name: (Lemongrass Oil)
Folkloric Use: Can be used in bath herbs, hair rinses. Said to normalize overactive oil glands, dandruff and similar skin problems.

3481:
Strength: F.E.
3082:
Strength: 1:1 PG
3289:
Strength: 5:1 PG
Common Name: Lemon Peel
Botanical Name: (Citrus limonum)
CTFA Name: (Lemon Peel Extract)
Folkloric Use: Antiseptic, healing, bacteriostat skin bleach. Is aromatic and astringent.

BIO-BOTANICA, INC.: Botanicals in Cosmetics(Continued):

Code:

3487:
Strength: F.E.
3024:
Strength: 1:1 PG
3297:
Strength: 5:1 PG
Common Name: Marigold
Botanical Name: (Calendula officinalis)
CTFA Name: (Calendula, Calendula Extract, Calendula Oil)
Folkloric Use: Calendula contains a volatile oil, Saponins, a yellow resin calendulin and a bitter principle. Calendula has been employed as an aromatic, anti-inflammatory, anti-septic, anti-haemorrhagic, styptic, spasmolytic, vulnerary, emmenagogue, mild diaphoretic, topically it has been used for crural ulcer, varicose veins, hemorroids, anal eczema, proctitis, lymphadenoma, inflamed cutaneous lesions, conjunctivitis, as eye lotion.

3493:
Strength: F.E.
3093:
Strength: 1:1 PG
3303:
Strength: 5:1 PG
Common Name: Mullein
Botanical Name: (Verbascum thapsus)
CTFA Name: (Mullein Extract)
Folkloric Use: Mullein has been used as a demucent, diuretic, astringent, anodyne, antispasmodic, pectoral. Herbalists have used it both internally and externally. In the form of a fomentation, applied to inflamed piles, tumors, ulcers, various swellings as a discutient. Recently several companies have employed it in smoking mixtures to relieve symptoms of asthma.

3495:
Strength: F.E.
3095:
Strength: 1:1 PG
3305:
Strength: 5:1 PG
Common Name: Nettles
Botanical Name: (Urtica dioical)
CTFA Name: (Nettle, Nettle Extract)
Folkloric Use: Nettles extract consists of the dried aerial parts of the plant. The plant grows up to 150 cm tall, upright, having stinging hairs, with ovate, usually cordate, toothed leaves, bearing insignificant auxillary heads of green flowers.

BIO-BOTANICA, INC.: Botanicals in Cosmetics(Continued):

Code:

Common Name: Nettles(Continued)
Folkloric Use: Constituents - contains indolic compounds including histamine and 5-hydroxytryptamine, and various acids, including ascorbic acid; mucilage and mineral salts.
Medicinal uses - Astringent, cutaneous, eruptions nervous eczema, epistaxis malaena.
Cosmetic uses - stimulates hair growth. As a hair tonic prevents hair from falling out and renders it soft and glossy.

3517:
Strength: F.E.
3117:
Strength: 1:1 PG
3329:
Strength: 5:1 PG
Common Name: Rose Hips
Botanical Name: (Rose canina)
CTFA Name: (Rose Hips Extract)
Folkloric Use: Rose Hips are widely used for their high Vitamin C content, and when combined with various other botanicals in hair preparations it will add highlights to light hair. It also acts as a fixative for various herbs.

3518:
Strength: F.E.
3118:
Strength: 1:1 PG
3330:
Strength: 5:1 PG
Common Name: Rosemary
Botanical Name: (Rosmarinus officinalis)
CTFA Name: (Rosemary Extract, Rosemary Oil)
Folkloric Use: Tonic, astringent, diaphoretic, stimulant. Its basic external use is in hair lotions, for its effect in stimulating the hair follicle to renewed activity and preventing premature baldness. An extract of the plant combined with sodium borate and applied cold is said to be one of the finest hair washes known. It is a surprisingly effective remedy for the control of scurf and dandruff.

BIO-BOTANICA, INC.: Botanicals for Cosmetics(Continued):

Code:

3519:
Strength: F.E.
3119:
Strength: 1:1 PG
3331:
Strength: 5:1 PG
Common Name: Sage
Botanical Name: (Salvia officinalis L)
CTFA Name: (Sage Extract, Sage Oil)
Folkloric Use: Sage is an extract of the dried leaf containing not more than 10% of stems.
Sage is a perennial low shrub, growing up to two feet in height with opposite, petiolate, ovate-lanceolate, crenulate, wrinkled leaves of a grayish-green color; sometimes tinged with red or purple, the flowers are blue.
Part used - the dried leaves are used for extraction purposes.
Constituents - contains volatile oil, (Thujone, camphor salvene, pinene, cineol) resin, tannin, bitter principles.
Analytical data - Ferrous Sulphate, strikes a black color with a dilute mixture of sage extract.
Cosmetically - an extract of sage is used to cleanse old ulcers and wounds. Massaged into the scalp, it will control dandruff, falling hair or loss of hair if the papilla is dormant and not destroyed.
Russian Folk Medicine - claim it to be aromatic, astringent, antiseptic, carminative, disinfectant (against inflammations).

3411:
Strength: F.E.
3012:
Strength: 1:1 PG
3212:
Strength: 5:1 PG
Common Name: Walnut
Botanical Name: (Juglans regia)
CTFA Name: (Walnut Extract, Walnut Leaves, Walnut Oil)
Folkloric Use: Juglone (5-Hydroxy-1,4-naphthoquinone), Alpha-hydrojuglone (1,4,5-Tri-hydroxy-naphthalene), Beta hydrojuglone. Tonic restorative, depurative, disinfectant, vermifuge, astringent. Used for skin complaints, eczema, scabbing, pruritis, glandular disturbances, scrofula, rheumatism, gout, irritation of the eye lid, opthalmia styes, blisters, varicose ulcers. Walnut Extract is an old fashioned hair dye.

BIO-BOTANICA, INC.: Botanicals for Cosmetics(Continued):

Code:

3534A:
Strength: F.E.
3133A:
Strength: 1:1 PG
3349:
Strength: 5:1 PG
Common Name: Watercress
Botanical Name: (Nasturtium officinale)
CTFA Name: (Watercress Extract)
Folkloric Use: Note: Watercress is extremely heat sensitive destroying most of its active principles. Copper, Iron, Calcium, manganese, phosphorus, Zinc, Potassium, Iodine, Sulfo-nitrogenized oil. Vitamins A, B2, C, P, E. Stimulates hair follicle, according to L. Binet. Injection of Watercress Extract. "Brake" experimental cancers, mineralizing and strengthening to the system. In the middle ages it was believed that rubbing the juice of watercress into the scalp, will strengthen and thicken the hair. It was also used as a Beauty Treatment for freckles and also for clearing the complexion in combination with honey. Externally Watercress was used as a poultice on atonic ulcers, tumors, lymphatic or oedematous swellings, its action being both detersive and healing.

3451:
Strength: F.E.
3138:
Strength: 1:1 PG
3357:
Strength: 5:1 PG
Common Name: Wood Betony (Bishopwort)
Botanical Name: (Betonica officinalis)
Folkloric Use: Alkaloids, including Betonicine, stachydrine. Sedative, bitter. Externally leaves were applied to purulent wounds and ulcers. It is said to be vulernary (healing wounds) and decongestant.

BIO-BOTANICA, INC.: Botanicals for Cosmetics(Continued):

Code:

3542:
Strength: F.E.
3139:
Strength: 1:1 PG
3358:
Strength: 5:1 PG
Common Name: Yarrow
Botanical Name: (Archillea mille-
CTFA Name: (Yarrow Extract, Yarrow Oil)
Folkloric Use: Aromatic volatile oil, which is blue in color, owing the presence of chamazulene, and achilleine, a glyco-alkaloid, silicic acid. Diaphoretic, antipyretic, hypotensive, astringent, diuretic and urinary antiseptic. Frequent applications locally as a skin lotion; is said to improve the complexion, remove pimples or scabbing. It is said to promote healing and is also cleansing. It has been used on cuts, injuries, varicose ulcers. Yarrow has a firming affect on the connective tissue, it is also used in hair care preparations.

BIO-BOTANICA, INC.: **Shampoo Herbal Complexes:**

3143:
Chemical/Trade Name: BIO-CHELATED NEUTRAL HENNA PLUS I (Bio-Botanica)
CTFA Adopted Name: Henna Extract (and) Rosemary Extract (and) Sage Extract (and) Nettle Extract
Approximate Use Levels: NEUTRAL HENNA PLUS I has been formula- for normal to dry dark hair, to condition, add highlights, and stimulate. Can be incorporated in shampoos, rinses, conditioners at a level of 1-5% in aqueous phase depending on formulation.

3146:
Chemical/Trade Name: BIO-CHELATED NEUTRAL HENNA PLUS II (Bio-Botanica)
CTFA Adopted Name: Henna Extract (and) Chamomile Extract (and) Calendula Extract (and) Mullein Extract (and) Rose Hips Extract
Approximate Use Levels: NEUTRAL HENNA PLUS II has been formulated for light hair, to add highlights, condition and stimulate. For use in shampoos, hair rinses, conditioners, at a level of 1-5% in aqueous phase depending on formulation.

BIO-BOTANICA, INC.: Shampoo Herbal Complexes(Continued):

Code:

3144:
Chemical/Trade Name: BIO-CHELATED NEUTRAL HENNA EXTRACT (Bio-Botanica)
CTFA Adopted Name: Henna Extract
Approximate Use Levels: 10-12% Alcohol...Lawsonia Alba. Use level 1-5%.

3145:
Chemical/Trade Name: BIO-CHELATED RED HENNA EXTRACT (Bio-Botanica)
CTFA Adopted Name: Henna Extract
Approximate Use Levels: 10-12% Alcohol...Lawsonia Inermis. Use level 1-5%.

3152:
Chemical/Trade Name: BIO-DANDRA PLEX (Bio-Botanica)
CTFA Adopted Name: Nettle Extract (and) Horsetail Extract (and) Hops Extract (and) Comfrey Extract (and) Chamomile Extract (and) Henna Extract (and) Rosemary Extract (and) Burdock Extract (and) Quassia (and) Witch Hazel Extract (and) Chapparel Extract (and) Crane's Bill Extract (and) Black Walnut Extract
Approximate Use Levels: This formula has been compounded to be of benefit in formulations used to control dandruff and condition hair. It can be incorporated in ointments, shampoos, conditioners, and hair rinses at a level of between 1-5% in the aqueous phase depending on formulation.

BIO-BOTANICA, INC.: Hair Complexes:

Code:

3147:
Chemical/Trade Name: BIO-CHELATED NUTRA PLANT I (Bio-Botanica)
CTFA Adopted Name: Nettle Extract (and) Horsetail Extract (and) Hops Extract (and) Comfrey Extract (and) Chamomile Extract (and) Henna Extract (and) Rosemary Extract
Approximate Use Levels: NUTRA PLANT I has been formulated for its stimulating, conditioning, highlighting and strengthening properties when incorporated in shampoos, hair rinses, conditioners, and treatments, usual level of introduction 1-5% in aqueous phase depending on formulation.

3149:
Chemical/Trade Name: BIO-CHELATED NUTRA PLANT II (Bio-Botanica)
CTFA Adopted Name: Nettle Extract (and) Chamomile Extract (and) Rosemary Extract (and) Horsetail Extract (and) Honey Extract (and) Comfrey Extract (and) Henna Extract (and) Hops Extract
Approximate Use Levels: NUTRA PLANT II has been formulated for incorporation into shampoos, hair rinses, conditioners, to add highlights, stimulate and strengthen. NUTRA PLANT II is for use in dry hair formulations usual level of introduction 1-5% in the aqueous phase depending on formulations.

BIO-BOTANICA, INC.: Herbal Skin Care Complexes:

Code:

3153:
Chemical/Trade Name: BIO-CHELATED DERMA-PLEX I (Bio-Botanica)
CTFA Adopted Name: Comfrey Extract (and) Plantain Extract (and) Elder Flower Extract (and) Horsetail Extract (and) Calendula Extract (and) Sage Extract (and) Crane's Bill Extract (and) Ginseng Root Extract (and) Wild Honey
Approximate Use Levels: DERMA-PLEX I has been formulated to aid in the control of dry skin, wrinkled skin. It is stimulating, hydrating, softening, emollient, and healing. Bio-Botanica has incorporated those herbs which are noted for their cell proliferating and healing qualities. This formula can be used in creams, lotions, foundations, lipstick, wherever moisturizing qualities will be of benefit at a rate of 0.5-5% in the aqueous phase depending on formulation.

3154:
Chemical/Trade Name: BIO-CHELATED DERMA-PLEX II
CTFA Adopted Name: Comfrey Extract (and) Witch Hazel Extract (and) Yarrow Extract (and) Hops Extract (and) Horsetail Extract (and) Calendula Extract (and) Lemon Peel Extract (and) Burdock Root Extract
Approximate Use Levels: DERMA-PLEX II has been formulated to aid in the control of acne, eczema, oily skin at a level between 0.5-5% depending on formulation in creams, lotions, and ointments, in the aqueous phase. This formula is astringent, antiseptic and healing.

3155:
Chemical/Trade Name: BIO-CHELATED SAUNA DERM I (Bio-Botanica)
CTFA Adopted Name: Comfrey Extract (and) Orange Peel Extract (and) Chamomile Extract (and) Althea Extract (and) Yarrow Extract (and) Fennel Extract (and) Licorice Extract
Approximate Use Levels: This formula has been compounded for use in facial steams, sauna's, face creams, lotions, soap, cleansing cream, for its hydrating qualities. It can be incorporated from 0.5-5% in aqueous phase depending on formulation.

3156:
Chemical/Trade Name: BIO-CHELATED SAUNA DERM II (Bio-Botanica)
CTFA Adopted Name: Yarrow Extract (and) Comfrey Extract (and) Witch Hazel Extract (and) Lavender Extract (and) Licorice (and) Lemongrass Extract
Approximate Use Levels: This formula has been compounded for use in facial steams, sauna's, face creams, lotions, soaps, cleansing creams for its astringent, antiseptic, decongesting action. It can be incorporated from 0.5-5% in aqueous phase, depending on formulation.

BROOKS INDUSTRIES INC.: Active Ingredients:

Essential Fatty Acids:

EFA-PLEX:
100% Stabilized Complex
contains omega 6-linoleic, omega 3-Linolenic, omega-6 Arachidonic
Linoleic Acid & Myristyl Myristate & Isopropyl Palmitate & Linolenic Acid & Oleic Acid & Arachidonic Acid & Tocopherol

Tissue Respiratory Factors - Live Yeast Cell Derivative:

BIODYNES TRF:
Active Powder
Approx 10 respiration units/mg (Warburg/Micro-biological Bioassay)
A Cosmetic Grade LYCD
Tissue Respiratory Factors

BIODYNES TRF 5% Solution:
500 Units/ml, Clear Low odor Liquid
Tissue Respiratory Factors

DERMASOME TRF
500 Units/ml, Liposome
Tissue Respiratory Factors & Soy Lecithin

"Bioengineered" Protein Mineral Complexes:

BIOMIN CINQUE
Powder, 5 Essential Minerals bound in a low Mwt Protein/Carbohydrate/Nucleotide Cellular Extract
Silicon & Zinc & Copper & Iron & Magnesium Glyconucleopeptides

BIOMIN ACQUACINQUE Liquid
Clear Filtered version of BIOMIN CINQUE

BIOMIN Se/P/C:
Powder, Selenium Protein Complex
Selenopeptides

BIOMIN Z/P/C:
Powder, Complex Peptide/Zinc/Nucleotide/Carbohydrate
Zinc Yeast Derivative

BIOMIN Z/P/C-20 Liquid:
Clear Filtered Version of BIOMIN Z/P/C (20% Liquid)

BROOKS INDUSTRIES INC.: Active Ingredients(Continued):

<u>"Bioengineered" Protein Mineral Complexes:</u>

BIOMIN Ca/P/C:
Powder, Complex Peptide/Calcium/Nucleotide/Carbohydrate
Calcium Protein Complex

BIOMIN Fe/P/C:
Powder, Complex Peptide/Iron/Nucleotide/Carbohydrate
Iron Protein Complex

BIOMIN Si/P/C:
Powder, Complex Peptide/Silicon/Nucleotide/Carbohydrate
Silicon Protein Complex

<u>Super-Oxide Dismutase-Protein Bound:</u>

BIOCELL SOD:
Free Radical Scavenging Enzyme - Activity 600 Units/g
Superoxide Dismutase

DERMASOME SOD:
600 Units SOD in a Liposome
Soy Lecithin & Superoxide Dismutase

Substantive Humectants

QUAMECTANT AM-50:
50% Active
Quaternized Acetamide MEA

Nucleic Acid Extract - Containing RNA & DNA

BIOPLEX RNA:
20% Liquid, soluble nucleic acids
Propylene Glycol & Hydrolyzed RNA & DNA

Cosmetic Preservatives

MIKROKILL:
15% Active
Polyaminopropyl Biguanide

MIKROKILL 2:
15% Active
Polyaminopropyl Biguanide & Chloroxylenol

Highly effective against gram negative organisms - Use levels 0.2% - Good toxicological profile

BROOKS INDUSTRIES INC.: Active Ingredients:

Emulsifiers:

Emulsifying Waxes:

BROOKSWAX D:
100% active flaked solid
Cetearyl Alcohol & Ceteareth 20

BROOKSWAX P:
100% active flaked solid
Emulsifying Wax NF

Relaxer Concentrate:

RELAXER CONCENTRATE NO. 2:
100% active flaked emulsifying agent
Add mineral oil, petrolatum and base (sodium hydroxide) and water and have an easy to manufacture, stable, effective finished product (hair relaxer).

Phosphate Esters:

BROPHOS OL-3:
100% Active, Liquid Surfactant
Oleth 3 Phosphate

BROPHOS OL-3N:
100% Active, Thick Paste
DEA Oleth 3 Phosphate

BROPHOS OL-3NPG:
Free flowing liquid - Easy to use
DEA Oleth 3 Phosphate & Propylene Glycol

BROOKS INDUSTRIES INC.: Lanolin & Lanolin Derivatives:

3 generations of MALMSTROM expertise are behind the manufacture of a comprehensive range of Woolgrease, refined Lanolins, and their many derivatives which are offered.

Pure Lanolins & Lanolin Oil:

IVARLAN 3000:
Cosmetic Anhydrous Lanolin USP
Lanolin

IVARLAN LIGHT:
High Quality Cosmetic Lanolin USP
Lanolin

IVARLAN 3100:
Fluid fraction of whole Lanolin
Lanolin Oil

Lanolin Absorption Bases:

IVARBASE 101:
Mineral Oil & Lanolin Alcohol

IVARBASE 3230:
"Lanolin Base No. 3"
Makes very stable milks & creams
Mineral Oil & PEG-30 Lanolin & Cetyl Alcohol

IVARBASE 3240:
"Lanolin Base A"
Isopropyl Lanolate & Lanolin Oil & Oleyl Alcohol

IVARBASE 3250:
"IP33"
Rich Light Emollient
Isopropyl Palmitate & Lanolin Oil

Lanolin Alcohol Derivatives:

IVARBASE 3210:
Light greaseless emollient
CTFA Formerly Acetylated Lanolin Alcohol
Cetyl Acetate & Acetylated Lanolin Alcohol

IVARBASE "98":
CTFA Formerly Laneth 10 Acetate
Polysorbate 80 & Cetyl Acetate & Acetylated Lanolin Alcohol

BROOKS INDUSTRIES INC.: Lanolin & Lanolin Derivatives(Continued):

Lanolin Alcohol Derivatives(Continued):

IVARLAN 3310:
Cosmetic Grade - Distilled
Lanolin Alcohol

IVARLAN C-24:
CTFA Formerly Choleth 24
Choleth 24 & Ceteth 24

Lanolin Ethoxylates:

IVARLAN 3400:
100% Active Water Soluble Lanolin
PEG-75 Lanolin

IVARLAN 3401:
50% Active - Easy to handle
PEG-75 Lanolin

IVARLAN 3406:
100% Active Water Soluble Lanolin
PEG-60 Lanolin

IVARLAN 3420:
100% Active Liquid Lanolin
PPG-12 PEG-50 Lanolin

IVARLAN AWS:
Alcohol/Water Soluble Lanolin Oil
PPG-12 PEG-65 Lanolin Oil

IVARLAN 3450:
White Water Soluble Lanolin
PEG-20 Hydrogenated Lanolin

BROOKS INDUSTRIES INC.: Proteins, Amino Acids & Derivatives:

1. Collagen Derived

Hydrolyzed Proteins:

a. Enzyme Hydrolyzed (EN): Excellent light color, very low odor, controlled Mwts.

HYDROCOLL EN-55:
55% Liquid, Low Salt, Mwt: 2,000
Hydrolyzed Animal Protein

HYDROCOLL EN-SD:
Powder, Low Salt, Mwt: 2,000
Hydrolyzed Animal Protein

b. Alkaline Hydrolyzed (AL): Broader Mwt Distribution.

HYDROCOLL AL-55:
55% Liquid, Low Salt, Mwt: 1,000
Hydrolyzed Animal Protein

HYDROCOLL G-40:
40% Liquid - Mwt: 100,000
Liquid Gelatin

c. Acid Hydrolyzed: High iP, Light color, low odor

HYDROCOLL AC-30:
30% Liquid, Low Salt, Mwt: 5,000
Cationic Animal Protein

Native Soluble Collagen:

SOLU-COLL:
1%, Triple Helix, Mwt: 300,000
Soluble Animal Collagen

Quaternary Hydrolyzed Proteins:

QUAT-COLL QS:
Powder, High iP, Mwt: 10,000
Steartrimonium Hydrolyzed Animal Protein

QUAT-COLL IP10-30:
30% Liquid, iP 10+, Mwt: 25,000
Polytrimonium Gelatin

BROOKS INDUSTRIES INC.: Proteins, Amino Acids & Derivatives (Continued):

1. Collagen Derived(Continued)

Alcohol Soluble Proteins:

ETHA-COLL AAS-20:
20% Alcoholic Liquid
Ethyl Ester of Hydrolyzed Animal Protein

ETHA-COLL A210-20:
20% Alcoholic Liquid
Myristoyl Hydrolyzed Animal Protein

Oil Soluble Proteins:

OLEO-COLL LP
100% Oily Liquid
Lecithin & Butyl Stearate & Coco-Hydrolyzed Animal Protein & Oleoyl Sarcosine & Sesame Oil & Lanolin Alcohol

Foaming Proteins:

FOAM-COLL 4C:
35% Liquid, mild surfactant
Potassium Coco-Hydrolyzed Animal Protein

FOAM-COLL 4CT:
38% Liquid, mild surfactant
TEA-Coco-Hydrolyzed Animal Protein

FOAM-COLL 5:
70% Liquid, mild surfactant
TEA-Coco-Hydrolyzed Animal Protein & Sorbitol

Moisturizing Protein Complexes:

COLLA-MOIST WS:
45% Liquid
Hydrolyzed Animal Protein & PPG-12-PEG-65 Lanolin Oil & Propylene Glycol

SOLU-COLL COMPLEX:
10% Active, Collagen Complex
Soluble Animal Collagen & Animal Collagen Amino Acids

COLLA-MOIST CG:
30% Active, Amino Acid Ester
Collagen Glycerides

BROOKS INDUSTRIES INC.: Proteins, Amino Acids & Derivatives (Continued):

1. Collagen Derived(Continued):

<u>Collagen Amino Acids:</u>

COLLAMINO 25:
25% Active, Mwt: 200
Animal Collagen Amino Acids

COLLAMINO COMPLEX:
50% Active, Moisture Complex
Animal Collagen Amino Acids & Acetamide MEA & Propylene Glycol

<u>Cationic Amino Acid Derivative:</u>

COLLAMINO COMPLEX L/O:
90%, Conditioning Complex
Lauryloleylmethylamine Animal Collagen Amino Acids

2. Elastin Derived:

SOLU-LASTIN 10:
10%, Low Odor, Mwt: 4,000
Hydrolyzed Elastin

SOLU-LASTIN 30:
30%, Low Odor, Mwt: 4,000
Hydrolyzed Elastin

3. Keratin Derived:

<u>Hydrolyzed Keratins:</u>

HYDROKERATIN AL-30:
30% Liquid, Mwt: 1,000
Hydrolyzed Keratin

HYDROKERATIN AL-SD:
Powder, Mwt: 1,000
Hydrolyzed Keratin

<u>Alcohol Soluble Keratin:</u>

ETHA-KERATIN A-20:
20% Alcoholic Liquid
Ethyl Ester of Hydrolyzed Keratin

BROOKS INDUSTRIES INC.: Proteins, Amino Acids & Derivatives (Continued):

3. Keratin Derived:

Oil Soluble Keratin:

OLEO-KERATIN ISO:
100% Oily Liquid
AMP-Isostearic Hydrolyzed Animal Keratin & Isostearic Acid & Myristyl Myristate & Isopropyl Palmitate

Keratin Amino Acids:

KERAMINO 25:
25% Liquid, Mwt: 200
Keratin Amino Acids

KERAMINO SD:
50% Active Powder, Mwt: 200
Keratin Amino Acids & Sodium Chloride

4. Silk Derived:

Powdered Silk Pigment:

FIBRO-SILK Powder:
Insoluble crushed silk
Silk

Hydrolyzed Silk Protein:

SOLUSILK Protein:
10% Liquid, Mwt: 1,000
Hydrolyzed Silk

Silk Amino Acids:

SOLUSILK 25:
30% Liquid, 15% Protein, Mwt: 100
Silk Amino Acids

SOLUSILK SF:
15% Active, Low Salt, Mwt: 100
Silk Amino Acids

Alcohol Soluble Silk:

ETHA-SILK EEHS:
50% Alcoholic Liquid
Ethyl Ester of Hydrolyzed Silk

BROOKS INDUSTRIES INC.: Proteins, Amino Acids & Derivatives (Continued):

5. Soyabean Derived:

Enzymatically Hydrolyzed Soyabean Protein:

SOLU-SOY EN-25:
25% Active, Low Salt, Mwt: 1,000
Hydrolyzed Soy Protein

Foaming Soy Protein Derivative:

FOAM-SOY C:
30% Active, Mild surfactant
Sodium Coco-Hydrolyzed Soy Protein

Alcohol Soluble Soy Protein Derivative:

ETHA-SOY ISO
25% Alcoholic Liquid
AMP-Isostearoyl-Hydrolyzed Soy Protein

Oil Soluble Soy Protein Derivative:

OLEO-SOY C:
100% Oily Liquid
Coco-Hydrolyzed Soy Protein

6. Vegetable Derived:

Enzymatically Hydrolyzed Vegetable Protein:

SOLU-VEG EN-35:
35% Active, Low Salt, Mwt: 2,000
Hydrolyzed Vegetable Protein

7. Single Cell Organism Derived:

HYDROCELL SCP-30:
30% Acellular Protein, Mwt: 4,000
Single Cell Protein

HYDROCELL AYP-30:
30% Enzymatic Yeast Protein Extr
Autolyzed Yeast

HYDROCELL YP-30:
30% Pure Yeast Protein Mwt: 1,000
Hydrolyzed Yeast Protein

BROOKS INDUSTRIES INC.: Proteins, Amino Acids & Derivatives (Continued):

8. Whole Milk Derived Proteins:

Enzymatically Hydrolyzed Whole Milk Protein:

HYDROMILK EN-20:
20% Active, Mwt: 4,000
Hydrolyzed Milk Protein

Milk Amino Acids:

MILKAMINO 20:
20% Active, Low Salt, Mwt: 200
Milk Amino Acids

9. Serum Proteins:

Enzymatically Hydrolyzed Nutritional Serum Protein:

SERUMPRO EN-10:
10% Active, Mwt: 7000
Serum Protein
Contains all Essential Amino Acids

10. Marine Proteins:

Enzymaticaly Hydrolyzed Cold Water Fish Protein:

SOLU-MAR EN-30:
30% Liquid, Mwt: 1,000
Hydrolyzed Marine Protein

11. Glycoproteins & Mucoproteins (Mucins)

Whole Native Globular Protein Extract of Glycoproteins:

GLYPROSOL 20
20% Liquid, Mwt: 10,000
Soluble Glycoprotein

GLYPROSOL SD
Powder
Soluble Glycoprotein

BROOKS INDUSTRIES INC.: Proteins, Amino Acids & Derivatives (Continued):

11. Glycoproteins & Mucoproteins (Mucins)(Continued):

Cell Wall Extract Consisting of Native Mucins:

MUSOL 20:
20% Viscous liquid, Mwt: 10,000
Soluble Mucoprotein

MUSOL SD:
Powder, cospraydried with Dextrin
Maltodextrin & Soluble Mucoprotein

12. Whole Dermal Tissue Extract

Cattle Skin Extract Consisting of Collagen, Elastin & Mucopolysaccharides:

BIOPOL TE:
2% Liquid, 1% Protein, 1% MPS
Animal Tissular Extract

CAPITAL CITY PRODUCTS CO.: Emulsifiers and Surfactants:

CAPMUL Mono- and Diglycerides:

CAPMUL mono- and diglycerides are nonionic emulsifiers made by reacting glycerine with fats, oils or fatty acids. They are lipophilic, insoluble in water and soluble in oils at elevated temperatures. They are used in fats and oils, often with other emulsifiers, to produce water in oil emulsions and to increase viscosity. All CAPMUL mono- and diglycerides use the highest quality raw materials in order to meet the rigid specifications of the cosmetic and pharmaceutical industries.

Typical Specifications:

CAPMUL GDL:
Color LOVIBOND Red Max.: 4
Form: Semi-Solid
Iodine Value Max.: 20
Acid Value Max: 3
Sap. Value: 215-230
M.P. C: 28-31
HLB: 3.6

CAPMUL GMS:
Color LOVIBOND Red Max.: 4
Form: Flake
Iodine Value Max.: 5
Acid Value Max.: 3
Free Glycerol % Max.: 1#
Sap. Value: 155-165
M.P. C: 57-62
HLB: 3.6

CAPMUL GMVS:
Color LOVIBOND Red Max.: 2.5
Form: Plastic
Iodine Value Max.: 75
Acid Value Max.: 3
Free Glycerol % Max.: 1#
Sap. Value: 155-165
M.P. C: 44-54
HLB: 3.5

CAPMUL GMO:
Color LOVIBOND Red Max.: 7
Form: Semi-Solid
Iodine Value Max.: 75
Acid Value Max.: 5
Free Glycerol % Max.: 1
Sap. Value: 160-170
M.P. C: 25 Max.
HLB: 3.4

Products having higher glycerol contents are available on request. Anionic, cationic and nonionic emulsifiers are available for self-emulsifying GMS.

CAPITAL CITY PRODUCTS CO.: Emulsifiers and Surfactants (Continued):

CAPMUL MCM:

CAPMUL MCM is a mono- and diglyceride of medium chain fatty acids (caprylic and capric). It has solvent properties and is useful as an oil-in-water emulsifier.

CAPMUL MCM:

Color LOVIBOND Red Max.: 4
Form: Liquid
Acid Value Max.: 2.5
Free Glycerol % Max.: 2.5
Alpha Mono % Min.: 70
Moisture % Max.: 0.5

Additional information on the uses of medium chain triglycerides and mono- and diglyceride esters in special dietary products, cosmetics and for use as a solubilizing agent is available upon request.

CAPMUL Sorbitan Esters:

CAPMUL sorbitan esters are made by reacting sorbitol with fats or fatty acids. They are lipophilic (W/O), insoluble in water or dispersible in oil. They are used in combination with higher HLB emulsifiers (polysorbates) to form oil-in-water emulsions.

CAPMUL S:

Color Gardner Max.: 6
Acid Value Max.: 10
Sap. Value: 147-157
Hydroxyl Value: 235-260
Moisture % Max.: 1.5
HLB: 4.7

CAPMUL O:

Color Gardner Max.: 6
Acid Value Max.: 10
Sap. Value: 145-160
Hydroxyl Value: 190-215
Moisture % Max.: 1.0
HLB: 4.3

CAPITAL CITY PRODUCTS CO.: Emulsifiers and Surfactants (Continued):

CAPMUL Ethoxylated Esters:

CAPMUL POE esters are ethoxylated sorbitan esters (polysorbates). CAPMUL EMG is an ethoxylated mono- and diglyceride. They are soluble or dispersible in water, and are used to form oil-in-water emulsions. They are also used to adjust the HLB of emulsion systems and as solubilizers and dispersants for oils into water.

CAPMUL POE-L:
Color Gardner Max.: 3.0
Acid Value Max.: 2.0
Sap. Value: 40-50
Hydroxyl Value: 96-108
Moisture % Max.: 3.0#
HLB: 16.7

CAPMUL POE-S:
Color Gardner Max.: 7.0
Acid Value Max.: 2.0
Sap. Value: 45-55
Hydroxyl Value: 81-96
Moisture % Max.: 3.0#
HLB: 14.9

CAPMUL POE-O:
Color Gardner Max.: 6.0
Acid Value Max.: 2.0
Sap. Value: 45-55
Hydroxyl Value: 65-80
Moisture % Max.: 3.0#
HLB: 15.0

CAPMUL EMG:
Color Gardner Max: 4.0##
Acid Value Max.: 2.0
Sap. Value: 65-75
Hydroxyl Value: 65-80
Moisture % Max.: 1.0#
HLB: 13.1

CAPMUL POE esters are also available with a 1.0% max. moisture content.
LOVIBOND Red Color

CAPITAL CITY PRODUCTS CO.: Emulsifiers and Surfactants (Continued):

CAPROL Polyglycerol Esters:

CAPROL polyglycerol esters have been developed to cover a wide hydrophylic and lipophylic range. They are generally prepared by esterification with fatty acids or by alcoholysis of a vegetable oil with a polyglycerol. The starting polyglycerol molecular weight may be low (diglycerol) or very high (decaglycerol). CAPROL esters may vary from viscous liquids to waxy, brittle solids. This wide range of physical and structural properties makes CAPROL products useful in both water-in-oil and oil-in-water systems. Polyglycerol ester products are currently in use in cosmetic, pharmaceutical, confectionary, toiletry, fragrance, food, synthetic fiber and plastic industries.

Capital City Products Company is prepared to offer other polyglycerol esters to fulfill specific customer requirements. Depending on type, chain length and amount of fatty acid used, polyglycerol esters may be:

1. Partial or complete
2. Solid, paste or liquid
3. Saturated, unsaturated or polyunsaturated
4. Aliphatic or aromatic
5. Simple or complex esters
6. Esters of mono-, di or polycarboxylic acids
7. Esters of hydroxy acids
8. Esters of single or mixed fatty acids
9. High, medium or low molecular weight

Some of the standard polyglycerols esters are listed below.

CAPROL 2G4S:
Form: Solid
Color Gardner Max.: 8
Acid Value Max.: 6
Sap. Value: 165-185
Iodine Value Max.: 3
HLB: 2.5

CAPROL 3GO:
Form: Liquid
Color Gardner Max.: 7
Acid Value Max.: 6
Sap. Value: 125-150
Iodine Value Max.: 78
HLB: 6.2

CAPITAL CITY PRODUCTS CO.: Emulsifiers and Surfactants (Continued):

CAPROL Polyglycerol Esters(Continued):

CAPROL 3GS:
Form: Solid
Color Gardner Max.: 8
Acid Value Max.: 6
Sap. Value: 120-135
Iodine Value Max.: 3
HLB: 6.2

CAPROL 6G20:
Form: Liquid
Color Gardner Max.: 10
Acid Value Max.: 6
Sap. Value: 105-125
Iodine Value Max.: 75
HLB: 8.5

CAPROL 6G2S:
Form: Solid
Color Gardner Max.: 10
Acid Value Max.: 6
Sap. Value: 105-125
Iodine Value Max.: 3
HLB: 8.5

CAPROL 10G20:
Form: Liquid
Color Gardner Max.: 7
Acid Value Max.: 6
Sap. Value: 100-120
Iodine Value Max.: 60
HLB: 10.0

CAPROL 10G40:
Form: Liquid
Color Gardner Max.: 7
Acid Value Max.: 6
Sap. Value: 125-150
Iodine Value Max.: 60
HLB: 6.2

CAPITAL CITY PRODUCTS CO.: Emulsifiers and Surfactants (Continued):

CAPROL Polyglycerol Esters(Continued):

CAPROL 10G100:
Form: Liquid
Color Gardner Max.: 9
Acid Value Max.: 10
Sap. Value: 155-185
Iodine Value Max.: 75
HLB: 3.5

CAPROL PGE 860:
Form: Liquid
Color Gardner Max.: 8
Acid Value Max.: 6
Sap. Value: 90-100
Iodine Value Max.: 60
HLB: 11.0

CAPROL ET:
Form: Solid
Color Gardner Max.: 5
Acid Value Max.: 3
Sap. Value: 190-197
Iodine Value Max.: 25
HLB: 2.5

CAPROL JB:
Form: Flakes
Color Gardner Max.: 8
Acid Value Max.: 6
Sap. Value: 167-173
Iodine Value Max.: 4
HLB: 2.5

CAPITAL CITY PRODUCTS CO.: Emulsifiers and Surfactants (Continued):

ACCOMEEN Ethoxylated Fatty Amines:

ACCOMEEN surfactants are a group of tertiary amines derived from fatty sources ranging in chain length from C8 to C18 and substituted with two or more polyoxyethylene groups. These products are essentially cationic, but with increasing polyoxyethylene content, POE-5 and above, they become nonionic in character. The following list represents typical products in this line. Other variations can be supplied by utilizing different types of fatty amines or by changing the polyoxyalkane content.

ACCOMEEN C2:
Form: Liquid
Color Gardner Max.: 3
Surface Tension: 28
Neutral Equivalent: 280-300
Moisture % Max.: 1.0

ACCOMEEN C5:
Form: Liquid
Color Gardner Max.: 3
Surface Tension: 33
Neutral Equivalent: 415-430
Moisture % Max.: 1.5

ACCOMEEN C10:
Form: Liquid
Color Gardner Max.: 3
Surface Tension: 39
Neutral Equivalent: 640-660
Moisture % max.: 1.5

ACCOMEEN C15:
Form: Liquid
Color Gardner Max.: 3
Surface Tension: 41
Neutral Equivalent: 840-890
Moisture % Max.: 1.5

ACCOMEEN T2:
Form: Semi-Solid
Color Gardner Max.: 3
Surface Tension: 29
Neutral Equivalent: 345-355
Moisture % Max.: 1.0

CAPITAL CITY PRODUCTS CO.: Emulsifiers and Surfactants (Continued):

ACCOMEEN Ethoxylated Fatty Amines(Continued):

ACCOMEEN T15:
Form: Semi-solid
Color Gardner Max.: 6
Surface Tension: 41
Neutral Equivalent: 895-955
Moisture % Max.: 1.0

ACCOMEEN S2:
Form: Liquid
Color Gardner Max.: 8
Surface Tension: 26
Neutral Equivalent: 345-360
Moisture % Max.: 1.0

ACCOMEEN S5:
Form: Liquid
Color Gardner Max.: 10
Surface Tension: 33
Neutral Equivalent: 480-505
Moisture % Max.: 1.0

ACCOMEEN S15:
Form: Liquid
Color Gardner Max.: 10
Surface Tension: 40
Neutral Equivalent: 900-950
Moisture % Max.: 1.0

Surface tension in dynes/cm2 of 0.1% solution in water.

ACCOMID Surfactants:

The ACCOMID line of products are alkanolamides manufactured for the chemical specialties and personal care markets. Primary uses for these products are viscosity building, foam boosting/stabilizing and emulsification. They find applications in shampoos, liquid soaps, dish detergents and bubble bath products. The ACCOMIDS are classified as 'Super Amides" of the 1:1 amide type which exhibit nonionic surface active characteristics.

ACCOMID C:
Form: Liquid
Color Gardner Max.: 3
pH 10% Aqueous: 9-10.5
Free Amine %: 4-8.5

ACCOMID PK:
Form: Liquid
Color Gardner Max.: 5
pH 10% Aqueous: 9-10.6
Free Amine %: 4-8.5

CAPITAL CITY PRODUCTS CO.: Emulsifiers and Surfactants (Continued):

ACCONON Surfactants:

The ACCONON series of products offers a range of nonionic surfactants from oil soluble to water soluble. They are used as emulsifiers, dispersants, solubilizers and viscosity control agents in many different cosmetic, pharmaceutical and industrial formulations. The ACCONON line ranges from ethoxylated fatty acids, oils or esters to ethoxylated/propoxylated products. The following list represents typical products available; many variations of these can be supplied to meet specific requirements.

ACCONON TGH:
Form: Liquid
Color Gardner Max.: 3
pH: 6.0-7.0
HLB: 16

ACCONON ETG:
Form: Liquid
Color Gardner Max.: 3
pH: 5.5-6.5
HLB: 15

ACCONON CON:
Form: Liquid
Color Gardner Max.: 2
pH: 6.0-7.0
HLB: 10

ACCONON 1300:
Form: Liquid
Color Gardner Max.: 3
pH: 6.0-7.0
HLB: 11

ACCONON CA-5:
Form: Liquid
Color Gardner Max.: 3
pH: 6.0-7.0
HLB: 16

ACCONON CA-8:
Form: Liquid
Color Gardner Max.: 3
pH: 6.0-7.0
HLB: 8

CAPITAL CITY PRODUCTS CO.: Emulsifiers and Surfactants (Continued):

ACCONON Surfactants(Continued):

ACCONON CA-15:
Form: Liquid
Color Gardner Max.: 3
pH: 6.0-7.0
HLB: 16

ACCONON E:
Form: Liquid
Color Gardner Max.: 3
pH: 6.0-7.0
HLB: 16

ACCONON W-230:
Form: Solid
Color Gardner Max.: 3
pH: 6.5-7.5
HLB: 15

ACCONON 200-MS:
Form: Solid
Color Gardner Max.: 4
pH: 5.5-6.5
HLB: 8

ACCONON 400-MS:
Form: Solid
Color Gardner Max.: 4
pH: 5.5-6.5
HLB: 12

ACCONON 400-MO:
Form: Liquid
Color Gardner Max.: 4
pH: 5.5-6.5
HLB: 12

CAPITAL CITY PRODUCTS CO.: Specialty Fats--Fractionated Non-Lauric Hard Butter:

ARKOPOL E:
Form: Solid
Partially Hydrogenated Soybean Oil with Lactic Acid Esters of Monoglycerides
Typical Data:
WMP: 96-100F
SFI: 50F = 72 Min
70F = 63 Min
80F = 55 Min
92F = 19 Min
100F = 1 Max
Applications: Non-Tempering Confectioner's Coating, Vegetable Dairy, Cosmetics, Pharmaceuticals

ARKOPOL R:
Form: Solid
Partially Hydrogenated Vegetable Oil (Soybean, Cottonseed)
Typical Data:
WMP: 97-101F
SFI: 50F = 69 Min
70F = 59 Min
80F = 52 Min
92F = 23 Min
100F = 5 Max
Applications: Non-Tempering Confectioner's Coating, Vegetable Dairy, Cosmetics, Pharmaceuticals

CAPITAL CITY PRODUCTS CO.: Specialty Fats - Standard Lauric Hard Butter:

HB-80:
Form: Solid
Palm Kernel Oil
Typical Data:
WMP: 80-84F
FFA: 0.05% Max
Color (Lovibond): 1.5R Max
Multi-Purpose Vegetable Oil

HB-92:
Form: Solid
Partially Hydrogenated Palm Kernel Oil
Typical Data:
WMP: 91-93F
SFI: 50F = 64-68
70F = 51-55
80F = 34-38
92F = 3- 6
100F = 0
Confectioner's Coating and Centers

HB-95:
Solid
Partially Hydrogenated Vegetable Oil (Palm Kernel, Coconut)
Typical Data:
WMP: 94-97F
SFI: 50F = 64-68
70F = 53-57
80F = 36-40
92F = 8-12
100F = 0
Confectioner's Coating, Drops and Centers

HB-102:
Solid Flake
Partially Hydrogenated Vegetable Oil (Palm Kernel, Coconut and Palm)
Typical Data:
WMP: 101-104F
SFI: 50F = 62-68
70F = 50-56
80F = 38-42
92F = 12-16
100F = 4 Max
Confectioner's Coating, Drops and Centers, Vegetable Dairy, Icings, Cosmetics, Pharmaceuticals

CAPITAL CITY PRODUCTS CO.: Specialty Fats - Standard Lauric Hard Butter(Continued):

HB-102-PKO:
Solid
Partially Hydrogenated Palm Kernel Oil
Typical Data:
WMP: 101-104F
SFI: 50F: 66-72
70F: 55-60
80F: 40-44
92F: 10-14
100F: 3 Max
Applications: Confectioner's Coating, Drops and Centers, Vegetable Dairy, Icings, Cosmetics, Pharmaceuticals

HB-103:
Solid
Partially Hydrogenated Vegetable Oils (Palm Kernel and Palm)
Typical Data:
WMP: 102-104F
SFI: 50F = 66-72
70F = 55-60
80F = 41-45
92F = 15-19
100F = 4 Max
Applications: Confectioner's Coating, Drops and Centers

HB-108:
Solid
Partially Hydrogenated Vegetable Oils (Palm Kernel, Coconut & Palm)
Typical Data:
WMP: 107-109F
SFI: 50F = 64-70
70F = 54-60
80F = 43-49
92F = 19-25
100F = 8-12
Applications: Confectioner's Coating, Drops and Centers

HB-108-PKO:
Solid
Partially Hydrogenated Palm Kernel Oil
Typical Data:
WMP: 107-109F
SFI: 50F = 67-74
70F = 57-64
80F = 43-49
92F = 16-22
100F = 4-10
Applications: Confectioner's Coating, Drops and Centers

CAPITAL CITY PRODUCTS CO.: Specialty Fats - Standard Lauric Hard Butter(Continued):

HB-112:

Solid/Flake

Partially Hydrogenated Vegetable Oils (Palm Kernel, Coconut & Palm)

Typical Data:

WMP: 112-114F

SFI: 50F = 64-70
70F = 55-63
80F = 45-52
92F = 19-27
100F = 11-17

Confectioner's Coating, Centers, Icing Stabilizer, and Vegetable Dairy

HB-118:

Solid/Flake

Partially Hydrogenated Vegetable Oils (Palm Kernel, Coconut & Palm)

Typical Data:

WMP: 117-120F

SFI: 50F = 66-72
70F = 56-62
80F = 50-56
92F = 28-34
100F = 18-24

Confectioner's Coating, Centers, Icing Stabilizer, Vegetable Dairy, Cosmetic & Pharmaceutical

CAPITAL CITY PRODUCTS CO.: Specialty Oils and Bases:

CAPTEX Oils:

The CAPTEX oils are esters of fractionated coconut oil fatty acids with glycerine or propylene glycol. They are commonly called MCT oils. They have been designed specifically for pharmaceutical, nutritional and cosmetic applications. They are useful as emollients, solvents, carriers, fixatives and extenders.

CAPTEX 200:
Color LOVIBOND Red Max.: 1.0
Acid Value Max.: 0.1
Cloud Point Max. C: -15
Iodine Value Max.: 0.5
Viscosity @ 25C cps: 8
Moisture % Max.: 0.1

CAPTEX 300:
Color LOVIBOND Red Max.: 1.0
Acid Value Max.: 0.1
Cloud Point Max. C: -5
Iodine Value Max.: 0.5
Viscosity @ 25C cps: 23
Moisture % Max.: 0.1

CAPTEX 350:
Color LOVIBOND Red Max.: 1.0
Acid Value Max.: 0.1
Cloud Point Max. C: 0
Iodine Value Max.: 1.5
Viscosity @ 25C cps: 33
Moisture % Max.: 0.1

CAPTEX 355:
Color LOVIBOND Red Max.: 1.0
Acid Value Max.: 0.1
Cloud Point Max. C: -5
Iodine Value Max.: 0.5
Viscosity @ 25C cps: 23
Moisture % Max.: 0.1

CAPTEX 810B:
Color LOVIBOND Red Max.: 1.0
Acid Value Max.: 0.1
Cloud Point Max. C: 10
Iodine Value Max.: 51.0
Viscosity @ 25C cps: 30
Moisture % Max.: 0.1

CAPITAL CITY PRODUCTS CO.: Specialty Oils and Bases(Continued):

HYDROKOTE Specialty Bases:

HYDROKOTE bases are made from selected fractions of vegetable oils that are hydrogenated, blended, refined and deodorized. They are used primarily as replacements for cocoa butter and in cosmetic and pharmaceutical applications where their textures and narrow melting range characteristics are essential.

HYDROKOTE bases are available in a variety of melting point ranges and in different Solid Fat Indexes (SFI's)#.

Some of the standard HYDROKOTES are listed below:

Cocoa Butter:
Color LOVIBOND Max. Red: 1.5
Melting Pt. C Capillary: 30.0-34.0
FFA Value % Max.: 1.00
Iodine Value Max.: 35-40
Melting Pt. C USP Class II: 27.0-30.0

HYDROKOTE S-7:
Color LOVIBOND Max. Red: 1.5
Melting Pt. C Capillary: 33.3-35.0
FFA Value % Max.: 0.05
Iodine Value Max.: 3-6
Melting Pt. C USP Class II: 32.8-34.4

HYDROKOTE 27:
Color LOVIBOND Max. Red: 1.5
Melting Pt. C Capillary: 38.3-40.0
FFA Value % Max.: 0.05
Iodine Value Max.: 1-5
Melting Pt. C USP Class II: 33.0-35.7

HYDROKOTE SP:
Color LOVIBOND Max. Red: 1.5
Melting Pt. C Capillary: 32.8-33.3
FFA Value % Max.: 0.05
Iodine Value Max.: 6-8
Melting Pt. C USP Class II: 31.2-32.2

HYDROKOTE 25:
Color LOVIBOND Max. Red: 1.5
Melting Pt. C Capillary: 35.6-36.7
FFA Value % Max.: 0.05
Iodine Value Max.: 1-5
Melting Pt. C USP Class II: 33.6-36.3

CAPITAL CITY PRODUCTS CO.: Specialty Oils and Bases(Continued):

HYDROKOTE Specialty Bases(Continued):

HYDROKOTE 79:
Color LOVIBOND Max. Red: 1.5
Melting Pt. C Capillary: 41.1-42.8
FFA Value % Max.: 0.05
Iodine Value Max.: 1-5
Melting Pt. C USP Class II: 35.8-41.0

HYDROKOTE 711:
Color LOVIBOND Max. Red: 1.5
Melting Pt. C Capillary: 45.6-47.8
FFA Value % Max.: 0.05
Iodine Value Max.: 1-5
Melting Pt. C USP Class II: 39.5-44.5

The Solid Fat Index (SFI) is an empirical measure of the amount of solid fat at a particular temperature. It may be detrmined by the American Oil Chemists' Society method Cd 10-57.

Typical Solid Fat Indexes:

Cocoa Butter:
50F: 62-66
70F: 53-58
80F: 35-40
92F: 0-1
100F: 0

HYDROKOTE S-7:
50F: 64-70
70F: 58-62
80F: 35-40
92F: 1-5
100F: 0

HYDROKOTE 27:
50F: 62-68
70F: 54-60
80F: 40-46
92F: 12-18
100F: 1-5

HYDROKOTE SP:
50F: 68-73
70F: 61-67
80F: 50-56
92F: 0-2
100F: 0

CAPITAL CITY PRODUCTS CO.: Specialty Oils and Bases(Continued):

HYDROKOTE Specialty Bases(Continued):

HYDROKOTE 25:
50F: 62-68
70F: 53-59
80F: 37-43
92F: 8-14
100F: 0- 2

HYDROKOTE 79:
50F: 64-70
70F: 55-61
80F: 41-47
92F: 14-20
100F: 3- 4

HYDROKOTE 711:
50F: 66-72
70F: 61-67
80F: 49-55
92F: 25-31
100F: 10-16

CAPITAL CITY PRODUCTS CO.: Specialty Oils and Bases(Continued):

STEROTEX Powdered Lubricants:

STEROTEX products are finely divided white powders made from food grade vegetable oils. STEROTEX powders are neutral, low in ash content and practically without metal contamination. STEROTEX products are widely used as tablet lubricants in the pharmaceutical industry. They can be made in a variety of particle size and melting point ranges.

STEROTEX#:
Melting Pt. C: 60-63
Acid Value Max.: 0.4
Moisture & Volatiles Max. %: 0.1
Iodine Value Max.: 5.0

STEROTEX HM:
Melting Pt. C: 67-70
Acid Value Max.: 0.4
Moisture & Volatiles Max. %: 0.1
Iodine Value Max.: 5.0

STEROTEX K:
Melting Pt. C: 81-84
Acid Value Max.: 1.0
Moisture & Volatiles Max. %: 0.1
Iodine Value Max.: 5.0

All Sterotex products have a 10 ppm max. specification on heavy metals.

Conforms to the Monograph for Hydrogenated Vegetable Oil in the National Formulary, XVI edition.

CAPITAL CITY PRODUCTS CO.: Specialty Oils and Bases(Continued):

Vegetable Oil Products:

Vegetable Oils:

Peanut:
Color LOVIBOND Max. Red: 1.5
Melting Pt. F: <70
Sap. Value Max.: 185-195
Iodine Value Max.: 84-100
AOM Stability (Hours): 16

Corn:
Color LOVIBOND Max. Red: 3.5
Melting Pt. F: <60
Sap. Value Max.: 187-193
Iodine Value Max.: 102-128
AOM Stability (Hours): 16

Specialty Oils:

1600 Oil:
Color LOVIBOND Max. Red: 1.0
Melting Pt. F: 75-80
Sap. Value Max.: 235-250
Iodine Value Max.: 25
AOM Stability (Hours): 200

STABLAND:
Color LOVIBOND Max. Red: 1.5
Melting Pt. F: <40
Sap. Value Max.: 189-195
Iodine Value Max.: 107-115
AOM Stability (Hours): 16

1200 Oil:
Color LOVIBOND Max. Red: 2.0
Melting Pt. F: <60
Sap. Value Max.: 188-195
Iodine Value Max.: 125-135
AOM Stability (Hours): 8

Flaked Stearines:

CAPITAL 5330:
Color LOVIBOND Max. Red: 2.5
Melting Pt. F: 142-146
Sap. Value Max.: 189-198
Iodine Value Max.: 5
AOM Stability (Hours): NA

CAPITAL CITY PRODUCTS CO.: Specialty Oils and Bases(Continued):

Vegetable Oil Products(Continued):

Flaked Stearines(Continued):

CAPITAL 5370:
Color LOVIBOND Max. Red: 2.0
Melting Pt. F: 153-157
Sap. Value Max.: 189-197
Iodine Value Max.: 5
AOM Stability (Hours): NA

CAPITAL 5380:
Color LOVIBOND Max. Red: 4.0
Melting Pt. F: 138-141
Sap. Value Max.: 196-202
Iodine Value Max.: 5
AOM Stability (Hours): NA

Lauric Oils:

Palm Kernel:
Color LOVIBOND Max. Red: 1.0
Melting Pt. F.: 84
Sap. Value Max.: 243-247
Iodine Value Max.: 15-22
AOM Stability (Hours): NA

PURECO 76:
Color LOVIBOND Max. Red: 1.0
Melting Pt. F.: 76-80
Sap. Value Max.: 248-264
Iodine Value Max.: 12
AOM Stability (Hours): 200

PURECO 92:
Color LOVIBOND Max. Red: 1.0
Melting Pt. F.: 100-104
Sap. Value Max.: 248-264
Iodine Value Max.: 4
AOM Stability (Hours): 300

PURECO 110:
Color LOVIBOND Max. Red: 1.0
Melting Pt. F.: 112-115
Sap. Value Max.: 246-262
Iodine Value Max.: 4
AOM Stability (Hours): 300

CAPITAL CITY PRODUCTS CO.: Specialty Oils and Bases(Continued):

CAPITAL Fatty Acids:

CAPITAL fatty acids are a line of high quality acids derived from vegetable oils.

Fatty acids and their derivatives have a place in practically all phases of modern living. They add needed qualities to pharmaceuticals, cosmetics and other personal care products. They improve the performance of paints, lubricants, textiles, detergents and rubber products.

One important advantage in the use of fatty acids, as compared to whole oils, is that type-for-type they are more reactive and permit faster and more complete saponification, esterification or other reactions.

CAPITAL 160:
Color Gardner Max.: 6
Sap. Value: 250-260
Iodine Value: 25 max.
Acid Value: 250-260
Titer C: 22-26
Source Oil: Coconut

CAPITAL 170:
Color Gardner Max.: 4
Sap. Value: 260-270
Iodine Value: 8-16
Acid Value: 258-263
Titer C: 22-26
Source Oil: Coconut

CAPITAL 180:
Color Gardner Max.: 2
Sap. Value: 262-272
Iodine Value: 8-15
Acid Value: 260-270
Titer C: 22-26
Source Oil: Coconut

CAPITAL 220 (low IV):
Color Gardner Max.: 1
Sap. Value: 266-276
Iodine Value: 6 max.
Acid Value: 264-274
Titer C: 23-27
Source Oil: Coconut

CAPITAL CITY PRODUCTS CO.: Specialty Oils and Bases(Continued):

Capital Fatty Acids(Continued):

CAPITAL 172:
Color Gardner Max.: 6
Sap. Value: 196-206
Iodine Value: 122 min.
Acid Value: 195-205
Titer C: 26-30
Source Oil: Soya

CAPITAL 182 (Alkyd):
Color Gardner Max.: 3
Sap. Value: 198-207
Iodine Value: 125 min.
Acid Value: 195-205
Titer C: 26-30
Source Oil: Soya

CAPITAL 184:
Color Gardner Max.: 5
Sap. Value: 196-216
Iodine Value: 85-100
Acid Value: 195-215
Titer C: 27-31
Source Oil: Mixed

Typical Fatty Acid Chain Distribution:

CAPITAL 160:
C8: 4
C10: 5
C12: 34
C14: 18
C16: 14
C18: 4
C18-1: 18
C18-2: 3

CAPITAL 170:
C8: 7
C10: 7
C12: 49
C14: 17
C16: 10
C18: 2
C18-1: 7
C18-2: 1

CAPITAL CITY PRODUCTS CO.: Specialty Oils and Bases(Continued):

CAPITAL Fatty Acids(Continued):

Typical Fatty Acid Chain Distribution(Continued):

CAPITAL 180:
C8: 7
C10: 7
C12: 51
C14: 16
C16: 9
C18: 2
C18-1: 6
C18-2: 2

CAPITAL 220:
C8: 5
C10: 6
C12: 55
C14: 19
C16: 7
C18: 2
C18-1: 5
C18-2: 1

CAPITAL 172:
C16: 17
C18: 4
C18-1: 28
C18-2: 49
C18-3: 2

CAPITAL 182:
C16: 13
C18: 4
C18-1: 28
C18-2: 52
C18-3: 3

CAPITAL 184:
C8: 1
C10: 1
C12: 7
C14: 3
C16: 14
C18: 7
C18-1: 33
C18-2: 31
C18-3: 3

CASCHEM INC.: Emollient Esters:

NATURECHEM Non-Comedogenic Fatty Acid Esters:

NATURECHEM esters are derived from castor oil fatty acid (ricinoleic acid) and possess many of the benefits found in pure castor oil:

- Mildness.
- Non-comedogenic.
- Resistance to rancidity.
- Wetting properties.
- Refatting properties.
- Velvety skin feel.

Emollients:

NATURECHEM CR:

CTFA Name: Cetyl Ricinoleate
Color (Gardner): 1
Pour Point C: 27
Sap Value: 204
Iodine Value: 50
OH Value: 98
Acid Value: .5

NATURECHEM CAR:

CTFA Name: Cetyl Acetyl Ricinoleate
Color (Gardner): 1
Pour Point C: 7
Sap Value: 189
Iodine Value: 46
OH Value: 5
Acid Value: 1

NATURECHEM PGR:

CTFA Name: Propylene Glycol Ricinoleate
Color (Gardner): 1
Pour Point C: -15
Sap Value: 159
Iodine Value: 76
OH Value: 293
Acid Value: 2

NATURECHEM GTH:

CTFA Name: Glyceryl Triacetyl Hydroxystearate
Color (Gardner): 1
Pour Point C: -1
Sap Value: 298
Iodine Value: 5
OH Value: 5
Acid Value: 1

CASCHEM INC.: Emollient Esters(Continued):

Emollients(Continued):

NATURECHEM OHS:
CTFA Name: Octyl Hydroxystearate
Color(Gardner): 1
Pour Point C: 30
Sap Value: 150
Iodine Value: 3
OH Value: 71
Acid Value: 1

NATURECHEM GTR:
CTFA Name: Glyceryl Triacetyl Ricinoleate
Color (Gardner): 2
Pour Point C: -40
Sap Value: 300
Iodine Value: 76
OH Value: 5
Acid Value: 1

NATURECHEM MAR:
CTFA Name: Methyl Acetyl Ricinoleate
Color (Gardner): 1
Pour Point C: -25
Sap Value: 301
Iodine Value: 75
OH Value: 5
Acid Value: 2

Emulsifiers:

NATURECHEM GMHS:
CTFA Name: Glyceryl Hydroxystearate
Color (Gardner): 2
Melting Point C: 69
Sap Value: 162
Iodine Value: 5
OH Value: 320
Acid Value: 6

NATURECHEM EGHS:
CTFA Name: Glycol Hydroxystearate
Color (Gardner): 1
Melting Point C: 66
Sap Value: 169
Iodine Value: 5
OH Value: 266
Acid Value: 3

NATURECHEM PGHS:
CTFA Name: Propylene Glycol Hydroxystearate
Color (Gardner): 1
Melting Point C: 53
Sap Value: 160
Iodine Value: 5
OH Value: 289
Acid Value: 2

CASCHEM INC.: Emollient Esters(Continued):

WICKENOL and WAXENOL Emollient Esters:

These esters range from the reliable isopropyl esters to the elegant porosity esters. This large selection of emollients fulfills a diverse number of formulators' needs. Listed below summarizes a few of the benefits you will find in this selection of emollients.

- imparts emolliency and skin softening to skin.
- excellent bodying agent.
- excellent wetting agent for pigments.
- provides cushioning in emulsion products.
- mildness.
- improves pay out in lipsticks.

Emollients:

WICKENOL 101:
CTFA Name: Isopropyl Myristate
Color (Gardner): -1
Pour Point C: 0
Ester Value: 205
Refractive Index: 1.433
Acid Value: .7

WICKENOL 105:
CTFA Name: Isopropyl Myristate/Palmitate
Color (Gardner): -1
Pour Point C: 7
Ester Value: 194
Refractive Index: 1.433
Acid Value: .7

WICKENOL 111:
CTFA Name: Isopropyl Palmitate
Color (Gardner): -1
Pour Point C: 7
Ester Value: 185
Refractive Index: 1.436
Acid Value: .7

WICKENOL 127:
CTFA Name: Isopropyl Stearate
Color (Gardner): -1
Pour Point C: 18
Ester Value: 180
Refractive Index: 1.438
Acid Value: .7

CASCHEM, INC.: Emollient Esters(Continued):

WICKENOL and WAXENOL Emollient Esters(Continued):

Emollients(Continued):

WICKENOL 131:
CTFA Name: Isopropyl Isostearate
Color (Gardner): 2+
Pour Point C: -10
Ester Value: 175
Refractive Index: 1.4418
Acid Value: .7

WICKENOL 139:
CTFA Name: Synthetic Jojoba Oil
Color (Gardner): 2+
Pour Point C: -20
Ester Value: 87
Refractive Index: 1.4355
Acid Value: .7

WICKENOL 141:
CTFA Name: Butyl Myristate
Color (Gardner): -1
Pour Point C: 2
Ester Value: 198
Refractive Index: 1.437
Acid Value: .7

WICKENOL 142:
CTFA Name: Octyldodecyl Myristate
Color (Gardner): 2+
Pour Point C: -5
Ester Value: 110
Refractive Index: 1.460
Acid Value: .7

WICKENOL 143:
CTFA Name: Oleyl Oleate
Color (Gardner): 2+
Pour Point C: 12
Ester Value: 107
Refractive Index: 1.463
Acid Value: .7

WICKENOL 144:
CTFA Name: Isodecyl Oleate
Color (Gardner): 2+
Pour Point C: -20
Ester Value: 135
Refractive Index: 1.45
Acid Value: .7

CASCHEM INC.: Emollient Esters(Continued):

WICKENOL and WAXENOL Emollient Esters(Continued):

Emollients(Continued):

WICKENOL 151:
CTFA Name: Isononyl Isononanoate
Color (Gardner): -1
Pour Point C: -15
Ester Value: 198
Refractive Index: 1.435
Acid Value: .7

WICKENOL 152:
CTFA Name: Isodecyl Isononanoate
Color (Gardner): -1
Pour Point C: -15
Ester Value: 188
Refractive Index: 1.438
Acid Value: .7

WICKENOL 153:
CTFA Name: Isotridecyl Isononanoate
Color (Gardner): -1
Pour Point C: -15
Ester Value: 165
Refractive Index: 1.444
Acid Value: .7

WICKENOL 155:
CTFA Name: Octyl Palmitate
Color (Gardner): -1
Pour Point C: -3
Ester Value: 150
Refractive Index: 1.445
Acid Value: .7

WICKENOL 156:
CTFA Name: Octyl Stearate
Color (Gardner): -1
Pour Point C: 5
Ester Value: 150
Refractive Index: 1.4465
Acid Value: .7

WICKENOL 158:
CTFA Name: Dioctyl Adipate
Color (Gardner): -1
Pour Point C: -15
Ester Value: 302
Refractive Index: 1.445
Acid Value: .7

CASCHEM INC.: Emollient Esters(Continued):

WICKENOL and WAXENOL Emollient Esters(Continued):

Emollients(Continued):

WICKENOL 159:
CTFA Name: Dioctyl Succinate
Color (Gardner): -1
Pour Point C: -15
Ester Value: 320
Refractive Index: 1.443
Acid Value: .7

WICKENOL 160:
CTFA Name: Octyl Perlargonate
Color (Gardner): 2+
Pour Point C: -15
Ester Value: 210
Refractive Index: 1.437
Acid Value: .7

WICKENOL 171:
CTFA Name: Octyl Hydroxystearate
Color (Gardner): 2+
Pour Point C: 5
Ester Value: 150
Refractive Index: 1.456
Acid Value: .7

WICKENOL 506:
CTFA Name: Myristyl Lactate
Color (Gardner): 2+
Pour Point C: 9
Ester Value: 175
Refractive Index: 1.4442
Acid Value: 1.0

WICKENOL 174:
CTFA Name: Myristyl Octanoate
Color (Gardner): 2+
Pour Point C: -10
Ester Value: 162
Refractive Index: 1.443
Acid Value: .7

CASCHEM INC.: Emollient Esters(Continued):

WICKENOL and WAXENOL Emollient Esters(Continued):

Emollients Blended Esters:

WICKENOL 136:
CTFA Name: Isopropyl Myristate (and) Isopropyl Palmitate (and) Isopropyl Stearate
Color (Gardner): -1
Pour Point C: 5
Ester Value: 195
Refractive Index: 1.435
Acid Value: .7

WICKENOL 161:
CTFA Name: Dioctyl Adipate (and) Octyl Palmitate (and) Octyl Stearate
Color (Gardner): -1
Pour Point C: -12
Ester Value: 268
Refractive Index: 1.446
Acid Value: .7

WICKENOL 163:
CTFA Name: Octyl Stearate (and) Octyl Palmitate (and) Dioctyl Adipate
Color (Gardner): -1
Ester Value: 188
Refractive Index: 1.446
Acid Value: .7

CASCHEM INC.: Emollient Esters(Continued):

WICKENOL and WAXENOL Emollient Esters(Continued):

Solid Emollients:

WAXENOL 801:
CTFA Name: Arachidyl Propionate
Color (Gardner): 2+
Melt Point C: 37
Ester Value: 113
Iodine Value: 12
Acid Value: .7

WAXENOL 810:
CTFA Name: Myristyl Myristate
Color (Gardner): 2+
Melt Point C: 38
Ester Value: 128
Iodine Value: 1.0 Max.
Acid Value: .7

WAXENOL 815:
CTFA Name: Cetyl Palmitate
Color (Gardner): 1+
Melt Point C: 50
Ester Value: 113
Iodine Value: 1.0 Max.
Acid Value: .7

WAXENOL 821:
CTFA Name: Synthetic Beeswax
Color (Gardner): 1+
Melt Point C: 63
Ester Value: 75
Iodine Value: 12
Acid Value: 21.0

WAXENOL 822:
CTFA Name: Arachidyl Behenate
Color (Gardner): 2+
Melt Point C: 65
Ester Value: 75
Iodine Value: 10
Acid Value: 22.0

CASCHEM INC.: Specialty Products:

CasChem's specialty ingredient line consists of the following product types:

- VITA-COS
- Humectant
- Cetyl & Lanolin Ethers
- Synthetic Beeswax
- Binder

Each specialty product provides the formulator with a benefit for a specific formulation requirement. Listed below are a few of the benefits contained within this specialty line.

- anti-irritancy
- liquid bodying agent
- water soluble binder
- moisturization

WICKENOL 535:
CTFA Name: Wheat Germ Glycerides
Color (Gardner): (2+)
Acid Value: 1.5 Max.
Iodine Value: 110 Min.

WICKENOL 545:
CTFA Name: Glucose Glutamate
Color (Gardner): 2+
Acid Value: N/A
Iodine Value: N/A

WICKENOL 707:
CTFA Name: PPG-30 Cetyl Ether
Color (Gardner): (2+)
Acid Value: 1.5 Max.
Iodine Value: 1.0 Max.

WICKENOL 727:
CTFA Name: PPG-30 Lanolin Ether
Color (Gardner): (4+)
Acid Value: 1.0 Max.
Iodine Value: 12

WICKENOL 550:
CTFA Name: Maltodextrin
Color (Gardner): (2+)
Acid Value: N/A
Iodine Value: N/A

CASCHEM INC.: Specialty Products(Continued):

Emulsifiers--SURFACTOL Non-ionic Surfactants:

The SURFACTOL series of nonionic emulsifiers are ethoxylated castor oils with varying amounts of ethylene oxide added to hydroxyl bearing fatty acid chain. They vary from self-emulsifiable to completely water soluble. The HLB values vary from 3.6 to 16.

SURFACTOL nonionic surfactants offer these benefits:
- Low odor.
- Low foaming.
- Excellent stability over broad pH range.
- Lubricity.
- Excellent Fragrance Solubilizer

Non-Ionics:

SURFACTOL 318:
CTFA Name: PEG-5 Castor Oil
Pour Point C: -25
Iodine Value: 70
Water Solubility: Dispersible

SURFACTOL 365:
CTFA Name: PEG-40 Castor Oil
Pour Point C: 10
Iodine Value: 36
Water Solubility: Soluble

NATURECHEM THS-200:
CTFA Name: PEG-200 Trihydroxystearin
Melting Point C: 53
Iodine Value: 5
Water Solubility: Soluble

SOLARSCREEN O Sunscreen:

SOLARSCREEN O is CasChem's trademark for Padimate O Octyl Dimethyl Para-aminobenzoate the most widely used sunscreening agent. SOLARCHEM O's ultra violet absorption range is between 290-315 nm i.e. the ideal range for screening out sunburning rays. Additional benefits provided by SOLARCHEM O are:
- High purity.
- Low odor.
- Light color.
- Easy to use liquid.

Specifications:
Chemical Name: 2-Ethyl Hexyl Paradimethylaminobenzoate
CTFA Name: Octyl diemethyl PABA
Generic Name: Padimate O
Acid Number: 1.0 Max.
Saponification Value: 195-215
Refractive Index @ 25C: 1.5390-1.5430
Specific Gravity @ 25C: 0.99-1.00
% Purity GC Method: 99.00 Min.

CASCHEM INC.: Specialty Products(Continued):

Emulsifiers--SURFACTOL Cationic Surfactants:

The SURFACTOL Q Series represents CasChem's new line of Castor base quaterniums. These dimethyl amino propylamine quaterniums are prepared from high purity ricinoleic acid and hydroxystearic acid. They offer superior substantivity to skin and hair without build up. SURFACTOL Q series quaterniums offer the following advantages:

- refatting agent.
- impart substantivity and emolliency to skin and hair.
- Foam boosting properties in shampoos.
- Water soluble.
- Broad compatibility with anionic surfactants.
- Mildness.

Cationics:

SURFACTOL Q1:
CTFA Name: Ricinoleamido-propyl Trimonium Chloride
Pour Point C: -15
Iodine Value: N/A
Water Solubility: Soluble

SURFACTOL Q2:
CTFA Name: Hydroxy Stearamidopropyl Trimonium Chloride
Melting Point C: 35
Iodine Value: 5
Water Solubility: Dispersible

SURFACTOL Q3:
CTFA Name: Hydroxy Stearamidopropyl Trimonium Methyl Sulfate
Melting Point C: 45
Iodine Value: 5
Water Solubility: Dispersible

Waxes--CASTORWAX Hydrogenated Castor Oils:

CasChem offers three types of hydrogenated castor oils for use in cosmetics and pharmaceutical products. CASTORWAX MP-80 and MP-70 are specialty products having lower melting points than the standard grade of hydrogenated castor oil i.e. CASTORWAX N.F. These materials provide ease of use in their flaked form and in finished products release no undesirable odors nor become rancid. Some of the outstanding performance benefits obtained from CASTORWAX products are:

- Broad compatibility with most natural and animal waxes.
- Enhance pay-off in stick products.
- Compatible with silicone fluids.
- Provide thickening and opacification.
- Odorless and tasteless.

CASCHEM INC.: Specialty Products(Continued):

Waxes--CASTORWAX Hydrogenated Castor Oils(Continued):

CASTORWAX MP80:
CTFA Name: Hydrogenated Castor Oil
Melting Point C: 80
Sap Value: 180
Iodine value: 29
OH Value: 158
Acid Value: 2

CASTORWAX NF:
CTFA Name: Hydrogenated Castor Oil
Melting Point C: 87
Sap Value: 180
Iodine Value: 5
OH Value: 158
Acid Value: 2

CASTORWAX MP70:
CTFA Name: Hydrogenated Castor Oil
Melting Point C: 70
Sap Value: 180
Iodine Value: 38
OH Value: 158
Acid Value: 2

Cosmetic Grade Castor Oils:

CasChem's CRYSTAL O and DIAMOND QUALITY are specialty refined U.S.P. castor oils. These oils are brilliant, light colored, odorless and tasteless and are manufactured in accordance with the Food and Drug Administration's Good Manufacturing Practices. For formulations that require castor oil with additional oxidative stability CasChem offers CRYSTAL CROWN and COSMETOL X Oils. Both oils contain a food grade anti-oxidant.

CRYSTAL O:
CTFA Name: Castor Oil
Pour Point C: -23
Sap Value: 180
Iodine Value: 86
OH Value: 164
Acid Value: 2

CRYSTAL CROWN:
CTFA Name: Castor Oil
Pour Point C: -23
Sap Value: 180
Iodine Value: 86
OH Value: 164
Acid Value: 2

COSMETOL X:
CTFA Name: Castor Oil
Pour Point C: -23
Sap Value: 180
Iodine Value: 86
OH Value: 164
Acid Value: 2

DIAMOND QUALITY:
CTFA Name: Castor Oil
Pour Point C: -23
Sap Value: 180
Iodine Value: 86
OH Value: 164
Acid Value: 2

COSTEC, INC.: COJOBA JoJoba Oil:

This naturally pure, stable, liquid wax-ester, obtained from the seeds of the desert shrub jojoba (simmondsia Chinensis) is unique among vegetable oils. Chemically, the "oil" is actually a liquid wax made up almost entirely of non-glyceride, unsaturated straight-chain acids and alcohols of high molecular weight, principally C20 and C22. It manifests excellent spreadability and rub-in characteristics with non-oily afterfeel making it an elegant emollient, lubricant and conditioner. Although unsaturated, the oil is non-drying and can be stored for years without becoming rancid.

Typical Chemical Composition:
Saturated acids: 1.5-1.7%
Oleic Acid: 0.66
Palmitoleic acid: 0.24
Eicosenoic acid (C20): 30.30
Docosenic acid (C22): 14.20
Eicosenol (C20): 14.60
Docosenol (C22): 33.70
Glycerol: none
Unsaponifiable matter: 37-48%

Character:
Clear, light yellow liquid of mild, pleasant odor.

Solubility:
Soluble in benzene, petroleum ether, chloroform, carbon tetrachloride, carbon disulfide and hexane; immiscible with alcohol and acetone.

Typical Properties:
Pour point: 10C
Solidifying point: Approximate 7C
Specific gravity (g/cc, 77F): 0.86-0.89
Viscosity (cs, 25C): 56-59
Refractive value:(25C): 1.464±.004
Saponification Value: 90 minimum
Iodine value (Hanus): 80 minimum
Acid Number: .23-.57
Acidity: 1.5%

Stability:
Unchanged by repeated heatings to 285C or by heating for 4 days at 370C. Non-volatile, non-drying.

Hydrogenation:
Easily hydrogenated to a hard white wax with nickel-copper catalyst and relatively mild temperatures and pressures. Melting point of wax 65-68C; hardness (trionic gauge 90).

CTFA Name: Jojoba oil

COSTEC, INC.: Collagen Proteins for Skin Care:

SOLLAGEN:

CTFA: Soluble Animal Collagen

SOLLAGEN brand soluble collagen and film former is a natural protein moisturizer that can bind and retain many times its own weight of water. SOLLAGEN is not chemically produced or modified but is rather one of nature's own skin components thought to be partly responsible for the soft, pliant texture of young skin; it is found in all animal skin and available in three varieties.

SOLLAGEN 1%:

At this level of protein exhibits a high viscosity due to its strong water binding ability. A nearly clear to slightly milky appearance with a mild, pleasant odor.

SOLLAGEN .5% "Extra Clear":

Is a solution of tropo-collagen or collagen monomer that has been further processed to produce the highest clarity possible. It is designed for those applications that require utmost clarity.

SOLLAGEN LA:

This is a 1% solution of natural soluble collagen which has reduced buffer strength to accomodate a wider range of emulsion applications. SOLLAGEN LA is non-tacky with complete rub in that leaves the skin smooth and soft.

COSTEC, INC.: Skin Lipids:

LIPITEIN P:

CTFA: Porcine Skin Lipids

These natural lipids are essentially clear at room temperature. They are high in monounsaturated fats and stabilized with tocopherol (Vitamin E). LIPITEIN P should be used in moisturizing cremes for its spreadability and soothing, supple feel. This natural skin lipid is noncomedogenic and low in irritancy.

LIPITEIN P with Vitamin E:

CTFA: Porcine Skin Lipids with Tocopherol

Same product as LIPITEIN P, but with 1% Vitamin E added. Use in moisturizing cremes, lotions, and hair care preparations like conditioners, relaxers, and other specialty products. Has natural UV absorption properties.

ELASTEIN 5000:

CTFA: Hydrolyzed elastin

ELASTIN 5000 is a enzymatically hydrolyzed elastin that is recognized as one of the three essential connective tissue proteins in skin; as a natural protein, elastin is non toxic, a film former, and aids in the binding of moisture and the forming of protective colloids.

COSTEC, INC.: COSEPT Parabens:

COSEPT parabens are extra fine, odorless, white powders that are non-volatile, non-hygroscopic, and generally recognized as safe (GRAS) for use as preservatives in foods as well as cosmetics. For over 50 years, these neutral esters of parahydroxybenzoic acid have functioned as excellent bacteriostatic and fungistatic agents at low concentrations against a wide range of organisms. The parabens are considered non-toxic and non-sensitizing. They are stable and active in the 3-8 pH range and are often used in conjunction with other additives such as imidazolidinyl urea, EDTA, etc., for synergistic effects. For personal care products, the COSEPTS are typically used at .2% methyl and .1% propyl in combination; these levels should be adjusted to specific circumstances. The preference for parabens continues due to their 50-year history of safety and reliability.

COSEPT M:

Methyl paraben, technical

Appearance: fine white powder
Solubility: meets NF tests
Identification: positive NF tests
Melting range: 124-128C
Acidity: passes NF tests
Loss of drying: .5% maximum
Residue on ignition: .05% maximum
Assay: .99% minimum

CTFA Name: methyl paraben

COSEPT P:

Propyl paraben, technical

Appearance: fine white powder
Solubility: meets NF tests
Identification: positive NF tests
Melting Range: 93-98C
Acidity: passes NF tests
Loss of drying: .5% maximum
Residue on ignition: .05% maximum
Assay: .99% minimum

CTFA Name: propyl paraben

COSTEC, INC.: COSEPT Preservatives:

COSEPT 200:

CTFA Name: Quaternium-15

Chemical Nomenclature: Cis isomer of 1-(3-chloroallyl)-3,5,7-triaza-1-azoniaadamantane chloride

Type of Compound: Adamantane

Physical Form: Fine, off-white hygroscopic powder

CAS Number: 51229-78-8

Optimum pH: 4.0 to 10.0

Usage Levels: 0.02-0.3%

Solubility: Water soluble (127g/100g)

Stability: Stable below 60C

Compatibility: Compatibile with common cosmetic and household product ingredients

Activity Spectrum: Active against Gram+ and Gram- bacteria, including Pseudomonas, yeast and mold.

Comments: Does not contain free formaldehyde. Easy to use. Very effective. All purpose preservative. Rapid kill time.

Application: Detergent systems
Emulsion products
Personal care and cosmetics
General purpose

Summary: A broad spectrum antimicrobial especially effective against pseudomonas as well as other bacteria, yeast and mold. COSEPT 200 is active in the 4-10 pH range and is compatible with anionic, nonionic, or cationic systems. It has excellent water solubility and offers better shelf life protection than most preservatives. Safe, excellent high performance preservative.

NUOSEPT C:

CTFA Name: Polymethoxy Bicyclic Oxazolidine

Chemical Nomenclature: 5-Hydroxymethoxymethyl-1-aza-3-dioxabicyclo (3.3.0) octane

Type of Compound: Oxazolidine

Physical Form: 50% Active Liquid in Water

CAS Number: 56709-13-8

Optimum pH: 4.5-9.5

Usage Levels: 0.1 to 0.5% w/w

Solubility: Water soluble; mineral oil: 0.18g/100g

Stability: Stable

Compatibility: Not inactivated by nonionics, compatible with common cosmetic and household product ingredients.

Activity Spectrum: Active against Gram+ and Gram- bacteria, yeasts and molds

Comments: Good toxicological profile at recommended use levels. Halogen free. Easy to use. Stable at elevated temperatures.

COSTEC, INC.: COSEPT Preservatives(Continued):

NUOSEPT C(Continued):

Application: Detergent systems
Emulsion products
Personal care and cosmetics
General purpose
Coatings & Industrial
Some indirect food uses.

Summary: An exciting, newer, liquid preservative with excellent credentials for most applications. It is a stable, nonionic, nonreactive product for use in the 4.5 to 9.5 pH range. NUOSEPT C is a clear, low odor liquid with excellent temperature stability when used between 32 to 200F. Excellent choice for a cost-effective, safe, high performance preservative.

COSEPT M:

CTFA Name: Methyl Paraben
Chemical Nomenclature: Methyl p-Hydroxy Benzoate
Type of Compound: Benzoic acid ester
Physical Form: Odorless white powder
CAS Number: 99-76-3
Optimum pH: 3.8 to 8.0
Usage Levels: 0.15 to 0.3%
Solubility: In water: 0.25g/100g @ 25C
Stability: Stable
Compatibility: Incompatible with nonionic and cationic surfactants
Activity Spectrum: Primarily fungi and Gram+ bacteria, poor vs pseudomonads
Comments: 50 year history of use. GRAS listed. Poor water solubility. Use in combination with other products. Available as N.F. grade.
Application: Detergent systems
Emulsion products
Personal care and cosmetic
Pharmaceuticals
Food systems
High oil systems

Summary: For over 50 years the parabens have been used as preservatives for food, pharmaceuticals and cosmetics. The methyl ester is the most water soluble, best against gram positive bacteria and mold, and the most commonly used paraben. COSEPT M is active in the 3-7 pH range and generally compatible with anionics and some nonionics.

COSTEC, INC.: COSEPT Preservatives(Continued):

COSEPT P:

CTFA Name: Propyl Paraben
Chemical Nomenclature: Propyl p-Hydroxy Benzoate
Type of Compound: Benzoic acid ester
Physical Form: Odorless white powder
CAS Number: 94-13-3
Optimum pH: 3.8 to 8.0
Usage Levels: 0.02 to 0.2%
Solubility: Soluble in: 2000 parts water
Stability: Stable
Compatibility: Incompatible with nonionics, and cationic surfactants
Activity Spectrum: Primarily Fungi and Gram+ bacteria; poor vs pseudomonads
Comments: Excellent safety. GRAS listed. Used in combination with methyl paraben. Available as N.F. grade.
Application: Detergent systems
Emulsion products
Personal care and cosmetics
Pharmaceuticals
Food systems
High oil systems
Summary: Because the parabens are more soluble in oils and organic solvents than in water, mixtures of these esters are normally used for better oil-water distribution and more efficient antimicrobial activity. COSEPT P is generally used in combination with M and often both will be used with COSEPT 115 or COSEPT 200. Propyl is effective against fungi and gram positive bacteria.

COSEPT B:

CTFA Name: Butyl Paraben
Chemical Nomenclature: Butyl p-Hydroxy Benzoate
Type of Compound: Benzoic acid ester
Physical Form: Odorless white powder
CAS Number: 94-26-8
Optimum pH: 3.8-7.0
Usage Levels: 0.001 to 0.2%
Solubility: Very slightly soluble in water
Stability: Stable
Compatibility: Incompatible with nonionic and cationic surfactants
Activity: Primarily fungi and Gram+ bacteria; poor vs pseudomonads
Comments: Safe. GRAS listed. Effective only in acidic pH. Anti fungal. Available as N.F. grade.

COSTEC, INC.: COSEPT Preservatives(Continued):

COSEPT B(Continued):

Application: Detergent systems
Emulsion products
Personal care and cosmetics
Pharmaceuticals
Food systems
High oil systems

Summary: More than the other parabens, COSEPT B provides the strongest action against fungi and yeast and is less water soluble. It is generally used as a co-preservative with other COSEPTS in order to increase the activity spectrum. It is effective only in acid pH range.

COSEPT E:

CTFA Name: Ethyl Paraben
Chemical Nomenclature: Ethyl p-Hydroxy Benzoate
Type of Compound: Benzoic acid ester
Physical Form: Odorless white powder
CAS Number: 120-47-8
Optimum pH: 3.8-8.0
Usage Levels: 0.1 to 0.25%
Solubility: In water: 0.075% @ 25C
Stability: Stable
Compatibility: Incompatible with nonionic and cationic surfactants
Activity Spectrum: Primarily fungi and Gram+ bacteria; poor vs pseudomonads
Comments: Safe. GRAS listed. Used in combinations. Not widely used. Available as N.F. grade.

Application: Detergent systems
Emulsion products
Personal care and cosmetics
Pharmaceuticals
Food systems
High oil systems

Summary: The ethyl ester of COSEPT E generally demonstrates the good performance of all the parabens against gram positive bacteria and has the same limitations regarding pH range, system compatibility and water solubility. It is most effective against mold and yeast and is used in combination with the other COSEPTS.

COSTEC, INC.: COSEPT Preservatives(Continued):

COSEPT 115:

CTFA Name: Imidazolidinyl Urea

Chemical Nomenclature: N.N-methylenebis (N-(1-(hydroxymethyl)-2.5-dioxo-4-imidazolidinyl) urea)

Type of Compound: Heterocyclic substituted urea

Physical Form: Odorless white powder

CAS Number: 39236-46-9

Optimum pH: 3.0 to 9.0

Usage Levels: 0.05 to 0.5%

Solubility: Water soluble

Stability: Stable

Compatibility: Compatible with common cosmetic ingredients

Activity Spectrum: Good spectrum when used in combination with the parabens

Comments: Well known co-preservative when used in combination with the parabens for broad spectrum activity. Good water solubility and stability.

Application: Emulsion products
Personal care and cosmetics
Some specialty systems

Summary: This well known co-preservative is used in combination with the parabens for broad spectrum activity. COSEPT 115 has good water solubility, is odorless and compatible with most ingredients. It is active in the 3-9 pH range and offers good stability. Generally not used alone.

COSEPT BNP:

CTFA Name: 2-bromo-2 nitropropane-1,3 diol

Chemical Nomenclature: 2-Bromo-2-nitro-1,3-propanediol

Type of Compound: Substituted aliphatic diol

Physical Form: White crystals; slight odor

CAS Number: 52-51-7

Optimum pH: 4.0 to 10.0

Usage Levels: 0.01 to 0.1%

Solubility: Soluble in water, alcohol

Stability: Unstable alkaline pH; high temp.

Compatibility: Inactivated by compounds containing (-SH) groups; cysteine, aluminum

Activity Spectrum: Broad spectrum; most effective against bacteria.

Comments: Effective against a broad range of microorganisms; good water solubility. Low use levels.

Application: Detergent systems
Emulsion products
Personal care and cosmetics
General purpose, but not in presence of amines

Summary: COSEPT BNP is effective against a broad range of microorganisms, but performs better against fungi and yeasts. It has good water solubility and is most effective at pH 6 or below. It is generally unaffected by nonionics, but is inactivated by sulfur containing compounds.

COSTEC, INC.: COVERA:

COVERA is the natural aloe vera juice obtained from the parenchyma of fresh leaves of mature, Texas-grown Aloe barbadensis Miller. This plant, a member of the lily family is commonly called Aloe Vera or true Aloe. The essentially clear inner pulp matrix is specially processed, pasteurized, and filtered. Sufficient amounts of anti-oxidants and preservatives have been included to help stabilize COVERA's natural state and color. Natural unpreserved aloe vera juice used for cosmetic or food purposes generally consists of 99.5% water and 0.5% total solids. Allowing for variations owing to climatic and seasonal growing conditions, the solids have been found, by various researchers, to contain sugars, mucopolysaccharides, amino acids, minerals, sterols, sapogenins, enzymes, fatty acids, and natural hydroxyquinones.

Typical Properties:
Appearance: Clear Liquid
Color: Light yellow to pale amber
Odor: Typical, light, vegetable/plant-like note
Specific Gravity: 1.000±.002 @ 25C
pH (as is): 4.0-5.5
Microbiology: Pathogenic organisms - none
Total Aerobic count - 200/g maximum

Solubility:
Soluble in common solvents such as water, propylene glycol and glycerine.

CTFA Name: Aloe Vera Gel

COSTEC, INC.: COVERA - Dry:

COVERA Dry is derived from the species Aloe Barbadensis Miller and represents a 199 fold concentrate of aloe vera gel. This powder form of COVERA is for manufacturing use in various health, cosmetic, and pharmacutical products where the high moisture levels of the liquid type can present formulation problems. COVERA Dry can be readily solubilized by adding the product at a 1:199 ratio by weight to well agitated water. Once reconstituted, the mixture should be used or protected with a suitable preservative system.

Typical Properties:
- Appearance: free-flowing off white to light tan
- Odor: slight plant-like
- pH (1%): 4.0-6.0
- Specific Gravity (0.5% solution): 1.002±0.002
- Water Insolubles: less than 0.05%
- Solution (.5%): water white to very slight yellow
- Microbiology: Less than 5000 per gram
- E. Coli: negative
- Salmonellae: negative

CTFA Name: Aloe Vera Gel

Implications for Cosmetic Utility:

As a Moisturizing Agent

Aloes are classed among natural gum-resins. Like so many other mucopolysaccharides, their mucilaginous gel extract is a film-former. This characteristic in combination with the high water content of the A. vera gel extract suggests that the mucilaginous liquid is potentially useful in aiding skin moisturizing effects. Additionally, the natural acidity in the range of pH 4-5 of the mucilaginous gel puts it in the same region as that of normal skin pH.

As a Sunscreen Agent:

Natural anthraquinones or anthracenic derivatives, such as those found in aloe sap in free state or combined as glycosides reportedly absorb U.V. light within the skin's erythemal range. For example, the parent compound of barbaloin, a highly conjugated glycoside, anthrone absorbs U.V. light between 250-290 nm with peak absorbance at 260-275 nm. Similarly, emodin is considered potentially useful as a U.V. absorber in the erythemal range of 290-300 nm.

As a Cosmetic Ingredient:

A. vera has had a long history of uses as topical treatment in home remedies for minor skin injuries and for beautification purposes with no apparently harmful effects reported. It is generally compatible with conventional cosmetic ingredients and can be incorporated into most cosmetic products.

COSTEC, INC.: Proteins for Hair Care:

PEPTEIN 2000:
CTFA: Hydrolyzed Animal Protein
PEPTEIN 2000 is an enzymatically hydrolyzed protein that is produced under closely controlled conditions from high collagen content raw materials. PEPTEIN 2000 has an extremely low odor, is remarkably uniform and very substantive to hair and skin; it is the product by which others measure quality and value. Available as a liquid or powder.

PF-6:
CTFA: Hydrolyzed Animal Protein
PF-6 is an enzymatically derived collagen available in liquid or dried forms. It is recommended for use in cost critical protein applications where substantivity is secondary in importance.

PEPTEIN AH:
CTFA: Hydrolyzed Animal Protein
PEPTEIN AH is an excellent liquid alkaline hydrolyzate, not enzymatic. The alkaline hydrolyzates have a diverse nature, so PEPTEIN AH contains a wide range of proteins from amino acids through oligo peptides. This diversity offers humectancy, water binding power and film forming functions.

POLYPRO 5000:
POLYPRO 15,000:
CTFA: Hydrolyzed Animal Protein
POLYPRO 5000 is a completely natural powdered protein; it does not contain chemical or microbial preservatives and can act as an anti-irritant and reduce the harshness of certain surfactants. A higher molecular weight product is also available named POLYPRO 15,000.

PROLAGEN MP-1:
CTFA: Propyltrimonium Hydrolyzed Animal Protein
PROLAGEN MP-1 is a unique liquid quaternary derivative of pure collagen which has controlled activating substantivity enhancement properties. Use MP-1 in shampoo and conditioner formulations for substantivity enhancement over a wide pH range. This cationic derivative has unique properties for the cosmetic formulator for use in soaps, waving lotions, shampoos and conditioners.

PEPTEIN CCA:
CTFA: Collagen amino acids
PEPTEIN CCA is a liquid, enzymatically produced product of excellent quality and uniformity. It is very hygroscopic which enhances its humectancy over other protein forms. This humectancy can be used to increase the softness and feel characteristic of shampoo and conditioners.

Cosmetic Gelatin:
Gelatin has numerous uses in cosmetics as a humectant, viscosity builder, film forming, anti-irritant, gel forming and sheen contributor. Gelatin is available in various grades from 125 bloom to 300 bloom.

CRODA INC.: Absorption Bases:

Absorption bases are the most convenient way of obtaining lanolin alcohol in a pre-dissolved, easy to use form.

Product:

CREMBA:
Mineral Oil (and) Petrolatum (and) Lanolin Alcohol (and) Lanolin
White to Yellow
Soft Solid
Developed for detergent surgical scrub emulsion polymerizations. Highly emollient moisturizer and auxiliary low HLB emulsifier. Typical use levels 2-10%

CREMBA B6:
Mineral Oil (and) Lanolin (and) Paraffin (and) Lanolin Alcohol (and) Beeswax
Pale Yellow
Soft Solid
Similar to CREMBA. Produces thicker emulsions and creams. Typical use levels 2-10%.

Liquid Absorption Base Type A:
Mineral Oil (and) Lanolin Alcohol
Clear Yellow Liquid
Used in liquid make-up preparation to improve dispersion and application properties of pigments. Also used as primary oil phase ingredients in oil-in-water emulsions. Typical use levels 2-10%

Liquid Absortion Base Type T:
Mineral Oil (and) Lanolin Alcchol
Clear Yellow Liquid
Similar to Type A. Better solvent for oil soluble dyes. Typical use levels 2-10%.

CRODA INC.: Alkanolamides(INCROMIDES):

100% Active 1:1 Ethanolamide Surfactants. Usually used with anionic surfactants in a ratio of approximately 4:1 active anionic: alkanolamide, to build viscosity, enhance the surfactant slip and feel and to provide foam stabilization in shampoos, bubble baths, soaps, liquid soaps and bath products.

INCROMIDE ALD:
Almondamide DEA
Clear Liquid
Nonionic
Improved conditioning in a natural based amide which builds viscosity and acts as a foam stabilizer. Typical use levels 1-10%

INCROMIDE AVD:
Avocadamide DEA
Clear Liquid
Nonionic
Viscosity builder and foam stabilizer mede from avoacado oil. Typical use levels 1-10%

INCROMIDE BAD:
Babassamide DEA
Clear Liquid
Nonionic
Viscosity buider and foam stabilizer with emulsifying properties. Typical use levels 1-10%

INCROMIDE BED:
Behenamide DEA
Pale Yellow Wax
Nonionic
A high melting alkanolamide useful as pearling and opacifying agent. Incorporated as a stick ingredient will give a creamy feel. Typical use levels 0.5-15%

INCROMIDE BEM:
Behenamide MEA
Pale Yellow Flake
Nonionic
An excellent structural wax for anti-perspirant and other stick products producing a smooth, creamy pay-off. Typical use levels 0.5-10%

CRODA INC.: Alkanolamides(INCROMIDES)(Continued):

INCROMIDE CA:
Cocamide DEA
Clear Liquid
Nonionic
Good foam stabilizing and emulsifying properties. Typical use levels 1-10%

INCROMIDE CAC:
Cocamide DEA Cocoyl Sarcosinate
Clear Liquid
Anionic
Low irritation with enhanced forming. Especially recommended as a clarifying agent for clear soap and shampoo bar preparations. Freely water soluble. Typical use levels 1-10%

INCROMIDE CM:
Cocamide MEA
Yellow Flake
Nonionic
Economical and efficient viscosity builder and foam stabilizer. Less soluble than the DEA amides. Especially useful in Alpha Olefin Sulfonates (AOS) formulations. Typical use levels 1-10%

INCROMIDE L90:
Lauramide DEA
White Solid
Nonionic
Excellent thickener and foam stabilizer. Typical use levels 1-10%

INCROMIDE LA:
Linoleamide DEA
Clear Liquid
Nonionic
Highly efficient thickener and superfatting agent. Good clarity and ease of use in surfactant systems. Especially recommended for low active shampoos and bubble baths. Typical use level 1-10%

CRODA INC.: Alkanolamides(INCROMIDES)(Continued):

INCROMIDE LM-70:
Lauramide DEA
White Solid
Nonionic
A viscosity builder and foam stabilizer especially recommended for shampoos and bath products. Typical use levels 1-10%

INCROMIDE LR:
Lauramide DEA
Clear Liquid
Nonionic
Similar to L90 but promotes higher viscosity. Typical use levels 1-10%

INCROMIDE MINK D:
Minkamide DEA
Clear Liquid
Nonionic
Naturally derived viscosity builder and foam stabilizer. Typical use levels 1-10%

INCROMIDE OD:
Oleamide DEA
Clear Liquid
Nonionic
Low color viscosity builder with good solubility in anionic surfactants. Typical use levels 0.5-5%

INCROMIDE OLD:
Olivamide DEA
Clear Liquid
Nonionic
Naturally derived viscosity builder and foam stabilizer. Typical use levels 0.5-10%

INCROMIDE OM:
Oleamide MEA
Yellow Solid
Nonionic
Low solubility but highly efficient viscosity building agent. Typical use levels 0.5-5%

INCROMIDE SED:
Sesamide DEA
Clear Liquid
Nonionic
Viscosity builder and foam stabilizer. Typical use levels 0.5-5%

CRODA INC.: Alkanolamides(INCROMIDES)(Continued):

INCROMIDE SM:
Stearamide MEA
Pale Yellow Flake
Nonionic
A softener and dye carrier for textiles. A pearling, opacifying and thickening agent for shampoos and bath gelees. Typical use levels 0.5-5%. Superior gellant for anti-perspirant stick formulations. Typical use levels 25-30% (in sticks).

INCROMIDE WGD:
Wheat Germamide DEA
Clear Liquid
Nonionic
Viscosity builder and foam stabilizer. Typical use levels 1-10%

CRODA INC.: Amine Amides(INCROMINES):

INCROMINES can be neutralized in-situ to produce unique cationic conditioning salts using any low molecular weight organic or inorganic acid. INCROMINES are soluble in alcohol but not water soluble until neutralized. Compatible with nonionic, anionic and cationic surfactants.

INCROMINE BB:
Behenamidopropyl Dimethylamine
Yellow Flakes
Nonionic/Cationic
A viscosity builder and conditioner that adds body and bulk to hair. Typical use levels 1-5%

INCROMINE CB:
Cocamidopropyl Dimethylamine
Pale Yellow Liquid
Nonionic/Cationic
A good foaming agent and conditioner that improves dry comb and reduces flyaway. Typical use levels 1-5%

INCROMINE IB:
Isostearamidopropyl Dimethylamine
Viscous Yellow Liquid
Nonionic/Cationic
Good foaming agent and conditioner that provides excellent wet combing properties. Typical use levels 1-5%

INCROMINE ISM:
Isostearamidopropyl Morpholine
Yellow Paste
Nonionic/Cationic
A good foaming agent and excellent wet combing conditioner. Typical use levels 1-5%

INCROMINE OPB:
Oleamidopropyl Dimethylamine
Pale Yellow Soft Solid
Nonionic/Cationic
This conditioning agent provides good detangling and wet comb. Typical use levels 1-5%

CRODA INC.: Amine Amides(INCROMINES)(Continued):

INCROMINE PB:
Palmitamidopropyl Dimethylamine
Off White Flakes
Nonionic/Cationic
A thickening agent and conditioner adding body and bulk to the hair. Typical use levels 1-5%

INCROMINE SB:
Stearamidopropyl Dimethylamine
Off White Flakes
Nonionic/Cationic
A viscosity builder and conditioning agent improving the body and bulk of the hair. Typical use levels 1-5%

CRODA INC.: Amine Oxides(INCROMINE OXIDES):

Nonionic surfactants that can act as cationic conditioning agents at low pH in hair care products and facial cleansers.

INCROMINE OXIDE AL:
Almondamidopropylamine Oxide
Pale Yellow Gel
Nonionic
Percent Active: 25
Viscosity builder with improved substantivity for good conditioning in hair care products. Typical use levels 0.5-10%

INCROMINE OXIDE AV:
Avocadamidopropylamine Oxide
Pale Yellow Gel
Nonionic
Percent Active: 25
Viscosity builder with improved substantivity for good conditioning in hair care products. Typical use levels 0.5-10%

INCROMINE OXIDE BA:
Babassamidopropylamine Oxide
Pale Yellow Liquid
Nonionic
Percent Active: 30
Produces a voluminous, lacey foam. Good foam stabilization and boosting. Typical use levels 0.5-10%

INCROMINE B-30P:
Behenamine Oxide
White Paste
Nonionic
Percent Active: 30
Good softener and conditioner for hair care products. Typical use levels 0.5-10%

INCROMINE OXIDE C:
Cocamidopropylamine Oxide
Pale Yellow Oxide
Nonionic
Percent Active: 30
Produces a voluminous, lacey foam. Typical use levels 0.5-20%

CRODA INC.: Amine Oxides(INCROMINE OXIDES)(Continued):

INCROMINE OXIDE I:
Isostearamidopropylamine Oxide
Essentially Colorless Gel
Nonionic
Percent Active: 25
Good conditioner, slip agent and viscosity builder. Recommended for clear non-quat type conditioners. Typical use levels 0.5-10%

INCROMINE OXIDE ISMO:
Isostearamidopropyl Morpholine Oxide
Essentially Colorless Gel/Liquid
Nonionic
Percent Active: 25
Very mild effective conditioner and viscosity builder. Typical use levels 0.5-10%

INCROMINE OXIDE L:
Lauramine Oxide
Essentially Colorless Liquid
Nonionic
Percent Active: 30

INCROMINE OXIDE L-40:
Lauramine Oxide
Essentially Colorless Liquid
Nonionic
Percent Active: 40
Excellent foaming agent, foam stabilizer and foam booster. An excellent degreaser. Also recommended for shampoos and light duty liquids. Typical use levels 0.5-20%.

INCROMINE OXIDE M:
Myristamine Oxide
Essentially Colorless Liquid
Nonionic
Percent Active: 30
Produces a dense, creamy foam. Improved viscosity builder. Can be used as primary surfactant or foam booster. Typical use levels 0.5-20%

CRODA INC.: Amine Oxides(INCROMINE OXIDES)(Continued):

INCROMINE OXIDE MC:
Myristamine Oxide (and) Cetamine Oxide
Colorless Liquid
Nonionic
Percent Active: 30
Viscosity builder with improved substantivity for good conditioning in hair care products. Typical use levels 0.5-10%

INCROMINE OXIDE MINK:
Minkamidopropylamine Oxide
Pale Yellow Gel
Nonionic
Percent Active: 25
Similar to M, but with denser foam and more efficient viscosity building. Typical use levels 0.5-20%.

INCROMINE OXIDE O:
Oleamidopropylamine Oxide
Essentially Colorless Liquid
Nonionic
Percent Active: 25
Good thickener and conditioner. Improved foaming and foam stabilization. Better dry comb. Typical use levels 0.5-10%

INCROMINE OXIDE OD-50:
Oleamine Oxide
Essentially Colorless Liquid
Nonionic
Percent Active: 50
Good thickener and conditioner. Improved foaming. Typical use levels 0.5-10%

INCROMINE OXIDE OL:
Olivamidopropylamine Oxide
Yellow Gel
Nonionic
Percent Active: 25
Good thickener. Recommended for clear systems. Typical use levels 0.5-10%

CRODA INC.: Amine Oxides(INCROMINE OXIDES)(Continued):

INCROMINE OXIDE S:
Stearamine Oxide
White Paste
Nonionic
Percent Active: 25
Auxiliary emulsifier in conditioners and cream rinses. Good viscosity builder. Good lotionizer with stearyl quats. Typical use levels 0.5-10%

INCROMINE OXIDE SE:
Sesamidopropylamine Oxide
Pale Yellow Gel
Nonionic
Percent Active: 25
Efficient visocosity builder and conditioner. Improved dry comb. Typical use levels 0.5-10%

INCROMINE OXIDE WG:
Wheat Germamidopropylamine Oxide
Yellow Gel
Nonionic
Percent Active: 25
Efficient viscosity builder and conditioning agent. Improved dry comb. Typical use levels 0.5-10%

CRODA INC.: Amino Acids:

The building blocks of proteins. Their low molecular weight and amphoteric salt-like character enables them to retain moisture and to be highly substantive to hair and skin.

AMINO GLUTEN MG:

Maize Gluten Amino Acids (and) Sodium Chloride
Type: Maize (Corn)
Amber Liquid
Approximate Molecular Weight: 150
Activity: 15%
Solubility: Water Soluble

A plant derived amino acid recommended for skin creams and lotions and hair conditioners. Typical use levels 0.2-3%.

CROSILK Liquid:

Silk Amino Acids
Type: Silk
Pale Amber Liquid
Approximate Molecular Weight: 90
Activity: 15%
Solubility: Water Soluble

Gives skin and hair a silky feel. The low molecular weight permits penetration of the cuticle in hair and of the surface layers of the epidermis. Typical use levels 0.2-3%.

CROTEIN CAA/SF:

Animal Collagen Amino Acids
Type: Collagen
Yellow Liquid
Approximate Molecular Weight: 150
Activity: 40%
Solubility: Water Soluble

The hygroscopic (moisture binding) nature of amino acids will improve the moisturizing efficiency of skin creams and lotions. Typical use levels 0.2-3%.

CRODA INC.: Amino Acids(Continued):

CROTEIN HKP Powder:

Hair Keratin Amino Acids (and) Sodium Chloride
Type: Keratin
Physical Form: Light Brown Powder
Approximate Molecular Weight: 150
Activity: 50%
Water Soluble

In shampoos, conditioners and cream rinses it will improve body, appearance and manageability of hair whose cuticle has been damaged by harsh chenical treatment since it will penetrate into exposed cortex where it is highly substantive. Typical use levels 0.5-2%

CROTEIN HKP SF:

Animal Keratin Amiino Acids
Type: Keratin
Pale Straw Liquid
Approximate Molecular Weight: 150
Activity: 25%
Compatible with Aqueous Alcohol

Used in conditioners and cream rinses, which cannot tolerate inorganic salts, keratin amino acids have a high affinity for hair keratin. Due to their low molecular weight, amino acids penetrate the hair cuticle. Protein conditioning for nail care products. Typical use levels 0.5-2%.

CRODA INC.: Betaines--Amphoterics(INCRONAMS):

Generally amohoteric surfactants are mild (nonirritating) and are often used in combination with other surfactants. They are recommended especially for frequent use shampoos as a foam booster, foam stabiizer and adjuvant surfactant for foam feel and viscosity modification.

INCRONAM AL-30:
Almondamidopropyl Betaine
Yellow Liquid
Percent Active: 30
Water and alcohol soluble. Good foaming to give light, lacey foam with a light, dry feel on the skin. Good rinsing. Typical use levels 2-20%

INCRONAM AV-30:
Avocadamidopropyl Betaine
Pale Yellow Liquid
Percent Active: 30
Water and alcohol soluble. Forms voluminous lacey foam with a light dry feel. Good foam booster. Typical use levels 2-20%

INCRONAM BA-30:
Babassamidopropyl Betaine
Yellow Liquid
Percent Active: 30
Alcohol and water soluble. Produces voluminous light lacey foam. Good foam boosting and rinsability. Typical use levels 2-20%

INCRONAM B-40:
Behenyl Betaine
White Paste
Percent Active: 40
Forms opaque gel in water, with moderate foaming. Excellent slip and conditioning. Good wetting and rinse aid. Good water sheeting. Typical use levels 2-20%

INCRONAM 30:
Cocamidopropyl Betaine
Thin Yellow Liquid
Percent Active: 30
Alcohol and water soluble. Good foam booster producing a light lacey foam with a light dry feel. Rinses well. Typical use levels 2-20%

CRODA INC.: Betaines--Amphoterics(INCRONAMS)(Continued):

INCRONAM I-30:
Isostearamidopropyl Betaine
Viscous Yellow Liquid
Percent Active: 30
Soluble in alcohol and water. Good viscosity builder, producing a denser, creamier lather. Moderate slip and rinses well. Typical use levels 2-20%

INCRONAM MINK 30:
Minkamidopropyl Betaine
Amber Liquid
Percent Active: 30
Alcohol and water soluble. Best conditioner of the INCRONAMS. Good foaming with nice volume and density. Excellent slip. A superb viscosity builder that leaves a nice smooth after feel. Typical use levels 2-20%

INCRONAM OL-30:
Olivamidopropyl Betaine
Amber Liquid
Percent Active: 30
Water and alcohol soluble. Good foaming to give a light, lacey foam with a light, dry feel on the skin. Good rinsing. Typical use levels 2-20%

INCRONAM OP-30:
Oleamidopropyl Betaine
Viscous Yellow Liquid
Percent Active: 30
Water and alcohol soluble. Good foam booster and stabilizer. Typical use levels 2-20%

INCRONAM SE-30:
Sesamidopropyl Betaine
Yellow Liquid
Percent Active: 30
Forms clear solutions in alcohol and clear gels in water. Produces a good volume of creamy, dense foam. Good slip and conditioning, leaving a smooth after feel. Typical use levels 2-20%

INCRONAM WG-30:
Wheat Germamidopropyl Betaine
Yellow Liquid
Percent Active: 30
Forms clear solutions in hot alcohol and clear solutions in water. Produces good foaming with a dry feel. Rinses well leaving a soft smooth after feel. Typical use levels 2-20%

CRODA INC.: Cationic Polymers(CRODACELS):

These quaternized celluloses are modified with fatty alkyl groups which differentiates CRODACELS from cationic celluloses. They are substantive to skin and hair. The fatty group enhances wet and dry combing, increases body and reduces flyaway. CRODACELS produce a lubricious, long lasting feel in skin care products. Their approximate molecular weight is 10,000 and their activity is 20%.

CRODACEL QL:
Laurdimonium Hydroxyethyl Cellulose
Pale Yellow Solution
Especially improves foaming and is the most water soluble. Typical use levels 1-4%

CRODACEL QM:
Cocodimonium Hydroxyethyl Cellulose
Clear Viscous Solution
Improves foaming and imparts body. Typical use levels 1-4%.

CRODACEL QS:
Steardimonium Hydroxyethyl Cellulose
Opaque Viscous Solution
Slightly less water soluble than QL and QM, and imparts the most body to the hair. Typical use levels 1-4%.

CRODA INC.: Cationic Salts(INCROMATES):

These cationic salts are compatible with anionic systems. Due to their cationic nature, INCROMATES are substantive, light conditioners that should be used at a neutral or acid pH. Can be used in hair and skin care products.

INCROMATE ALL:
Almondamidopropyl Dimethylamine Lactate
Viscous Amber Liquid
Percent Active: 75
Based on natural almond oil. Provides slip and improvements in wet comb. Good rinsing and dry feel. Typical use levels 0.5-5%

INCROMATE AVL:
Avocadamidopropyl Dimethylamine Lactate
Viscous Amber Liquid
Percent Active: 75
Based on natural avocado oil. Produces slip and improvements in wet comb. Good rinsing and dry feel. Typical use levels 0.5-5%

INCROMATE BAL:
Babassamidopropyl Dimethylamine Lactate
Viscous Amber Liquid
Percent Active: 68
Based on natural babassu oil. Good foam with large lacey bubbles. Nice clean rinse. Improved dry comb. Typical use levels 0.5-5%

INCROMATE CDL:
Cocamidopropyl Dimethylamine Lactate
Viscous Yellow Liquid
Percent Active: 95
Highly concentrated, good foamer and conditioner for clear rinses and shampoos. Typical use levels 0.5-5%

INCROMATE CDP:
Cocamidopropyl Dimethylamine Propionate
Yellow Liquid
Percent Active: 40
Moderate foaming conditioning agent for clear rinses and shampoos. Good detangling. Typical use levels 1-10%

CRODA INC.: Cationic Salts(INCROMATES)(Continued):

Product:

INCROMATE IDL:
Isostearamidopropyl Dimethylamine Lactate
Viscous Yellow Liquid
Percent Active: 95
Improved substantivity. Good conditioner to improve slip, wet comb and manageability. In hand creams and lotions improves rub-in feel and after feel. Typical use levels 0.5-5%

INCROMATE ISML:
Isostearamidopropyl Morpholine Lactate
Viscous Yellow Liquid
Percent Active: 25
Improved substantivity. Good conditioner to improve slip, wet comb and manageability. In hand creams and lotions improves rub-in feel and after feel. Improved viscosity builder. Typical use levels 1-10%

INCROMATE MINK L:
Minkamidopropyl Dimethylamine Lactate
Viscous Amber Liquid
Percent Active: 75
Based on natural mink oil. Excellent foam and conditioning. Very good slip and detangling. Good after feel and dry comb. Also recommended for skin creams. Typical use levels 0.5-5%

INCROMATE ODL:
Oleamidopropyl Dimethylamine Lactate
Viscous Yellow Liquid/Gel
Percent Active: 30
Improved viscosity builder and conditioner, moderate foamer for clear rinses and shampoos with good detangling properties. Typical use levels 1-10%

INCROMATE OLL:
Olivamidopropyl Dimethylamine Lactate
Viscous Amber Liquid
Percent Active: 75
Very similar properties to MINK L. Based on natural olive oil. Typical use levels 0.5-5%

INCROMATE SDL:
Stearamidopropyl Dimethylamine Lactate
Pale Yellow Slurry
Percent Active: 25
Viscosity builder and opacifier for cream rinses and conditioners. Good emulsifier for hand creams, cleansers and lotions. Typical use levels 1-10%

CRODA INC.: Cationic Salts(INCROMATES)(Continued):

INCROMATE SEL:
Sesamidopropyl Dimethylamine Lactate
Viscous Amber Liquid
Percent Active: 75
Forms clear solutions in alcohol and water. Provides good slip, wet combing and detangling. Good foamer. Typical use levels 0.5-10%

INCROMATE WGL:
Wheat Germamidopropyl Dimethylamine Lactate
Viscous Amber Liquid
Percent Active: 75
Based on natural wheat germ oil. Superb conditioner. Gives clear solutions in alcohol. Excellent slip, wet comb and dry feel. Typical use levels 0.5-5%

CRODA INC.: Emollient Esters(CRODAMOLS):

CRILL 6:
Sorbitan Isostearate
Clear Yellow Amber Viscous Liquid
Mobility: Wetting Agent
Melting Point C: 10
Solubility: Oil Soluble
Water Insoluble

A liquid sorbitan ester due to the complex branching of the fatty acid. This powerful water-in-oil emulsifier (HLB=4.6) with excellent oxidation resistance is highly recommended in creams, lotions and aerosols. Typical use levels 1-5%

CRODALAN LA:
Cetyl Acetate (and) Acetylated Lanolin Alcohol
Clear Pale Yellow Liquid
Mobility: Spreading Agent
Melting Point C: 20
Solubility: Oil Soluble

A clear, low viscosity emollient possessing excellent spreading properties on the skin. Good dye and lipid solvent. Typical use levels 2-10%.

CRODAMOL AC:
Cetyl Acetate
Clear Colorless Liquid
Mobility: Spreading Agent
Melting Point: 19
Solubility: Oil Soluble

A low cost alternative to CRODALAN LA. An occlusive emollient ester that is easily emulsified. Good dye solvent and pigment dispersion properties. Typical use levels 2-10%.

CRODAMOL BE:
Behenyl Erucate
Light Amber Waxy Solid
Mobility: Viscosity Builder
Melting Point: 43-46
Solublity: Oil Soluble
Water Insoluble

A thickening and opacifying agent that forms dense white emulsions. Improves stability of pigment and powder suspenions in sticks and lipsticks, also modifies feel and application to provide a creamier pay-off. Typical use levels 1-5%

CRODA INC.: Emollient Esters(CRODAMOLS)(Continued):

CRODAMOL CAP:
Cetearyl Octanoate
Clear Colorless Liquid
Mobility: High Wetting and Spreading
Melting Point C: 0
Solubility: Oil Soluble
Water Insoluble

Designed to simulate the properties of preen gland oil, this branched chain emollient spreads quickly imparting a high degree of water repellency. Typical use levels 5-10%

CRODAMOL CP:
Cetyl Palmitate
White Flakes
Mobility: Improves Body of Emulsions
Melting Point C: 50-54
Solubility: Oil Soluble
Water Insoluble

Similar to CSP, produces more viscous emulsions. Typical use levels 3-10%.

CRODAMOL CSP:
Cetearyl Palmitate
White Flakes
Mobility: Improves Body of Emulsions
Melting Point C: 48-52
Solubility: Oil Soluble
Water Insoluble

A useful replacement for spermaceti wax. CSP produces a stable white cream, imparting a velvety feel to the skin. Typical use levels 3-10%.

CRODAMOL MM:
Myristyl Myristate
Waxy Off-White Solid
Mobility: Liquifies at Body Temperature
Melting Point C: 36-39
Solubility: Oil Soluble
Water Insoluble (Studies above melting point)

This dry emollient improves emulsion texture and stability Typical use levels 3-10%

CRODA INC.: Emollient Esters(CRODAMOLS)(Continued):

CRODAMOL PMP:
PPG-2 Myristyl Ether Propionate
Colorless Liquid
Mobility: Improves Spreadability
Melting Point C: -5
Solubility: Oil Soluble
Water Insoluble
The outstanding choice for non-oily emollience. This superb light, dry, greaseless ester is a good solvent for sunscreen actives. It is easy to emulsify, imparts better freeze/thaw stability to emulsions and can partially replace mineral oil to reduce the greasy feel. Typical use levels 5-20%

CRODAMOL PTC:
Pentaerythritol Tetra Caprate/Caprylate
Light Yellow Liquid
Mobility: Cushion Ester
Melting Point C: 10
Solubility: Oil Soluble
Ethanol Soluble
A lubricating ester with "cushion" properties gives long lasting feel after application of cream or lotion. Reduces tack in clear gel micro-emulsion systems. Typical use levels 1-12%.

CRODAMOL PTIS:
Pentaerthyritol Tetra Isostearate
Light Amber Liquid
Mobility: Cushion Ester
Melting Point C: 0
Solubility: Oil Soluble
Ethanol Soluble
Similar in properties to CRODAMOL PTC but with a heavier, though still dry, feel. Typical use levels 1-12%

CRODAMOL SS:
Cetyl Esters
Nearly White Crystalline Solid
Mobility: Viscosity Builder
Melting Point C: 43-47
Solubility: Oil Soluble
Water Insoluble
Synthetic spermaceti wax, a lubricious emollient imparts body to emulsions and improves stability. Typical use levels 3-10%.

CRODAMOL W:
Stearyl Heptanoate
Waxy Solid
Mobility: Melts Rapidly on Skin with Cooling
Melting Point C: 23-27
Solubility: Oil Soluble Water Insoluble
This synthetic preen gland wax, a non-greasy emollient, produces highly water-repellent films. Excellent solvency properties. Especially good in stick formulations.

CRODA INC.: Ethoxylated (Nonionic) Emulsifiers:

Mixtures of VOLPO emulsifiers act synergistically. They are the industry standards for solubilizing and emulsifying.

CETOMACROGOL 1000 BP:
Ceteth-20
White to Off White Solid
HLB: 15.7
Solubility: Water Soluble
Oil Insoluble
Excellent emulsifier for oil-in-water creams and lotions and wetting agent in stick formulations. Typical use levels 0.5-5%

CRILL 6:
Sorbitan Isostearate
Clear Yellow Amber Viscous Liquid
HLB: 4.6
Solubility: Oil Soluble
Water Insoluble
A liquid sorbitan ester due to the complex branching of the fatty acid. This powerful water-in-oil emulsifier with excellent oxidation resistance is highly recommended in creams, lotions and aerosols. Typical use levels 1-5%

PROCETYL AWS:
PPG-5 Ceteth-20
Clear to Slightly Hazy Liquid
HLB: 16
Soluble: Water Soluble
Alcohol Soluble
Fragrance solubilizer, emollient and humectant in bath oils, shampoos, clear gels and hydroalcoholic systems. Recommended in anti-perspirant sticks to reduce fabric staining. Typical use levels 0.5-5%.

VOLPO 3:
Oleth-3
Off White Hazy Liquid
HLB: 6.6
Solubility: Oil Soluble
Water Insoluble

Recommended as a solubilizer for bromo acid dyes in lipsticks and liquid rouge. A superlative spreading agent for bath oils. Forms the primary co-emulsifier with CRODAFOS phosphate esters in clear, colloidal micro-emulsion gels. Typical use levels 0.5-5%

CRODA INC.: Ethoxylated (Nonionic) Emulsifiers(Continued):

VOLPO 5:
Oleth-5
Off White Hazy Liquid
HLB: 8.8
Solubility: Oil Soluble
Water Insoluble

Recommended as a solubilizer for bromo acid dyes in lipsticks and liquid rouge. A superlative spreading agent for bath oils. Forms the primary co-emulsifier with CRODAFOS phosphate esters in clear, colloidal micro-emulsion gels. Typical use levels 0.5-5%.

VOLPO 10:
Oleth-10
Off White Semi Solid
HLB: 12.4
Solubility: Water Soluble
Oil Insoluble

High HLB non ionic emulsifier, especially useful where the oil phase contains unsaturated fatty moieties. Excellent solubilizer. Typical use levels 0.5-5%

VOLPO 20:
Oleth-20
Off White Soft Solid
HLB: 15.4
Solubility: Water Soluble
Oil Insoluble

Fragrance solubilizer especially useful in clear CARBOPOL gel systems. Typical use levels 0.5-5%.

VOLPO CS-10:
Ceteareth-10
White Soft Solid
HLB: 14.2
Solubility: Alcohol Soluble
Oil Insoluble

Recommended as emulsifier in oil-in-water systems where the oil phase consists of saturated fatty moieties. Typical use levels 0.5-5%

VOLPO S-2:
Steareth-2
White Soft Solid
HLB: 4.9
Solubility: Oil Soluble
Water Insoluble

Auxiliary low HLB emulsifier for oil-in-water systems. Typical use levels 0.5-5%

CRODA INC.: Ethoxylated (Nonionic) Emulsifiers(Continued):

VOLPO S-10:
Steareth-10
White Soft Solid
HLB: 12.4
Solubility: Alcohol Soluble
Oil Insoluble

VOLPO S-20:
Steareth-20
White Solid
HLB: 15.3
Solubility: Water Soluble
Oil Insoluble

Useful primary oil-in-water emulsifiers for formulating ethnic hair straighteners and relaxers. Typical use levels 0.5-5%.

VOLPO S-100:
Steareth-100
White Waxy Solid Flakes
HLB: 18.8
Solubility: Water Soluble
Oil Insoluble

Good solubilizer for creams and lotion systems. Can be used as a wetting agent in stick systems to improve anti-perspirant efficacy. Typical use levels 0.5-5%

CRODA INC.: Fatty Acids:

A unique group of fatty acids.

CRODACID B(PG-3440):
Behenic Acid
White Solid
This long chain (C22) fatty acid forms more viscous emulsions than stearic acid. Useful gellant for stick formulations when neutralized. Typical use levels 0.5-5%

SYNCROWAX AWIC:
C18-36 Acid
Pale Cream Wax Flakes
This long chain acid may be neutralized to give interesting soap emulsions and gels with high viscosity and excellent stability. Typical use levels 1-15%

SKLIRO Distilled:
Lanolin Acid
Yellow Waxy Solid
Superfatting agent in aerosol shave foams and hand soaps. Has the ability to form soaps for commercial oil-in-water emulsions. Typical use levels 0.5-4.0%.

CRODA INC.: Fatty Alcohols (CRODACOLS):

CRODACOLS act as secondary emulsifiers, thickeners, opacifiers and structural agents in anhydrous stick systems.

Cetostearyl Alcohol BP:
Cetearyl Alcohol
White Flakes
Typical G.C. Distribution: %C16: 68
%C18: 24
This cetostearyl alcohol forms dense emulsions and conforms to the BP specification. Typical use levels 2-30%

Cetostearyl Alcohol NF:
Cetearyl Alcohol
White Flakes
Typical G.C. Distribution: %C16: 41
%C18: 49
This cetostearyl alcohol conforms to the N.F. specification. Typical use levels 2-30%.

CRODACOL C70:
Cetyl Alcohol
White Flakes
Typical G.C. Distribution: %C16: 67
%C18: 23
Confers a smooth, dense appearance resulting in emulsions of enhanced stability. Typical use levels 2-30%.

CRODACOL C95 NF:
Cetyl Alcohol
White Flakes
Typical G.C. Distribution: %C16: 92
%C18: 2
High purity cetyl alcohol complying to NF specifications. Used as a primary structural agent in anti-perspirant sticks. Typical use levels 2-30%

CRODACOL CS-50:
Cetearyl Alcohol
White Flakes
Typical G.C. Distribution: %C16: 41
%C18: 49
An economical mixed alcohol with excellent color and odor. Produces dense emulsions without graininess. Typical use levels 2-30%

CRODA INC.: Fatty Alcohols(CRODACOLS)(Continued):

CRODACOL S-70:
Stearyl Alcohol
White Flakes
Typical G.C. Distribution: %C16: 28
%C18: 65
The optimum C16/C18 ratio for many emulsion systems. Typical use levels 2-30%.

CRODACOL S-95NF:
Stearyl Alcohol
White Flakes
Typical G.C. Distribution: %C16: 2
%C18: 95
A primary structural surfactant in cosmetic anti-perspirant stick formulations. Typical use levels 2-30%.

NOVOL:
Oleyl Alcohol
Clear Liquid
NOVOL consists mainly of mono unsaturated C18 alcohol
This extremely low color and odor oleyl alcohol is an easily emulsified emollient, penetrant, super-fatting agent amd pigment suspending agent. The coupling agent and emollient of choice in lipstick formulas. Perfume and fragrance fixative. Typical use levels 2-30%

CRODA INC.: Humectants(INCROMECTANTS):

INCROMECTANTS show superior water absorption and retention when compared to commonly used humectants such as glycerin. INCROMECTANTS can replace these less effective humectants, and are especially recommended in shampoos, cream rinses, conditioners and skin creams and lotions.

INCROMECTANT AQ:
Acetamidopropyl Trimonium Chloride
Liquid
Percent Active: 75

INCROMECTANT LQ:
Lactamidopropyl Trimonium Chloride
Liquid
Percent Active: 75

Magnetic Moisturizing Factors. Cationic antistatic agents in shampoos and conditioners. Super effective humectants, non tacky glycerin replacement in ethnic hair care systems. AQ is the most hygroscopic INCROMECTANT, retaining as much as 200% more moisture than glycerin. Plasticizers for hair conditioning/setting polymers. Typical use levels 0.5-3%

INCROMECTANT AMEA-70:
Acetamide MEA
Liquid
Percent Active: 70

INCROMECTANT AMEA-100:
Acetamide MEA
Liquid
Percent Active: 100

Clarifying detangling agents for shampoos, conditioners and cream rinses. Humectants in creams and lotions. Typical use levels 0.5-15%

INCROMECTANT LAMEA:
Acetamide MEA (and) Lactamide MEA
Liquid
Percent Active: 100

A blend of AMEA and LMEA that out performs glycerin as a moisturizing agent. Recommended in both hair and skin care products. Typical use levels 0.5-15%

INCROMECTANT LMEA:
Lactamide MEA
Liquid
Percent Active: 100

Similar to AMEA-70 except lower in odor, more concentrated and more stable over a broader pH range. Typical use levels 0.5-15%

CRODA INC.: Lanolin:

Lanolin is a natural, tenacious soft wax, super-fatting emollient with some emuulsifying properties. This complex mixture of chemically stable esters and polyesters is effective due to its similarity to sebum, the skin's natural lipid.

SUPER CORONA Lanolin Anhydrous USP:
Lanolin
Yellow Soft Solid
Maximum Gardner Color: 8½
Melting Point C: 38-44
A highly purified low odor luxury lanolin.

SUPERFINE Lanolin Anhydrous USP:
Lanolin
Yellow/Amber Soft Solid
Maximum Gardner Color: 9½
Melting Point C: 38-44
A low odor lanolin, slightly darker than SUPER CORONA

Cosmetic Lanolin Anhydrous USP:
Lanolin
Light Amber Soft Solid
Maximum Gardner Color: 10½
Melting Point C: 38-44
Recommended for creams, lotions, lipsticks, make-up and sunscreen products

Pharmaceutical Lanolin Anhydrous USP:
Lanolin
Amber Soft Solid
Maximum Gardner Color: 11½
Melting Point C: 38-44
Specially developed for the pharmaceutical industry.

CRODA INC.: Lanolin Derivatives:

A range of derivatives with a wide variety of solubility developed to enhance the multifunctionality of lanolin.

ACYLAN:
Acetylated Lanolin
Light Yellow Soft Solid
Solubility: Oil Soluble
Acylan is more hydrophobic and oily in character than lanolin and helps form water repellant films. Typical use levels 1-10%.

Cholesterol NF:
Cholesterol
White to Pale Yellow Powder
Solubility: Soluble in 1% Ethyl Alcohol
Turbidity after 2 hrs.
A moisturizer and emollient that acts as a powerful primary emulsifier in water-in-oil systems. Typical use levels 0.1-1%.

CRODALAN AWS:
Polysorbate 80 (and) Cetyl Acetate (and) Acetylated Lanolin Alcohol
Clear Liquid
Solubility: Water Soluble
Alcohol Soluble
This emollient is used as a solubilizer and emulsifier in aqueous and alcoholic systems. Typical use levels 1-5%

CRODALAN LA:
Cetyl Acetate (and) Acetylated Lanolin Alcohol
Clear Pale Yellow Liquid
Solubility: Oil Soluble
Ester Soluble
A clear low viscosity emollient possessing excellent spreading properties on the skin. Typical use levels 2-10%

FLUILAN:
Lanolin Oil
Clear Viscous Liquid
Solubility: Oil Soluble
Ester Soluble
Improved application and solubility properties for lipsticks, baby oils, brilliantines and cleansing lotions compared to lanolin. Good dye solvent and suspending agent. Reduced bleed and improved pay-off in lipsticks. Typical use levels 2-10%.

CRODA INC.: Lanolin Derivatives(Continued):

FLUILAN AWS:
PPG-12-PEG-65 Lanolin Oil
Amber Viscous Liquid
Solubility: Water Soluble
Alcohol Soluble

Used as a plasticizer and film modifier in hair sprays. Also used as a water and alcohol soluble emollient. Excellent solubilizer. Typical use levels 0.1-2%.

HARTOLAN:
Lanolin Alcohol
Light Brown Solid Wax
Solubility: Oil Soluble
Ester Soluble

In water-in-oil systems HARTOLAN absorbs a considerable quantity of water which is slowly released for moisturization. A cost effective source of cholesterol. Powerful water-in-oil emulsifier. Typical use levels 0.5-3%.

SUPER HARTOLAN:
Lanolin Alcohol
Golden Solid Wax
Solubility: Oil Soluble

The premier lanolin alcohol for cosmetic and pharmaceutical use. The most powerful non-ionic water-in-oil emulsifier. SUPER HARTOLAN has a much lighter color and a milder odor than HARTOLAN. Both are rich in cholesterol. A molecularly distilled lanolin alcohol. Typical use levels 0.5-3%

LANEXOL AWS:
PPG-12-PEG-50 Lanolin
Amber
Viscous Liquid
Solubility: Water Soluble
Alcohol Soluble

This nonionic lanolin derived emollient forms films from hydroalcoholic and aqueous solutions which are soft, glossy silky and tack free. Excellent solubilizer. Typical use levels 1-5%.

CRODA INC.: Lanolin Derivatives(Continued):

LANPOLAMIDE 5:
PEG-5 Lanolinamide (and) PEG-5 Lanolate
Soft Amber Solid
Solubility: Water Insoluble
Slightly Hazy at 1% In Mineral Oil

A highly efficient primary water-in-oil emulsifier forming surprisingly stable emulsions containing over 90% water. Used in high water content aerosols as an emulsion stabilizer and can corrosion inhibitor. Typical use levels 0.1-0.5%

POLYCHOL 5:
Laneth-5
Soft Golden Yellow Wax
Solubility: Water Dispersible

The most lipophilic POLYCHOL is an excellent bromo dye solvent that imparts emolliency. Typical use levels 1-10%.

POLYCHOL 10:
Laneth-10
Soft Golden Yellow Wax
Solubility: Water Dispersible

POLYCHOL 15:
Laneth-15
Semi-hard Yellow Wax
Solubility: Opalescent Solutions in Water

POLYCHOL 20:
Laneth-20
Semi-hard Yellow Wax
Solubility: Almost Totally Water Soluble

POLYCHOL 40:
Laneth-40
Hard Yellow Wax
Solubility: Water Soluble

Combinatons of the POLYCHOLS are excellent oil-in-water emulsifiers especially effective for emulsifying lanolin and branched fatty moieties. More effective thickening action than other ethoxylated alcohols. Can be used to produce mild clear micro-emulsion gels with soft, honey-like consistency. POLYCHOL 15 especially is an excellent emulsifier in hair straighteners and relaxers. Typical use levels 1-10%.

CRODA INC.: Lanolin Derivatives(Continued):

SKLIRO Distilled:
Lanolin Acid
Yellow Waxy Solid
Solubility: Oil Soluble
Ester Soluble
Superfatting agent in aerosol shave foams and hand soaps. Has the ability to form soaps for commercial oil-in-water emulsions. Typical use levels 1-10%

SOLAN:
PEG-60 Lanolin
Yellow Wax
Water Soluble
A mild water soluble lanolin with emollient, conditioning and super fatting properties. Efficient solubilizer. Typical use levels 1-10%

SOLAN 50:
PEG-60 Lanolin
Viscous Yellow Liquid
Solubility: Water Soluble
A convenient liquid form of SOLAN. Typical use levels 1-10%

SUPER SOLAN Flaked:
PEG-75 Lanolin
Yellow Flake
Solubility: Water Soluble
Easy to use dry flaked form of SOLAN. Efficient solubilizer. Produces more viscous solutions in water. Typical use levels 1-10%

SUPER SOLANGEL 25:
PEG-75 Lanolin
Yellow Liquid
Solubility: Water Soluble
A convenient liquid form of SOLAN. Typical use levels 2-20%

SUPER STEROL Ester:
C10-30 Cholesterol/Lanosterol Esters
Soft White Solid
Solubility: Oil Soluble
Ester Soluble
The white, odorless ester complex is comparable to sebum and is an excellent dry skin emollient and moisturizer. Typical use levels 1-10%

CRODA INC.: Nonionics# (INCROCAS and INCROPOL):

INCROCAS and INCROPOL surfactants are nonionic, ethoxylated emulsifiers for oil-in-water systems, solubilizers, wetting agents and lubricants.

INCROCAS 10:
PEG-10 Castor Oil
Liquid
HLB: 6.3
Percent Active: 100

INCROCAS 30:
PEG-30 Castor Oil
Liquid
HLB: 11.7
Percent Active: 100

INCROCAS 40:
PEG-40 Castor Oil
Soft Paste
HLB: 13.0
Percent Active: 100

INCROCAS 60:
PEG-60 Castor Oil
Soft Paste
HLB: 14.7
Percent Active: 100

INCROCAS 100:
PEG-100 Castor Oil
Pale Yellow Waxy Solid
HLB: 16.5
Percent Active: 100

Excellent low color oil-in-water solubilizers recommended for fragrance oils and other oils which may be difficult to solubilize. Typical use levels 0.5-5%. Good starting point for fragrance solubilization is to use 1:1 fragrance: INCROCAS. The highly ethoxylated INCROCAS' are more liquid and easier to handle than other high HLB ethoxylates.

INCROPOL CS-20:
Ceteareth-20
Solid
HLB: 15.5
Percent Active: 100
Good solubilizer/emulsifier. Typical use levels 0.5-10%

Croda Inc. will custom ethoxylate materials to customer specifications

CRODA INC.: Nonionics#(INCROCAS and INCROPOL)(Continued):

INCROPOL CS-50:
Ceteareth-50
Flake
HLB: 17.9
Percent Active: 100
Good foamer and wetting agent for toilet bowl cleansers, and in cistern blocks. Typical use levels 0.5-10%

INCROPOL L-7:
Laureth-7
Liquid
HLB: 11.8
Percent Active: 100
INCROPOL L-12:
Laureth-12
Paste
HLB: 14.2
Percent Active: 100

Good wetting agents for hard surface cleaners. Good oil emulsifiers. Also used as emulsifier in emulsion polymerization. Typical use levels 0.5-10%

INCROPOL L-23:
Laureth-23
Solid
HLB: 16.5
Percent Active: 100
L23 is a good solubilizer. Recommended as a lubricant in shaving creams.

Croda Inc. will custom ethoxylate materials to customer specifications

CRODA INC.: Pearling Agents(CRODAPEARLS):

These agents are typically added to shampoos, cream rinses and lotions after the emulsion has cooled to produce rich pearling effects.

CRODAPEARL Liquid:

Sodium Laureth Sulfate (and) Hydroxyethyl Stearamide MIPA

White Paste

For use in shampoo systems and bubble baths where a stable irridescent pearl is needed, without heating. Typical use levels 1-5%

CRODAPEARL N.I. Liquid:

Hydroxyethyl Stearamide MIPA (and) PPG-5 Ceteth-20

White Paste

A non-ionic based cold process pearl for the pearlisation of lotions, gels, clear rinses and bath products. Typical use levels 1-5%

CRODA INC.: Phosphate Ester Emulsifiers(CRODAFOS):

CRODAFOS esters are extraordinarily powerful emulsifiers. The neutral phosphate esters are easy to use and produce stable oil-in-water emulsions. The acid phosphate esters are more flexible and may be neutralized with a wide range of organic and inorganic bases to modify the pH and the thickening, emulsifying and suspending properties.

CRODAFOS CAP:
PPG-10 Cetyl Ether Phosphate
Clear Yellow Liquid
Solubility: Oil Soluble
Water Soluble
Acid or Neutral: Acid
CRODAFOS CAP is more lipophilic than CRODAFOS SG and provides enhanced conditioning. Recommended for micro emulsion systems. Typical use levels 0.5-3%

CRODAFOS N3 Acid:
Oleth-3 Phosphate
Viscous Amber Liquid
Solubility: Oil Soluble
Water Soluble
Powerful oil-in-water emulsifier. Especially useful for clear micro-emulsion gels. Typical use levels 0.5-5%

CRODAFOS N10 Acid:
Oleth-10 Phosphate
Viscous Yellow Liquid
Solubility: Oil Soluble
Water Soluble
Acid or Neutral: Acid
Oil-in-water emulsifier. Especially useful for clear micro-emulsion gels. Typical use levels 0.5-10%

CRODAFOS N3 Neutral:
DEA Oleth-3 Phosphate
Viscous Amber Liquid
Solubility: Oil Soluble
Water Soluble
Acid or Neutral: Neutral
General purpose oil-in-water emulsifier for clear micro-emulsion gels, and hair relaxers. Typical use levels 0.5-5%

CRODA INC.: Phosphate Ester Emulsifiers(CRODAFOS)(Continued):

CRODAFOS N10 Neutral:
DEA Oleth-10 Phosphate
Viscous Yellow Liquid
Solubility: Oil Soluble
Water Soluble
Acid or Neutral: Neutral
Oil-in-water emulsifier. Typical use levels 0.5-10%

CRODAFOS SG:
PPG-5-Ceteth-10 Phosphate
Viscous Amber Liquid
Solubility: Oil Soluble
Water Soluble
Acid or Neutral: Acid
Provides superior wet combing and manageability in shampoos and cream rinses. Used as thickener, gellant and pH adjuster in shampoos. Anionic fragrance co-solubilizer. Typical use levels 0.5-3%

CRODA INC.: Propoxylates Couplers and Emollients:

Their branched nature confers fluidity, while the ether links impart solubilizing properties.

PROCETYL 10:
PPG-10 Cetyl Ether
Clear Liquid
Solubility: Oil Soluble
Alcohol Soluble
Lanolin Oil Soluble

An excellent coupling agent for the incorporation of mineral oil and esters into hydroalcoholic systems. Typical use levels 5-30%

PROCETYL 50:
PPG-50 Cetyl Ether
Yellow to Slight Yellow Clear Liquid
Solubility: Polyethylene Glycol 200 Soluble
Alcohol Soluble
Lanolin Oil Soluble

This very mild emollient is particularly useful in the formulation of three layer bath oils. Improves lubricity in oil based hair dressings. Typical use levels 5-30%

PROMYRISTYL PM-3:
PPG-3 Myristyl Ether
Clear Liquid
Solubility: Oil Soluble--Alcohol Soluble--Lanolin Oil Soluble

An effective low viscosity emollient for clear analgesic, deodorant and fragrance sticks. Typical use levels 5-15%.

PROSTEARYL 15:
PPG-15 Stearyl Ether
Clear Liquid
Solubility: Alcohol Soluble--Water Insoluble

Mobile emollient and oil coupler for bath oils, sunscreens, hair products, aerosol anti-persirants and hand and body lotions. Typical use levels 5-15%.

PROVOL 10:
PPG-10 Oleyl Ether
Clear Liquid
Solubility: Oil Soluble--Lanolin Oil Soluble--Ester Soluble

PROVOL 50:
PPG-50 Oleyl Ether
Clear Liquid
Solubility: Oil Soluble--Lanolin Oil Soluble--Ester Soluble

Aid combing and improve lubricity of hair gels and ethnic hair pomades. Useful coupling agents for unsaturated materials. After bath/shower emollients with low irritation. Aid spreading and pigment dispersion in make-up systems. Typical use levels 5-20%.

CRODA INC.: Proteins:

Proteins are natural, substantive film forming moisturizers that impart a smooth feel to the skin or a high degree of conditioning/body to the hair.

CLEARCOL:
Soluble Animal Collagen
Type: Collagen
Clear to Slightly Hazy Viscous Liquid
Approximate Molecular Weight: 300,000
Activity: 1%
Recommended for low solids facial systems where clarity is important. The aqueous tropocollagen lubricates and moisturizes skin and hair. Typical use levels 2-20%.

COLLASOL:
Soluble Animal Collagen
Type: Collagen
Off-White Dispersion
Approximate Molecular Weight: 300,000
Activity: 3%
Undenatured soluble collagen is a highly effective film forming agent which enhances moisture retention and improves skin feel. The highest activity soluble collagen currently available on the market. Typical use levels 2-10%.

CRODYNE BY19:
Gelatin
Type: Collagen
Buff Crystalline Powder
Approximate Molecular Weight: 25,000
Activity: 85%
Pharmaceutical gelatin is the classic protective colloid for skin and hair care products. In shaving creams, mousses and soaps will improve foam body and creaminess. Helps ameliorate irritation. Typical use levels 1-5%.

CROLASTIN:
Hydrolyzed Animal Elastin
Type: Elastin
Yellow Liquid
Approximate Molecular Weight: 4,000
Activity: 30%
Applied topically, CROLASTIN helps enhance flexibility, moisturizes and improves skin tone and appearance. Especially recommended in skin cleansers. Typical use levels 1-5%.

CRODA INC.: Proteins(Continued):

CROMOIST CS:

Chondroitoin Sulfate (and) Hydrolyzed Animal Protein
Type: Mucopolysaccharide
Clear to Slightly Hazy Viscous Solution
Approximate Molecular Weight: 115,000
Activity: 20%

Increased water binding properties enhance the moisturizing effects for face and body creams and lotions. Typical use levels 0.5-5%.

CROMOIST HYA:

Hydrolyzed Animal Protein (and) Hyaluronic Acid
Type: Mucopolysaccharide
Amber Liquid
Approximate Molecular Weight: 500,000
Activity: 15%

A proteoglycan that shows improved moisturizing over pure hyaluronic acid. This important glycosoaminoglycan improves elasticity, reduces dryness and can increase the moisture content of the skin by 33%. Highly recommended in facial creams. Typical use levels 0.5-2%.

CRONECTIN H:

Hydrolyzed Fibronectin
Type: Fibronectin
Opalescent Low Viscosity Liquid
Approximate Molecular Weight: 20,000
Activity: 3%

A humectant and moisturizing agent for skin creams and lotions. Especially effective in low solids lotions. Typical use levels 0.5-6%.

CROSILK 10,000:

Hydrolyzed Silk
Type: Silk
Amber Liquid
Approximate Molecular Weight: 10,000
Activity: 15%

This water soluble protein will improve manageability, gloss and texture in hair care products and will moisturize and protect in skin care products. Very effective in eye wrinkle cream. Typical use levels 0.5-2%.

CRODA INC.: Proteins(Continued):

CROSILK Powder:
Silk Powder
Type: Silk
Lustrous Grey/White Powder
Approximate Molecular Weight: Over 500,000
Activity: 100%
Recommended in solid make-up systems where it absorbs oil, improves levelling, modifies application properties and provides a silky, lustrous appearance. Typical use levels 1-5%.

CROTEIN K:
Hydrolyzed Animal Keratin
Type: Keratin
Light Tan Powder
Approximate Molecular Weight: 2,000
Activity: 60%
Helps replace protein in damaged hair. Its high substantivity and hygroscopicity (moisture binding) improve hair condition, body and manageability. It is also substantive to nail keratin in nail care products. Typical use levels 0.5-3%.

CROTEIN SPA:
Hydrolyzed Animal Protein
Type: Collagen
White Powder
Approximate Molecular Weight: 4,000
Activity: 93%
The quality benchmark protein for personal care products. Clearly soluble in 50/50 ethanol/water at 1%. Improves flexibility, manageability, gloss, shine and body of hair. Typical use levels 0.2-2%.

CROTEIN SPA55:
Hydrolyzed Animal Protein
Type: Collagen
Amber Liquid
Approximate Molecular Weight: 4,000
Activity: An easy to use, light color, low odor protein solution. Typical use level 0.5-5%.

CRODA, INC.: Proteins(Continued):

CROTEIN SPC:

Hydrolyzed Animal Protein
Type: Collagen
White Powder
Approximate Molecular Weight: 10,000
Activity: 93%

Clearly soluble in water and 80/20 water/ethanol solutions at 1%. A general purpose protein conditioner. Binder, excipient and encapsulant for pharmaceuticals. Typical use levels 0.2-2%.

CROTEIN SPO:

Hydrolyzed Animal Protein
Type: Collagen
White Powder
Approximate Molecular Weight: 1,000
Activity: 93%

High solubility in aqueous alcohol especially useful in hair spray formulations. Forms a hard clear non-tacky film. Typical use levels 0.2-2%.

CROTEIN WKP:

Hydrolyzed Animal Keratin
Type: Keratin
Pale Yellow Liquid
Approximate Molecular Weight: 600
Activity: 22%

The polypeptides in CROTEIN WKP improve lustre and feel, also replace cystine in damaged hair. Low molecular weight results in enhanced substantivity to keratinous substrates. Typical use levels 0.2-3%.

ELASTOSOL:

Soluble Animal Elastin (and) Soluble Animal Collagen
Type: Elastin/Collagen
White to Off-White Viscous Liquid
Approximate Molecular Weight: 200,000
Activity: 2%

Leaves a colloidal film which mimics the natural protein responsible for skin elasticity. An effective moisturizer with superb results in eye wrinkle creams. Typical use levels 2-10%.

CRODA INC.: Proteins(Continued):

HYDROSOY 2000/SF:
Hydrolyzed Soy Protein
Type: Soy
Clear Amber Liquid
Approximate Molecular Weight: 4,000
Activity: 20%
A protein derived from a natural vegetable source that is substantive and has similar moisturizing properties to the collagen and keratin derived proteins. Typical use levels 0.2-3%

KERASOL:
Soluble Animal Keratin
Type: Keratin
Slightly Hazy Amber Liquid
Approximate Molecular Weight: 125,000
Activity: 15%
Studies show that this unique keratin protein conditioner can be covalently bonded to the hair improving body and sheen with permanent conditioning even after successive shampooings. Also conditions in nail care products. Typical use levels 1-10%.

RETICUSOL:
Soluble Animal Reticulin
Type: Reticulin
Pale Straw Liquid
Approximate Molecular Weight: 3,000
Activity: 20%
The third structural protein present in the skin. Dries to give a unique feel without tackiness, unlike other protein hydrolysates of similar molecular weight. Helps increase the water content of the outer skin layers plumping the epidermis leaving the skin smoother and softer. Typical use levels 0.5-2%.

CRODA INC.: Protein Derivatives:

A range of derivatives with widely varying properties.

AMINOFOAM C:

TEA-Lauroyl Animal Collagen Amino Acids
Type: Modified Collagen
Light Amber Liquid
Approximate Molecular Weight: 550
Activity: 40%
Solubility: Water Soluble

Very mild, creates foams of small bubble size producing extremely rich lather in shampoos, foaming conditioners, and facial cleansers. Good feel on wash off. Typical use levels 2-10%.

AMINOFOAM K:

TEA-Lauroyl Animal Keratin Amino Acids
Type: Modified Keratin
Light Amber Liquid
Approximate Molecular Weight: 550
Activity: 40%
Solubility: Water Soluble

The keratin amino acid version of Aminofoam C. Typical use levels 2-10%.

ATP Nucleotides:

Propylene Glycol (and) Animal Collagen Amino Acids (and) Adenosine Triphosphate
Type: Nucleotides
Clear Liquid
Approximate Molecular Weight: 150/500
Activity: 5%
Solubility: Water Soluble

ATP is the universal energy currency of the cell. An excellent moisturizer. Typical use levels 1-5%.

CROQUAT L:

Laurdimonium Hydrolyzed Animal Protein
Type: Cationic Collagen
Clear Yellow Liquid
Approximate Molecular Weight: 2,500
Activity: 40%
Solubility: Essentially Water Soluble

An enhanced substantivity protein. Recommended for clear rinses, shampoos and conditioners. Improves wet and dry combability without overconditioning. Typical use levels 0.5-2.5%.

CRODA INC.: Protein Derivatives(Continued):

CROQUAT M:
Cocodimonium Hydrolyzed Animal Protein
Type: Cationic Collagen
Clear Yellow Liquid
Approximate Molecular Weight: 2,500
Activity: 40%
Solubility: Essentially Water Soluble

An enhanced substantivity protein. Effective in shampoos and conditioners. Especially useful in formulating conditioning perms and hair relaxers. Good substantive foamer. Typical use levels 0.5-2.5%.

CROQUAT S:
Steardimonium Hydrolyzed Animal Protein
Type: Cationic Collagen
Yellow Paste/Gel
Approximate Molecular Weight: 2,700
Activity: 40%
Solubility: Slight Haze, Partly Water Soluble

An enhanced substantivity protein. Provides good conditioning especially from conventional cream rinses. Typical use levels 0.5-2.5%.

CROQUAT WKP:
Cocodimonium Hydrolyzed Animal Keratin
Type: Cationic Keratin
Clear Amber Liquid
Approximate Molecular Weight: 1,000
Activity: 30%
Solubility: Water Soluble

A permanent conditioning protein which can be covalently bonded to the hair permanently attaching a fatty quat and producing excellent combability, body and shine. Recommended for shampoos, conditioners, cream rinses, conditioning perms and hair relaxing systems. Also effective in nail care products. Typical use levels 0.25-2%.

CROTEIN AD Anhydrous:
AMP Isostearoyl Hydrolyzed Animal Protein
Type: Modified Collagen
Clear Yellow Liquid
Approximate Molecular Weight: 2,000
Activity: 30%
Solubility: Alcohol Soluble

Used in alcohol products, such as aerosol hair sprays, skin tonics, pump hair sprays, setting lotions and after shave lotions. Typical use levels 0.1-2.0%

CRODA INC.: Protein Derivatives(Continued):

CROTEIN ASC:
Ethyl Ester of Hydrolyzed Animal Protein
Type: Modified Collagen
Clear Liqht Amber Liquid
Approximate Molecular Weight: 2,000
Activity: 20%
Solubility: Water Soluble
Alcohol Soluble

A protein conditioning ingredient and film modifier for alcoholic and hydro-alcoholic lotions and hair care products. Typical use levels 0.1-0.5%.

CROTEIN ASK:
Hydrolyzed Animal Keratin
Type: Keratin
Clear Amber Liquid
Approximate Molecular Weight: 2,000
Activity: 12%
Soluble: Water Soluble
Alcohol Soluble

Film modifier for aerosol and non-aerosol hair setting/ conditioning systems. Typical use levels 0.1-0.5%.

CROTEIN IP:
Isostearoyl Hydrolyzed Animal Protein
Type: Modified Collagen
Viscous Liquid
Approximate Molecular Weight: 2,200
Activity: 100%
Solubility: Oil/Solvent Soluble

Conditioning agent for incorporation into solvent based nail polish and remover preparations. Typical use levels 0.1-0.5%.

CROTEIN Q:
Steartrimonium Hydrolyzed Animal Protein
Type: Cationic Collagen
Off-White Free-Flowing Powder
Approximate Molecular Weight: 12,000
Activity: 90%
Solubility: Water Soluble

The benchmark quaternized collagen which provides body and gloss for shampoos and conditioners. Its cationic nature enhances substantivity. Good anionic compatibility. Typical use levels 0.5-2.5%.

CRODA INC.: Quaternary Cosmetic Conditioners(INCROQUATS):

INCROQUATS are cationic quaternary amine salts, substantive to hair. These conditioners improve wet and dry combing, reducing flyaway and enhance manageability.

INCROQUAT AL-85:
Almondamidopropalkonium Chloride
Amber Liquid
Percent Active: 85
Based on almond oil, and soluble in water and alcohol to give clear solutions. Good foamer. Typical use levels 0.5-5%

INCROQUAT AV-85:
Avocadamidopropalkonium Chloride
Amber Liquid
Percent Active: 85
Based on avocado oil. Clearly soluble in alcohol and water. Foams and specifically imparts good slip and wet comb. Typical use levels 0.5-5%

INCROQUAT BA-85:
Babassamidopropalkonium Chloride
Amber Liquid
Percent Active: 85
Gives good foam. Improves dry combing, manageability and static control. Provides soft feel and good shine. Typical use levels 0.5-5%

INCROQUAT B65C:
Behenalkonium Chloride (and) Cetyl Alcohol
White Flake
Percent Active: 65
Excellent substantive conditioner adding body and manageability to hair. Good rinsability with a smooth after feel. Typical use levels 1-5%

INCROQUAT BEHENYL TMS:
Behenyl Trimonium Methosulfate (and) Cetearyl Alcohol
White Flake
Percent Active: 25
Especially improves wet comb and leaves hair with excellent body and spring. Typical use levels 1-10%

CRODA INC.: Quaternary Cosmetic Conditioners(INCROQUATS) (Continued):

INCROQUAT CR Concentrate:
Cetearyl Alcohol (and) PEG-40 Castor Oil (and) Stearalkonium Chloride
White Flake
Percent Active: 100
One part formulating aid for the production of effective cream rinses and conditioners. Typical use levels 5-7%

INCROQUAT CTC-30:
Cetrimonium Chloride
Colorless Liquid
Percent Active: 30
A light conditioner that gives very good antistatic properties, moderate detangling and wet slip. Typical use levels 0.5-5%

INCROQUAT DBM-90:
Debehenyldimonium Methosulfate
White Flake
Percent Active: 90
Soluble in hot alcohol. Insoluble in cold alcohol and water. Low foamer. Excellent conditioner giving good wetting and slip and superb after feel. Typical use levels 0.5-5%

INCROQUAT I-85:
Isostearamidopropalkonium Chloride
Yellow Liquid
Percent Active: 85
Good solubility in water and good foaming. Good slip and detangling. Typical use levels 0.5-5%

INCROQUAT MINK-85:
Minkamidopropalkonium Chloride
Amber Liquid
Percent Active: 85
One of the best quaternary cosmetic conditioners. Based on natural mink oil, MINK 85 gives lots of foam and excellent slip. Very good wet comb/dry comb balance. Typical use levels 0.5-5%

CRODA INC.: Quaternary Cosmetic Conditioners(INCROQUATS) (Continued):

INCROQUAT O-50:
Olealkonium Chloride
Pale Yellow Liquid
Percent Active: 50
Clearly soluble in alcohol and water. Moderate foam. Good slip and improved wet comb. Typical use levels 0.5-5%

INCROQUAT OL-85:
Olivamidopropalkonium Chloride
Amber Liquid
Percent Active: 85
Based on natural olive oil and clearly soluble in alcohol and water. Foams and imparts good slip and wet comb. Typical use levels 0.5-5%

INCROQUAT S-75CG:
Quaternium-27
Yellow Liquid
Percent Active: 75
Low foamer, with dry feel on hair, good bodying and dry comb. Typical use levels 0.5-5%

INCROQUAT S-85:
Stearalkonium Chloride
White Flake
Percent Active: 85
Soluble in water and alcohol. Good foam, wetting and rinsing. This conditioner provides good slip, wet comb and antistatic properties. Typical use levels 0.5-5%

INCROQUAT SDQ-25:
Stearalkonium Chloride
White Paste
Percent Active: 25
Superb all round conditioning, wet slip and detangling. Good dry feel, antistatic properties and manageability. Typical use levels 2-10%

CRODA INC.: Quaternary Cosmetic Conditioners(INCROQUATS) (Continued):

INCROQUAT SE-85:
Sesamidopropalkonium Chloride
Amber Liquid
Percent Active: 85
Gives good foam and a dry feel. Excellemt dry manageability and static control. Typical use levels 0.5-5%

INCROQUAT WG-85:
Wheat Germamidopropalkonium Chloride
Amber Liquid
Percent Active: 85
Excellent conditioner. Soluble in alcohol and water. Very good foam, slip and wet comb. Typical use levels 0.5-5%

INCROQUAT 100#:
Methyl Bis (Hydrogenated Tallow amido ethyl) 2-Hydroxyethyl Ammonium Chloride
Flake or Powder
Percent Active: 100
Soluble in hot alcohol, precipitates when cold. Insoluble in water. Does not foam. Strongly substantive, gives good body, manageability and after feel. Good antistatic properties. Typical use levels 1-5%.

INCROQUAT 248:
Quaternium-72
Yellow Solid
Percent Active: 90
Economical hair conditioner. Typical use levels 0.5-5%

#INCROQUAT 100 is available in a powdered or flaked form.

CRODA INC.: Self-emulsifying Waxes:

Complete emulsifying agents for oil-in-water emulsions combining body, opacity and conditioning with ease of formulation.

COSMOWAX J:
Cetearyl Alcohol (and) Ceteareth 20
White to Off White Flakes
Melting Point C: 46.5-52.5
Nonionic
General emulsifying agent for creams and lotions. Viscosity building characteristics help improve stability especially for emulsions thickened with gums. Stable at extremes of alkalinity and acidity. Typical use levels 2-15%.

COSMOWAX K:
Stearyl Alcohol (and) Ceteareth 20
White to Off White Flakes
Melting Point C: 55-60
Nonionic
General emulsifying agent that improves opacity and produces more viscous emulsions than COSMOWAX J. Typical use levels 2-10%.

COSMOWAX P:
Ceteareth 20 (and) Cetearyl Alcohol
White to Off White Flakes
Melting Point C: 46-53
Nonionic
Emulsifying agent providing improved emulsion stability in lotions. CRODACOL CS50 can be added to increase viscosity when formulated in creams. Stable at extremes of alkalinity and acidity. Typical use levels 2-15%.

INCROQUAT BEHENYL TMS:
Behenyl Trimonium Methosulfate (and) Cetearyl Alcohol
White Flakes
Melting Point C: 56-62
Cationic
A novel cationic self emulsifying wax that produces conditioning benefits for hair and skin care products. Improves wet combing and reduces static in cream rinses. Produces a soft "powdery" feel on the skin. Typical use levels 1-5%

CRODA, INC.: Self-emulsifying Waxes(Continued):

INCROQUAT CR Concentrate:
Cetearyl Alcohol (and) PEG-40 Castor Oil (and) Stearalkonium Chloride
White Flakes
Melting Point C: 58-62
Cationic
One-step additive for the formulation of effective cream rinses and conditioners. Typical use levels 5-7%

POLAWAX:
Emulsifying Wax NF
Creamy White Flakes
Melting Point C: 48-52
The premier nonionic emulsifier for cosmetics and pharmaceuticals. Emulsifying agent of choice for ethnic hair straighteners and relaxers. Produces stable emulsions even at pH 13. "Official" emulsifier for pharmaceutical emulsions where an ionic emulsifier may reduce drug activity. Typical use levels 2-15%.

POLAWAX A31:
Emulsifying Wax NF
Creamy White Flakes
Melting Point C: 48-52
Nonionic
Specially developed emulsifying agent for quick breaking aerosol foams and mousses. Typical use levels 1-8%.

CRODA INC.: Sugar Esters(CRODESTAS):

CRODESTAS are esters of sucrose and selected fatty acids giving mild, edible, ethylene oxide free emulsifiers with a wide range of HLB values.

CRODESTA A-10:
Acetylated Sucrose Distearate
Yellow/White Wax
Solubility: Oil Soluble
Water Insoluble

Confers a smooth feel, good spreading properties, pigment dispersion and thermal properties to anhydrous sticks and hot poured products. Typical use levels 2-12%

CRODESTA F-10:
Sucrose Distearate
White Creamy Powder
HLB: 3
Solubility: Oil Soluble (Translucent)
Water Insoluble

Produces emulsions which resist rinsing off and produces an emollient feel on the skin. Useful for sun tan and baby lotions. Typical use levels 1-3%

CRODESTA F-50:
Sucrose Distearate
White Powder
HLB: 6.5
Solubility: Water Soluble (Translucent)
Oil Insoluble

Unique structural properties impart exceptional stability and a creamy feel to the lips in lipstick formulations. Typical use levels 3-6%

CRODESTA F-110:
Sucrose Stearate (and) Sucrose Distearate
White Powder
HLB: 12.0
Solubility: Water Soluble (Translucent)
Oil Insoluble

Useful in producing non-alkaline, non-soap containing emulsions with much reduced irritational scores. Recommended in skin creams and lotions. Typical use levels 3-6%

CRODA INC.: Sugar Esters(CRODESTAS)(Continued):

CRODESTA F-160:

Sucrose Stearate
White Powder
HLB: 14.5
Solubility: Water Soluble (Translucent)
Oil Insoluble

Enables the formulation of clear gel micro-emulsion systems with reduced physiological effect on the skin and eyes. A good foaming agent, wetting agent and emulsifier for ingestible pharmaceutical products. Typical use levels 3-6%.

CRODESTA SL-40:

Sucrose Cocoate
Amber Liquid
HLB: 15.0
Solubility: Water Soluble
Oil Insoluble

This high foaming non-ionic emollient is recommended for mild shampoos and skin cleansers. Typical use levels 5-20%

CRODA INC.: Sulfosuccinates(INCROSULS):

Mild, non-irritating, high foaming anionic surfactants with light color used in baby shampoos and bath products.

INCROSUL LAFS:
Disodium Laneth-5 Sulfosuccinate
Yellow Liquid/Slurry
Anionic
Percent Active: 50
A mild conditioning surfactant for cream/emulsion shampoos. Low foaming with good emulsifying properties. Typical use levels 5-50%

INCROSUL LMA:
Diammonium Lauramido MEA Sulfosuccinate
Colorless Liquid/Slurry
Anionic
Percent Active: 40
Recommended for clear shampoo formulations, very water soluble with high foam. Typical use levels 5-50%

INCROSUL LMS:
Disodium Lauramido MEA Sulfosuccinate
Yellow Liquid/Slurry
Anionic
Percent Active: 40
High foaming, recommended for mild shampoos. Also recommended for carpet cleaners. Typical use levels 5-50%

INCROSUL LS:
Disodium Lauryl Sulfosuccinate
White Paste
Anionic
Percent Active: 40
Mild conditioning surfactant and viscosity modifier for creams, lotions, shampoos and bath products. Typical use levels 0.5-50%

INCROSUL LSA:
Diammonium Lauryl Sulfosuccinate
Colorless Liquid
Anionic
Percent Active: 40
Mild, high foaming surfactant and lime soap dispersant for shampoos and bubblebaths. Typical use levels 5-50%

CRODA INC.: Sulfosuccinates(INCROSULS)(Continued):

INCROSUL TLS:
Disodium Laureth Sulfosuccinate
Colorless Liquid
Anionic
Percent Active: 40
Very low eye irritation with high foam and very light color. Especially useful for baby shampoos. Typical use levels 5-50%

INCROSUL OMS:
Disodium Oleamido MEA Sulfosuccinate
Yellow to Amber Liquid
Anionic
Percent Active: 35
Recommended for baby shampoos due to extreme mildness, good conditioning and rich foam. Typical use levels 5-50%.

INCROSUL TS:
Disodium Tridecyl Sulfosuccinate
Colorless Liquid
Anionic
Percent Active: 40
Very low color material that produces good flash foam. Typical use levels 5-50%

CRODA INC.: SUPER REFINED Oils:

SUPER REFINED oils are produced by exhaustive chromatographic purification yielding oils of exceptional color, odor and oxidative stability.

Almond Oil, NF SUPER REFINED:
Almond Oil
Typical APHA Color: 40
Fatty Acid Distribution:
Myristic: 7
Oleic: 66
Linoleic: 27
Provides an elegant skin feel and promotes spreading in creams, lotions and bath oils.

Apricot Kernel Oil, NF SUPER REFINED:
Apricot Kernel Oil
Typical APHA Color: 30
Fatty Acid Distribution:
Palmitic: 5
Oleic: 62
Linoleic: 33
Lubricant and softener in nail oils. Conditioner for hair care products.

Avocado Oil, SUPER REFINED:
Avocado Oil
Typical APHA Color: 30
Fatty Acid Distribution:
Myristic: 11
Palmitic: 3
Oleic: 69
Linoleic: 15
Provides an elegant skin feel and promotes spreading in creams, lotions and bath oils. Emollient for hair care products.

Babussu Oil, SUPER REFINED:
Babussu Oil
Typical APHA Color: 20
Fatty Acid Distribution:
Lauric: 46
Myristic: 17
Palmitic: 10
Stearic: 4
Oleic: 15
Linoleic: 3
Superior emollient melts close to body temperature. Less "raspy" than coconut oil. Exceptional in sunscreen products.

CRODA INC.: SUPER REFINED Oils(Continued):

Coconut Oil, SUPER REFINED:
Coconut Oil
Typical APHA Color: 20
Fatty Acid Distribution:
Lauric: 50
Myristic: 20
Palmitic: 10
Stearic: 3
Oleic: 8
Linoleic: 2
Color and odor dramatically better than typical commercial product. Used in sunscreen products.

CROSSENTIAL EPO, SUPER REFINED:
Evening Primrose Oil
Typical APHA Color: 20
Fatty Acid Distribution:
Palmitic: 6
Stearic: 2
Oleic: 9
Linoleic: 68
Linolenic: 15
Contains essential fatty acids (EFA) which are vital to the metabolism of healthy skin.

Grapeseed Oil, SUPER REFINED:
Grapeseed Oil
Typical APHA Color: 50
Fatty Acid Distribution:
Palmitic: 6
Stearic: 3
Oleic: 27
Linoleic: 64
Contains a high degree of linoleic acid.

Menhaden Oil, SUPER REFINED:
Menhaden Oil
Typical APHA Color: 100
Fatty Acid Distribution:
Myristic: 8
Palmitic: 19
Palmitoleic: 10
Oleic: 11
Eicosapentaenoic: 15
Docosahexaenoic: 12
Contins essential fatty acids (EFA) which are vital to the metabolism of healthy skin.

CRODA INC.: SUPER REFINED Oils(Continued):

Mink Oil, SUPER REFINED:
Mink Oil
Typical APHA Color: 30
Fatty Acid Distribution:
Myristic: 4
Palmitic: 16
Palmitoleic: 18
Stearic: 2
Oleic: 42
Linoleic: 18
A gentle effective oil for make-up remover systems evoking luxury.

Olive Oil, SUPER REFINED:
Olive Oil
Typical APHA Color: 30
Fatty Acid Distribution:
Myristic: 10
Stearic: 3
Oleic: 80
Linoleic: 7
The standard for cosmetic emollience. This super stable version eliminates the hassles. A lubricant for hair care products.

Orange Roughy Oil, SUPER REFINED:
Orange Roughy Oil
Typical APHA Color: 30
Fatty Acid Distribution:
Myristic: 1
Palmitic: 1
Palmitoleic: 12
Oleic: 56
Linoleic: 2
In skin care preparations shows superior spreading properties and skin softening ability to Jojoba Oil as it is more "fluid".

Peanut Oil, SUPER REFINED:
Peanut Oil
Typical APHA Color: 30
Fatty Acid Distribution:
Palmitic: 10
Stearic: 2
Oleic: 48
Linoleic: 40
An emollient in skin care preparations. Also recommended for nutritional supplements and pharmaceutical delivery systems.

CRODA INC.: SUPER REFINED Oils(Continued):

Safflower Oil, USP Super Refined:
Safflower Oil
Typical APHA Color: 30
Fatty Acid Distribution:
Palmitic: 5
Stearic: 4
Oleic: 13
Linoleic: 76
The polyunsaturate content yields an oil of extreme mobility.

Sesame Oil, SUPER REFINED:
Sesame Oil
Typical APHA Color: 30
Fatty Acid Distribution:
Palmitic: 9
Stearic: 5
Oleic: 40
Linoleic: 47
In skin care preparations, nutritional supplements and pharmaceutical delivery systems.

Shark Oil, SUPER REFINED:
Shark Liver Oil
Typical APHA Color: 40
Fatty Acid Distribution:
Myristic: 2
Palmitic: 12
Palmitoleic: 7
Stearic: 4
Oleic: 30
Linoleic: 6
Linolenic: 5
Docosahexaenoic: 10
Very moisture repellent and may be used for skin protection. Contains 13% C20 unsaturated, 18% C22 trisaturated and 8% C24 unsaturated.

Soybean Oil, SUPER REFINED:
Soybean Oil
Typical APHA Color: 60
Fatty Acid Distribution:
Palmitic: 11
Stearic: 4
Oleic: 25
Linoleic: 54
Linolenic: 6
In skin care preparations, nutritional supplements and pharmaceutical delivery systems.

Wheat Germ Oil, SUPER REFINED:
Wheat Germ Oil
Typical APHA Color: 250
Fatty Acid Distribution:
Palmitic: 13
Stearic: 3
Oleic: 14
Linoleic: 58
Linolenic: 8
An economical source of gamma linolenic acid.

CRODA INC.: Synthetic Waxes(SYNCROWAXES):

SYNCROWAXES are specially engineered to provide stable uniform properties in sticks and emulsions.

SYNCROWAX AW1-C:
C18-36 Acid
Pale Cream Wax Flakes
Melting Point C: 69-74
Compatibility: Mineral Oil
Esters

This long chain acid may be neutralized yielding soap emulsions and gels with high viscosity and excellent stability. Typical use levels 1-15%.

SYNCROWAX BB4:
Synthetic Beeswax
Pale Cream Wax Flakes
Melting Point C: 60-65
Compatibility: Similar to Beeswax

A synthetic beeswax designed to avoid the problems caused by the natural batch to batch variations of Beeswax. Typical use levels 1-5%

SYNCROWAX ERL-C:
C18-36 Acid Glycol Ester
Light Tan Wax Flakes
Melting Point C: 70-75
Compatibility: Mineral Oil

Used with HGL-C it provides stability with fine texture and good strength in stick formulations due to its micro-crystalline nature. Also used in suppositories and powder make-ups. Typical use levels 1-4%.

SYNCROWAX HGL-C:
C18-36 Acid Triglyceride
Light Yellow Wax Flakes
Melting Point C: 70-75
Compatibility: Castor Oil

The most crystalline SYNCROWAX. With ERL-C provides rigidity and gives excellent gloss and molding characteristics in stick formulations. Typical use levels 4-12%

CRODA INC.: Synthetic Waxes(SYNCROWAXES)(Continued):

SYNCROWAX HR-C:
Glyceryl Tribehenate
Off White Wax Pastilles/Powder
Melting Point C: 60-65
Compatibility: Castor Oil
A more amorphous wax that yields a creamy payoff and ease of application in anti-perspirant sticks, lipsticks and lip balms. Also tolerates high levels of powder in anhydrous make-up. Typical use levels 4-12%.

SYNCROWAX HRS-C:
Glyceryl Tribehenate (and) Calcium Behenate
Pale Cream Solid Flakes
Melting Point C: 100-115
Compatibility: Volatile Silicone
Mineral Oil
A gellant for relatively non polar systems. Its high gelling capacity enebles cosmetic oils and esters to gel without tack. Can form firm gels with mineral oil and volatile silicone. Typical use levels 2-10%

CYPRUS INDUSTRIAL MINERALS: CYPRUS Talc Products for Cosmetic Applications:

Pressed and loose powders require finer talc products in the 300 to 500 mesh size range, as do antiperspirants, creams, lotions and nail polishes. Consistent, high color, lack of oversize particles, good compression and dispersion characteristics are desired.

SUPREME:
Color: 88
Economical 325 mesh Montana talc

ALTALC 400:
Color: 87
400 mesh floated blend

ALTALC 500:
Color: 8
500 mesh floated blend

ACT II 400:
Color: 90
400 mesh, ultrasoft, floated Alabama talc

ACT II 500:
Color: 91
500 mesh, ultrasoft, floated Alabama talc

BRILLIANTE:
Color: 86
200 mesh, translucent Italian talc

SUPRAFINO:
Color: 88
325 high slip, platy blend

ULTRAFINO:
Color: 89
Ultrafine, high slip, Italian talc

TOPNOTE 300:
Color: 89
300 mesh, platy, Italian talc

CYPRUS INDUSTRIAL MINERALS: Quality CYPRUS Talc Products for Cosmetics:

Baby Powder requires a 200 mesh talc product with good slip, a consistent loose bulk density in the 22-32 lb/ft 3 range, good color and purity to meet CTFA specifications. Bath and dusting powders also require similar properties. Cyprus products include:

VERTAL C-O:
Color (GEB): 78
Loose Bulk (#/ft3): 32
Economical, floated Vermont talc

SUPREME:
Color (GEB): 86
Loose Bulk (#/ft3): 26
Well known dry milled Montana talc

ALTALC 200:
Color (GEB): 86
Loose Bulk (#/ft3): 24
Mainstream, floated talc blend

REGAL:
Color (GEB): 89
Loose Bulk (#/ft3): 24
Ultrasoft, floated Alabama talc

CASTLE 65:
Color (GEB): 82
Loose Bulk (#/ft3): 28
Platy, floated, Vermont talc

AURA 200:
Color (GEB): 88
Loose Bulk (#/ft3): 28
Platy, floated, Australian talc

SUPRA A:
Color (GEB): 87
Loose Bulk (#/ft3): 24
Platy, high sheen blend

TOPNOTE 200:
Color (GEB): 87
Loose Bulk (#/ft3): 22
The ultimate premium platy Italian talc

The last four products are used for bath and dusting powder applications.

DESERT KING CORP.: FOAMATION 50 Natural Foaming Agent:

FOAMATION 50 is a concentrated natural saponin# extract of the Mohave Yucca plant (Yucca schidigera) which is native to the Southwestern deserts of North America. Only select plants grown in the Baja California region of Mexico are used in the making of FOAMATION 50. FOAMATION 50 is cold processed to ensure maximum activity and functionality in a variety of foaming applications.

Appearance: Pourable liquid
Color: Dark Brown
Odor: Typical
Organoleptic: Sl. Bittersweet
Water Solubility: 99.6%
Density: 10 lbs/gal
Viscosity (Brookfield): 25.5
Concentration: 50 Brix
pH: 2.8-3.8
lower concentration levels are available upon request

Proximate Analysis:
Moisture: 49.75%
Protein: 0.8%
Fat: 0.6%
Crude Fiber: 0.3%
Carbohydrate: 45.95%
Sodium Benzoate: .1% (added to preserve freshness)
Ash (Mineral Matter): 2.5%

Microbiological Standards:
Standard Plate count: 5000/g max.
Yeast: 10/g max.
Staphylcoccus aureus: Negative
Coliform: 10/g. max.
E. Coli.: Negative
Salmonella: Negative

Regulatory Status:
Mohave yucca is approved for use as a natural food adjuvant under Title 21 CFR 172.510.

Yucca saponin is a sapogenin glycoside. Upon hydrolysis, the saponin yields sapogenin and simple sugars.

DESERT KING JMC, LTD.: ESTRAN-Series Cosmetic Grade Jojoba Oils:

The ESTRAN cosmetic oils contain approximately 98% liquid wax esters which are predominantly composed of cis-monounsaturated w-9 C-18 to C-26 linear fatty alcohols and cis-monounsaturated w-9 C-16 to C-24 linear fatty acids. Over 90% of the wax esters in ESTRAN cosmetic oils are C-40 to C-44 monoesters.

The physical properties of the ESTRAN cosmetic oils are listed below:

Acid Value: 0.3 max
Peroxide value (mcg/kg): 2.0 max
Freezing point (C): 10.6-7.0
Melting point (C): 6.8-7.0
Boiling point (757 mm under N2) (C): 389
Heat of fusion by DSC (cal/g): 21
Refractive index (at 25C): 1.4650
Dielectric constant (27C): 2.680
Specific conductivity (27C) (mho/cm): 8.86 x 10 -13
Specific gravity 25/25C: 0.863
Smoke point (C): 195
Flash point (C): 295
Fire point (C): 338
Iodine value: 82
Saponification value: 92
Acetyl value: 2
Unsaponifiable matter (%): 51
Total acids: 52
Iodine value of alcohols: 77
Iodine value of acids: <76
Average molecular weight: 606

Because of the chemical composition of Jojoba oil, its high molecular weight and stable, low viscosity, the ESTRAN oils deliver a package of functional properties and aesthetic qualities superior to those of any other natural oil. Some of the functional values of the ESTRAN oils are:

- nongreasiness
- superior transpirational water control in skin
- enhanced skin elasticity
- high skin penetration
- enhanced cellular regeneration
- excellent keratoplastic effect
- high miscibility with sebum
- effective control of exfoliation, flaking, and dryness
- mildness (very low acid value)
- hypoallergenicity
- unsurpasssed resistance to temperature fluctuations, oxidation, hydrolysis and rancidity
- biodegradability

DESERT KING JMC, LTD.: ESTRAN-Series Cosmetic Grade Jojoba Oils: (Continued):

Skin Care Applications:
- Hand care preparations
- Body care preparations
- Face care preparations
- Leg care
- Moisturizers
- Day creams and lotions
- Skin oils
- Night creams
- Healing & soothing creams
- Sport preparations
- Vanishing creams
- Sun screens
- Sun tanners
- Soaps
- Lip care
- Bath oils
- Perfume oil
- Massage oil

Hair and Scalp Applications:
- Liquid shampoos
- Cream shampoos
- Conditioners for stressed hair
- Hair dressing creams
- Hair dressing gels
- Hair mousses
- Dry hair emulsions and gels
- Scalp preparations for babies and adults
- Hair treatment fixatives
- Hair treatment creams and lotions for prophylaxis of hair loss

Miscellaneous Uses:
- Eye balsams
- Regenerative eye creams
- Eye make-up removers
- Eyebrow pencils
- Eye liners
- Mascara
- Eyeshadow
- Lip gloss
- Rouges
- Lip balm
- Lip stick
- Liquid soap (hand cleaner creams)
- Bars soap
- Pomades
- Nail polishes
- Nail polishes removers

ESTRAN Product Line:

ESTRAN-PURE:
A cold extracted, single-pressed Jojoba oil. The equivalent of "extra-virgin" grades of vegetable oils. ESTRAN-PURE is typified by a virtual lack of odor, acid value less than .3%, and a pleasing yellow color. ESTRAN-PURE can be used for most all cosmetic preparations.

ESTRAN-LITE:
A virtually colorless refined Jojoba oil with a very slight odor. ESTRAN-LITE possesses all the functional properties of ESTRAN-PURE and is used in preparations requiring colorless ingredients.

ESTRAN-ULTRA:
A molecularly distilled Jojoba oil that is the "creme-de-la-creme" of all natural oils. ESTRAN-ULTRA is completely devoid of odor and flavor and is a crystal-clear (white) oil. ESTRAN-ULTRA is used in delicate preparations, such as pharmaceutical ointments and fine perfume preparations.

DOW CHEMICAL U.S.A.: DOWICIL 200 Preservative:

Reliable, broad-spectrum antimicrobial protection for cosmetics and personal care formulations

DOWICIL 200 offers a unique combination of desirable properties:

- Effective at low concentrations
- Broad spectrum effectiveness
- Excellent formulation compatibility
- Well documented, favorable toxicologic profile
- High water solubility, virtually insoluble in oil
- Shelf life: Highly effective for two or more years
- Cost effective
- Easy to work with

Product Types:

- Hair care products
- Lotions
- Powders
- Shaving products
- Raw materials
- Miscellaneous
- Eye area products
- Baby products
- Creams
- Suntanning products
- Ethnic market products

Physical Properties:
Formula: C6H12N4(CH2CHCHCl)Cl
Molecular Weight: 251.2
Bulk Density lb/ft3: 25.0
CAS#: 51229-78-8

Sales Specifications:
Description: Off-white powder
Active ingredient# (minimum): 94%
Color, Gardner, maximum: 2
Sieve analysis:
Through No. 20 (U.S. Standard Sieve): 100%

Cis isomer 1-(3-chloroallyl)-3,5,7-triaza-1-azoniaadamantane chloride

DOW CHEMICAL U.S.A.: DOW POLYGLYCOLS:

Polyethylene Glycols E-Series:
CAS #25322-68-3:

E200:
Average Molecular Weight: 200
Average Viscosity, Centistokes: 77F: 40
Flash Point PMCC, F: 340
Refractive Index at 25C: 1.459
Specific Gravity 25/25C: 1.124
Viscosity Index: 111
CTFA: PEG-4

E300:
Average Molecular Weight: 300
Average Viscosity, Centistokes: 77F: 69
Flash Point PMCC, F: >400
Refractive Index at 25C: 1.463
Specific Gravity 25/25C: 1.125
Viscosity Index: 118
CTFA: PEG-6

E400:
Average Molecular Weight: 400
Average Viscosity, Centistokes: 77F: 90
Flash Point PMCC, F: >450
Refractive Index at 25C: 1.465
Specific Gravity 25/25C: 1.125
Viscosity Index: 124
CTFA: PEG-8

E600:
Average Molecular Weight: 600
Average Viscosity, Centistokes: 77F: 131
Flash Point PMCC, F: >450
Refractive Index at 25C: 1.466
Specific Gravity 25/25C: 1.126
Viscosity Index: 154
CTFA: PEG-12

E900:
Average Molecular Weight: 900
Average Viscosity, Centistokes: 100F: 100
Flash Point PMCC, F: >450
Specific Gravity 25/25C: 1.204
Viscosity Index: 182

DOW CHEMICAL U.S.A.: DOW POLYGLYCOLS(Continued):

Polyethylene Glycols E-Series(Continued):

E1000:
Average Molecular Weight: 1000
Average Viscosity, Centistokes: 210F: 18
Flash Point PMCC, F: >450
Specific Gravity 25/25C: 1.214
CTFA: PEG-20

E1450:
Average Molecular Weight: 1450
Average Viscosity, Centistokes: 210F: 29
Flash Point PMCC, F: >450
Specific Gravity 25/25C: 1.214
CTFA: PEG-6-32

E3350:
Average Molecular Weight: 3350
Average Viscosity, Centistokes: 210F: 93
Flash Point PMCC, F: >450
Specific Gravity 25/25C: 1.224
CTFA: PEG-75

E4500:
Average Molecular Weight: 4500
Average Viscosity, Centistokes: 210F: 180
Flash Point PMCC, F: >450
Specific Gravity 25/25C: 1.224
CTFA: PEG-100

E8000:
Average Molecular Weight: 8000
Average Viscosity, Centistokes: 210F: 800
Flash Point PMCC, F: >500
Specific Gravity 25/25C: 1.224
CTFA: PEG-150

DOW CHEMICAL U.S.A.: DOW POLYGLYCOLS(Continued):

Methoxypolyethylene Glycols MPEG:
CAS #9004-74-4

MPEG 350:
Average Molecular Weight: 350
Average Viscosity, Centistokes: 77F: 27
Flash Point PMCC, F: >350
Refractive Index at 25C: 1.455
Specific Gravity 25/25C: 1.097
Viscosity Index: 138
CTFA: PEG-6 Methyl Ether

MPEG 550:
Average Molecular Weight: 550
Average Viscosity, Centistokes: 77F: 56
Flash Point PMCC, F: >400
Refractive Index at 25C: 1.461
Specific Gravity 25/25C: 1.102
Viscosity Index: 181
CTFA: PEG-10 Methyl Ether

MPEG 750:
Average Molecular Weight: 750
Average Viscosity, Centistokes: 100F: 53
Flash Point PMCC, F: >450
Refractive Index at 25C: 1.463
Specific Gravity 25/25C: 1.096
CTFA: PEG-16 Methyl Ether

Polypropylene Glycols L-Series:
CAS #9003-13-8

L910:
Average Molecular Weight: 910
Average Viscosity, Centistokes: 77F: 83
Flash Point PMCC, F: 345
Refractive Index at 25C: 1.444
Specific Gravity 25/25C: 0.9833
Viscosity Index: 181
CTFA: PPG-14 Butyl Ether

DOW CHEMICAL U.S.A.: DOW POLYGLYCOLS(Continued):

Polypropylene Glycols P-Series:
CAS #29434-03-5

P425:
Average Molecular Weight: 425
Average Viscosity, Centistokes: 77F: 70
Flash Point PMCC, F: 330
Refractive Index at 25C: 1.447
Specific Gravity 25/25C: 1.007
CTFA: PPG-9

P1200:
Average Molecular Weight: 1200
Average Viscosity, Centistokes: 77F: 175
Flash Point PMCC, F: 345
Refractive Index at 25C: 1.448
Specific Gravity 25/25C: 1.007
Viscosity Index: 161
CTFA: PPG-20

P2000:
Average Molecular Weight: 2000
Average Viscosity, Centistokes: 77F: 300
Flash Point PMCC, F: 390
Refractive Index at 25C: 1.449
Specific Gravity 25/25C: 1.002
Viscosity Index: 183
CTFA: PPG-26

P4000:
Average Molecular Weight: 4000
Average Viscosity, Centistokes: 77F: 800
Flash Point PMCC, F: 365
Refractive Index at 25C: 1.450
Specific Gravity 25/25C: 1.005
Viscosity Index: 191
CTFA: PPG-30

DOW CHEMICAL U.S.A.: DOW POLYGLYCOLS(Continued):

Polyglycol Copolymers:

CAS #51258-15-2:
15-200:
Average Molecular Weight: 2600
Average Viscosity, Centistokes: 77F: 420
Flash Point PMCC, F: >450
Refractive Index at 25C: 1.460
Specific Gravity 25/25C: 1.060
Viscosity Index: 200
CTFA: PPG-24 Glycereth-24

CAS #9082-00-2:
112-2:
Average Molecular Weight: 4900
Average Viscosity, Centistokes: 77F: 1000
Flash Point PMCC, F: 455
Refractive Index at 25C: 1.455
Specific Gravity 25/25C: 1.028
Viscosity Index: 200
CTFA: PPG-66 Glycereth-12

CAS #53637-25-5:
EP530:
Average Molecular Weight: 2000
Average Viscosity, Centistokes: 77F: 321
Flash Point PMCC, F: >420
Refractive Index at 25C: 1.452
Specific Gravity 25/25C: 1.017
Viscosity Index: 192
CTFA: Poloxamer-181

DOW CHEMICAL U.S.A.: METHOCEL Cellulose Ethers:

METHOCEL cellulose ethers are water-soluble polymers derived from cellulose, the most abundant polymer in nature.

Personal Care Products:
Shampoos:
METHOCEL cellulose ether is widely used as a thickener in shampoos.
Creams and lotions:
METHOCEL can contribute film forming and secondary thickening properties which improve afterfeel and other sensory characteristics in creams and lotions.

Properties of solutions of METHOCEL:

Specific gravity, 4C, all types:
1% solutions: 1.0012
5% solutions: 1.0117
10% solutions: 1.0245

Refractive index, 2% solutions, all types: 1.336 nD20

Partial specific volume:
4,000 mPa-s METHOCEL A: 0.725 cc/g (0.087 gal/lb)
4,000 mPa-s METHOCEL E: 0.767 cc/g (0.092 gal/lb)
4,000 mPa-s METHOCEL F: 0.734 cc/g (0.087 gal/lb)
5,000 mPa-s METHOCEL J: 0.725 cc/g (0.087 gal/lb)
4,000 mPa-s METHOCEL K: 0.717 cc/g (0.086 gal/lb)
15,000 mPa-s METHOCEL K: 0.724 cc/g (0.087 gal/lb)

Freezing point, 2% solutions, all types:
0.0C at 2% concentration

Surface tension, 25C, 0.05% concentrations:
Water: 72-74 x 10 -3 Newton/meter (72-74 dynes/cm)
Methylcellulose: 53-59 x 10 -3 Newton/meter (53-59 dynes/cm)
Hydroxypropyl methylcellulose: 43-55 x 10 -3 Newton/meter (43-55 dynes/cm)

Viscosities of METHOCEL cellulose ethers#:

METHOCEL Product:	Viscosity, cP
E4M Premium	4,000
F4M Premium	4,000
K4M Premium	4,000
K15M Premium	15,000
K100M Premium	100,000
40-100	12,000
40-101	75,000

Expressed as 2% aqueous solutions at 20C.
Note: All 40-series METHOCEL products are surface-treated for cold water dispersibility

DOW CORNING CORP.: Silicones:

Amodimethicone (and) Tallowtrimononium Chloride (and) Nonoxynol-10:

DOW CORNING 929 cationic emulsion is an emulsion of an amine functional polymer. In hair care products it provides substantivity by acting as a conditioning agent that dries to a film by crosslinking. It is especially effective on damaged hair, where benefits include improved wet and dry combing, a soft feel, and durability without buildup.

Cyclomethicones:

These volatile fluids are well-suited to a variety of applications. In hair care products, they improve wet combing, yet evaporate without a residue. In antiperspirants and deodorants, they act as transient carriers for active salts. Similarly, in skin care products, cyclomethicones deliver active ingredients and serve as noncomedogenic emollients and carriers. As vehicles for delivering fragrance, cyclomethicones provide a smooth, silky, noncooling feel.

DOW CORNING 244 fluid and DOW CORNING 344 fluid are composed primarily of cyclic silicone tetramer: DOW CORNING 244 fluid contains more than 95-percent cyclic tetramer; DOW CORNING 344 fluid typically contains 85-percent cyclic tetramer and 15-percent cyclic pentamer. Both silicones readily volatilize from the skin and hair and are appropriate choices for products such as aerosol and roll-on antiperspirants and deodorants, where rapid evaporation is desirable.

DOW CORNING 245 fluid and DOW CORNING 345 fluid evaporate more slowly and have lower freezing points than cyclomethicones that are composed primarily of cyclic tetramer. DOW CORNING 245 fluid contains more than 95-percent cyclic pentamer; DOW CORNING 345 fluid typically contains 80-percent cyclic pentamer and 20-percent cyclic hexamer. Because of their more controlled volatility, these silicones are the choice for solid stick products because they help avoid product shrinkage over the stick product's usable life.

Cyclomethicone (and) Dimethicone Copolyol:

DOW CORNING 3225C formulation aid is a self-emulsifying cyclomethicone that allows a greater level of compatibility with other ingredients than does cyclomethicone alone. Used primarily for formulating water-in-silicone emulsions, the formulation aid is also useful for making various types of oil-in-water emulsions by acting as a coemulsifier, and has been effective in the preparation of various transparent products. In hair care products, it reduces tackiness in formulations and improves spreadability and wet combing, while providing a nongreasy feel. It also provides a drier feel to aqueous-based antiperspirants and acts as an emulsifier for aqueous aluminum salts. In skin care products, cyclomethicone (and) dimethicone copolyol adds an elegant feel and reduces the tackiness of moisturizing ingredients.

DOW CORNING CORP.: Silicones(Continued):

Cyclomethicone (and) Dimethiconol:

DOW CORNING Q2-1401 fluid is a blend of cyclomethicone and an ultra-high molecular weight silicone polymer. This gumlike silicone imparts long-lasting, substantive properties to skin and hair care products. It provides for uniform spreading and a dry, emollient feel in antiperspirant/deodorant sticks. In skin care products, cyclomethicone (and) dimethiconol acts as a carrier to deliver active ingredients, offering durability and a soft, nongreasy feel.

Dimethicone:

200 fluid by Dow Corning and DOW CORNING 225 fluid are dimethyl silicone fluids whose properties are related to a variety of viscosities and molecular weights. In hair care applications, the fluids provide lubrication, gloss and softness, along with reduced tackiness and resistance to humidity. In antiperspirants and deodorants, low-viscosity dimethicone helps alleviate the whitening effects of antiperspirant salts and can act as an emollient and lubricant. Dimethicone also reduces valve clogging in aerosol products. In skin lotions and creams, dimethicones are excellent emollients that reduce the "soaping" effect of emulsifiers. As active ingredients, they can act as skin protectants, a claim that may be included on product labels based on the proposed FDA monograph on OTC skin-protectant drugs. Dimethicones improve the flow or pay-out of cosmetic powders, aid spreading and rub-in, and act as a water barrier.

Dimethicone (and) Trimethylsiloxysilicate:

DOW CORNING 593 fluid is a resin-modified polysiloxane that acts as a breathable, protective barrier in skin care and cosmetic applications. This nongreasy, tenacious silicone fluid also provides excellent lubricating properties and is an effective water-proofing agent for sunscreen creams and lotions.

DOW CORNING CORP.: Silicones(Continued):

Dimethicone Copolyol:
DOW CORNING 190 surfactant, DOW CORNING 193 surfactant, and DOW CORNING Q2-5220 resin modifier are copolymers of a polydimethylsiloxane and polyoxyalkylene ether that are water- and ethanol-soluble. In hair care applications, these silicones aid spreading, reduce irritation of surfactants, and can be used as resin plasticizers and detackifiers. In skin care products, dimethicone copolyols provide a very soft feel, reduce soap irritation and offer excellent foam control. In men's shaving products, they improve foaming and add lubrication and glide.

Phenyl Trimethicone:
DOW CORNING 556 fluid is similar to dimethicone, but its phenyl substitution provides broader compatibility with organic waxes and oils. This silicone provides gloss and sheen in hair care products as well as protection from the harsh effect of heat. In antiperspirants and deodorants, phenyl trimethicone can be formulated with wax, oil or alcohol-based stick products to provide the benefits of a nonoily emollient; it also greatly reduces the white residue of antiperspirant salts. In skin care applications, it acts as a water barrier, leaving the skin smooth and soft. In cosmetics, phenyl trimethicone adds emolliency and reduces tackiness.

Methicone:
DOW CORNING 1107 fluid is a reactive, methyl hydrogen functional fluid used to develop free-flowing cosmetic powders.

Stearoxytrimethylsilane (and) Stearyl Alcohol:
DOW CORNING Q5-0158A wax is a solid silicone well-suited for use in stick products. It offers excellent lubricating properties and is compatible with most waxes used in cosmetics.

Trimethylsilylamodimethicone:
DOW CORNING Q2-8220 conditioning additive is an amine-functional fluid used primarily in hair care products. It improves wet and dry combing and adds softness and gloss. This fluid is very substantive and reduces damage caused by reactive chemicals.

Trimethylsilylamodimethicone (and) Octoxynol-40 (and) Isolaureth-6 (and) Propylene Glycol:
DOW CORNING Q2-7224 conditioning agent is a cationic emulsion of a noncrosslinking, amine functional polymer. Like DOW CORNING 929 cationic emulsion, DOW CORNING Q2-7224 conditioning agent is used primarily in hair care products, where it forms a film that improves wet and dry combing, adds softness and reduces drying time. It is substantive, yet does not build up on hair, and is particularly effective on damaged hair.

DOW CORNING CORP.: Silicones(Continued):

Cyclomethicone:

DOW CORNING 244:
Viscosity (Centistokes): 2.5
Specific Gravity @ 25C: 0.953
Flash Point, F/C (Closed Cup): 130/55
Melting Point, F/C: 63/17
Boiling Point, F/C: 342/172
Refractive Index @ 25C: 1.394
Appearance: Clear

DOW CORNING 344:
Viscosity (Centistokes): 2.5
Specific Gravity @ 25C: 0.953
Flash Point F/C (Closed Cup): 130/55
Melting Point F/C: 50/10
Boiling Point F/C: 351/178
Refractive Index @ 25C: 1.394
Appearance: Clear

DOW CORNING 245:
Viscosity (Centistokes): 4.2
Specific Gravity @ 25C: 0.956
Flash Point F/C (Closed Cup): 170/76
Melting Point F/C: <-40/-40
Boiling Point F/C: 402/205
Refractive Index @ 25C: 1.397
Appearance: Clear

DOW CORNING 345:
Viscosity (Centistokes): 4.5
Specific Gravity @ 25C: 0.957
Flash Point F/C (Closed Cup): 170/76
Melting Point F/C: <-40/-40
Boiling Point F/C: 423/217
Refractive Index @ 25C: 1.398
Appearance: Clear

Dimethicone (and) Trimethylsiloxysilicate:

DOW CORNING 593:
Viscosity (Centistokes): 650
Specific Gravity @ 25C: 1.034
Flash Point F/C (Closed Cup): 570/299
Melting Point F/C: <-40/-40
Boiling Point F/C: >400/200
Refractive Index @ 25C: 1.409
Appearance: Clear

DOW CORNING CORP.: Silicones(Continued):

Cyclomethicone (and) Dimethiconol:

DOW CORNING Q2-1401:
Viscosity (Centistokes): 5000
Specific Gravity @ 25C: 0.960
Flash Point F/C (Closed Cup): 130/55
Melting Point F/C: 0/-18
Boiling Point F/C: 360/182
Refractive Index @ 25C: 1.397
Appearance: Clear

Dimethicone:

DOW CORNING 200:
Viscosity (Centistokes): 0.65
Specific Gravity @ 25C: 0.761
Flash Point F/C (Closed Cup): 30/-2
Melting Point F/C: <-40/-40
Boiling Point F/C: 212/100
Refractive Index @ 25C: 1.375
Appearance: Clear

DOW CORNING 200:
Viscosity (Centistokes): 5
Specific Gravity @ 25C: 0.920
Flash Point, F/C (Closed Cup): 275/135
Melting Point, F/C: <-40/-40
Boiling Point, F/C: >400/200
Refractive Index @ 25C: 1.397
Appearance: Clear

DOW CORNING 200:
Viscosity (Centistokes): 10
Specific Gravity @ 25C: 0.934
Flash Point, F/C (Closed Cup): 325/163
Melting Point, F/C: <-40/-40
Boiling Point, F/C: >400/200
Refractive Index @ 25C: 1.399
Appearance: Clear

DOW CORNING 225:
Viscosity (Centistokes): 10
Specific Gravity @ 25C: 0.934
Flash Point, F/C (Closed Cup): 325/163
Melting Point, F/C: <-40/-40
Boiling Point, F/C: >400/200
Refractive Index @ 25C: 1.399
Appearance: Clear

DOW CORNING CORP.: Silicones(Continued):

Dimethicone(Continued):

DOW CORNING 200:
Viscosity (Centistokes): 20
Specific Gravity @ 25C: 0.949
Flash Point F/C (Closed Cup): 450/232
Melting Point F/C: <-40/-40
Boiling Point F/C: >400/200
Refractive Index @ 25C: 1.401
Appearance: Clear

DOW CORNING 200:
Viscosity (Centistokes): 50
Specific Gravity @ 25C: 0.960
Flash Point F/C (Closed Cup): 545/285
Melting Point F/C: <-40/-40
Boiling Point F/C: >400/200
Refractive Index @ 25C: 1.402
Appearance: Clear

DOW CORNING 200:
Viscosity (Centistokes): 100
Specific Gravity @ 25C: 0.960
Flash Point, F/C (Closed Cup): 600/315
Melting Point, F/C: <-40/-40
Boiling Point, F/C: >400/200
Refractive Index @ 25C: 1.403
Appearance: Clear

DOW CORNING 200:
Viscosity (Centistokes): 350
Specific Gravity @ 25C: 0.970
Flash Point, F/C (Closed Cup): 600/315
Melting Point, F/C: <-40/-40
Boiling Point, F/C: >400/200
Refractive Index @ 25C: 1.403
Appearance: Clear

DOW CORNING 200:
Viscosity (Centistokes): 500
Specific Gravity @ 25C: 0.971
Flash Point F/C (Closed Cup): 600/315
Melting Point, F/C: <-40/-40
Boiling Point, F/C: >400/200
Refractive Index @ 25C: 1.403
Appearance: Clear

DOW CORNING CORP.: Silicones(Continued):

Dimethicone(Continued):

DOW CORNING 200:
Viscosity (Centistokes): 1000
Specific Gravity @ 25C: 0.971
Flash Point, F/C (Closed Cup): 610/321
Melting Point, F/C: <-40/-40
Boiling Point, F/C: >400/200
Refractive Index @ 25C: 1.403
Appearance: Clear

DOW CORNING 200:
Viscosity (Centistokes): 12500
Specific Gravity @ 25C: 0.975
Flash Point, F/C: <-40/-40
Boiling Point, F/C: >400/200
Refractive Index @ 25C: 1.403
Appearance: Clear

Dimethicone Copolyol:

DOW CORNING 190:
Viscosity (Centistokes): 1500
Specific Gravity @ 25C: 1.035
Flash Point, F/C (Closed Cup): 130/54
Melting Point, F/C: NA
Boiling Point, F/C: >400/200
Refractive Index @ 25C: 1.448
Appearance: Hazy

DOW CORNING 193:
Viscosity (Centistokes): 465
Specific Gravity @ 25C: 1.070
Flash Point, F/C (Closed Cup): 198/92
Melting Point, F/C: NA
Boiling Point, F/C: >400/200
Refractive Index @ 25C: 1.454
Appearance: Hazy

DOW CORNING Q2-5220:
Viscosity (Centistokes): 1000
Specific Gravity @ 25C: 1.030
Flash Point, F/C (Closed Cup): 140/60
Melting Point, F/C: NA
Boiling Point, F/C: >400/200
Refractive Index @ 25C: 1.457
Appearance: Hazy

DOW CORNING CORP.: Silicones(Continued):

Cyclomethicone (and) Dimethicone Copolyol:

DOW CORNING 3225C:
Viscosity (Centistokes): 500
Specific Gravity @ 25C: 0.963
Flash Point, F/C (Closed Cup): 140/60
Melting Point, F/C: 50/10
Boiling Point, F/C: 351/178
Refractive Index @ 25C: 1.398
Appearance: Hazy

Phenyl Trimethicone:

DOW CORNING 556:
Viscosity (Centistokes): 22.5
Specific Gravity @ 25C: 0.980
Flash Point, F/C (Closed Cup): 250/121
Melting Point, F/C: <-40/-40
Boiling Point, F/C: >400/200
Refractive Index @ 25C: 1.460
Appearance: Clear

Stearoxytrimethylsilane (and) Stearyl Alcohol:

DOW CORNING Q5-0158A:
Viscosity (Centistokes): Solid
Specific Gravity @ 25C: 0.852
Flash Point, F/C (Closed Cup): 162/72
Melting Point, F/C: 104/40
Boiling Point, F/C: >400/200
Refractive Index @ 25C: Solid
Apprearance: Ivory

Trimethylsilylamodimethicone:

DOW CORNING Q2-8220:
Viscosity (Centistokes): 145
Specific Gravity @ 25C: 0.970
Flash Point, F/C (Closed Cup): 170/76
Melting Point, F/C: <-40/-40
Boiling Point, F/C: >400/200
Refractive Index @ 25C: 1.407
Appearance: Clear

DOW CORNING CORP.: Silicones(Continued):

Silicone Functions in Formulations:

CTFA Designation:

Amodimethicone (and) Tallowtrimonium Chloride (and) Nonoxynol-10:
- Substantive: 929
- Conditions Hair: 929

Cyclomethicone:
- Spreading, Wetting: 244, 245, 344, 345
- Reduces Tackiness: 244, 245, 344, 345
- Conditions Hair: 244, 245, 344, 345
- Lubricate: 244, 245, 344, 345
- Add Shine, Gloss: 244, 245, 344, 345
- Emolliency: 244, 245, 344, 345
- Non-Comedogenic: 244, 245, 344, 345
- Carrier for Actives: 244, 245, 344, 345

Cyclomethicone (and) Dimethicone Copolyol:
- Spreading, Wetting: 3225C
- Reduce Tackiness: 3225C
- Emulsify: 3225C
- Lubricate: 3225C
- Moisturize: 3225C
- Anti-irritancy: 3225C
- Non-Comedogenic: 3225C

Cyclomethicone (and) Dimethiconol:
- Protection, Water Barrier: Q2-1401
- Spreading, Wetting: Q2-1401
- Substantive: Q2-1401
- Condition Hair: Q2-1401
- Lubricate: Q2-1401
- Moisturize: Q2-1401
- Emolliency: Q2-1401
- Non-Comedogenic: Q2-1401
- Carrier for Actives: Q2-1401

Dimethicone:
- Protection, Water Barrier: 200, 225
- Spreading, Wetting: 200, 225
- Reduce Tackiness: 200, 225
- Substantive: 200
- Condition Hair: 200
- Lubricate: 200, 225
- Moisturize: 200, 225
- Add Shine, Gloss: 200, 225
- Emolliency: 200, 225
- Anti-Whitening: 200, 225
- Non-Comedogenic: 200, 225
- Antifoaming: 200, 225

DOW CORNING CORP.: Silicones(Continued):

Silicone Functions in Formulations(Continued):

CTFA Designation:

Dimethicone (and) Trimethylsiloxysilicate:
Protection, Water Barrier: 593
Substantive: 593
Add Shine, Gloss: 593
Non-Comedogenic: 593

Dimethicone Copolyol:
Spreading, Wetting: 190, 193, Q2-5220
Reduce Tackiness: 190, 193, Q2-5220
Emulsify: 190, 193, Q2-5220
Lubricate: 190, 193, Q2-5220
Anti-Irritancy: 190, 193, Q2-5220
Non-Comedogenic: 190, 193, Q2-5220
Modify Foam: 190, 193, Q2-5220

Phenyl Trimethicone:
Protection, Water Barrier: 556
Reduce Tackiness: 556
Lubricate: 556
Moisturize: 556
Add Shine, Gloss: 556
Emolliency: 556
Anti-Whitening: 556

Methicone:
Protection, Water Barrier: 1107
Substantive: 1107
Lubricate: 1107

Stearoxytrimethylsilane (and) Stearyl Alcohol:
Reduce Tackiness: Q5-0158A
Lubricate: Q5-0185A

Trimethylsilylamodimethicone:
Substantive: Q2-8220
Condition Hair: Q2-8220
Add Shine, Gloss: Q2-8220

Trimethylsilylamodimethicone (and) Octoxynol-40 (and) Isolaureth-6 (and) Propylene Glycol:
Substantive: Q2-7224
Condition Hair: Q2-7224

DOW CORNING CORP.: Silicones(Continued):

Emulsions and Compounds:

Amino-Functional Silicones:

DOW CORNING Q2-7224:
Emulsifier Type: Nonionic
Color: White
Percent Silicone: 35
pH: 10.5
Suitable Diluent: Water

DOW CORNING 929:
Emulsifier Type: Cationic
Color: White
Percent Silicone: 35
pH: 7.6
Suitable Diluent: Water

Antifoam Compounds:

DOW CORNING Antifoam AF Emulsion:
Emulsifier Type: Nonionic
Color: White
Percent Silicone: 30
pH: 3.5
Suitable Diluent: Water

DOW CORNING Antifoam 1520-US Emulsion:
Emulsifier Type: Nonionic
Color: White
Percent Silicone: 30
pH: 3.5
Suitable Diluent: Water

DRAGOCO, INC.: DRAGOCO EXTRAPONES:

EXTRAPONE Alpine Herbs Special:
Recommended for Use In:
Skin Products
Hair Products
Number: 2/789350
CTFA Adopted Name: TEA-Dodecylbenzene Sulfonate (and) Ethoxydiglycol (and) Propylene Glycol (and) Butylene Glycol (and) Gentian Extract (and) Hayflower Extract (and) Hypericum Extract (and) Balm Mint Extract (and) Juniper Extract (and) Pine Needle Extract (and) Arnica Extract.

EXTRAPONE Arkin Special:
Recommended for Use In:
Hair Products
Number: 2/789360
CTFA Adopted Name: Ethoxydiglycol (and) Propylene Glycol (and) Butylene Glycol (and) Matricaria Extract (and) Nettle Extract (and) Birch Sap Extract (and) Arnica Extract (and) Cinchona Extract (and) Birch Leaf Extract.

EXTRAPONE Biopollin Special:
Recommended for Use In:
Skin Products
Number: 2/789370
CTFA Adopted Name: Ethoxydiglycol (and) Propylene Glycol (and) Butylene Glycol (and) Witch Hazel Extract (and) Wheat Germ Extract (and) Calcium Pantothenate (and) Vitamin A (and) Boric Acid (and) Niacinamide (and) Inositol (and) Thiamine HCL (and) Biotin (and) Methionine (and) Folic Acid (and) Ascorbic Acid (and) Pyridoxine HCL (and) Colloidal Sulfur (and) Allantoin.

EXTRAPONE Witchhazel Special:
Recommended for Use In:
Hair Products
Skin Products
Number: 2/789420
CTFA Adopted Name: Propylene Glycol (and) Witch Hazel Extract.

EXTRAPON Chamomile Special:
Recommended for Use In:
Skin Products
Hair Products
Number: 2/789450
CTFA Adopted Name: Ethoxydiglycol (and) Propylene Glycol (and) Butylene Glycol (and) Matricaria Extract.

DRAGOCO, INC.: DRAGOCO EXTRAPONES(Continued):

EXTRAPONE Phytostimulin Special:

Recommended for Use In:

Skin Products

Number: 2/789440

CTFA Adopted Name: Ethoxydiglycol (and) Propylene Glycol (and) Butylene Glycol (and) Witch Hazel Extract (and) Wheat Germ Extract (and) Calcium Pantothenate (and) Vitamin A (and) Hops Extract (and) Allantoin (and) Boric Acid (and) Calendula Extract (and) Arnica Extract (and) Inositol (and) Niacinamide (and) Thiamine HCL (and) Colloidal Sulfur (and) Pyridoxine HCL (and) Ascorbic Acid (and) Biotin (and) Methionine (and) Folic Acid.

EXTRAPONE #3 Special:

Recommended for Use In:

Hair Products
Skin Products

Number: 2/789490

CTFA Adopted Name: Ethoxydiglycol (and) Propylene Glycol (and) Butylene Glycol (and) Sage Extract (and) Yarrow Extract (and) Coltsfoot Extract (and) Matricaria Extract (and) Balm Mint Extract (and) Wild Thyme Extract (and) Horsetail Extract (and) Rosemary Extract (and) Althea Extract.

EXTRAPONE #4 Special:

Recommended for Use In:

Skin Products

Number: 2/788400

CTFA Adopted Name: Ethoxydiglycol (and) Propylene Glycol (and) Butylene Glycol (and) Matricaria Extract (and) Sage Extract (and) Coltsfoot Extract (and) Restharrow Extract (and) Horsetail Extract (and) Yarrow Extract (and) Rosemary Extract (and) Dandelion Extract (and) Althea Extract (and) Witch Hazel Extract.

EXTRAPONE #5 Special:

Recommended for Use In:

Hair Products

Number: 2/789500

CTFA Adopted Name: Ethoxydiglycol (and) Propylene Glycol (and) Butylene Glycol (and) Nettle Extract (and) Yarrow Extract (and) Coltsfoot Extract (and) Rosemary Extract (and) Birch Sap (and) Clover Blossom Extract (and) Birch Leaf Extract (and) Horsetail Extract (and) Sage Extact (and) Calcium Pantothenate (and) Inositol.

DRAGOCO, INC.: DRAGOSANTOL:

Product Number: 2/012681

DRAGOSANTOL is a fluid which contains at least 85% of d,1-α-bisabolol. Ingredients of secondary importance are mainly farnesol isomers.

d-1-α-bisabolol is a monocyclic unsaturated sesquiterpene alcohol.

Empirical formula: C15H26O
Molecular weight: 222.36

Directions for use:

DRAGOSANTOL has been shown to possess anti-inflammatory properties. Moreover, bacteriostatic effectiveness has been demonstrated both for d,1-α-bisabolol and DRAGOSANTOL.

Physico-chemical Properties:

Consistency: viscous liquid
Colour: colourless to straw yellow
Odour: slight inherent odour
Density D 20/4: 0.922 to 0.928
Refractive Index n 20/D: 1.492 to 1.498

Soluble in lower alcohols (ethanol, isopropanol), in fatty alcohols, glycerin esters and paraffin.

Practically insoluble in water and glycerin. Aqueous solutions can be prepared by adding solubilizers at the rate 1:4.

The use of DRAGOSANTOL is dermatogically safe.

Aqueous solutions can be prepared by means of solubilizers at the rate 1:4.

Applications and dosage:

DRAGOSANTOL is suitable for use in cosmetic products due to its stability and good compatibility with the skin. Unlike azulene or camomile oil, DRAGOSANTOL does not alter its colour after long storage periods nor diffuse through plastic containers and thus can be used without any problems.

The main areas of application for DRAGOSANTOL are skin protective and skin-care cosmetic preparations, in particular ointments, creams or lotions for sensitive skin. DRAGOSANTOL is also suitable for sunscreen products, after sun preparations, baby care and after-shave formulations. Furthermore, DRAGOSANTOL is recommended for oral hygiene preparations such as toothpastes or mouthwashes.

FDA-CRMCS-NO.: R 00 12527
CTFA-Adopted Name: Bisabolol

DUPONT: Chemicals and Pigments:

ALBONE 35, 50, 70 CG Hydrogen Peroxide:

Available as aqueous solutions in concentrations of 35%, 50%, and 70%

Properties:
- Clear, colorless liquid
- Slightly sharp odor
- Miscible in water in all proportions
- Liberates oxygen when decomposed

Uses:

Special grade for preparing and bottling dilute solutions for cosmetic use (hair coloring and bleaching), pharmaceuticals, reagent grade, and liquid laundry bleach

When properly diluted to 3% and 6% with deionized water, it produces stable dilute hydrogen peroxide solutions that meet USP specifications without needing added stabilizers

AVITEX Surface Active Agents/Softeners/Static Control Agents:

Sodium salts of sulfated oleyl acetates

Properties:
- Modified fatty alcohol sulfate sodium salts, cationic softeners, antistatic agents, and bath stabilizers
- A reddish-brown liquid (50% A.I.)
- Does not inhibit moisture absorption
- Durable static protection

Uses:
- Textile softener
- Dyeing assistant for acid and direct dyes on paper
- Penetrating and leveling agent for textile dyeing
- Beam and packaging dyeing of cottons
- In the preparation of hair grooming aids
- Alone, or in combination, with bodying and stiffening agents and thermosetting resins to reduce the development of static electricity
- To increase the pliability and lubricity of natural and synthetic fibers and fabrics
- Bath stabilizer for aqueous application baths
- May be co-applied with TEFLON MF Carpet Protector

Types:
- AVITEX DN Softener
- AVITEX E Static Control Agent
- AVITEX ML Softener
- AVITEX NA Softener
- AVITEX R Softener
- AVITEX 2153 Surface Active Agent

DUPONT: Chemicals and Pigments(Continued):

DUPONOL C Surface Active Agents:

Sodium lauryl sulfate

NF Grade--Anionic, white to cream-colored powder (90% to 96% A.I.)

Uses:
As an emulsifying, foaming, and dispersing agent for the cosmetic, dental, and medical preparation industries

DUPONOL EP Surface Active Agent:

Diethanolamine salt of lauryl sulfate

Properties:
- Anionic, pale yellow, slightly viscous paste (33% to 36% A.I.)

Uses:
- Particularly suitable for shampoos and cosmetics
- Excellent solubility, color stability, and foaming properties
- As a detergent, wetting, dispersing, and emulsifying agent

DUPONOL QC Surface Active Agent:

Sodium lauryl sulfate

Properties:
- An anionic, slightly viscous, almost colorless liquid (29% to 31% A.I.)

Uses:
- Specifically designed for shampoo industry use in all three major types of shampoos (clear liquid, cream lotion, and paste)

DUPONOL WA Paste Surface Active Agent:

Sodium lauryl sulfate

Properties:
- An anionic, pale yellow viscous liquid (29% to 31% A.I.)

Uses:
- In liquid cream-lotion and solid cream or paste shampoos. Good foaming, desired body, and controlled degreasing action
- With cosmetics, metal processing, leather, building materials, paper, pigments, and cleaning processes

DUPONT: Chemicals & Pigments(Continued):

DUPONOL WAQ Surface Active Agent:

Sodium lauryl sulfate

Properties:
- An anionic, pale yellow, noticeably viscous liquid (28% to 30% A.I.)

Uses:
- Lower viscosity form of DUPONOL WA
- Detergent, wetting, foaming, dispersing, and emulsifying agent
- For cosmetics, metal processing, leather, building materials, paper, pigments, and cleaning processes

DUPONOL XL Surface Active Agent:
Amine salt of lauryl sulfate plus amphoteric salt.

Properties:
- An anionic, yellow, moderately viscous liquid (36% A.I.)

Uses:
- As a shampoo base for high quality, clear liquid, and gel shampoos

OXONE Monopersulfate Compound:
A nonchlorine oxidizer used in a variety of cleaning applications.

Properties:
- An acidic, white, granular, free-flowing solid containing the active ingredient potassium peroxymonosulfate
- Readily soluble in water
- 1% solution has pH of 2.3
- Minimum active oxygen content 4.5%
- Strong oxidizing agent

Uses:
- Denture cleaners
- Hair wave neutralizers
- Bleaching nylon textiles
- Etching electronic circuit boards
- Swimming pool shocking agent
- Synthesis of specialty chemicals
- Dry bleaches, detergent-bleach washing compounds, scouring powders

EASTMAN CHEMICAL PRODUCTS, INC.: EASTMAN Distilled Monoglycerides:

High-Purity Products for Specialty Applications

Type:

18-00:
Fat Source: Hydrogenated lard or tallow
Monoester Content, Minimum %: 90
Glycerol Content, Maximum %: 1.2
Acid Value, Maximum: 3
Iodine Value: 5 max
Specific Gravity at 80C: 0.91
Melting Point, Approx. C (F): 68 (154)
Physical Form: Small beads

18-04K:
Fat Source: Hydrogenated palm
Monoester Content, Minimum %: 90
Glycerol Content, Maximum %: 1.2
Acid Value, Maximum: 3
Iodine Value: 5 max
Specific Gravity at 80C: 0.94
Melting Point, Approx. C (F): 66 (151)
Physical Form: Small beads

18-06(K):
Fat Source: Hydrogenated soybean oil
Monoester Content, Minimum %: 90
Glycerol Content, Maximum %: 1.2
Acid Value, Maximum: 3
Iodine Value: 5 max
Specific Gravity at 80C: 0.92
Melting Point, Approx. C(F): 69 (156)
Physical Form: Small beads

18-07K:
Fat Source: Hydrogenated vegetable oil
Monoester Content, Minimum %: 90
Glycerol Content, Maximum %: 1.2
Acid Value, Maximum: 3
Iodine Value: 5 max
Specific Gravity at 80C: 0.92
Melting Point, Approx. C(F): 68 (154)
Physical Form: Small beads

EASTMAN CHEMICAL PRODUCTS, INC.: EASTMAN Distilled Monoglycerides(Continued):

MYVEROL Distilled Monoglycerides(Continued):

Type:

18-30:
Fat Source: Edible beef tallow
Monoester Content, Minimum %: 90
Glycerol Content, Maximum %: 1.2
Acid Value, Maximum: 3
Iodine Value: 27-40
Specific Gravity at 80C: 0.92
Melting Point, Approx. C(F): 60 (140)
Physical Form: Plastic

18-35K:
Fat Source: Refined palm
Monoester Content, Minimum %: 90
Glycerol Content, Maximum %: 1.2
Acid Value, Maximum: 3
Iodine Value: 36-45
Specific Gravity at 80C: 0.94
Melting Point, Approx. C(F): 60 (140)
Physical Form: Plastic

18-40:
Fat Source: Edible lard
Monoester Content, Minimum %: 90
Glycerol Content, Maximum %: 1.2
Acid Value, Maximum: 3
Iodine Value: 43-55
Specific Gravity at 80C: 0.92
Melting Point, Approx. C(F): 58 (136)
Physical Form: Plastic

18-50K:
Fat Source: Partially hydrogenated soybean oil
Monoester Content, Minimum %: 90
Glycerol Content, Maximum %: 1.2
Acid Value, Maximum: 3
Iodine Value: 50-60
Specific Gravity at 80C: 0.94
Melting Point, Approx. C(F): 54 (129)
Physical Form: Plastic

EASTMAN CHEMICAL PRODUCTS, INC.: EASTMAN Distilled Monoglycerides(Continued):

MYVEROL Distilled Monoglycerides(Continued):

Type:

18-85K:
Fat Source: Cottonseed oil
Monoester Content, Minimum %: 90
Glycerol Content, Maximum %: 1.2
Acid Value, Maximum: 3
Iodine Value, Maximum: 85-95
Specific Gravity at 80C: 0.95
Melting Point, Approx. C(F): 46 (115)
Physical Form: Plastic

18-92(K):
Fat Source: Sunflower oil
Monoester Content, Minimum %: 90
Glycerol Content, Maximum %: 1.2
Acid Value, Maximum: 3
Iodine Value: 105-115
Specific Gravity at 80C: 0.90
Melting Point, Approx. C(F): 41 (106)
Physical Form: Semiplastic

18-99K:
Fat Source: Low erucic rapeseed oil
Monoester Content, Minimum %: 90
Glycerol Content, Maximum %: 1.2
Acid Value, Maximum: 3
Iodine Value: 90-95
Specific Gravity at 80C: 0.93
Melting Point, Approx. C(F): 35 (94)
Physical Form: Semiplastic

MYVEROL Distilled Propylene Glycol Monoester:

P-06(K):
Fat Source: Hydrogenated soybean oil
Monoester Content, Minimum %: 90
Glycerol Content, Maximum %: 1.2
Acid Value, Maximum: 3
Iodine Value: 5 max.
Specific Gravity at 80C: 0.89
Melting Point, Approx. C(F): 45 (113)
Physical Form: Beads

EASTMAN CHEMICAL PRODUCTS, INC.: EASTMAN Distilled Monoglycerides(Continued):

MYVEROL Succinylated Monoglyceride:

Type:

SMG-VK:
Fat Source: Hydrogenated palm
Monoester Content, Minimum: 12-20
Glycerol Content, Maximum %: 55
Acid Value, Maximum: 70-120
Iodine Value: 3 max
Specific Gravity @ 80C: 3
Melting Point, Approx. C(F): 58(136)
Physical Form: Beads

MYVAPLEX Concentrated Glyceryl Monostearates:

600(K):
Fat Source: Hydrogenated soybean oil
Monoester Content, Minimum %: 90
Glycerol Content, Maximum %: 1.2
Acid Value, Maximum: 3
Iodine Value: 5 max
Specific Gravity at 80C: 0.92
Melting Point, Approx. C(F): 69 (156)
Physical Form: Small beads

600P(K):
Fat Source: Hydrogenated soybean oil
Monoester Content, Minimum %: 90
Glycerol Content, Maximum %: 1.2
Acid Value, Maximum: 3
Iodine Value: 5 max
Specific Gravity at 80C: 0.92
Melting Point, Approx. C(F): 69 (156)
Physical Form: Powder

EASTMAN CHEMICAL PRODUCTS, INC.: EASTMAN Distilled Monoglycerides(Continued):

MYVACET Distilled Acetylated Monoglycerides:

Type:

5-07(K):
Fat Source: Hydrogenated vegetable oil
Acetylation %: 48.5-51.5
Melting Point, Approx. C (F): 41-46 (105-114)
Specific Gravity at 80C: 0.94
Physical Form: Waxy solid

7-00:
Fat Source: Hydrogenated lard
Acetylation %: 66.5-69.5
Melting Point, Approx. C (F): 37-40 (99-104)
Specific Gravity at 80C: 0.94
Physical Form: Waxy solid

7-07K:
Fat Source: Hydrogenated vegetable oil
Acetylation %: 66.5-69.5
Melting Point, Approx. C (F): 37-40 (99-104)
Specific Gravity at 80C: 0.94
Physical Form: Waxy solid

9-08K:
Fat Source: Hydrogenated coconut oil
Acetylation %: 96 min
Melting Point, Approx. C (F): 4-12 (40-54)
Specific Gravity at 80C: 0.94
Physical Form: Liquid

9-40:
Fat Source: Edible lard
Acetylation %: 96 min
Melting Point, Approx. C (F): 4-12 (40-54)
Specific Gravity at 80C: 0.94
Physical Form: Liquid

9-45K:
Fat Source: Partially hydrogenated soybean oil
Acetylation %: 96 min
Melting Point, Approx. F (C): 4-12 (40-54)
Specific Gravity at 80C: 0.94
Physical Form: Liquid

EASTMAN CHEMICAL PRODUCTS, INC.: EASTMAN Distilled Monoglycerides(Continued):

MYVATEX Emulsifiers:

Type:

3-50K:
Fat Source(s): Soybean oil
Composition: Distilled monoglycerides and distilled propylene glycol monoesters
Monoester Content, Minimum %: 90
Acid Value: <3
Iodine Value: 5 max
Specific Gravity at 80C: 0.91
Melting Point, Approx. C (F): 58 (136)
Physical Form: Beads

8-06K:
Fat Source(s): Soybean oil, sunflower oil
Composition: Distilled monoglycerides with 20% hydrogenated soybean oil
Monoester Content, Minimum %: 72
Acid Value: <3
Iodine Value: 24-30
Specific Gravity at 80C: 0.93
Melting Point, Approx. C (F): 67 (153)
Physical Form: Beads

8-20:
Fat Source(s): Lard, tallow, soybean oil
Composition: Distilled monoglycerides with 20% hydrogenated soybean oil
Monoester Content, Minimum %: 72
Acid Value: <3
Iodine Value: 19-24
Specific Gravity at 80C: 0.93
Melting Point, Approx. C (F): 58 (136)
Physical Form: Beads

8-20E:
Fat Source: Lard, tallow, sunflower oil, soybean oil
Composition: Distilled monoglycerides with 20% hydrogenated soybean oil
Monoester Content, Minimum %: 72
Acid Value: <3
Iodine Value: 24-30
Specific Gravity at 80C: 0.93
Melting Point, Approx. C (F): 57 (135)
Physical Form: Beads or powder

EASTMAN CHEMICAL PRODUCTS, INC.: EASTMAN Distilled Mono-Glycerides(Continued):

MYVATEX Emulsifiers:

Type:

Peanut butter stabilizer K:
Fat Source(s): Palm oil
Composition: Distilled monoglycerides with 50% hydrogenated palm oil
Monoester Content, Minimum %: 45
Acid Value: <6
Iodine Value: 5 max
Specific Gravity at 80C: 0.93
Melting Point, Approx. C (F): 61 (142)
Physical Form: Beads

7-85K:
Fat Source(s): Cottonseed oil
Composition: Distilled monoglycerides with 30% cottonseed oil
Monoester Content, Minimum %: 63
Acid Value: <3
Iodine Value: 91-101
Specific Gravity at 80C: 0.94
Melting Point, Approx. C (F): 49 (120)
Physical Form: Plastic

MIGHTY SOFT emulsifier:
Fat Source(s): Soybean oil
Composition: Distilled monoglyceride
Monoester Content, Minimum %: 90
Acid Value: <3
Iodine Value: 19-36
Specific Gravity at 80C: 0.94
Melting Point, Approx. C (F): 67 (153)
Physical Form: Powder

DO CONTROL strengthener (K):
Fat Source(s): Palm oil
Composition: Distilled succinylated monoglycerides and distilled monoglycerides
Monoester Content, Minimum %: 41
Acid Value: >74
Iodine Value: 4-7
Specific Gravity at 80C: 0.94
Melting Point, Approc. F (C): 53 (127)
Physical Form: Powder

EASTMAN CHEMICAL PRODUCTS, INC.: EASTMAN Distilled Monoglycerides(Continued):

MYVATEX Emulsifiers(Continued):

Type:

TEXTURE LITE emulsifier (K):
Fat Source(s): Soybean oil
Composition: Distilled monoglycerides, distilled propylene glycol monoesters, sodium stearoyl lactylate with silicon dioxide
Monoester Content, Minimum %: 80
Acid Value: <13
Iodine Value: 3 max
Specific Gravity at 80C: 0.97
Physical Form: Powder

LIQUID LITE emulsifier:
Fat Source(s): Soybean oil
Composition: Distilled propylene glycol monoester and distilled acetylated monoglyceride
Monoester Content, Minimum %: 80
Acid Value: <3
Iodine Value: 3 max
Specific Gravity at 80C: 0.97
Melting Point, Approx. C (F): 44-48 (111-118)
Physical Form: Beads

MONOSET emulsifier K:
Fat Source(s): Rapeseed oil, cottonseed oil
Composition: Distilled monoglyceride, rapeseed and cottonseed oils
Monoester Content, Minimum %: 18
Acid Value: <4
Iodine Value: 5 max
Melting Point, Approx. C (F): 63 (146)
Physical Form: Beads

40-06S:
Fat Source(s): Soybean oil
Composition: Distilled propylene glycol monoesters, distilled monoglycerides, and lactylic esters of fatty acids (stearic), water, and potassium sorbate
Monoester Content, Minimum %: 25
Acid Value: <13
Iodine Value: 1 max
Physical form: Soft plastic

EASTMAN CHEMICAL PRODUCTS, INC.: EASTMAN Distilled Monoglycerides(Continued):

MYVATEX Emulsifiers(Continued):

Type:

25-07(K):
Fat Source(s): Soybean oil
Composition: Distilled monoglyceride, lecithin, water, propionic acid, and sodium propionate
Monoester Content, Minimum %: 25
Acid Value: <7.3
Iodine Value: 1 max
Physical Form: Soft plastic

SMG 30K **emulsifier:**
Fat source(s): Vegetable oil
Composition: Distilled succinylated monoglyceride, distilled monoglycerides, lecithin, water, and propionic acid
Monoester Content, Minimum %: 30
Acid Value: 30-36
Iodine Value: 1 max
Physical Form: Soft plastic

SSH **emulsifier:**
Fat source(s): Soybean oil
Composition: Distilled monoglycerides, lecithin, water, and propionic acid
Monoester Content, Minimum %: 45
Acid Value: 7.3
Iodine Value: 6 min
Physical Form: Soft plastic

MYVATEM Dispersing Agents:

Type:

06(K):
Fat Source: Hydrogenated soybean oil
Physical Form: Solid
Melting Point, Approx. C (F): 47 (117)

30:
Fat Source: Edible tallow
Physical Form: Semisolid
Melting Point, Approx. C (F): 33 (91)

35K:
Fat Source: Refined palm oil
Physical Form: Semisolid
Melting Point, Approx. C (F): 26 (79)

92K:
Fat Source: Refined sunflower oil
Physical Form: Liquid
Melting Point, Approx. C (F): <0 (<32)

EASTMAN CHEMICAL PRODUCTS, INC.: Products for Cosmetics and Personal Care:

Alcohols:

Ethanol:
CTFA: Alcohol
CAS/RD: 64-17-5

Ethyl Alcohol, USP Undenatured(Anhydrous and 95%):

SDA Formulas:
(Anhydrous and 95%):

SDA-1:
CTFA: SD Alcohol 1
CAS/RD: 977021-58-1

SDA-2B:
CTFA: SD Alcohol 2-B

SDA-3A:
CTFA: SD Alcohol 3-A
CAS/RD: 977021-59-2

SDA-3C:
CTFA: SD Alcohol 3-C

SDA-4:
CTFA: SD Alcohol 4

SDA-12A:
CTFA: SD Alcohol 12-A
CAS/RD: 977021-62-7

SDA-23A:
CTFA: SD Alcohol 23-A
CAS/RD: 977021-64-9

SDA-23H:
CTFA: SD Alcohol 23-H
CAS/RD: 977021-66-1

SDA-29-2:
CTFA: SD Alcohol 29-2

SDA-29-3:
CTFA: SD Alcohol 29-3

SDA-29-5:
CTFA: SD Alcohol 29-5

EASTMAN CHEMICAL PRODUCTS, INC.: Products for Cosmetics and Personal Care(Continued):

Alcohols(Continued):

SDA-29-6:
CTFA: SD Alcohol 29-6

SDA-29-8:
CTFA: SD Alcohol 29-8

SDA-29-9:
CTFA: SD Alcohol 29-9

SDA-30:
CTFA: SD Alcohol 30
CAS/RD: 977021-68-3

SDA-32:
CTFA: SD Alcohol 32

SDA-35A:
CTFA: SD Alcohol 35-A

SDA-36:
CTFA: SD Alcohol 36
CAS/RD: 977021-70-7

SDA-39B:
CTFA: SD Alcohol 39-B
CAS/RD: 977021-78-5

SDA-39C:
CTFA: SD Alcohol 39-C
CAS/RD: 977021-79-6

SDA-40:
CTFA: SD Alcohol 40
CAS/RD: 977021-81-0

SDA-40A:
CTFA: SD Alcohol 40-A
CAS/RD: 977021-82-1

SDA-40B:
CTFA: SD Alcohol 40-B
CAS/RD: 977021-83-2

SDA-40C:
CTFA: SD Alcohol 40-C
CAS/RD: 977021-84-3

EASTMAN CHEMICAL PRODUCTS, INC.: Products for Cosmetics and Personal Care(Continued):

Alcohols(Continued):

Glycols:

CHDM:
CTFA: 1,4-Cyclohexanedimethanol
CAS/RD: 105-08-8

DEG:
CTFA: Diethylene Glycol
CAS/RD: 111-46-6

DMCD:
CTFA: Dimethyl 1,4-Cyclohexanedicarboxylate
CAS/RD: 94-60-0

EG:
CTFA: Glycol
CAS/RD: 107-21-1

NPG Glycol:
CTFA: Neopentyl Glycol
CAS/RD: 126-30-7

TEG:
CTFA: Triethylene Glycol
CAS/RD: 112-27-6

TMPD Glycol:
CTFA: 2,2,4-Trimethyl-1,3-Pentanediol
CAS/RD: 144-19-4

Oxo Alcohols:

n-Butanol:
CTFA: n-Butyl Alcohol
CAS/RD: 71-36-3

Isobutanol:
CTFA: Isobutyl Alcohol
CAS/RD: 78-83-1

Propanol:
CTFA: n-Propyl Alcohol
CAS/RD: 71-23-8

EASTMAN CHEMICAL PRODUCTS, INC.: Products for Cosmetics and Personal Care(Continued):

Antioxidants:

2,5-DTBHQ:
CTFA: Di-t-Butylhydroquinone
CAS/RD: 88-58-4

TENOX BHA:
CTFA: BHA
CAS/RD: 25013-16-5

TENOX GT-1 and GT-2:
CTFA: Natural Tocopherols
CAS/RD: N/A

TENOX PG:
CTFA: Propyl Gallate
CAS/RD: 121-79-9

TENOX TBHQ:
CTFA: t-Butyl Hydroquinone
CAS/RD: 1948-33-0

TENOX 2:
CTFA: Propylene Glycol, BHA, Propyl Gallate, Citric Acid
CAS/RD: N/A

TENOX 4:
CTFA: Corn Oil, BHA, BHT
CAS/RD: N/A

TENOX 6:
CTFA: Corn Oil, Glyceryl Oleate, Propylene Glycol, BHA, BHT, Propyl Gallate, Citric Acid
CAS/RD: N/A

TENOX 20:
CTFA: Propylene Glycol, t-Butyl Hydoquinone, Citric Acid
CAS/RD: N/A

TENOX 26:
CTFA: Corn Oil, Glyceryl Oleate, Propylene Glycol, BHA, BHT, t-Butyl Hydroquinone, Citric Acid
CAS/RD: N/A

EASTMAN CHEMICAL PRODUCTS, INC.: Products for Cosmetics and Personal Care(Continued):

Chemical Intermediates:

Acetic Acid:
CTFA: Acetic Acid
CAS/RD: 64-19-7

Acetic Anhydride:
CTFA: Acetic Anhydride
CAS/RD: 108-24-7

n-Butyraldehyde:
CTFA: Butanal
CAS/RD: 123-72-8

Ethyl Acetoacetate:
CTFA: EAA
CAS/RD: 141-97-9

Isobutyraldehyde:
CTFA: 2-Methylpropanal
CAS/RD: 78-84-2

Isobutyric Acid:
CTFA: Isobutyric Acid
CAS/RD: 79-31-2

Methyl Acetoacetate:
CTFA: MAA
CAS/RD: 105-45-3

Propionaldehyde:
CTFA: Propanal
CAS/RD: 123-38-6

EASTMAN CHEMICAL PRODUCTS, INC.: Products for Cosmetics and Personal Care(Continued):

Emulsifiers:

Concentrated Glyceryl Monostearate:

MYVAPLEX 600P (K):
CTFA: Hydrogenated Soy Glycerides
CAS/RD Number: 61789-08-0

Dispersing Agents:

MYVATEM 06K:
CTFA: Diactyl Tartaric Acid Ester of Distilled Hydrogenated Soy Glycerides
CAS/RD Number: N/A

MYVATEM 30:
CTFA: Diacetyl Tartaric Acid Ester of Distilled Tallow Glycerides
CAS/RD Number: N/A

MYVATEM 35K:
CTFA: Diacetyl Tartaric Acid Ester of Distilled Palm Glycerides
CAS/RD Number: N/A

MYVATEM 92K:
CTFA: Diacetyl Tartaric Acid Ester of Distilled Sunflower Glycerides
CAS/RD Number: N/A

Distilled Acetylated Monoglycerides:

MYVACET 5-07(K):
CTFA: Acetylated Hydrogenated Cottonseed Glyceride
CAS/RD Number: 977055-83-6

MYVACET 7-00:
CTFA: Acetylated Hydrogenated Lard Glyceride
CAS/RD Number: 8029-91-2

MYVACET 7-07K:
CTFA: Acetylated Monoglycerides
CAS/RD Number: N/A

MYVACET 9-08K:
CTFA: Acetylated Monoglycerides
CAS/RD Number: N/A

MYVACET 9-40:
CTFA: Acetylated Lard Glycerides
CAS/RD Number: 8029-92-3

MYVACET 9-45K:
CTFA: Acetylated Vegetable Glycerides
CAS/RD Number: N/A

EASTMAN CHEMICAL PRODUCTS, INC.: Products for Cosmetics and Personal Care(Continued):

Emulsifiers(Continued):

Distilled Monoglycerides:

MYVEROL 18-00:
CTFA: Hydrogenated Animal Glyceride
CAS/RD Number: 977036-91-1

MYVEROL 18-04K:
CTFA: Hydrogenated Palm Oil Glyceride
CAS/RD Number: 67784-87-6

MYVEROL 18-06(K):
CTFA: Hydrogenated Soy Glyceride
CAS/RD Number: 61789-08-0

MYVEROL 18-07K:
CTFA: Hydrogenated Cottonseed Glyceride
CAS/RD Number: 61789-07-9

MYVEROL 18-30:
CTFA: Tallow Glyceride
CAS/RD Number: 61789-13-7

MYVEROL 18-35K:
CTFA: Palm Oil Glyceride
CAS/RD Number: 977013-38-9

MYVEROL 18-40:
CTFA: Lard Glyceride
CAS/RD Number: 61789-10-4

MYVEROL 18-50K:
CTFA: Hydrogenated Vegetable Glyceride
CAS/RD Number: 69028-36-0

MYVEROL 18-85K:
CTFA: Cottonseed Glyceride
CAS/RD Number: 8029-44-5

MYVEROL 18-92(K):
CTFA: Sunflower Seed Oil Glyceride

MYVEROL 18-99K:
CTFA: Canola Monoglyeride
CAS/RD Number: N/A

Distilled Propylene Glycol Monoester:

MYVEROL P-06(K):
CTFA: Propylene Glycol Stearate
CAS/RD Number: 1323-39-3

EASTMAN CHEMICAL PRODUCTS, INC.: Products for Cosmetics and Personal Care(Continued):

Emulsifiers(Continued):

Emulsifier Blends:

MYVATEX TEXTURE LITE Emulsifier:
CTFA: Glyceryl Monostearate, Propylene Glycol Monostearate, Sodium Stearoyl Lactylate
CAS/RD Number: N/A

MYVATEX Emulsifiers:
CTFA: Blends of Distilled Monoglycerides with one or more of the following: Distilled Monoglycerides, Water, Triglycerides, Propylene Glycol Monostearate
CAS/RD Number: N/A

Succinylated Monoglycerides:

MYVEROL SMG VK:
CTFA: Distilled Succinylated Monoglyerides
CAS/RD Number: N/A

Plasticizers:

KODAFLEX DBP Plasticizer:
CTFA: Dibutyl Phthalate
CAS/RD Number: 84-74-2

KODAFLEX DEP Plasticizer:
CTFA: Diethyl Phthalate
CAS/RD Number: 84-66-2

KODAFLEX DMP Plasticizer:
CTFA: Dimethyl Phthalate
CAS/RD Number: 131-11-3

KODAFLEX DOA Plasticizer:
CTFA: Dioctyl Adipate
CAS/RD Number: 103-23-1

KODAFLEX DOTP Plasticizer:
CTFA: Dioctyl Terephthalate
CAS/RD Number: 422-86-2

KODAFLEX TOTM Plasticizer:
CTFA: Trioctyl Trimellitate
CAS/RD Number: 3319-31-1

KODAFLEX TXIB Plasticizer:
CTFA: 2,2,4-Trimethyl-1,2-Pentanediol, Diisobutyrate
CAS/RD Number: 6846-50-0

Triacetin USP:
CTFA: Tracetin
CAS/RD Number: 102-76-1

EASTMAN CHEMICAL PRODUCTS, INC.: Products for Cosmetics and Personal Care(Continued):

Polyethylene Waxes:

EPOLENE C Waxes:
(Nonemulsifiable, medium molecular weight)
CTFA: Polyethylene
CAS/RD Number: 9002-88-4

EPOLENE N Waxes:
(Nonemulsifiable, low molecular weight)
CTFA: Polyethylene
CAS/RD Number: 9002-88-4

EPOLENE E Waxes:
(Emulsifiable)
CTFA: Polyethylene
CAS/RD Number: 12616-32-9

Polymers:

Cellulose Esters:
(Available in various grades)

Cellulose Acetate:
CTFA: Cellulose Acetate
CAS/RD Number: 9004-35-7

Cellulose Acetate Butyrate:
CTFA: Cellulose Acetate Butyrate
CAS/RD Number: 9004-36-8

Cellulose Acetate Propionate:
CTFA: Cellulose Acetate Propionate
CAS/RD Number: 9004-39-1

Water-Dispersible Polymers:
(Available in solid and dispersed forms)

EASTMAN AQ 29 Polymer:
CTFA: Diglycol Isophthalates Sulfoisophthalates Copolymer

EASTMAN AQ 38 & 55 Polymers:
CTFA: Diglycol Cyclohexanedimethanol Isophthalates Sulfoisophthalates Copolymer

EASTMAN CHEMICAL PRODUCTS, INC.: Products for Cosmetics and Personal Care(Continued):

Solvents:

Esters:

n-Butyl Acetate:
CTFA: Butyl Acetate
CAS/RD Number: 123-86-4

Ethyl Acetate:
CTFA: Ethyl Acetate
CAS/RD Number: 141-78-6

2-Ethylhexyl Acetate:
CTFA: 2-Ethylhexyl Acetate
CAS/RD Number: 103-09-3

Isobutyl Acetate:
CTFA: Isobutyl Acetate
CAS/RD Number: 110-19-0

Isobutyl Isobutyrate:
CTFA: IBIB
CAS/RD Number: 97-85-8

Isopropyl Acetate:
CTFA: Isopropyl Acetate
CAS/RD Number: 108-21-4

n-Propyl Acetate:
CTFA: Propyl Acetate
CAS/RD Number: 109-60-4

Glycol Ethers:

EE Solvent:
CTFA: Ethoxyethanol
CAS/RD Number: 110-80-5

EKTASOLVE DB Solvent:
CTFA: Butoxydiglycol
CAS/RD Number: 112-34-5

EKTASOLVE DE Solvent:
CTFA: Ethoxydiglycol
CAS/RD Number: 111-90-0

EASTMAN CHEMICAL PRODUCTS, INC.: Products for Cosmetics and Personal Care(Continued):

Solvents(Continued):

Glycol Ethers(Continued):

EKTASOLVE DM Solvent:
CTFA: Methoxydiglycol
CAS/RD Number: 111-77-3

EKTASOLVE DP Solvent:
CTFA: Diethylene Glycol Monopropyl Ether
CAS/RD Number: 6881-94-3

EKTASOLVE EEH Solvent:
CTFA: Ethylene Glycol 2-Ethylhexyl Ether
CAS/RD Number: 1559-35-9

EKTASOLVE EP Solvent:
CTFA: Ethylene Glycol Monopropyl Ether
CAS/RD Number: 2807-30-9

Glycol Ether Esters:

EE Acetate:
CTFA: Ethoxyethanol Acetate
CAS/RD Number: 111-15-9

EKTASOLVE DB Acetate:
CTFA: Diethylene Glycol Monobutyl Ether Acetate
CAS/RD Number: 124-17-4

EKTASOLVE DE Acetate:
CTFA: Ethoxydiglycol Acetate
CAS/RD Number: 112-15-2

EKTASOLVE PM Acetate:
CTFA: Propylene Glycol Monomethyl Ether Acetate
CAS/RD Number: 108-65-6

Ketones:

Diisobutyl Ketone:
CTFA: DIBK
CAS/RD Number: 108-83-8

Methyl n-Amyl Ketone:
CTFA: MAK
CAS/RD Number: 110-43-0

EASTMAN CHEMICAL PRODUCTS, INC.: Products for Cosmetics and Personal Care(Continued):

Solvents:

Ketones(Continued):

Methyl Isoamyl Ketone:
CTFA: MIAK
CAS/RD Number: 110-12-3

Methyl n-Propyl Ketone:
CTFA: MPK
CAS/RD Number: 107-87-9

Methyl Isobutyl Ketone:
CTFA: MIBK
CAS/RD Number: 108-10-1

Specialty Chemicals:

SAIB:
CTFA: Sucrose Acetate Isobutyrate
CAS/RD Number: 126-13-6

TECQUINOL Hydroquinone:
CTFA: Hydroquinone
CAS/RD Number: 123-31-9

EASTMAN CHEMICAL PRODUCTS, INC.: Products for Cosmetics and Personal Care(Continued):

Vitamin E (Natural Source):

Mixed Tocopherols:
EASTMAN Vitamin E 4-50
EASTMAN Vitamin E 4-80
CTFA: Mixed Tocopherols
CAS/RD Number: N/A

d-Alpha Tocopherol:
EASTMAN Vitamin E 5-40
EASTMAN Vitamin E 5-67
CTFA: Tocopherol
CAS/RD Number: 59-02-9

d-Alpha Tocopherol Acetate:
EASTMAN Vitamin E 6-40
EASTMAN Vitamin E 6-81
EASTMAN Vitamin E 6-100
CTFA: Tocopheryl Acetate
CAS/RD Number: 7695-91-2

Dry d-Alpha Tocopheryl Acetate:
EASTMAN Vitamin E 700
CTFA: Tocopheryl Acetate
CAS/RD Number: 7695-91-2

Dry d-Alpha Tocopheryl Succinate:
EASTMAN Vitamin E Succinate
CTFA: Tocopheryl Succinate
CAS/RD Number: 4345-03-3

d-Alpha Tocopheryl Polyethylene Glycol 1000 Succinate:
EASTMAN Vitamin E TPGS (Water Soluble)
CTFA: Tocopheryl Polyethylene Glycol
CAS/RD Number: 9002-96-4

ECC AMERICA INC.: BENTOLITE Bentonites:

BENTOLITE viscosifiers are white bentonite produced by a special Southern Clay Products process which eliminates the major portion of iron and other color producing impurities.

Grades:

BENTOLITE L:
Low viscosity, calcium form. Recommended where increased loading of clay is desired without significantly increasing the viscosity.
BENTOLITE H:
A modified, high sodium form. Recommended for applications requiring higher viscosities at lower loading levels.

Typical Physical Properties:

BENTOLITE L:
Color: White
Brightness (1): 86
Moisture, %: 10
Swelling, ml (2): 8
pH (3): 7.0
Sieve Fineness (4): 98

BENTOLITE H:
Color: White
Brightness (1): 82
Moisture, %: 10
Swelling, ml (2): 16
pH (3): 9.5
Sieve Fineness (4): 98

(1): GE % of MgO
(2): 2 grams in 100 ml. distilled water
(3): 10% solids in distilled water
(4): % thru 200 mesh

Typical Chemical Analysis:

BENTOLITE L:	BENTOLITE H:
SiO_2: 71.7	65.8
Al_2O_3: 15.7	14.1
MgO: 3.6	3.2
CaO: 1.7	1.5
Fe_2O_3: 0.3	0.4
TiO_2: 0.3	0.2
Na_2O: 0.2	4.1
K_2O: 0.1	0.1

CTFA Name: Bentonite

ECC AMERICA INC.: CLAYTONE XL Bentonite:

A Rheological Control Additive for Roll-On Antiperspirants

CLAYTONE XL is a refined bentonite which has been beneficiated to provide you with a high performance organoclay.

Organoclays are the preferred rheological control additives for roll-on antiperspirants. They provide high viscosity under low shear conditions and thin readily with increased shear. These characteristics impart good shelf stability and easy application of the finished product.

Organoclays are excellent stabilizing and suspending agents in roll-on antiperspirant formulations. Recommended concentrations of 1.5-2.5% should be sufficient for syneresis control.

Features of CLAYTONE XL:

- Significantly reduces syneresis
- Improves suspension stability
- Imparts a reproducible thixotropic viscosity

Typical Properties:

Composition: Quaternary Ammonium Bentonite
Color: Light Buff
Form: Fine Powder
Free Moisture: 1.5%
Density: 14.1 lbs./gal.
Particle Size: 99.8% min. through 200 mesh
95.0% min. through 325 mesh

ECC AMERICA INC.: FLOWTONE Organic Rheological Additives:

FLOWTONE additives are recommended for non-aqueous systems to improve viscosity, for pigment suspension, and to provide sag control with optimum leveling characteristics.

FLOWTONE R:
A white organic powder, recommended for systems based on aliphatic solvents.

FLOWTONE GR:
A modified FLOWTONE R designed for faster and more efficient dispersion.

FLOWTONE ST:
A white organic powder, recommended for systems based on aromatic, aliphatic or blends of each solvent.

FLOWTONE GST:
A modified FLOWTONE ST designed for faster and more efficient dispersion.

FLOWTONE 4:
A liquid rheological additive designed to be incorporated at any step of the plant manufacturing process.

Typical Physical Properties:

FLOWTONE R:
Density Lbs/Gal: 8.50
Specific Gravity: 1.02
Bulking Value Gal/Lb: 0.118
White Powder

FLOWTONE GR:
Density Lbs/Gal: 12.25
Specific Gravity: 1.47
Bulking Value Gal/Lb: 0.082
White Powder

FLOWTONE ST:
Density Lbs/Gal: 8.50
Specific Gravity: 1.02
Bulking Value Gal/Lb: 0.118
White Powder

FLOWTONE GST:
Density Lbs/Gal: 12.16
Specific Gravity: 1.46
Bulking Value Gal/Lb: 0.082
White Powder

FLOWTONE 4:
Density Lbs/Gal: 8.66
Specific Gravity: 1.04
Bulking Value Gal/Lb: 0.115
Liquid

ECC AMERICA INC.: GELWHITE Montmorillonite:

Features of GELWHITE:
- Pure and natural
- White color (GE Brightness of 80)
- Suspension aid/emulsion stabilizer
- Thixotropic flow properties
- Performance over a wide pH range
- Synergism with organic gums
- Performance unaffected by microbial activity
- Smooth, non-tacky feel in topicals
- High exchange capacity
- Ease of dispersion

Description: Purified colloidal montmorillonite

Typical Properties:

GELWHITE L:
Color: White
Brightness, % G.E.: 80
Free Moisture, %: 10
Sieve Fineness, dry, % finer than 200 mesh: 80
pH, 2% Slurry: 9.0
Viscosity, % solids required to obtain 800 cps in distilled water: 14.0

GELWHITE H, NF:
Color: White
Brightness, % G.E.: 80
Free Moisture, %: 8
Sieve Fineness, dry, % finer than 200 mesh: 80
pH, 2% Slurry: 9.5
Viscosity, % solids required to obtain 800 cps in distilled water: 5.0

GELWHITE GP:
Color: White
Brightness, % G.E. 80
Free Moisture, %: 8
Sieve Fineness, dry, % finer than 200 mesh: 80
pH, 2% Slurry: 10.0
Viscosity, % solids required to obtain 800 cps in distilled water: 5.0

CTFA Name: Montmorillonite
FDA Status: Confirmed as GRAS

ECC AMERICA INC.: MINERAL COLLOID Montmorillonite:

Features of MINERAL COLLOID:
- Improved gel formation
- Interacts with both inorganic and organic cations
- Exceptional film formations--nonflammable, stable at elevated temperatures and resistant to microbial degradation
- Stabilizing agent for emulsions, emulsion polymerization and suspended solids
- Tightly controlled viscosity character makes it a reliable additive for bodying and thickening systems.

Description: Pure Montmorillonite

Typical Properties:

MINERAL COLLOID MO:
Color: Light Amber
Free Moisture, %: 10
pH, 2% Slurry: 9.5
Viscosity (Brookfield, Cps 50 RPM): 2,000 A
A 3% Solids

MINERAL COLLOID BP:
Color: Light Amber
Free Moisture, %: 10
pH, 2% Slurry: 9.0
Viscosity (Brookfield, Cps 50 RPM): 550 B
B 5% Solids

CTFA Name: Montmorillonite

FINETEX, INC.: FINQUAT CT Highly Cationic Conditioner:

FINQUAT CT (CTFA Name: QUATERNIUM-75) is a highly cationic, non-polymeric quaternary ammonium ethosulfate. It is suggested for use in peramanent waving systems to give a more even distribution of anionic reducing agent, viz. thioglycolate or bisulfite, on undamaged hair and a selective concentration gradient at damaged hair sites. It has been reported that the use of FINQUAT CT in thioglycolate systems imparts a silky conditioning to hair. FINQUAT CT is recommended for use in permanent waving solutions at suggested concentrations of 1-3%. FINQUAT CT has also been found to be an effective stable conditioner in depilatory formulations. It is recommended for use in hair conditioners when clear aqueous solutions are required.

Typical Properties:
- Appearance: Light yellow clear to hazy liquid
- Total Solids: 25%-27%
- pH as is: 6.2-6.8
- Color (Gardner): 1-2
- Specific Gravity @ 25C: 1.01-1.03

Concentrations up to 5% are clear at room temperature. Although many quaternary ammonium compounds precipitate in an alkaline environment, 5% FINQUAT CT remains clear even up to a pH of 12.5.

FINQUAT CT is compatible with the commonly used anionic surfactants (e.g. the alkyl sulfate salts).

Toxicology (3% solids FINQUAT CT aqueous solution)

Primary Dermal Irritation Index (Rabbits)	0.0
Primary Ocular Irritation (Rabbits)	Mildly irritating, with no corneal involvement

FINETEX, INC.: FINSOLV TN#/FINSOLV SB# Ester Emollients:

FINETEX TN and FINSOLV SB are two of the unique, low toxicity emollient/solubilizers offered by Finetex. The CTFA names for the products are:

FINSOLV TN - C12-15 Alcohols Benzoate
FINSOLV SB - Isostearyl Benzoate

These ester emollients are used to perform many functions in a formulation.

As emollients they confer a non-greasy, sophisticated feel to skin, and on hair they provide lubricating conditioning and excellent wet and dry combability. They are also useful in attenuating the undesirable greasiness often associated with other emollients such as mineral oil and isopropyl myristate. In these combinations, they will also enhance spreadability.

As solvent/solubilizers, both FINSOLV TN and FINSOLV SB serve as excellent emollient vehicles for sunscreens, insect repellents and fragrances. In the latter case, the FINSOLVS also serve as extenders and fixatives without altering the characteristics of the fragrance.

As a suspending agent, FINSOLV TN is particularly useful in anti-perspirant formulations based on aluminum and related salts. In this application, the dry emolliency which FINSOLV TN confers is an additional advantage.

Typical Physical Properties:

FINSOLV TN:
Actives: 100%
Appearance: Clear, almost colorless liquid
Odor: Practically odorless
Viscosity @ 25C: 40-45 cps
Freezing Point: -3 to -12 C
FINSOLV SB:
Actives: 100%
Appearance: Clear, yellow liquid
Odor: Practically odorless
Viscosity @ 25C: 63 cps
Freezing Point: below -10C

Patented

Toxicology:
(Offered here are the toxicological data for FINSOLV TN, which is indicative of the class).
Acute Oral Toxicity (Rats), LD50: 34.5 g/kg
Ocular Irritation (Rabbits), No Wash: 0.0
Primary Dermal Irritation Index (Rabbits): 0.08
Acute Dermal Toxicity (Rabbits), LD50: >2.0 g/kg Non-toxic
Acute Inhalation Toxicity (Rats), LD50: >200 mg/l Non-toxic
Guinea Pig Sensitization: Not a potential sensitizer
Comedogenicity Assay (Rabbits): Non-comedogenic
Repeated Insult Patch Test: No sensitization, no irritation

FINETEX, INC.: SURFINE Carboxylated Surfactants:

The SURFINE series of carboxylated surfactants is represented by the general structure:

R-O-(EO)x-CH2COO--M+

Within the homologous series,

R generally represents C6-C20
x generally represents 1-30, and
M+ can be any mono- or polyvalent cation, including H+

A true class of specialty surfactants, the SURFINES display a combination of physical properties unlike any other common or commercially available anionic surface active compounds. Among these properties are included:

- High lime soap dispersion activity
- Variable HLB with pH
- Formulation and process stability
- Compatibility with most cationic conditioners
- Mild to skin and hair
- Effective cleansers/emulsifiers/coupling agents

<u>Product Name:</u>

SURFINE AZI-A:
R: NP
x: 9
M+: H+
CTFA Name: Nonoxynol-10 Carboxylic Acid
SURFINE T-A:
R: C13
x: 6.5
M+: H+
CTFA Name: Trideceth-7 Carboxylic Acid
SURFINE WNT:
R: C12-15
x: 6.5
M+: Na+
CTFA Name: Na C12-15 Pareth-7 Carboxylate
SURFINE WNG-A:
R: C14-15
x: 7
M+: H+
CTFA Name: C14-15 Pareth-8 Carboxylic Acid
SURFINE WLL:
R: C12
x: 12
M+: Na+
CTFA Name: Sodium Laureth-13 Carboxylate

FINETEX, INC.: TAURANOL Sodium Cocoyl Isethionate:

TAURANOL I-78 is the FINETEX designation for the coco ester of sodium isethionate (CTFA Name: Sodium Cocoyl Isethionate).

A mild, hard-water tolerant detergent, TAURANOL I-78 has traditionally been used in syndet bars or in conjunction with soap where it modifies/enhances the soap lather, serves as an extremely effective lime-soap dispersant, and can lower the irritancy level of the soap alone.

More recently, the use of TAURANOL I-78 as the prime surfactant in skin and hair cleaning products has gained increasing popularity. Its own high foaming properties, gentle and effective detergency, and compatibility with hard water lead the list of desirable properties which make TAURANOL I-78 useful in formulations.

Finetex manufactures TAURANOL I-78 in several forms designed to suit your production requirements.

TAURANOL I-78:
Form: Powder
Activity: 78-83%

TAURANOL I-78 Flakes:
Form: Flakes
Activity: 78-83%

TAURANOL I-78-3:
Form: Paste
Activity: 24-25%

TAURANOL I-78-6:
Form: Hydrogel
Activity: 49-50%

TAURANOL I-78-6, the newest addition to the line, was specifically developed as an easy-to-handle, high concentration form of TAURANOL I-78. Shipped as an off-white paste, TAURANOL I-78-6 easily melts at 60-70C and can be handled as a liquid. In this form, TAURANOL I-78-6 finds particular utility in the manufacture of combo (syndet-soap) products where the miscibility of the TAURANOL I-78-6 in molten Neat Soap improves production rates and final product quality.

FINETEX, INC.: TAURANOL Sulfoalkyl Amido Surfactants:

TAURANOL Taurates are sulfonated fatty acid amides whose foaming and detergency properties mimic those of soap, but do not have many of the drawbacks of soap.

The sulfo-amido combination gives the TAURANOL Taurates good solubility, close textured and persistant foam, mildness, and renders these molecules excellent lime-soap dispersants. They are unusually stable in alkaline and moderately acid media and display excellent detergency and emulsification properties.

Finetex currently manufactures and markets two sulfo-amido surfactants under the TAURANOL Taurate label in a number of physical forms designed to best suit your applications.

TAURANOL WS Series (CTFA Name: Sodium Methyl Cocoyl Taurate):

TAURANOL WS:
Form: Soft Paste
Activity: 24-25%

TAURANOL WS Conc.:
Form: Soft Paste
Activity: 30-31%

TAURANOL WSP:
Form: Powder
Activity: 72-75%

TAURANOL WS H.P.:
Form: Powder
Activity: 96-97%

TAURANOL M Series (CTFA Name: Sodium Methyl Oleyl Taurate):

TAURANOL M-35:
Form: Viscous Liquid
Activity: 23-25%

TAURANOL ML:
Form: Viscous Liquid
Activity: 32-33%

TAURANOL MS:
Form: Soft Paste
Activity: 32-33%

The TAURANOL Taurates provide many benefits in hair and skin care preparations.

FINETEX, INC.: Taurate/Carboxylate Gels:
Novel Clear Delivery Systems

In certain specific ratios with water, it has been found that TAURANOL WS Conc. and the TEA salts of two SURFINE carboxylated surfactants, SURFINE AZI-A and SURFINE WNG-A, will form clear, rigid gels. The coupling capabilities of the SURFINES enables these systems to deliver, in clear, aesthetic gel form, many active ingredients which heretofore could only be formulated into emulsions or otherwise opaque products.

Designated Experimantal Taurate/Carboxylate Gels 3500, the trio presented below are examples of the basic gel systems. Of particular interest is Gel B, which demonstrates the ability of SURFINE AZI-A to couple in a significant level of FINSOLV TN# (Actually, levels of up to 10% FINSOLV TN are possible). FINSOLV TN--itself an excellent solubilizer of many skin and hair care 'actives'--then serves to further potentiate the system's utility as a unique clear gel delivery composition. In addition, the FINSOLV TN also confers its own special emolliency benefits.

Many useful and desirable active ingredients have been included in one or more of these systems without effecting either the clarity or stability of the gels. Among the compounds which have been successfully incorporated are:

- Sunscreens--all types
- Foam boosters--generally lauryl based
- Cationic conditioners--including high cationic density types like FINQUAT CT
- Ethanol
- Fragrances
- Benzoyl Peroxide
- Biological Actives

Experimental Taurate/Carboxylate Gels 3500:

Component: A:

TAURANOL WS Conc.	30.3
SURFINE AZI-A	30.3
Ammonium Chloride	1.0
TEA	5.0
Water	33.4

Component: B:

TAURANOL WS Conc.	30.3
SURFINE AZI-A	30.3
FINSOLV TN	4.0
Ammonium Chloride	1.0
TEA	5.0
Water	29.4

Component: C:

TAURANOL WS Conc.	30.3
SURFINE WNG-A	30.3
Ammonium Chloride	1.0
TEA	5.0
Water	33.4

Patented

FLORIDA FOOD PRODUCTS, INC.: Aloe Vera Gel:

ALOE-CON:
Aloe Vera Concentrates
Applies to types WG & UP

Cosmetic:

Aloe Vera Gel is a natural moisturizer at effective levels in cosmetics. Its incorporation into skin care products improves their effectiveness in the moisturizing and "healing" of dry damaged skin caused by overexposure to sun and wind. Aloe is excellent in reducing skin irritations caused by diaper rash or insect bites. It also reduces the irritating effects of metallic salts such as aluminum chloride in antiperspirants.

In hair care products, because of Aloe's highly substantive nature, it provides effective conditioning at low usage levels.

Recommended Levels#:

Creams and lotions:	0.125-1.0%
Shampoos and rinses:	0.025-0.25%
Therapeutic products:	0.625-1.875%
Clear gel products:	1.25-2.425%

The levels listed are of Aloe concentrate. To convert to percent Aloe Vera Gel, simply multiply by 40. The level of gel on the ingredient listing is of the single strength.

ALOE-CON WG-40:

ALOE-CON WG-40 is a 40 fold concentrate of natural Aloe gel obtained by low temperature evaporation. Only pure Aloe gel is concentrated. No preservatives are added beforehand. The Aloe gel is kept cold to prevent spoilage before concentration.

ALOE-CON WG-40 is available in both cosmetic and food grades.

Advantages:
- Liquid concentrate, 1 qt. makes 10 gallons gel
- Ease of storage and handling
- Cold processed
- Immediate dispersion in product
- Full enzyme activity
- Can be added directly to batch, there is no need for pre-extension

Typical Properties:
Consistency: Semi-Turbid solution
Color (1:39 dilution): 6 Gardner max.
Total Solids: 20%±0.5
Specific gravity: 1.1±0.05
pH (1:39 dilution): 3.5-5.0
Total Microbial Count: <10 cfu/ml
Pathogenic Bacteria: 0
Dispersion rate: Immediate

FLORIDA FOOD PRODUCTS, INC.: Aloe Vera Gel(Continued):

ALOE-CON UP-40:
Aloe Vera Gel concentrate-purified

ALOE-CON UP-40 is a purified 40 fold concentrate of natural Aloe gel obtained by low temperature evaporation. This purification is obtained by an exclusive adsorption method, which selectively removes much of the natural impurities found in Aloe, which may cause color or chemical instability.

ALOE-CON UP-40 is available in both cosmetic and food grades.

Advantages:
- Liquid concentrate, 1 qt. makes 10 gallons gel
- Ease of storage and handling
- Cold processed
- Immediate dispersion in product
- Full enzyme activity
- Cold stability
- Can be added directly to batch, there is no need for pre-extension
- Less reactive with cosmetic ingredients than natural Aloe gel

Typical Properties:
Consistency: Semi-Turbid solution
Color (1:39 dilution): 3 Gardner max.
Total Solids: 20%±0.5
Specific Gravity: 1.1±0.05
pH(1:39 dilution): 3.5-5.5
Total Microbial Count: <10 cfu/ml
Pathogenic Bacteria: 0
Dispersion Rate: Immediate
Magnesium: 40-90 ppm
Calcium: 150-400 ppm
Aloe Profile Analysis (1:39 dilution): 100 percent Aloe Gel
X-Ray Diffraction: Match Standard
CTFA Designation: Aloe Vera Gel

Aloe-Vera Oil Soluble Extracts:
These Aloe oils are made by extracting the Aloe's oil soluble fractions onto an oil base. It finds use where Aloe is desired in products where using the gel is not possible, such as suntan oils or lipstick.

Aloe Oil Extracts:

Product #	Solvent Base
101	Light Mineral Oil
102	White Petrolatum
103	Refined Lanolin Oil
104	Isopropyl Myristate
105	Octyl Palmitate

FLORIDA FOOD PRODUCTS, INC.: Aloe Vera Gel(Continued):

ALOE-CON 200:
Freeze Dried Aloe Vera Gel

ALOE-CON freeze dried Aloe is the culmination of the concentration process. Only pure Aloe is freeze dried, there is no matrix or preservatives added to the Aloe beforehand. After FFP initially concentrates the Aloe to approximately 20% solids via their cold evaporation procedure the concentrate is then freeze dried, milled and packaged in a humidity controlled environment. Freeze drying is the process whereby you can achieve water removal from a product while it is in a frozen state.

Advantages:
- Highly concentrated, 1 lb makes 200 lbs gel
- Ease of storage and handling
- Long shelf life
- Full enzyme activity
- Readily water soluble
- 100% pure Aloe solids, no matrix added
- Available in both natural and ultra purified grades

Typical Properties:
WG-200:
Form: Brown-Tan Powder
Color: 6 Gardner max.
Moisture: 7% max.
pH: 3.5-5.0
Specific Gravity: 1.00
Magnesium: 40-90 ppm
Calcium: 150-400 ppm
Pathogenic Bacteria: 0
Total Microbial count: <500 cfu/ml
X-ray Diffraction: Match std
Aloe Profile analysis: 100%
Dispersion rate: 5 minutes
CTFA Designation: Aloe Vera Gel

UP-200:
Form: Off White powder
Color: 3 Gardner max.
Moisture: 7% max.
pH: 3.5-5.5
Specific Gravity: 1.00
Magnesium: 40-90 ppm
Calcium: 150-400 ppm
Pathogenic Bacteria: 0
Total Microbial count: <500 cfu/ml
X-Ray Diffraction: Match std.
Aloe Profile analysis: 100%
Dispersion rate: 5 minutes
CTFA Designation: Aloe Vera Gel

FREEMAN INDUSTRIES, INC.: Products for the Cosmetics Industry:

Products of United Guardian Inc.:

Trade Name:

AQUATHIX:
Modified Xanthan Gum
Uses: Water gelling agent for low pH products.

ALCOLITE:
High molecular weight fatty acid soap (C32-C36)
Uses: For gelling high concentrations of alcohol

BETA CAROTENE (stabilized):
Stabilized form of Beta Carotene pro-Vit. A
Uses: Natural yellow color for tinting cosmetic products

B-122:
Special Grade Acid Calcium Stearate
Uses: Binding agent in pressed powders; reduces breakage on dropping.

DERMAFORM:
Composed of acrylic polymers complexed with PVP Polyglceryl-methacrylate and PVP/DEA
Uses: Unique skin-forming water-based white emulsion with high stability. When dry, forms clear invisible film with breathes, yet resists tears, perspiration and rinsing with water.

KARAJEL:
Polyglyceryl alginate with excipients
Uses: Partially sea-derived moisturizing lubricant and mild surfactant; non-tacky, compatible with surfactants and salts.

LUBRAJEL:
Clear gel moisturizer; CTFA Name: Polyglyceryl methacrylate Propylene Glycol
Uses: Non-drying, water-soluble lubricating jelly for moisturizing skin; burn creams, prelubricated thermometers, catheters, etc. For cosmetic use as a moisturizing ingredient in liquids and creams.

Types:
LUBRAJEL MS-Maximum stability of viscosity
LUBRAJEL CG-Cosmetic grade, high moisturizing properties
LUBRAJEL DV-Highest viscosity material, retains excellent viscosity upon dilution
LUBRAJEL TW-Least tacky of all the Lubrajels
LUBRAJEL WA-Gel turns to "oil" on contact with skin
LUBRAJEL OIL-Free-flowing form with very low viscosity
LUBRAGEL WP-CTFA Name: Polyglycerylmethacrylate

FREEMAN INDUSTRIES, INC.: Products for the Cosmetics Industry (Continued):

Products of United Guardian Inc.(Continued):

Trade Name:

LUBRASLIDE:
Coated Micronized Teflon
Uses: Pressed powders, makeup, foot and shoe powders, gloves, creams

LUBRAQUAT CONC.:
Quaternized Polyjel
Uses: For hair conditioning

OIL OF ORCHIDS, WS:
OIL OF ORCHIDS, OS:
Orchid Extract
Uses: Moisturizers, creams, lotions

POLYJEL HV:
CTFA Name: Polyglyceryl-citrate, Hydroxypropyl methyl cellulose
Uses: Non-drying, water-soluble lubricating gel, to be used where highly ionized substance or water-soluble solvents are required. Will not affect viscosity.

SHEERSKIN:
Flexible film-former; copolymer of acrylic acid & PVP ester
Uses: Unique clear skin-forming water-base liquid; dries to an invisible film which forms a barrier impervious to water, tears, etc.

SKIN LITE:
Copolymer of N-butyl polyester, maleic acid & partially hydrolyzed ethyl cellulose, plasticized with acetyl tributyl citrate, the copolymer dissolved in 70% IPA
Use: For unique film-forming products where the presence of IPA would be desirable

SUPERTI POWDER: (0.1 - 0.3 microns) particle
ULTRATI POWDER: (0.01-0.03 microns) size
Especially fine grades of Titanium Dioxide
Uses: For face powders and rouges when hiding quality is needed; an important UV blocker

UNITWIX:
Glyceryl and Glycol esters of 24-40 carbon chain fatty acids
Uses: Thickener for lipo-soluble ingredients. Can convert oil, fatty alcohol liquids to ointments or sticks.

FREEMAN INDUSTRIES, INC.: Products for the Cosmetics Industry (Continued):

Products of Paninkret Chemicals & Pharmaceuticals GmbH and Dr. F.K Marcus GmbH:

Chlorophyll and other natural colors
Exotic Oils
Protein Derivatives:
- Elastin from calf placenta
- Elastin Hydrolysate
- Keratin (whole protein powder)
- CUTICULIN: Keratin Hydrolysate Powder
- Calf Thymus peptides
- Amino Acid/Peptide Complex
- Collagen Powder
- Collagen Hydrolysate Powder

Natural Colors:
- Green: Chlorophyll and Chlorophyllins
- Purple: Grapeskin Extract
- Yellow: Annatto, Beta Carotene
- Orange: Annatto
- Red: Beet Powders

Pectin & Derivatives:
- Apple Pectin USP
- Citrus Pectin USP

Proteins & Derivatives:
Whole Proteins, Hydrolysates & Autolysates; Corn Glutelin, Keratin, Skin, ZEIN.

Proteins:
- KERAMOIS (from wool)
- SILKPRO (amino acids)
- SILKPRO CM1000 (water-soluble)
- SILKPRO A.S. (alcohol-soluble)

Exotic Plant Oils:
- Borage Oil
- Evening Primrose Oil
- Camellia Oil
- Oil of Orchids

Rice Derivatives:
- Rice Bran Wax
- Rice Bran Extract, Defatted
- Rice Bran Oil

Sodium Bromate:
Cosmetic Grade, for cold wave neutralizers

FREEMAN INDUSTRIES, INC.: Products for the Cosmetics Industry (Continued):

Products of Paninkret Chemicals & Pharmaceuticals GmbH and Dr. F.K. Marcus GmbH(Continued):

Vitamins, Custom Designed Potencies:

Vitamin A, Acetate & Palmitate Pure concentrates in vegetable oil or aqueous solutions

Vitamin A & D

Vitamin D2 USP (ergocalciferol) Crystalline, resin, stabilized powder or pure concentrates in vegetable oil or aqueous solutions

Vitamin D3 USP (cholecalciferol)

Other Specialties:

Sericites: Powdered Mica, coated and uncoated: for pressed powder and makeups

Sodium Hyaluronate Powder: 95 & 98% purities

Sodium Hyaluronate: Bio type, powder & gel

Silk-coated TiO_2

SILKALL 100: Ultrafine silk powder

Aloe vera gel, 1X color stabilized gel and decolorized gel

Aloe Vera gel, 10X " "

Aloe Vera gel, 40X " "

Aloe Vera powder, freeze-dried or spray-dried 1:199X

Aloe Vera oil extract

H.B. FULLER: Specialty Waxes:

LUNACERA M:
microcrystalline wax of ozokerite structure
Colour: ivory
Form of supply: pellets
Congealing point: 68-72C
Penetration: 15-19
CTFA reference: microwax
DAB 9: corresponds to purity requirements
BGA recommendation XXV: corresponds

LUNACERA M distinguishes itself by its good absorption of mineral and vegetable oils. Stable and very fine crystals are formed.

LUNACERA M has a well balanced proportion of straight and branched-chain hydrocarbon waxes. LUNACERA M has consistency regulating and gel forming properties.

Fields of application:
- cosmetic and pharmaceutical preparations
- absorption bases
- fat bases
- emulsifier bases
- preparations in stick form

LUNACERA MW:
LUNACERA MWN:
microcrystalline waxes of ozokerite structure
LUNACERA MW:
Colour: ivory
Form of supply: slabs
Congealing point: 68-73C
Penetration: 20-30
Viscosity at 98,9C: mPa.s: 8,5-11,5
CTFA reference: microwax
BGA recommendation XXV: corresponds
LUNACERA MWN:
Colour: ivory
Form of supply: slabs
Congealing point: 65-70C
Penetration: 25-30
Viscosity at 98,9C: mPa.s: 11-14
CTFA reference: microwax
BGA recommendation XXV: corresponds

LUNACERA MW and LUNACERA MWN have good absorption of mineral and vegetable oils. Stable and very fine crystals are formed.

These LUNACERA types can be distinguished by their percentage of branced-chain hydrocarbon waxes. Both waxes give the same high consistency regulation and gel ability.

Fields of application:
- cosmetic and pharmaceutical preparations
- absorption bases
- preparations in stick form

H.B. FULLER: Specialty Waxes(Continued):

LUNACERA S:

microcrystalline wax of ozokerite structure
Colour: ivory
Form of supply: slabs
Congealing point: 70-75C
Penetration: 22-27
CTFA reference: microwax
BGA recommendation XXV: corresponds

LUNACERA S has excellent oil absorption and stabilizing properties because of its high content of branched-chain hydrocarbon waxes. With mineral and natural oils fine crystals are formed. LUNACERA S can be used alone, but also in combination with e.g. LUNACERA M.

Fields of application:
- cosmetic and pharmaceutical preparations
- absorption bases
- sticks

LUNACERA W 80:
LUNACERIN H 1038:

Hardwaxes
LUNACERA W 80:
Colour: white
Form of supply: pellets
Congealing point: 76-81C
Penetration: 9-13
CTFA reference: ceresin + microwax
BGA recommendation XXV: corresponds
LUNACERIN H 1038:
Colour: white
Form of supply: pellets
Congealing point: 92-97C
Penetration: <5
Viscosity at 120C: 6,5-9,5 mPa.s
CTFA reference: ceresin
BGA recommendation XXV: corresponds

LUNACERA W 80 and LUNACERIN H 1038 are hard, high-melting waxes of dry consistency. They are used for increasing the melting point and as regulator in preparations for decorative cosmetics.

LUNACERA W 80 also has an oil binding effect.

Fields of application:
- lipsticks
- mascara
- eye-liner

H.B. FULLER: Specialty Waxes(Continued):

LUNACERA 256:
LUNACERA 4776:
LUNACERA 5376:
ointment waxes

LUNACERA 256:
Colour: light-ivory
Form of supply: paste
Congealing point: 56-61C
Penetration: 70-90
Viscosity at 98,9C: mPa.s: 3,5-5,5
CTFA reference: petrolatum

LUNACERA 4776:
Colour: ivory
Form of supply: paste
Congealing point: 62-67C
Cone-penetration: 50-80
Viscosity at 98,9C: mPa.s: 7-10
CTFA reference: microwax
mineral oil

LUNACERA 5376:
Colour: ivory
Form of supply: paste
Congealing point: 58-63C
Cone-penetration: 30-50
Viscosity at 98,9C: mPa.s: 3,5-5,5
CTFA reference: ceresin
microwax
mineral oil

LUNACERA 256, LUNACERA 4776 and LUNACERA 5376 are mineral waxes of microcrystalline structure and with a balanced proportion of branched-chain and linear hydrocarbons.

Based on LUNACERA ointment waxes, stable absorption bases and so-called softcreams can be produced. They can easily be emulsified to give a cream-like consistency.

Field of application:
- cosmetic and pharmaceutical preparations

H.B. FULLER: Specialty Waxes(Continued):

LUNACERA 5396:
LUNACERA PA 5493:
polyethylene pastes

LUNACERA 5396:
Colour: light-ivory
Form of supply: paste
Congealing point: 78-82C
Cone-penetration: 70-90
Viscosity at 98,9C: 8-11 mPa.s
CTFA reference: microwax
mineral oil
polyethylene

LUNACERA PA 5493:
Colour: light-ivory
Form of supply: paste
Congealing point: 80-85C
Cone-penetration: 250-300
Viscosity at 98,9C: 9,2-10,2 mPa.s
CTFA reference: mineral oil

LUNACERA PA 5493 is a polyethylene wax dispersed in paraffin oil (per liquidum) DAB 9. LUNACERA 5396 contains in addition an ozokerite. Both products are of paste-like consistency. LUNACERA polyethylene pastes improve remarkably the heat stability and oil binding properties in cream formulations.

LUNACERA PA 5493 can be used as a substitute for vaseline.

Fields of application:
- cosmetic and pharmaceutical preparations
- decorative cosmetics

LUNACERA MWH:
ozokerite-resin-blend

Colour: light-ivory
Form of supply: blocks
Congealing point: 68-73C
Penetration: 24-31
Viscosity at 98,9C: 19-25 mPa.s
CTFA reference for the wax component: microwax
BGA recommendation XXV: corresponds

LUNACERA MWH is a microcrystalline wax, containing a hydrogenized hydrocarbon resin for improving adhesion properties on the skin without being tacky.

LUNACERA MWH is particularly used in baby-, sun- and skin-care-creams.

Field of application:
- cosmetic and pharmaceutical preparations

H.B. FULLER: Specialty Waxes(Continued):

LUNACERA P:

Paraffinum solidum
Colour: white
Form of supply: pellets
Congealing point: 58-62C
Penetration: 10-14
CTFA reference: Paraffin
DAB 9: corresponds
BGA recommendation XXV: corresponds

LUNACERA P is used in ointment and cream formulas for improving consistency. Its good oil absorption permits the addition of high proportions of effective substances to the fat phase.

Fields of application:
- cosmetic and pharmaceutical preparations
- absorption bases
- sticks

LUNACERA C 44:

Carnaubawax blend
Colour: ivory
Form of supply: pellets
Congealing point: 69-74C
Penetration: <8
CTFA reference: Carnaubawax + Ceresin

LUNACERA C44 is a blend of carnaubawax and ceresin. It is of hard consistency and has a faint odour which is typical for carnaubawax. LUNACERA C 44 has good oil absorption properties and increases the "hardness" of sticks.

Fields of application:
- decorative cosmetic preparations
- cream formulas
- coating of tablets

LUNACERA BW 568:

Beeswax substitute
Colour: yellowish
Form of supply: slabs
Congealing point: 64-68C
Penetration: 16-22
Acid Number: 20-30 mg KOH/g
Saponification number: 20-40 mg KOH/g

LUNACERA BW 568 is a modified acid wax, which is equivalent in a number of properties to natural beeswax. This depends on formulations and application.

Fields of application:
- cosmetic preparations
- fat- and emulsification bases

H.B. FULLER: Specialty Waxes(Continued):

LUNACERA JW:
Japanwax substitute

JW 187:
Colour: ivory
Form of supply: slabs
Congealing point: 56-61C
Penetration: 17-25
Viscosity at 98,9C: mPa.s: 8,5-10,5
Acid number: mg KOH/g: 15-20
Saponification number: mg KOH/g: 180-210

JW 189:
Colour: ivory
Form of supply: slabs
Congealing point: 45-50C
Penetration: 40-50
Viscosity at 98,9C: mPa.s: 6-8
Acid number: mg KOH/g: 5-8
Saponification number: mg KOH/g: 200-240

JW 642:
Colour: ivory
Form of supply: slabs
Congealing point: 42-47C
Penetration: 30-40
Viscosity at 98,9C: mPa.s: <3-5
Acid Number: mg KOH/g: <1,5
Saponification number: mg KOH/g: 65-85

LUNACERA JW are waxes based on fatty-acid glycerides. They are easily emulsifiable and have distinct dye solution properties. In addition they effect gloss.

Fields of application:
- preparation of decorative cosmetics
- crayons and pencils

COSMERIN 961:
wax-oil-compound
Colour: light-ivory
Form of supply: blocks of 8 kos each
Congealing point: approx. 76C
Penetration: approx. 100
Viscosity at 98,9C: approx. 11 mPa.s

COSMERIN is a compound to produce decorative cosmetic articles. Consistency and melt behaviour are adjusted in such a way, that e.g. for production of lipsticks about 15% of oily substances have to be added to absorb the colouring pigments.

Adding up to 10% of LUNACERA M yields a firmer consistency.

Fields of application:
- lipsticks
- eye-make-up sticks
- eye-shadow sticks
- rouge sticks and creams

H.B. FULLER: Specialty Waxes(Continued):

LUNACERA ALBA:

Beeswax LUNACERA ALBA meets the requirements of DAB 9 for "Cera alba".

Colour: white, yellowish-white, thin layer translucent
Consistency: soft and formable when hand warm
Odour: slight, not stinging
Form of supply: pellets
Drop/Melting Point: 61-65C
Acid Number: 17-24 mg KOH/g
Ester Number: 70-80 mg KOH/g
Proportion Ester/Acid: 3,3-4,3
Saponification Number: 87-104 mg KOH/g
Ceresins: not provable
Glycerole: not provable

Important notice: Store protected from light!

GAF CHEMICALS CORP.: Copolymer Series:

Poly(vinylpyrrolidone/dimethylaminoethyl methacrylate).

CTFA name: PVP/Dimethylamino-ethylmethacrylate Copolymer

Physical form and Description:

Copolymer 845 supplied as a 20% aqueous solution (average molecular weight = 10 6).

Copolymer 937 supplied as a 20% aqueous solution (average molecular weight = 10 6).

Copolymer 958 supplied as a 50% ethanol solution (average molecular weight = 10 5).

Properties:

Film-forming resin, providing good hold and curl retention even at high humidity.

Slight cationic charge provides for conditioning effects without build-up.

Compatible with alcoholic systems and with many cosmetic additives and thickeners.

Uses:

Styling and conditioning resin for mousses, gels, glazes, and sprays.

Conditioning agent for cream rinses, shampoos, hair colorants, skin creams, and lotions.

GAF CHEMICALS CORP.: GAFFIX VC-713 cationic fixative polymer:

CTFA name: Vinylcaprolactam/PVP/Dimethylaminoethylmethacrylate Copolymer

Physical form and description:
Supplied as a 37% ethanol solution

Properties:
Superior curl retention performance at low solids level.
Cationic functionality--can be used in combination with cationic conditioning resins.
Superior hair holding ability in high humidity with negligible tack.
Water soluble--easily removed by shampoo in either hot or cold water.
Compatible with a wide range of surfactants, silicones, thickeners, propellant systems, and conditioning resins.
No neutralization required.

Uses:
Film-former and styling aid in mousses, gels, glazes, lotions, sprays, hair colorants and peelable face masks.

GAF CHEMICALS CORP.: POLECTRON 430 opacifier:

Vinylpyrrolidone/Styrene Copolymer Emulsion
CTFA name: Styrene/PVP Copolymer

Physical Form and Description:
Fluid, milky-white emulsion, with solids content of 40%.
Sub-micron particle size (<0.5 micron).
pH (as is): 2.0-4.0.

Properties:
Unusually stable, moderately viscous emulsion, with excellent freeze-thaw and shear stability.
High acid, base, and salt tolerance.
Forms strong, light-stable films with high water resistance.

Uses:
Opacifier used in:
Shampoos.
Conditioners.
Acid rinses.
Permanent wave preparations (acid and high pH).
Setting lotion.
Dye acceptor in hair color preparations.

GAF CHEMICALS CORP.: GAFQUAT series quaternary copolymers:

Quaternized Copolymers of Vinylpyrrolidone and Dimethylaminoethylmethacrylate

CTFA name: Polyquaternium 11

Physical form and description:

GAFQUAT 734 is supplied as a 50% ethanol solution (average molecular weight = 10 5).

GAFQUAT 755 and 755N are supplied as a 20% aqueous solution (average molecular weight = 10 6).

Properties:

Form clear, non-tacky, continuous films. Cationic nature imparts substantivity to hair, providing conditioning and bodying effects with minimum build-up.

Contribute improved combability, gloss, smoothness and manageability to hair.

Compatible with nonionic, anionic, cationic and amphoteric surfactants.

Not an eye or skin irritant.

Uses:

Film-former and conditioning agent for shampoos and conditioners, colorants and acid perms. Styling agent for mousses, gels, lotions, and sprays.

Additive for improved skin feel in:

Deodorants and anti-perspirants.

Shaving products.

Skin creams and lotions.

Toilet soaps.

GAF CHEMICALS CORP.: GAFQUAT HS-100 Resin:

GAFQUAT HS-100 resin (formerly Polymer AT-1269) is a new cationic conditioning resin designed especially for use in personal care products. It is a copolymer of vinylpyrrolidine and methacrylamidopropyl trimethylammonium chloride (MAPTAC), supplied as a 20% aqueous solution. GAFQUAT HS-100 resin possesses a unique combination of substantive and film-forming properties which provide benefits in a wide range of products:

- shampoos
- conditioners
- permanent waving systems
- mousses
- gels
- glazes
- hand and body lotions
- moisturizing creams

Product Data:
Appearance @ 25C: Hazy, highly viscous liquid
Color: Light pink to straw
% Solids: 19-21
pH (as is): 5-8
Viscosity: 60,000 cps minimum

GAF CHEMICALS CORP.: GANTREZ SP-215 Resin:

GANTREZ SP-215 # resin (formerly Polymer ACV-4001) is a new product specifically designed to meet the growing need for stiffer, harder holding products. The most obvious way to increase holding is by increasing the solids content. However, due to limitations of current pump packaging technology, many commercially available resins can only be used at up to 5 or 6% solids before spray patterns deteriorate and drying times increase to unacceptable levels.

GANTREZ SP-215, ethyl ester of a Poly (methyl vinyl ether/ maleic anhdydride) (PVM/MA) copolymer, has been engineered to overcome these limitations.

- High humidity resistance
- Negligible tack
- No flaking
- Easily removed by shampoo
- Good combability
- High gloss/sheen
- Natural feel
- Compatible with additives (plasticizer)

Typical Product Data:
CTFA Name: Ethyl Ester of PVM/MA Copolymer
Appearance @ 25C: Clear viscous liquid
Color: Light yellow to straw
% Solids: 48-52

Patent Pending

GAF CHEMICALS CORP.: GANEX Alkylated Polyvinylpyrrolidone:

Typical Properties of GANEX Resins:

GANEX V-216 Polymer:
- CTFA Designation: PVP/Hexadecene Copolymer
- Physical Form: liquid
- % Activity: 100
- Appearance @ 25C: pale yellow viscous fluid
- % Nitrogen: 2.0-3.0
- HLB requirement: 10
- Solubilities:
  - Kerosene: S
  - Mineral Oil: S
  - Castor Oil: S (5%)
  - Ethanol: PS
  - n-propanol: I
  - Water: I

GANEX V-220 Polymer:
- CTFA Designation: PVP/Eicosene Copolymer
- Physical Form: solid
- % Activity: 100
- Appearance @ 25C: off white waxy solid
- Solidification Point: 35-40C
- % Nitrogen: 2.9-3.6
- HLB requirement: 8
- Solubilities:
  - Kerosene: S
  - Mineral Oil: S
  - Castor Oil: I
  - Ethanol: I
  - n-propanol: I
  - Water: I

S = Soluble to 20% GANEX, except at noted
PS = Partially Soluble
I = Insoluble

Benefits and Applications:
GANEX V-216 and GANEX V-220 resins have been used to impart four primary benefits to skin care and cosmetic products:

Water and wear resistance
Pigment dispersion
Rich feel
Improved stick integrity

The additional benefits of low toxicity, non-comedogenicity and broad compatibility further support their use in these applications.

GAF CHEMICALS CORP.: GANEX series copolymers:

Copolymers of Vinylpyrrolidone and long-chain α - Olefins.

P-904:
CTFA name: Butylated PVP
V-516, V-216:
CTFA name: PVP/Hexadecene Copolymer
V-220:
CTFA name: PVP/Eicosene Copolymer

Physical form and description:
Properties and solubility vary with series.

P-904:
Supplied as: 45% in Isopropanol or powder (100%)
Solubility: water, alcohol

V-516:
Supplied as: 55% in Isopropanol
Solubility: alcohol, hydro-alcohol

V-216:
Supplied as: viscous liquid (100%)
Solubility: oil

V-220:
Supplied as: waxy solid# (100%)
Solubility: oil
melting point approximately 35C.

Properties:

Film-formers with exceptional surface activity. GANEX V series provides water resistance and moisture barrier properties.

Differing hydrophobic/hydrophilic characteristics of series provides a wide range of solubility.

Excellent pigment dispersant and suspending agent.

Substantive to keratin materials (skin).

Excellent toxicological profile for skin care products.

Uses:

Waterproofing agents for sunscreen products.

Non-irritating color dispersants for make-up, mascara, eye liner, and lipstick.

Gloss additive for lipstick.

Moisture barrier additives for skin care products.

Conditioning agents for skin care products.

Provides unique "after-feel" characteristics in skin care products.

GAF CHEMICALS CORP.: GANTREZ ES anionic copolymers:

Monoalkyl Esters of Poly (methyl vinyl ether/maleic acid).

ES-225:
CTFA name: Ethyl ester of PVM/MA Copolymer
ES-335:
CTFA name: Isopropyl ester of PVM/MA Copolymer
ES-425, ES-435:
CTFA name: Butyl ester of PVM/MA Copolymer

Physical form and description:

Series of Copolymers with varying ester groups, supplied as clear, viscous solutions.

ES-225:
R = Ethyl, 50% in ethanol
ES-335:
R = Isopropyl, 50% in isopropanol.
ES-425:
R = n-Butyl, 50% in ethanol.
ES-435:
R = n-Butyl, 50% in isopropanol.

Properties:

Form tough, clear glossy films, tack-free, with excellent substantivity, hair-holding properties and moisture resistance.

Soluble in alcohols, esters, ketones and glycol ethers.

Good compatability with aerosol propellants, including hydrocarbons.

Film properties and solubility can be modified by type and degree of neutralization.

Uses:

Film-former in hairsprays, mousses, gels and lotions (especially "all weather" humidity-resistant products).

Emulsion stabilizer in creams and lotions.

GAF CHEMICALS CORP.: PVP/VA series copolymers:

Vinylpyrrolidone/vinyl acetate copolymers.
CTFA name: PVP/VA Copolymer

Physical form and description:

Series of copolymers covering a range of VP/VA ratios, supplied as either solutions or solid:

Ethanol Solutions (50%):
- PVP/VA E-735(70/30)#
- PVP/VA E-635(60/40)
- PVP/VA E-535(50/50)
- PVP/VA E-335(30/70)

Isopropanol Solutions (50%):
- PVP/VA I-735(70/30)
- PVP/VA I-535(50/50)
- PVP/VA I-335(30/70)

Mole Ratio VA/VP

Solid(100%):
- PVP/VA S-630(60/40)

Aqueous Solution (50%):
- PVP/VA W-735(70/30)

Properties:

Form hard, glossy, water-removable films.

Viscosity, softening point and water sensitivity vary with VP/VA ratio.

Good compatibility with many modifiers and plasticizers permits further variation of hygroscopicity and film flexibility.

Soluble in most common organic solvents.

Good compatibility with aerosol spray propellants.

Uses:

Film-former used in hairsprays, gels, mousses, lotions, hair thickeners, tints and dyes.

GE SILICONES: AF-72 Silicone Antifoam Emulsion:

AF-72 antifoam is an aqueous emulsion of polydimethylsiloxane.

When evaluating AF-72 antifoam, it is suggested that the evaluation be started at 10 ppm silicone. This starting point represents an average level of silicone found to be effective. Except in food processing applications, this concentration can be adjusted in either direction for the specific system requiring foam control.

AF-72 can be premixed in the batch during initial product formulation, or it can be metered into continuous process systems by the use of metering pumps.

Pre-diluted solutions should be kept under mild agitation.

Typical Applications:
- Personal Care Products
- Soaps and Detergents
- Food Processing

Property:
- Total Solids, %: 44.2
- Silicone Content, %: 30
- Density, lbs/gal.: 8.4
- Specific Gravity @ 25/25C (77/77F): 1.01
- Viscosity, cps max @ 25C (77F): 1500
- Color: White
- Heat Stability: Stable at 45C (114F)
- Dilution Stability: Less than 2% creaming and no settling after 24 hours at 10% silicone content.
- Dispersibility: Readily dispersible in warm or cold water with mild agitation.
- Emulsifier Type: Non-ionic
- Storage Stability: No visible settling in 60 days.

GE SILICONES: AF-75 Silicone Antifoam Emulsion:

AF-75 antifoam is a 10% antifoam emulsion of polydimethylsiloxane useful for defoaming aqueous systems. It is a low-viscosity emulsion, assuring ease of handling and dilution.

Typical Applications:
- Personal Care Products
- Soaps and Detergents
- Food Additives
- Food Packaging Materials

Property:
- Silicone Content, %: 10
- Density, lbs/gal.: 8.4
- Specific Gravity @ 25/25C (77/77F): 1.02
- Viscosity, cps max @ 25C (77F): 2500
- Color: White
- Heat Stability of Emulsion: 45C (114F)
- Heat Stability of Fluid: 315C (600F)
- Emulsifier Type: Non-ionic

GE SILICONES: AF-9020 Silicone Antifoam Emulsion:

AF-9020 antifoam is a 20% aqueous emulsion of polydimethylsiloxane useful as an antifoam in many industrial and food processing systems. It is recommended as a defoamer for highly alkaline or acid aqueous systems, exhibiting improved efficiency and longevity over conventional silicone antifoams. This emulsion also provides excellent defoaming properties in non-ionic, cationic, and anionic systems. Its low viscosity and easy dispersibility assure optimum foam control and handling ease.

Property:
- Silicone Content, %: 20
- Density, lbs/gal.: 8.4
- Specific Gravity @ 25/25C (77/77F): 1.01
- Viscosity, cps @ 25C (77F): 2500 max
- Color: White
- Heat Stability: Stable to 45C (114F)
- Dilution Stability: Less than 2% creaming and no settling after 24 hours at 10% silicone content.
- Dispersibility: Readily dispersible in warm or cold water with mild agitation.
- Emulsifier Type: Non-ionic
- Storage Stability: No visible settling in 60 days

GE SILICONES: SF-18(350)/SF-96/VISCASIL Fluids:

SF-96 polydimethylsiloxane fluids are available in 50, 100, 200, 350, 500, and 1000 centistoke viscosities.

VISCASIL fluids are identical in chemical structure to the SF-96 series of silicone fluids, but are higher in viscosity (5000, 10,000, 12,500, 30,000 and 60,000 centistoke viscosities.)

SF-18 (350) can be used in the same applications as SF-96 series fluids; in addition, it meets FDA regulations for direct food additives.

Typical Applications:
- Lotions & Creams
- Shampoos
- Soaps
- Antiperspirants

Property:

SF-18 Fluid:

Silicone Content, %: 100
Density, lbs/gal.: 8.0
Specific Gravity @ 25/25C (77/77F): 0.97
Viscosity, ctks @ 25C: 350
Flash Point C (F) (closed cup): 204 (400)

SF-96 Fluid:

Silicone Content, %: 100
Density, Lbs/gal.: 8.0
Specific Gravity @ 25/25C (77/77F): 0.96-0.97
Viscosity, ctks @ 25C: 50-1000
Flash Point C (F) (closed cup): 204 (400)

VISCASIL Fluid:

Silicone Content, %: 100
Density, lbs/gal.: 8.0
Specific Gravity @ 25/25C (77/77F): 0.97
Viscosity, ctks @ 25C: 5000-60,000
Flash Point C (F) (closed cup): 204 (400)

GE SILICONES: SF-96 (5) Silicone Fluid:

SF-96 (5) is a low-viscosity dimethylpolysiloxane fluid that is soluble in lower alcohols. It also has a greater compatibility with organic materials than other dimethicone products.

It is less volatile than the cyclomethicone fluids and improves lubricity in many formulations. It also improves gloss in hair-spray systems.

Typical Applications:

- Hair Sprays (enhanced gloss)
- Skin Care Products (alcohol solubility, improved lubricity)
- Fragrance Stick (lubricity)

Property:

Silicone Solids, %: 100
Viscosity, Ctks @ 25C (77F): 5
Specific Gravity @ 25/25C (77/77F): 0.916
Flash Point C (F) (closed cup): 135 (275)
Weight Loss, % [24 Hrs. @ 150C (302F)]: 90

GE SILICONES: SE-30 Silicone Gum:

SE-30 is a very-high-molecular-weight dimethicone polymer that is gum-like in nature. For ease of formulation, it is usually dispersed in another silicone system such as cyclomethicone fluid or low-viscosity dimethicone fluid.

Typical Applications:

- Hair Conditioners (conditioning; improved wet-combing)
- Lotions (lubricant; detackifier)
- Color Cosmetics (base)

Property:

Appearance: Clear, colorless
Specific Gravity @ 25/25C (77/77F): 0.95
Silicone Content, %: 100

GE SILICONES: SF-1188 Silicone Fluid:

SF-1188 silicone fluid is a stable copolymer of a polydimethylsiloxane and a polyoxyalkylene ether. It is soluble in cold water, but exhibits inverse solubility above 43C (110F).

SF-1188 is also soluble in lower alcohols and some hydrocarbons. Its use is primarily in hair care products as a conditioning aid and resin plasticizer. It also is a pro-foamer and can be used to improve lubricity in shave creams.

Typical Applications:

- Hair Conditioning Products
- Hair Styling Products
- Shave Creams

Property:

Viscosity, cps @ 25C (77F): 1000
Specific Gravity @ 25/25C (77/77F): 1.04
Flash Point C (F) (closed cup): 82 (180)
Density, lbs/gal. avg.: 8.65
Surface Tension, dynes/cm @ 25C (77F) avg.: 25.5
Appearance: Clear-amber

GE SILICONES: SF-1173/SF-1202/SF-1204 Silicone Fluids:

SF-1173, SF-1202 and SF-1204 are low-viscosity silicone fluids with relatively high volatility. All are cyclic structures: SF-1173 is primarily octamethyl-cyclotetrasiloxane (tetramer), SF-1202 is primarily decamethyl-cyclopentasiloxane, and SF-1204 is 85% SF-1173 and 15% SF-1202.

Non-polar and insoluble in water, these fluids are completely miscible in the lower alcohols as well as in typical aliphatic, aromatic, and halogenated hydrocarbon solvents.

The CTFA name for these products is cyclomethicone.

Typical Applications:

- Antiperspirants
- Skin Care Products
- Sun Screen Products
- Hair Conditioners
- Facial Make-up
- Particle Treatment

Typical Product Data:

SF-1173:

Viscosity, cps @ 25C (77F): 2.4
Boiling Point C: 175
F: 347
Flash Point C: 55
(closed cup) F: 130
Flash Point C: 57
(open cup) F: 135
Refractive Index 25C (77F): 1.394
Specific Gravity @ 25/25C (77/77F): 0.954
Freezing Point C: 17
F: 63

SF-1202:

Viscosity, cps @ 25C (77F): 4.0
Boiling Point C: 210
F: 410
Flash Point C: 82
(closed cup) F: 170
Flash Point C: 88
(open cup): 190
Refractive Index 25C (77F): 1.395
Specific Gravity @ 25C/25C (77/77F): 0.955
Freezing Point C: -40
F: -40

SF-1204:

Viscosity, cps @ 25C: 2.5
Boiling Point C: 175-210
F: 347-410
Flash Point C: 55
(closed cup) F: 130
Flash Point C: 57
(open cup) F: 135
Refractive Index 25C (77F): 1.394
Specific Gravity @ 25/25C (77/77F): 0.954
Freezing Point C: 11
F: 52

GE SILICONES: SF-1214 Silicone Fluid:

This product is a blend of cyclic silicone and silicone gum. Specifically, it is 15% (by weight) SE-30 gum blended with 85% (by weight) SF-1202 (pentamer). It provides several benefits as an ingredient in personal care formulations.

Typical Applications:
- Hair Conditioners (improves wet-combing, conditioning)
- Antiperspirants (anti-whitener)
- Skin Care/Sun Screen Products (protectant, water resistant, feel)

Property:
Silicone Solids, %: 100
Viscosity, ctks @ 25C (77F): 5000-8000
Acidity: 5 ppm maximum
Weight Loss, %: [24 hrs. @ 150C(302F)]: 85+2.0
Color: Clear

Note: This blend can be diluted to a lower percent gum solids using either SF-1202 (pentamer) or SF-1173 (tetramer). It is not soluble in alcohol systems.

GE SILICONES: SF-1228 Silicone Fluid:

SF-1228 is a 10% silicone-polyether copolymer emulsifier dispersed in 90% SF-1204 cyclic silicone. This product can be used to formulate stable water-in-oil emulsions, particularly water-in-silicone emulsions.

Typical Applications:
- Antiperspirants
- Lotions and Creams
- Sun Screen Products
- Facial Make-up
- Hair Care Products

Property:
Appearance: Clear to slight haze
Viscosity, ctks @ 25C (77F): 2500 max.
Silicone/Polyether Ratio: 75/25
Polyether: All ethylene oxide
Active solids, %: 10

Instructions for Use:
As the primary emulsifier in water-in-oil systems, concentrations from 6 to 12% as received (0.6 to 1.2% active ingredient) is recommended. Usually, an auxiliary hydrophilic emulsifier is also required at levels of 10-20% SF-1228 active ingredient. Although good emulsions can be formulated without ionic salts present, the addition of 2% sodium chloride aids emulsification. High shear is necessary to form most stable systems. In antiperspirants, active salt replaces the need for sodium chloride.

GE SILICONES: SF-1236 Silicone Fluid:

SF-1236 is a 100% silicone blend of silicone gum (SE-30) dispersed in 5 cps dimethyl silicone fluid [SF-96 (5)] in a ratio of 15:85. SF-1236 is substantive to skin and hair through mechanical deposition.

Typical Applications:

- Skin Care/Sun Screen Products (protectant, water resistant, feel)
- Facial Make-up (durability)
- Hair Care Products (conditioner, bodying agent)

Property:

Appearance: Clear
Viscosity, cps @ 25C (77F): 3000-5500
Specific Gravity @ 25C/25C (77/77F): .914-.934
Silicone Content %: 100

GE SILICONES: SS-4230 Silicone Fluid:

SS-4230 is a unique blend of a silicone MQ resin (trimethylsiloxysilicate) and SF-1202 cyclic silicone. It is a film former which provides increased water resistance, a "dry" emollient feel, and improved durability. A usage level of 3 to 5% SS-4230 is recommended for the various skin care products.

Typical Applications:

- Sun Screen Products
- Lotions
- Facial Make-up

Property:

Silicone Solids, %: 100
Appearance: Clear
Viscosity, ctks @ 25C (77F): 100-300
Specific Gravity @ 25/25C (77/77F): 1.05

GE SILICONES: SS-4267 Silicone Fluid:

SS-4267 silicone fluid is a unique product combining a resin product with a polydimethylsiloxane fluid to provide increased resistance to water and detergent removal from a surface. In addition, it retains the excellent characteristics of dimethyl fluids often used for detackification and anti-whitening in a variety of lotion cosmetic products.

Typical Applications:

- Sun Screen Products
- Protective Hand Lotions
- Color Cosmetics

Property:

Viscosity, ctks @ 25C (77F): 300-700
Color (Gardner): 1 max
Flash Point C (F) (closed cup): >191(>375)
Specific Gravity @ 25/25C(77/77F): 1.025

GE SILICONES: SM-2059 Silicone Emulsion:

SM-2059 cationic emulsion contains 35% aminofunctional silicone polymer. As the water evaporates, the silicone polymer begins to cure. Once the cure begins, the process is irreversible.

Typical Applications:
- Hair Conditioners
- Hair Styling Aids

Property:
Silicone Solids, %: 35
Color: Off-white
Density, lbs/gal.: 8.25
Viscosity, cps @ 25C(77F): 20
Emulsifier Type: Cationic
Diluent: Water
Shelf Life, months: Up to 6

Instructions for Use:

Because SM-2059 emulsion is reactive, it may not be compatible with certain materials. It is deactivated by acidic materials and may cause gelation of materials containing hydroxy (-OH) groups. Other materials such as anionics, lower alcohols, salts, etc. may break the silicone emulsion. It is recommended that any mixture of SM-2059 emulsion with other materials be thoroughly checked before it is put into factory production.

GE SILICONES: SM-2101 Silicone Emulsion:

SM-2101 is a 35% emulsion of a non-reactive aminofunctional silicone fluid. It is a unique conditioner for difficult-to-condition hair because of its high level of substantivity. It also enhances shine in a variety of hair care products. Because of its non-reactivity, it can be easily removed.

Typical Applications:
- Hair Conditioners
- Shampoos
- Hair Styling Aids

Property:
Silicone Solids, %: 35
Amine Equivalent (meq/g): 0.2
Emulsifier System: Non-ionic
pH Range: 9-11
Color: White
Appearance: Milky
Viscosity, cps @ 25C (77F): 200 max

GIVAUDAN: PARSOL MCX:

The effective non-paba UV-B filter

PARSOL MCX is a recognized safe and effective UV-B filter. PARSOL MCX and PARSOL 1789 combine to extend the UV protection range in the elaboration of sunblocks and protective cosmetic preparations.

Technical data:
C18H26O3
MW: 290.4
2-Ethylhexyl-p-methoxycinnamate
CTFA nomenclature: Octyl methoxycinnamate
CAS No. 5466-77-3

Analytical Specifications:
Appearance: Pale yellow, limpid, slightly oily liquid
Odor: Practically odorless
Boiling point: 198-200C at 3 mm Hg.
Congealing point: Below -25C
Flash point: Above 100C (T.C.C.)
Refractive index: n20/D: 1.542-1.548
Acid value: 1.0 max.
Solubility in methanol: Limpid at 10%
Purity, GLC: 98% min.

Regulatory status: TSCA/EEC registered

Cosmetic Applications:
The application of properly formulated broad spectrum protective sunscreening preparations enable people to enjoy the physiological advantages of sunlight whilst avoiding the physical disadvantages.

PARSOL MCX's excellent safety record, stability, light color, low odor, oil solubility, low viscosity, low freezing point, and ease of incorporation into emulsions makes it a choice filter to formulate NON-PABA sunscreening or protective cosmetic preparations. In a given formulation, the SPF value generally increases with the concentration of the filter (e.g. 4% PARSOL MCX may give an SPF of 4+ and 6% PARSOL MCX may give an SPF of 6+). It should always be kept in mind that the SPF value is formulation-dependant.

GOLDSCHMIDT CHEMICAL CORP.: ABIL Silicones:

ABIL S201:
CTFA: Sodium Poly PG-Propyl Dimethicone Thiosulfate
ABIL S255:
CTFA: Disodium Poly PG-Propyl Dimethicone Thiosulfate
Hair conditioner. Improves gloss and sheen on hair. Hydrophobic on hair. Has thermoset properties. Used in styling aids, mousses and gels.
ABIL B9950:
CTFA: Dimethicone Propyl PG-Betaine
Excellent conditioner for baby products and hair and skin care products. Antistat. Enhances gloss. Very mild to eyes/ skin. The benefits of a betaine plus the properties of a silicone. Used in many ethnic hair care products for gloss and conditioning.
ABIL B8842:
B8843:
B8851:
B8852:
B8863:
B8873:
B88183:
B88184:
CTFA: Dimethicone Copolyol
A series of eight (8) nonionic copolyols that vary in molecular weight and EO/PO adducts. Water soluble to water dispersible. Surface active. Good wetting/lubricating properties. Used in soaps, shampoos, shower gels, creams, lotions, and aerosols. Function as refatting agents. Contribute to compatibility/sheen in shampoos. Improve foam.
ABIL B9806:
CTFA: Cetyl Dimethicone Copolyol
Surfactant. Emulsifier for cyclomethicone. Functions as an emulsifier and emollient in creams and lotions. Forms water-in-oil emulsions. HLB 4-6.
ABIL B9808:
CTFA: Cetyl Dimethicone Copolyol (and) Hexyl Laurate
Surfactant. Water-in-oil emulsifier. Low odor. HLB 4-6.
ABIL WE-09:
CTFA: Cetyl Dimethicone Copolyol (and) Polyglyceryl-4 Isostearate (and) Hexyl Laurate
Water-in-oil emulsifier system. Forms stable emulsions with both hydrocarbons and polysiloxanes. Nonionic. Can be used in cold mixed emulsions. HLB 4-6.
ABIL QUAT 3270:
3272:
CTFA: Quaternium-80
Provides conditioning for skin care and hair care products. Improves sheen on hair. Used in both shampoos and conditioners. Antistat. Improves wet and dry combing.

GOLDSCHMIDT CHEMICAL CORP.: ABIL Silicone Oils:

Methyl silicone oils:

ABIL 10-10000:
Polydimethyl siloxanes
Liquid/colourless
Type: Nonionic
Specific Gravity: 0,94-0,98
Barrier against aqueous media, skin protection, perfume and active substance fixative, improved rub and spreadability, faster penetration, no sticking, good emollients.
Prevent aerosol valve clogging, sticking

Phenylmethyl silicone oils:

ABIL AV 20-1000:
Phenylmethyl polysiloxanes
Liquid/colourless
Nonionic
Specific Gravity: 0,97-1,078

ABIL AV 8853:
Phenylmethyl polysiloxanes
Liquid/colourless
Type: Nonionic
Specific Gravity: 0,90-0,94
Barrier against aqueous media, skin protection, perfume and active substance fixative, improved rub and spreadability, faster penetration, no sticking, good emollients.
Prevent aerosol valve clogging, non-sticky.

Volatile silicone oils:

ABIL K4:
Octamethyl cyclotetrasiloxane
Liquid/colourless
Nonionic
Specific Gravity: 0,94-0,97

ABIL B8839:
Decamethyl cyclopentasiloxane
Liquid/colourless
Type: Nonionic
Specific Gravity: 0,94-0.97
Residual free application of active ingredients; improvement of spraying; gives hair a silky shine; improved spreadability of emulsions.

GOLDSCHMIDT CHEMICAL CORP.: ABIL Silicone Surfactants:

Polyether polysiloxane copolymers:

ABIL B 8843:
Polyether polysiloxane copolymers
Liquid/pale yellow
Type: nonionic
Specific Gravity: 1,070

ABIL B 8851:
Polyether polysiloxane copolymers
Liquid/pale yellow
Type: nonionic
Specific Gravity: 1,050

ABIL B 8852:
Polyether polysiloxane copolymers
Liquid/pale yellow
Type: nonionic
Specific Gravity: 1,011

ABIL B 8863:
Polyether polysiloxane copolymers
Liquid/amber
Type: nonionic
Specific Gravity: 1,036

ABIL B 88183:
Polyether polysiloxane copolymers
Liquid/pale yellow
Type: nonionic
Specific Gravity: 1,024
Water soluble silicone surfactants.
Widely used in hair care products, deodorants, antiperspirants, creams and lotions.
Prevent aerosol valve clogging, plasticize hairspray resins.

ABIL B 88184:
Polyether polysiloxane copolymers
Liquid/pale yellow
Type: nonionic
Specific Gravity: 1,04
Anticracking agent for bar soaps.

W/O Emulsifiers:

ABIL WS 08:
Polysiloxan polyalkylen polyether copolymers
Liquid/pale yellow
Type: nonionic
Specific Gravity: 0,87-0,91

ABIL WE 09:
Polysiloxan polyalkylen polyether copolymers
Liquid/pale yellow
Nonionic
Specific Gravity: 0,89-0,93
For the manufacture of W/O creams or lotions.
Hot or cold emulsification possible.

GOLDSCHMIDT CHEMICAL CORP.: ABIL Special Products:

Cosmetic oils/waxes:

ABIL-Wax 9800:
Polysiloxane polyalkylene copolymers
Waxy/white
Type: nonionic
Specific Gravity: 0,86

ABIL-Wax 9801:
Polysiloxane polyalkylene copolymers
Liquid/pale yellow
Type: Nonionic
Specific Gravity: 0,86

ABIL-Wax 2434:
Dialkyl dimethyl polysiloxanes
Waxy/pale yellow
Type: nonionic
Specific Gravity: 0,88

ABIL-Wax 2440:
Dialkyl dimethyl polysiloxanes
Waxy/pale yellow
Type: nonionic
Specific Gravity: 0,90

Cosmetic oils/waxes, high lubricity, water protective, high spreadability, fast penetration.

Quaternary compounds:

ABIL-Quat 3270:
Diquaternary polydimethyl siloxanes
Liquid/amber
Type: cationic
Specific Gravity: 1,014

ABIL-Quat 3272:
Diquaternary polydimethyl siloxanes
Liquid/amber
Type: cationic
Specific Gravity: 1,008
Haircare: Conditioning agent, hair gets silky shine.
Skin cleansing: Film former, refatter.

Silicone betaine:

ABIL-B 9950:
Polysiloxane polyorganobetaine copolymer
Liquid/yellow
Type: amphoteric
Specific Gravity: 1,078-1,092
Hair care: Conditioning agent.
Skin cleansing: Film former.

GOLDSCHMIDT CHEMICAL CORP.: ABIL Waxes:

ABIL-Wax 9800:

CTFA: Stearyl Dimethicone

Improves the color, luster, and spreadability of pigmented products. In skin care products their low viscosity, spreading and penetrating properties are desirable. Waxy liquid.

ABIL-Wax 9801:

CTFA: Cetyl Dimethicone

Liquid wax useful in pigmented products and skin care products. Good pigment solubilizer. Soluble in cyclomethicone. Provides emolliency and application benefits for antiperspirants.

ABIL-Wax 2434:

CTFA: Stearoxy Dimethicone

Uses similar to ABIL-Waxes 9800/9801. Melting point 25C. Provides water barrier for creams and lotions. Soluble in cyclomethicone. Waxy liquid.

ABIL-Wax 2440:

CTFA: Behenoxy Dimethicone

Reduces whitening during application of creams and lotions, good pigment disperser. Melting point 35C. Soluble in mineral oil, cosmetic esters and oils. Thickens low molecular weight dimethicones. Glossing agent for cosmetics.

ABIL-Wax 9809:

CTFA: Stearyl Methicone

Provides water barrier for night creams and protection lotions. Useful in long lasting pigmented products. Soluble in mineral oil. Melting point 35C. Conditions hair.

ABIL-Wax 9810:

CTFA: C24-28 Alkyl Methicone

High molecular weight silicone wax useful as a thickening agent for mineral oils and cosmetic esters. Reduces greasiness. Provides water barrier for night creams and protecting lotions. Provides gloss and smoothness for lipsticks. Gels mineral oil, cosmetic esters and oils. Melting point 65C.

ABIL-Wax 9811:

CTFA: C30-45 Alkyl Methicone

High melting point silicone wax. Melting point 75C. Gels mineral oil, cosmetic esters and oils. Increases and stabilizes viscosity in creams and lotions. Has application in hot pour pigmented products.

GOLDSCHMIDT CHEMICAL CORP.: Additional Products:

LACTIL:

CTFA: Sodium Lactate (and) Sodium PCA (and) Hydrolyzed Animal Protein (and) Fructose (and) Urea (and) Niacinamide (and) Inositol (and) Sodium Benzoate (and) Lactic Acid

Moisturizer and humectant for skin and hair care products. A natural moisturizing factor.

TEGO-Pearl B-48:

CTFA: Cocamidopropyl Betaine (and) Glycol Distearate (and) Cocamide DEA (and) Cocamide MEA

Pearlizes and opacifies hair care products. Cold mixed.

TEGAMINE-18:

CTFA: Stearamidopropyl Dimethylamine

A cationic surfactant. Functions as a conditioner in hair care products. Effective auxiliary emulsifier for creams and lotions. Provides skin feel.

ANTIL 141 Solid:

CTFA: PEG-55 Propylene Glycol Oleate

ANTIL 141 Liquid:

CTFA: Propylene Glycol (and) PEG-55 Propylene Glycol Oleate

ANTIL 141 is used as a thickening agent for aqueous solutions of surfactants such as shampoos, foam baths, shower preparations and liquid soaps. It is compatible with all standard anionic, cationic, amphoteric and nonionic surfactants. ANTIL 141 may be used in a pH range of pH 5 up to pH 8. ANTIL 141 shows good dermatological compatibility and was especially developed for particularly mild formulations based on anionic/TEGO Betaine mixes. In addition to the thickening effect, ANTIL 141 is a good solubilizer. By increasing the solubilizing effect of the surface active substances, it contributes to forming clear solutions with perfume oils or essential oils and natural oils.

ANTIL 141 liquid was developed for cold process systems.

GOLDSCHMIDT CHEMICAL CORP.: Betaines/Amine Oxides:

TEGO-Betaine L-7:
TEGO-Betaine C:
TEGO-Betaine S:
CTFA: Cocamidopropyl Betaine
Amphoteric surfactants produced from coconut oil derived fatty acids. They act as mollifiers, viscosity builders and foam boosters in hair and skin cleansing products. TB L-7 is especially mild with current use in many baby bath and baby shampoo products.

TEGO-Betaine L5351:
CTFA: Cocamidopropyl Betaine
A low salt amphoteric surfactant for use in salt critical formulations; such as hair dyes, fixatives and reactive products.

TEGO-Betaine L-90:
CTFA: Lauramidopropyl Betaine
Mild amphoteric. Surfactant based in lauric acid. Used in shampoos and conditioners. Foam booster and viscosity builder.

TEGO-Betaine HS:
CTFA: Cocamidopropyl Betaine (and) Glyceryl Laurate
Mild amphoteric surfactant system designed for shampoos, conditioners and personal care products.

TEGAMINE Oxide WS-35:
CTFA: Cocamidopropylamine Oxide
Surfactant stable over a wide pH range. Used in hair and skin cleansing. Good foamer and viscosity builder.

GOLDSCHMIDT CHEMICAL CORP.: Esters and Emulsifiers:

TEGIN 515 VA:
CTFA: Glyceryl Stearate
A glyceryl monostearate emulsifier and opacifier for skin care products. Can be used in nonionic, cationic and anionic systems.
TEGIN VA:
CTFA: Glyceryl Stearate S.E.
Anionic. Self emulsifying glyceryl monostearate. Used in systems with a pH 6-12. Forms O/W emulsions for creams and lotions.
TEGACID Regular VA:
CTFA: Glyceryl Stearate (and) Stearamidoethyl-diethylamine
Cationic. Self emulsifying glyceryl monostearate that forms stable emulsions for craems and lotions. Reduces need for auxiliary emulsifiers in the acid pH range. Acid stable.
TEGACID Special:
CTFA: Glyceryl Stearate (and) Sodium Lauryl Sulfate
Anionic. Self emulsifying glyceryl monostearate for creams and lotions. Broad pH range--acid to alkaline. Very stable emulsions can be formed. Acid stable.
Ethylene Glycol Distearate VA:
CTFA: Glycol Distearate
Hair Care. Opacifier and pearling agent.
Ethylene Glycol Monostearate VA:
CTFA: Glycol Stearate
Stabilizes and thickens creams and lotions. Pearlizes and opacifies.
PROTEGIN:
PROTEGIN X:
CTFA: Mineral oil (and) Petrolatum (and) Ozokerite (and) Glyceryl Oleate (and) Lanolin Alcohols
A series of absoprtion bases for W/O emulsions. Produce soft stable creams compatible with a wide range of organic and inorganic compounds. Used as cosmetic and pharmaceutical bases.
PROTEGIN W:
PROTEGIN WX:
CTFA: Petrolatum (and) Ozokerite (and) Hydrogenated castor oil (and) Glyceryl Isostearate (and) Polyglyceryl-3-Oleate
A series of absorption bases for W/O emulsions. Produce soft stable creams compatible with a wide range of organic and inorganic compounds. Used as cosmetic and pharmaceutical bases.
TEGO Care 150:
CTFA: Glyceryl Stearate (and) Steareth-25 (and) Ceteth-20 (and) Stearyl Alcohol
Nonionic O/W emulsifier system for creams and esters. HLB 12. Produces cosmetically elegant and very stable emulsions.

GOLDSCHMIDT CHEMICAL CORP.: O/W-Emulsifiers for Acid, Alkali or Salt Stable Emulsions:

TEGO-CARE 150:
Blend of emulsifying and consistency-promoting substances, nonionic.
Powder/ivory
HLB: 12
Rise Melting Point C: 52-58
Acid Value: max. 3
Saponification Value: 90-106
Emulsifier for O/W creams with good hot and cold stability.
Extraordinary broad application properties.

TEGINACID, se:
Glycerol mono distearates mixed with other nonionics
Powder/ivory
Total Monoester: 50-60%
HLB: 12
Rise Melting Point C: 58-63
Acid Value: max. 3
Saponification Value: 153-165
Iodine Value: max. 4

TEGINACID X, se:
Glycerol mono distearates mixed with other nonionics
Powder/ivory
Total Monoester: 45-55%
HLB: 12
Rise Melting Point C: 55-61
Acid Value: max. 3
Saponification Value: 145-160
Iodine Value: max. 4

TEGINACID H, se:
Glycerol mono distearates mixed with other nonionics
Powder/ivory
Total Monoester: 20%
HLB: 11
Rise Melting Point C: 45-51
Acid Value: max. 3
Saponification Value: 55-70
Iodine Value: max. 3

TEGINACID Spezial, se:
Glycerol mono distearates mixed with fatty alcohol sulfates
Powder/ivory
Total Monoester: 35-45%
HLB: 12
Rise Melting Point C: 53-59
Acid Value: 16-18
Saponification Value: 160-172
Iodine Value: max. 3
Used for acid and salt-resisting emulsions of O/W type of liquid or creamy consistency.

GOLDSCHMIDT CHEMICAL CORP.: O/W-Emulsifiers for Acid, Alkali or Salt Stable Emulsions(Continued):

EMULGATOR E 2149, se:
Blends of nonionics
Waxy/white/ivory
HLB: 11
Rise Melting Point C: 41-47
Acid Value: max. 2
Saponification Value: max. 3
Iodine Value: max. 2

EMULGATOR E 2155, se:
Blends of nonionics
Waxy/white/ivory
HLB: 11
Rise Melting Point C: 49-55
Acid Value: max. 2
Saponification Value: max. 3
Iodine Value: max. 2

EMULGATOR E 2568, se:
Blends of nonionics
Waxy/white/ivory
HLB: 16
Rise Melting Point C: 43-49
Acid Value: max. 2
Saponification Value: max. 3
Iodine Value: max. 1

Emulsifiers and stabilizers for cosmetic and pharnaceutical emulsions of O/W type. Resistant to alkaline and acid substances.

GOLDSCHMIDT CHEMICAL CORP.: Special Products:

Aminoxide:

AMINOXID WS35:
Alkyl amido aminoxide
Liquid/amber
Type: nonionic
pH Value: 5-7
Specific Gravity: 1,000-1,010
Active Matter: 35%
Nonionic surfactant for use in shampoos and skin cleansing products.

Conditioning agent:

TEGO-Amid S18:
Stearamidopropyl dimethylamine
Flakes/ivory
Type: tertiary amine
Acid Value: max. 4
Cationic emulsifier for creams and lotions, conditioner for hair care products.

Pearlizing agents:

TEGO-Pearl B48:
Mixture of pearlescent substances with surfactants
Liquid/white
Type: amphoteric
pH Value: 5-7
Active Matter: 44%
Pearlescent agent for shampoos, bath and shampoo preparations with excellent superfatting properties, cold processible.

TEGO-Pearl S33:
Mixture of pearlescent substances with surfactants
Liquid/white
Type: Anionic/amphoteric
pH value: 5,5-7,5
Active Matter: 33%
Pearlescent agent for shampoos, bath and shower preparations with excellent superfatting properties; cold processible.

GOLDSCHMIDT CHEMICAL CORP.: Special Products:

Thickening agent:

ANTIL 141 liquid:
Polyoxyethylene propylene glycol dioleate
Low viscosity liquid/pale yellow
Type: Nonionic
Acid Value: max. 5
Saponification Value: 10-22
Thickening agent for surfactant systems.

Cosmetic oils/wax esters:

JOJOBA OIL:
Natural wax ester
Liquid/yellow
Natural liquid wax esters for high quality cosmetics.

PUR-OBA:
Natural wax ester
Liquid/nearly colourless
Natural liquid wax esters for high acidity cosmetics.

Humectants:

LACTIL:
Blend of humectants
Liquid/yellow
Spec. gravity: 1,27
pH Value: 6,7-7,6
Dry residue: 50-55%
For use in hydro-regulative cosmetics like moisturizing creams, anti-aging-preparations or after-swim products.

GOLDSCHMIDT CHEMICAL CORP.: Surfactants: Amido Alkyl Betaines:

TEGO Betain L 7, spray dried:
Fatty acid amido alkyl betaines
Powder/ivory
Sodium chloride content: 15%
Type: amphoteric
pH Value: 5
Active Matter: 75-85%
Betaine used in cosmetic formulations without water.

TEGO Betain L7:
Fatty acid amido alkyl betaines
Liquid/yellow
Sodium chloride content: 5%
Type: amphoteric
pH Value: 5
Active matter: 30%
For non-irritating shampoos, foam baths, hygiene, shower and baby care preparations.

TEGO Betain F:
Fatty acid amido alkyl betaines
Liquid/yellow
Sodium chloride content: 5%
Type: amphoteric
pH Value: 5-7
Active matter: 30%
For non-irritating shampoos, foam baths, hygiene, shower and baby care preparations.

TEGO Betain HS:
Fatty acid amido alkyl betaines
Liquid/yellow
Sodium chloride content: 5%
Type: amphoteric
pH Value: 6-7
Active matter: 30%
Especially suitable for cosmetic products for sensitive skin. Good refatting properties.

TEGO Betain L5351:
Fatty acid amido alkyl betaine, low salt
Liquid/yellow
Sodium chloride content: max. 1,2%
Type: amphoteric
pH Value: 6
Active matter: 30%
Amphoteric surfactant for aerosols and electrolytesensitive formulations.

GOLDSCHMIDT CHEMICAL CORP.: TAGAT Polyoxyethylene Glycerol Fatty Acid Esters:

TAGAT L:
Polyoxyethylene glycerol monolaurate
Liquid/pale yellow
Moles Ethylene Oxide: 30
HLB: 17,0
Hydroxyl Value: 52-68
Acid Value: max. 2
Saponification Value: 35-55
Iodine Value: max. 4

TAGAT L2:
Polyoxyethylene glycerol monolaurate
Liquid/ivory
Moles Ethylene Oxide: 20
HLB: 15,7
Hydroxyl Value: 60-80
Acid Value: max. 2
Saponification Value: 50-70
Iodine Value: max. 4

TAGAT S:
Polyoxyethylene glycerol monostearate
Solid, partially liquid/ivory
Moles Ethylene Oxide: 30
HLB: 16,4
Hydroxyl Value: 53-70
Acid Value: max. 2
Saponification Value: 30-47
Iodine Value: max. 2

TAGAT S2:
Polyoxyethylene glycerol monostearate
Solid, partially liquid/ivory
Moles Ethylene Oxide: 20
HLB: 15,0
Hydroxyl Value: 65-85
Acid Value: max. 2
Saponification Value: 40-60
Iodine Value: max. 2

TAGAT I:
Polyoxyethylene glycerol monoisostearate
Liquid/pale yellow
Moles Ethylene Oxide: 30
HLB: 15,6
Hydroxyl Value: 90-105
Acid Value: max. 2
Saponification Value: 25-40
Iodine Value: max. 4

Solubilizers for water-insoluble substances such as flavours, perfumes, vitamin oils, etc.

In combination with lipophilic emulsifiers for the preparation of O/W emulsions.

GOLDSCHMIDT CHEMICAL CORP.: TAGAT Polyoxyethylene Glycerol Fatty Acid Esters(Continued):

TAGAT I2:
Polyoxyethylene glycerol monoisostearate
Liquid/pale yellow
Moles Ethylene Oxide: 20
HLB: 14,2
Hydroxyl Value: 110-130
Acid Value: max. 2
Saponification Value: 45-65
Iodine Value: max. 4

TAGAT O:
Polyoxyethylene glycerol monooleate
Liquid/yellow
Moles Ethylene Oxide: 30
HLB: 16,4
Hydroxyl Value: 50-65
Acid Value: max. 2
Saponification Value: 30-45
Iodine Value: 15-19

TAGAT O2:
Polyoxyethylene glycerol monooleate
Liquid/yellow
Moles Ethylene Oxide: 20
HLB: 15,0
Hydroxyl Value: 70-85
Acid Value: max. 2
Saponification Value: 40-55
Iodine Value: 21-27

TAGAT TO:
Polyoxyethylene glycerol trioleate
Liquid/amber
Moles Ethylene Oxide: 25
HLB: 11,3
Hydroxyl Value: 18-33
Acid Value: max. 12
Saponification Value: 75-90
Iodine Value: 34-40

TAGAT R40:
Ethylene oxide derivative of castor oil
Solid/ivory
Moles Ethylene Oxide: 40
HLB: 13,0
Hydroxyl Value: 55-75
Acid Value: max. 2
Saponification Value: 45-65
Iodine Value: max. 2

Solubilizers for water-insoluble substances such as flavours, perfumes, vitamin oils, etc.

In combination with lipophilic emulsifiers for the preparation of O/W emulsifiers.

GOLDSCHMIDT CHEMICAL CORP.: TAGAT Polyoxyethylene Glycerol Fatty Acid Esters(Continued):

TAGAT R60:
Ethylene oxide derivative of castor oil
Solid/ivory
Moles Ethylene Oxide: 60
HLB: 15,0
Hydroxyl Value: 38-58
Acid Value: max. 2
Saponification Value: 35-52
Iodine Value: max. 2

TAGAT R63:
Ethylene oxide derivative of castor oil
Liquid/pale yellow
Moles Ethylene Oxide: 60
HLB: 15,0
Acid Value: max. 2

Solubilizers for water-insoluble substances such as flavours, perfumes, vitamin oils, etc.

In combination with lipophilic emulsifiers for the preparation of O/W emulsions.

GOLDSCHMIDT CHEMICAL CORP.: TEGIN Edible Acid Esters of Mono-Diglycerides:

1,2-Propylene glycol ester:

TEGIN P:
Propylenglykolmonostearat, se
Waxy/white/ivory
Total Monoester: 50%
HLB: 12
Rise Melting Point C: 39-46
Acid Value: 16-18
Saponification Value: 150-170
Iodine Value: max. 3
Used for cosmetic and pharmaceutical emulsions of O/W type.

Citric acid ester:

TEGIN C 62, se:
Powder/ivory
HLB: 10
Rise Melting Point C: 58-64
Saponification Value: 215-265
Iodine Value: max. 3
Hydrophilic emulsifier for cosmetic O/W emulsions.

GOLDSCHMIDT CHEMICAL CORP.: TEGIN--Glycerol Fatty Acid Partial Esters:

Palmitic stearic acid esters:

TEGIN, se:
Powder/white/ivory
Total Monoester: 32-40%
HLB: 12
Rise Melting Point C: 57-62
Acid Value: 32-36
Saponification Value: 145-160
Iodine Value: max. 4

TEGIN Spezial, se:
Powder/white/ivory
Total Monoester: 40-48%
HLB: 12
Rise Melting Point C: 54-60
Acid Value: 16-20
Saponification Value: 148-158

Used for cosmetic and pharmaceutical emulsions of O/W type; self-emulsifying.

TEGIN M, nse:
Powder/white/ivory
Total Monoester: 60%
HLB: 3,8
Rise Melting Point C: 58-63
Acid Value: max. 2
Saponification Value: 165-180
Iodine Value: max. 4

TEGIN 4100, nse:
Powder/white/ivory
Total Monoester: 45%
HLB: 3,8
Rise Melting Point C: 58-63
Acid Value: max. 2
Saponification Value: 164-180
Iodine Value: max. 3

TEGIN 90, nse:
Powder/white/ivory
Total Monoester: >90%
HLB: 4,5
Rise Melting Point C: 67-72
Acid Value: max. 3
Saponification Value: 155-170
Iodine Value: max. 2

Used as stabilizers for creamy and liquid emulsions of O/W type in cosmetic and pharmaceutical fields.

GOLDSCHMIDT CHEMICAL CORP.: TEGIN--Glycerol Fatty Acid Partial Esters(Continued):

Oleic acid ester:

TEGIN O, nse:
Pasty/pale yellow
Total Monoester: 60%
HLB: 3,3
Acid Value: max. 2
Saponification Value: 158-175
Iodine Value: 70-76
Used for cosmetic and pharmaceutical W/O emulsions. Listed in the Ph. Hevetica.

Isostearic acid ester:

TEGIN ISO, nse:
Pasty/ivory
Total Monoester: 50%
HLB: 3,4
Acid Value: max. 2
Saponification Value: 150-165
Iodine Value: max. 15
Used for cosmetic W/O emulsions.

GOLDSCMIDT CHEMICAL CORP.: TEGIN Glycol Fatty Acid Esters:

Ethylene glycol esters:

TEGIN G:
Ethylenglykolmonostearat, se
waxy/white/ivory
Total Monester: 50%
HLB: 12
Rise Melting Point C: 48-53
Acid Value: 36-38
Saponification Value: 150-165
Iodine Value: max. 3
Used for cosmetic and pharmaceutical emulsions of O/W type.

TEGIN G 6100:
Ethylenglykolmonostearat, nse
Powder/white/ivory
Total Monoester: 50-60%
HLB: 3,2
Rise Melting Point C: 53-58
Acid Value: max. 3
Saponification Value: 180-195
Iodine Value: max. 3
Lipophilic co-emulsifier for O/W type emulsions, promotes silky shine when used in shampoos etc.

TEGIN G 1100:
Ethylenglykoldistearat, nse
Powder/white/ivory
Total Monoester: 90% Diester
Rise Melting Point C: 59-63
Acid Value: max. 6
Saponification Value: 192-208
Iodine Value: max. 3
Pearlizing agent for shampoos, shower and bath preparations.

TEGIN D 6100:
Diethylenglykolmonostearat, nse
Waxy/white/ivory
Total Monoester: 50-60%
HLB: 2,8
Rise Melting Point C: 40-46
Saponification Value: 145-160
Iodine Value: max. 3
Lipophilic co-emulsifier for O/W type emulsions.

TEGIN D 1102:
Triethylenglykoldistearat, nse
Powder/white/ivory
Total Monoester: 90% Diester
HLB: 3,8
Rise Melting Point C: 47-52
Acid Value: 5-7
Saponification Value: 168-178
Iodine Value: max. 3
Pearlizing agent for shampoos, shower and bath preparations.

GOLDSCHMIDT CHEMICAL CORP.: W/O Emulsifiers:

PROTEGIN, se:
Blends of nonionics, sterols, aliphatic alcohols and hydrocarbons.
Waxy/ivory
HLB: 3
Thawpoint acc. to Ubbelohde C: 58-65
Acid Value: max. 1
Saponification Value: 8-12
Hydroxyl Value: 18-30

PROTEGIN X, se:
Blends of nonionics, sterols, aliphatic alcohols and hydrocarbons.
Waxy/ivory
HLB: 3
Thawpoint acc. to Ubbelohde C: 58-65
Acid Value: max. 1
Saponification Value: 10-16
Hydroxyl Value: 25-38
W/O absorption bases. Use: cosmetic and pharmaceutical fields.

PROTEGIN W, se:
Blends of nonionic fatty acid esters of polyols with waxes and purified saturated hydrocarbons.
Waxy/ivory
HLB: 3
Thawpoint acc. to Ubbelohde C: 75-82
Acid Value: max. 1
Saponification Value: 18-28
Hydroxyl Value: 18-28

PROTEGIN WX, se:
Blends of nonionic fatty acid esters of polyols with waxes and purified saturated hydrocarbons.
Waxy/ivory
HLB: 3
Thawpoint acc. to Ubbelohde C: 76-83
Acid Value: max. 1
Saponification Value: 27-37
Hydroxyl Value: 32-42
W/O absorption bases for especially soft and white creams. Use: cosmetic and pharmaceutical fields.

B.F. GOODRICH: CARBOPOL Water Soluble Resins:

CARBOPOL resins are a family of water-soluble resins used to:
...thicken--producing a wide range of viscosities and flow properties
...suspend insoluble ingredients
...stabilize emulsions

CARBOPOL resins offer these benefits:
Thickening efficiency-high viscosities at low concentrations
Uniform performance-reproducibility unattainable with natural gums.
History-over 25 years of successful use.
Temperature stability-only slight affect on mucilage properties after extended heating or cooling.
Unaffected by aging-excellent shelf life.
Microbial resistance-resists bacterial attack and does not support mold growth.
Versatility-although primarily used in aqueous systems with neutralization, it can also be used in solvent systems, with or without neutralization.

CARBOPOL resins are acrylic acid polymers offered as fluffy, dry powders (100% effective). The carboxyl groups in these polymers are responsible for many of the product benefits. CARBOPOL resins have an average equivalent weight of 76.

Types of CARBOPOL Resins:

907:
Molecular Weight (Approx.): 450,000
Special Properties: Exceptional water solubility.
Lubricity without viscosity.
Suggested Uses:
Applications requiring high carboxylic content without appreciable increase in viscosity. Builds solids in compounded formulations. Drilling mud, photo-sensitive emulsions, water treatment, fiber retention aid.

910:
Molecular Weight (Approx.): 750,000
Special Properties: Good ion tolerance, long rheology.
Effective at low (ppm) concentrations.
Suggested Uses:
Where long rheology is required such as in flocking, dip coating, back coating in textile applications. Brush-on-paints. Can be used to modify rheology of high-molecular-weight CARBOPOL resins.

B.F. GOODRICH: CARBOPOL Water Soluble Resins(Continued):

Types of CARBOPOL Resins(Continued):

941:

Molecular Weight (Approx.): 1,250,000

Special Properties: High stable emulsions and suspensions at relatively low viscosities--even with ionic systems. More efficient than CARBOPOL 934 and 940 at low to moderate concentrations.

Suggested Uses:

Emulsion stabilization of cosmetics, such as shampoos, lotions, thin gels with good clarity. Industrial uses: paper manufacturing and treatment, fiberglass and natural and synthetic fiber processing. Effective in moderately ionic system.

934:

Molecular Weight (Approx.): 3,000,000

Special Properties: Excellent stability at high viscosity. Produces thick formulations such as heavy gels, emulsions and suspensions.

Suggested Uses:

Stable lubricating, quenching and silicone emulsions. Stable graphite, polyethylene, fiber and paper suspensions. Water and solvent-based gels.

934P:

Molecular Weight (Approx.): 3,000,000

Special Properties: High purity grade of CARBOPOL 934 resin.

Suggested Uses:

Especially for the pharmaceutical industry. For thickening, suspending and emulsifying; for topical lotions, formulating sustained release tablets. More extensive toxicity studies have been conducted on CARBOPOL 934P than on the other CARBOPOL resins. A master file has been estabilished with the United States Food and Drug Administration.

940:

Molecular Weight (Approx.): 4,000,000

Special Properties: Excellent thickening efficiency at high viscosities.

Suggested Uses:

In cosmetic applications, for sparkling clear, water or hydro-alcoholic gels. Efficient solvent thickening, with or without neutralizing. Industrial uses: die casting and forging lubricants, thixotropic paints, can coating formulating. Used in water and solvent systems.

W.R. GRACE & CO.: SYLOID Silicas:

Industry: Pharmaceuticals, Cosmetics, and Household Products:
Function: Thickening and gelation
Suspension agents
Tabletting agents
Glidants
Protection against moisture
Carriers for active ingredients

Physical Properties:
Color: White powder; transparent and colorless in liquids
Refractive Index: 1.46
Odor: none
Hardness: 5 on Moh Scale
Solubility: Insoluble except in HF and strong bases, such as NaOH
Reactivity: Completely inert

63FP:
Loss on Ignition @ 1750F (%): 6.5
pH (5% Slurry in H2O): 4.3
SiO2 (%, ignited basis): 99.4
Avg. Particle Size (microns): 9.0
Color (Hunter): 95.0

72FP:
Loss on Ignition @ 1750F (%): 6.0
pH (5% Slurry in H2O): 6.5
SiO2 (%, ignited basis): 99.4
Avg. Particle Size (microns): 4.0
Color (Hunter): 96.0

74FP:
Loss on Ignition @ 1750F (%): 6.0
pH (5% Slurry in H2O): 6.5
SiO2 (%, ignited basis): 99.4
Avg. Particle Size (microns): 6.0
Color (Hunter): 96.0

244FP:
Loss on Ignition @ 1750F (%): 8.5
pH (5% Slurry in H2O): 7.0
SiO2 (%, ignited basis): 99.4
Avg. Particle Size (microns): 3.0
Color (Hunter): 96.0

W.R. GRACE & CO.: SYLOX Silicas:

Typical Properties:

SYLOX 2:

Chemical:
- Total Volatiles: 10%
- pH (5% in H_2O): 6-8
- Silica: 99%
- Na_2O: 0.3%
- SO_4: 0.2%

Physical:
- Surface Area: 225 m^2/g
- Oil Absorption: 230 lb./100 lb.
- Bulk Density: 4-6 lb/ft^3
- Particle Size: 2.0

SYLOX 15:

Chemical:
- Total Volatiles: 10%
- pH (5% in H_2O): 6-8
- Silica: 99%
- Na_2O: 0.3%
- SO_4: 0.2%

Physical:
- Surface Area: 225
- Oil Absorption: 220
- Bulk Density: 7-10
- Particle Size: 12

HAARMAN & REIMER: Perfumery Components:

Acetanisole cryst.

Acetophenone pure

Agrumex HC
o-tert.-Butylcyclohexylacetate

Alcohol C 8

Alcohol C 9

Alcohol C 10

Alcohol C 11

Alcohol C 12

Aldehyde C 8

Aldehyde C 9

Aldehyde C 10

Aldehyde C 11
Undecylene

Aldehyde C11MOA

Aldehyde C12MNA

Aldehyde C14 so-called

Aldehyde C16 so-called

Aldehyde C16 so-called, special

Aldehyde C18 so-called/Abricolin

Allyl caproate

Allyl cyclohexylpropionate

Allyl heptoate

Allyl phenoxyacetate

Ambrettia C
Substitute for Musk Ambrette

HAARMAN & REIMER: Perfumery Components(Continued):

alpha-Amylcinnamic aldehyde/Jasminal H&R

n-Amyl salicylate

Anethol NPU 21/22

Anethol supra 21/5

Anisaldehyde pure

Anisole

Anisyl acetate

Anisalcohol

Aurantesin abs.

Benzaldehyde ffc

Benzophenone cryst.

Benzyl acetate

Benzylacetone

Benzyl alcohol ffc double dist.

Benzyl alcohol FR

Benzyl benzoate H&R

Benzyl butyrate

Benzyl cinnamate

Benzyl formate

Benzyl propionate

Benzyl salicylate

Butyl butyrate

Butyl salicylate

HAARMAN & REIMER: Perfumery Components(Continued):

Cedryl acetate liquid

Cetodecanolide 50%

Cinnamic acid FCC 60 Mesh

Cinnamic aldehyde 100% ffc

Cinnamic alcohol

Cinnamyl acetate

Citral H&R pure

Citronellal supra

Citronellol supra

Citronellyl acetate supra

Citronellyl formate supra

Citronitril

Citral diethyl acetal

Citrylal

p-Cresyl methyl ether

Cyclopentadecanolide supra

Cyclorosan

beta-Decalyl acetate

beta-Decalyl formiate

Decyl acetate

Diacetyl pure

Dibenzyl ether

Diethyl phthalate (DEP)

HAARMAN & REIMER: Perfumery Components(Continued):

Dimethyl anthranilate

Dimethyl anthranilate extra

Dimethyl benzyl carbinyl butyrate

Diphenyl oxide

Ethyl benzoate

Ethyl butyrate

Ethyl caproate

Ethyl caprylate

Ethyl cinnamate

Ethyl formate pure

Ethyl heptanoate

Ethyl isobutyrate

Ethyl isovalerate

Ethyl phenyl acetate

Ethyl phenyl carbinol

Ethyl propionate

Ethyl salicylate

Ethyl vanillin/Bourbonal H&R

Eucalyptol

Eugenol ex Clove oil

Eugenol methyl ether extra

Farenal

Floropal

Freesiol

Frescolat

HAARMAN & REIMER: Perfumery Components (Continued):

Geraniol supra

Geranyl acetate

Geranyl formate

Heliofolal

Heliopan
o-Methoxy cinnamic aldehyde

Herbaflorat [4,7-Methano-3a,4,5,6,7,7a-hexahydro-5(or 6)-indenyl acetate]

Hexyl acetate

Hexyl salicylate

Hydroxycitronellal

Indoflor H&R cryst.

Indole

Ionone alpha 100%

Ionone pure 100%

Iraldein 100%

Iraldein delta H&R

Iraldein gamma H&R

Iraldein gamma pure

Isoamyl acetate

Isoamyl butyrate

Isoamyl isovalerate

Isoamyl salicylate

Isonanat

Isobornyl acetate

HAARMAN & REIMER: Perfumery Components(Continued):

Isobutyl phenyl acetate

Isoeugenol

Isoeugenol methyl ether

Isononyl acetate

Jasmaprunat

Lactojasmon

Limettene

Limonene pure O

Magnolan

I-Menthol H&R

Menthol Oil

Menthol rac.

I-Menthone

I-Menthyl acetate

Menthyl acetate rac.

RF Menthyl acetate

I-Menthyl isovalerate

I-Menthyl isovalerate EB 6

p-Methylacetophenone

Methyl anthranilate

Methyl benzoate H&R

alpha-Methylcinnamic aldehyde

Methyl cinnamate

Methyl-beta-naphtyl-ketone cryst.

Methyl phenyl acetate

HAARMAN & REIMER: Perfumery Components(Continued):

Methyl salicylate

Methyl-para-tolyl carbinol

Musk Ambrette H&R

Musk Ambrette lumped

Musk Ketone

Musk Xylol

Neocyclocitral

NEO HELIOPAN Type AV UV-B-Filter

NEO HELIOPAN, Type E1000 UV-B-Filter

NEO HELIOPAN, Type OS UV-B-Filter

NEO HELIOPAN, Type Hydro UV-B-Filter/water soluble

NEO HELIOPAN, Type BB UV-Broad-spectrum absorber

NEO HELIOPAN, Type MA UV-A-Filter

Neononyl acetate

Nerol N supra

Nerolin Bromelia cryst.
2-Ethoxy naphthalene

Nerolin Bromelia dist. fused
2-Ethoxy naphthalene

Nerolin Yara Yara cryst.
2-Methoxy naphthalene

Nerolin Yara Yara dist. flakes
2-Methoxy naphthalene

Oryclon extra
p-tert.-Butylcyclohexylacetate

Oryclon spec.
p-tert.-Butylcyclohexylacetate

Ozonil

HAARMAN & REIMER: Perfumery Components(Continued):

Palisandal

Palisandin

Phenirat
2-Phenoxyethyl isobutyrate

Phenoxyethyl alcohol/Arosol

Phenylacetaldehyde 50% in DEP

Phenylacetaldehyde 50% in PEA

Phenylacetaldehyde dimethyl acetal

Phenylacetic acid

beta-Phenylethyl acetate

beta-Phenylethyl alcohol pure

beta-Phenylethyl alcohol N

beta-Phenylethyl isobutyrate

beta-Phenylethyl phenyl acetate

Phenylpropyl alcohol

Profarnesal

Profarnesol

Projasmon P

Projasmon X

Resedafol

Resorcin dimethyl ether

Root Body

Rose acetate
1-Phenyl-2,2,2-trichloroethyl acetate

HAARMAN & REIMER: Perfumery Components(Continued):

Sandel H&R

Sandel SP

Sandel N

Sandel extra

Sandel 80

Sandolen H&R

Styrallyl acetate

Styrallyl alcohol

Styrallyl propionate

Terpineol pure

Terpineol N

Thymol cryst. H&R

2,6,10-Trimethyl-9-undecylenaldehyde

Vanillin pure

Vertocitral

Vertocitral C

Vertomugal

Vertosine

HEXCEL CHEMICAL PRODUCTS: Quaternary Ammonium Salts:

Tradename:

SUMQUAT 2355:
Chemical Name: Benzyl Triethyl Ammonium Chloride
CAS Reg No. 56-37-1
Synonym: Triethyl Benzyl Ammonium Chloride
BTEACI
White to off-white crystalline powder
M.W. 277.78
Applications: Phase transfer catalyst; catalyst for epoxy polyester resin polymerization; conditioning agent for personal care products.

ANTI-STAT Antistatic Agent 106G-90%:
SUMQUAT 5106:
Chemical Name: bis(2-hydroxyethyl) Octyl-methyl ammonium para-toluene sulfonate
CAS Reg. No. 58767-50-3
Viscous liquid
M.W. 403.57
Applications: Phase transfer catalyst, particularly in reduction reactions; anti-static agent for polystyrene or other thermoplastics and engineering resins.

CETATS:
SUMQUAT 5316:
Chemical Name: Cetyl Trimethyl Ammonium para-Toluene Sulfonate
CAS Reg. No. 138-32-9
Endostat, Hexadecyl Trimethyl Ammonium para-Toluene Sulfonate
Free flowing white powder
Assay: 99.0% Min.
M.W. 455.72
Applications: Personal care products, conditioning agent, anti-static agent.

BRETOL:
SUMQUAT 6020:
Chemical Name: Cetyl Dimethyl Ethyl Ammonium Bromide
CAS Reg. No. 124-03-8
EPA Reg. No. 1457-13
Cetethyldimonium Bromide Hexadecyl Dimethyl Ethyl Ammonium Bromide
Free flowing white powder
Assay: 99.0% Min.
M.W. 378.47
Applications: Medical disinfectant; automatic Lysing

HEXCEL CHEMICAL PRODUCTS: Quaternary Ammonium Salts (Continued):

Tradename:

BROMAT:
SUMQUAT 6030:
Chemical Name: Cetyl Trimethyl Ammonium Bromide
CAS Reg. No. 57-09-0
EPA Reg. No. 1457-15
Cetrimide B.P., Cetrimonium Bromide
Hexadecyl Trimethyl Ammonium Bromide
Free flowing white powder
Assay: 98.0% Min.
M.W.: 364.44
Applications: Cream rinses, as an emulsifier, as a preservavive, wetting agent in solder flux.

SUMQUAT 6045:
Distearyl Dimethyl Ammonium Methosulfate
DDAMS
Free flowing white powder.
Assay: 98.0% Min.
M.W. 661
Applications: Phase transfer catalyst, conditioning agent, personal care products.

CPC SUMQUAT 6060:
Chemical Name: Cetylpyridinium Chloride, U.S.P.
CAS Reg. No. 6004-24-6
EPA Reg. No. 1457-60
1-Hexadecyl-pyridinium Chloride monohydrate
Free flowing white powder
Assay: 99.0%-102.0%
M.W. 358
M.P. 80-84C
conforms to U.S.P. specifications
Applications: FDA approved cationic surfactant for mouthwash, feminine hygiene and cream rinse products.

TBAB SUMQUAT 6065:
Chemical Name: Tetra N-Butyl Ammonium Bromide
CAS Reg. No. 1643-19-2
White to off-white crystalline powder.
Assay: 99.0% Min.
M.W. 322.28
Applications: Phase transfer catalyst.

HEXCEL CHEMICAL PRODUCTS: Quaternary Ammonium Salts(Continued):

Tradename:

MYTAB SUMQUAT 6110:
Chemical Name: Myristyl Trimethyl Ammonium Bromide
CAS Reg. No. 1119-97-7
Tetradecyl Trimethyl Ammonium Bromide
Free flowing white powder.
Assay: 99.0% Min.
M.W. 336.40
Applications: Lysing agent in blood analysis; wetting agent in cosmetic and hair rinse preparations; anti-static agent.

CETOL SUMQUAT 6050:
Chemical Name: Cetyl Dimethyl Benzyl Ammonium Chloride
CAS Reg. No. 122-18-9
EPA Reg. No. 1457-16
Cetalkonium Chloride Hexadecyl Dimethyl Benzyl Ammonium Chloride
Free flowing white powder.
Assay: 98.0% Min.
M.P. 55-60C
M.W. 414.10
Applications: Phase transfer catalyst.

CPB:
Chemical Name: Cetyl Pyridinium Bromide
Hexadecyl Pyridinium Bromide
Off-white powder.
Applications: Phase transfer catalyst.

CTAC:
Chemical Name: Cetyl Trimethyl Ammonium Chloride
Hexadecyl Trimethyl Ammonium Chloride
White to off-white powder.
Applications: Phase transfer catalyst, conditioner for personal care products.

STEDBAC SUMQUAT 6210:
Chemical Name: Stearyl Dimethyl Benzyl Ammonium Chloride
Octadecyl Dimethyl Benzyl Ammonium Chloride
Free flowing white powder
Applications: Phase transfer catalyst, personal care products.

HOECHST CELANESE CORP.: Cosmetics Specialties:

Detergent Bases:

Alkyl Ether Sulfates:

GENAPOL LRO Liquid:
CTFA Name: Sodium Laureth Sulfate
28%/Pale yellow liq.
Base: C12 C14 Natural
Good foaming/cleansing

GENAPOL LRO Paste:
CTFA Name: Sodium Laureth Sulfate
90%/Slightly yellow mobile paste
Base: C12 C14 Natural
Good foaming/cleansing

GENAPOL ZRO Liquid:
CTFA Name: Sodium Laureth Sulfate
28%/Pale yellow clear liquid
Base: C12 C14 Synthetic
Good foaming/cleansing

GENAPOL ZRO Paste:
CTFA Name: Sodium Laureth Sulfate
70%/Slightly yellow mobile paste
Base: C12 C14 Synthetic
Good foaming/cleansing

Amide Ether Sulfates:

GENAPOL AMS:
CTFA Name: TEA-PEG-3 Cocamide Sulfate
40%/Clear, yellow low-viscous liquid
Base: Coconut fatty acid
Superior skin compatibility/mildness

Secondary Alkane Sulfonate:

HOSTAPUR SAS 60:
CTFA Name: Sodium C14 C17 Sec Alkyl Sulfonate
60%/Yellowish mobile paste
Base: C14 C12
Excellent cleansing
cost effective

HOECHST CELANESE CORP.: Cosmetic Specialties(Continued):

Detergent Bases(Continued):

Isethionate:

HOSTAPON KA Pdr.:
CTFA Name: Sodium Cocoyl Isethionate
78%/White powder
Base: Palm kernel/coco. fatty acid
For synthetic soaps, unique lather feel

Sarcosinates:

MEDIALAN KA:
CTFA Name: Sodium Cocoyl Sarcosinate
53%/Yellow paste
Base: Palm kernel/coco. fatty acid
Mildness, skin compatibility

MEDIALAN LD:
CTFA Name: Sodium Lauroyl Sarcosinate
30%/Clear yellowish liquid
Base: Lauric acid

MEDIALAN KF:
CTFA Name: TEA-Palm Kernel Sarcosinate
40%/Clear, pale yellow liquid
Base: Palm Kernel/coco. fatty acid

Amine Oxide:

GENAMINOX KC:
CTFA Name: Cocamine Oxide
30%/Clear, colorless liquid
Base: C10 C18
Foam Booster/stabilizer
Wide pH range

Methyl Taurides:

HOSTAPON T Powder:
CTFA Name: Sodium Methyl Oleoyl Taurate
63%/white powder
Base: Oleic acid
Good foaming/skin compatibility

HOSTAPON CT Paste:
CTFA Name: Sodium methyl cocoyl taurate
30%/white powder
Base: Palm Kernel/coco. fatty acid
Good foaming/skin compatibility

HOECHST CELANESE CORP.: Cosmetic Specialties(Continued):

Cosmetic Thickeners:

Polymers:

HOSTACERIN PM 73:
CTFA Name: Acrylamide/Sodium Acrylate Copolymer
White Powder
Pre-neutralized water soluble thickener

Cellulose Ethers:

TYLOSE Grades:

TYLOSE C, CB:
CTFA Name: Cellulose Gum
Granules
C-normally etherfied

TYLOSE C-p, CB-p:
CTFA Name: Cellulose Gum
Powders
CB-highly etherfied

TYLOSE H:
CTFA Name: Hydroxyethylcellulose
Granules
Normally etherfied

TYLOSE H-p:
CTFA Name: Hydroxyethylcellulose
Powders, readily soluble
Normally etherfied

TYLOSE H-yp:
CTFA Name: Hydroxyethylcellulose
Powders with retarded swelling
Normally etherfied

TYLOSE MH, MHB:
CTFA Name: Methylhydroxyethyl-cellulose
Granules
MH-normally etherified

TYLOSE MH-p, MHB-p:
CTFA Name: Methylhydroxyethyl-cellulose
Powders
MHB-highly etherified

HOECHST CELANESE CORP.: Cosmetic Specialties(Continued):

Emulsifiers for Creams and Lotions:

Oil-in-Water Emulsifiers:

HOSTAPHAT KL 340 N:
CTFA Name: Trilaureth-4 Phosphate
95%/clear pale yellowish liquid
Base: Anionic HLB 11.5
Effective at low levels

HOSTAPHAT KW 340 N:
CTFA Name: Triceteareth-4 Phosphate
96%/White wax
Base: Anionic HLB 10.5
Fast absorption into skin

HOSTACERIN DGS:
CTFA Name: Polyglyceryl-2-PEG-4-Stearate
100%/Flake
Base: Nonionic HLB 7.5
Co-emulsifier and thickener

GENAMIN DSAC:
CTFA Name: Distearyldimonium Chloride
97%/Yellowish white powder
Base: Cationic
For cationic emulsifiers good skin compatibility

HOSTACERIN CG:
CTFA Name: Cetearyl Alcohol (and) Triceteareth-4 Phosphate (and) PEG-6 Oleamide (and) Sodium C14 C17 Sec Alkane Sulfonate
100%/Yellowish white powder
Base: Anionic HLB 11
Self-emulsifying base

Water-in-Oil Emulsifiers:

HOSTACERIN WO:
CTFA Name: Polyglyceryl-2-Sesqui-isostearate (and) Beeswax (and) Mineral Oil (and) Magnesium Stearate (and) Aluminum Stearate
100%/Soft, yellowish white paste
Base: Anionic HLB 4.5
Emulsion concentrate; produces products with high stability

HOE S 2721:
CTFA Name: Polyglyceryl-2 Sesqui-isostearate
100%/Pale yellow, clear/low viscosity
Base: Nonionic HLB 3.5
Produces products with good high temperature stability

HOSTACERIN DGO:
CTFA Name: Polyglyceryl-2 Sesquioleate
100%/Yellow Liquid
Base: Anionic HLB 4.0

HOECHST CELANESE CORP.: Cosmetic Specialties(Continued):

Quaternaries:

GENAMIN KDM-F:
CTFA Name: Behentrimonium Chloride
80%/Flakes
Base: C20 C22
Improves Softness/gloss to thick dark hair

GENAMIN KSL:
CTFA Name: PEG-5 Stearyl Ammonium Lactate
21%/Clear Yellowish
Base: C18
Adds gloss to hair

GENAMIN DSAC:
CTFA Name: Distearyldimonium Chloride
97%/Yellowish White powder
Base: C18
Good wet combing, softening/skin comp.

HOE S 2650:
CTFA Name: Dilaureth-4 Dimonium Chloride
90%/Soft, light-colored paste
Clear conditioners clear cond. shampoos

GENAMIN CTAC:
CTFA Name: Cetrimonium Chloride
29%/Pale Yellowish Liquid
Base: C16
Good anti-static effect

GENAMIN KSE:
CTFA Name: Distearyl Dimonium Chloride (and) Cetyl Alcohol (and) Ceteareth-15 (and) Ceteareth-3 (and) PEG-3 Distearate
93%/Yellowish White Powder
Characteristics: One-step conditioner base with thickeners

Pearlescing Agents:

GENAPOL PGM Conc.:
CTFA Name: Sodium Laureth Sulfate (and) Glycol Distearate (and) Cocamide MEA
40%/White dispersion of high viscosity
Pearl lustre agent negligible effect on foam of systems

GENAPOL TS Powder:
CTFA Name: PEG-3 Distearate
100%/White Powder
Produce silk lustre effect

HOECHST CELANESE CORP.: OCTOPIROX Antidandruff Agent:

Chemical Name: 1-Hydroxy-4-methyl-6-(2,4,4-trimethylpentyl)-2(1H)-pyridone ethanolamine salt

Nonproprietary Name: Piroctone olamine (USAN, WHO)

Molecular Weight: 298,4

Description: White to slightly yellowish, fine crystalline, odorless powder

Melting Range: 130-135C (decomposition)

Solubility: Lightly soluble in alcohols and water/alcohol mixtures [>10% (w/w)]; moderately soluble in water containing surfactants [1-10% (w/w)]; slightly soluble in water [0.05% (w/w)] and oils [0.05-0.1% (w/w)].

Stability: OCTOPIROX incorporated in a shampoo base within the pH range of 5 to 9 is stable at 80C for at least 14 days and is, therefore, stable at temperatures normally used in formulating hair care products.

Substantivity: OCTOPIROX exhibits considerable substantivity to skin. The amount adsorbed onto keratin is dependent on time, temperature and concentration; it is independent of the pH within the range of 5 to 8.

Application: On account of its good solubility properties OCTOPIROX is easy to incorporate into various cosmetic formulations of any viscosity. OCTOPIROX permits the manufacture of clear truly effective antidandruff hair tonics and transparent shampoos. To Hoechst Celanese's knowledge OCTOPIROX is compatible with ingredients commonly used in hair care products; e.g. anionic, cationic and amphoteric surfactants. However, incompatibilities with some components of fragrance might occasionally occur. Preferred concentration of OCTOPIROX in rinsed off products is 0,5 to 1,0% (w/w). In formulations remaining on the hair and scalp after application a concentration between 0,05 and 0,1% (w/w) is recommended.

HOECHST CELANESE CORP.: Products for Cosmetics:

GENAPOL PGS:
Pumpable pearl lustre concentrate for the cosmetics industry

Chemical composition:
Dispersion of nonionic, pearl-lustring substances in an alkyl ether sulphate sodium salt solution with consistency-regulating and foam-stabilizing additives.

Appearance (20C):
Slightly yellowish, medium to high viscosity dispersion with pearl-lustre effect.

Odour: Faint

Chemical data:
Solids content: 37.0±1.0%
Alkyl ether sulphate (molar mass 382): 13.0±1.0%
Electrolytes: max. 1.5%
Water: 63.0±1.0%

Physical data:
pH (1% aqueous dispersion of the active ingredient, 20C): 7.2±0.8
Density (20C): approx. 1.0g/cm3

GENAPOL PMS:
Raw material with pearl luster properties for the manufacture of shampoos and cleansers used in cosmetics

Chemical composition: fatty acid mono glycol ester

Typical Properties:
Appearance: almost white powder
Odor: weak

Chemical data:
Concentration: approx. 100%
Water: max. 1%
Acid number: max. 6
Saponification number: 200±10

Physical data:
Dropping point: 62±3C
Iodine color number (70C): max. 5
Flash point: approx. 260C

HOECHST CORPORATION CORP.: Products for Cosmetics(Continued):

GENAPOL TS Powder:
Clouding agent for the manufacture of cosmetic washing and cleansing preparations

Chemical composition: Fatty acid polyglycol ester

Typical Properties:
Appearance: white powder
Odor: weak

Chemical data:
Active ingredient conc.: approx. 100%
Acid value: approx. 7
Saponification value: 165-175

Physical data:
Color (iodine color number 60C): max. 5
Viscosity (60C): approx 15 cP (mPa s)
Density (60C): approx. 0.90 g/cm3
Drop point (C): approx. 50
Flash point (C): approx. 270

GENAPOL TSM:
Pearl-luster concentrate for the cosmetics industry

Chemical composition: Dispersion of nonionic, pearl-lustering substances in an alkyl ether sulfate

Typical Properties:
Appearance (20C): white, homogeneous medium-viscosity dispersion
Odor: Faint, characteristic

Chemical data:
Active ingredient content: 38.0±1.0%
Alkyl ether sulfate: 17.0±1.0%
Electrolytes: max. 2.0%
Water: 62.0±2.0%

Physical data:
pH (1% aqueous solution of active ingredient, 20C): 7.2±0.8

HOECHST CELANESE CORP.: Products for Cosmetics(Continued):

GENAPOL 3520:
Biodegradable surfactant combination

Composition:
mixture of surface-active substances

Properties:
Appearance: yellowish liquid
Density at 20C: 0.96 g/cm3
pH (1% in water): about 7
Low temperature behaviour: freezing point below -30C
Storage stability: no restrictions
Compatability: compatible with water, acids, alkalis, alcohols and glycols

HOE S 2721:
W/O emulsifier for the cosmetic industry

Chemical composition: fatty acid polyglyceryl ester
CTFA designation: Polyglyceryl-2-Sesquiisostearate

Typical Properties:
Appearance: pale yellow, clear, low viscosity
Odor: weak

Chemical data:
Concentration: approx. 100%
Water: max. 1%
Acid Value: max. 3
Saponification Value: 170±10

Physical data:
Iodine Color Number: max. 5
pH Value (1% aqueous dispersion): 7.0±1.0
HLB Value: 3-4

HOE S 3267:
Amphoteric surfactant for cosmetic field
CTFA-adopted name: Cocamidopropyl Betain
Typical Properties:
Appearance (20C): clear light yellow low viscous liquid
Odor: bitter
Chemical and physical data:
Solid content (%): approx. 35
Active matter content (%): approx. 30
Water content (%): approx. 65
Salt content (%): approx. 5
Iodine value (20C): <3
Density (20C): approx. 1.04 g/cm3
pH (1% active matter content): 5-7

HOECHST CELANESE CORP.: Products for Cosmetics(Continued):

HOE S 3495:

Emulsifier for the manufacture of oil-in-water emulsions for the cosmetics and pharmaceutical industry.

Chemical composition: Ethoxylated fatty acid polyglycerol ester

CTFA designation: PEG 10 polyglyceryl-2 laurate

Typical Properties:
Appearance (20C): yellowish, clear, liquid
Odor: weak

Chemical data:
Active ingredient: approx. 100%
Water: max. 1%
Acid value: max. 3
Saponification value: 73±3
Iodine value: max. 1

Physical data:
Iodine color number (20C): max. 2
Viscosity (20C): approx. 200 mPas
Density (20C): 1.05 g/cm3
pH (1% in water, 20C): 6.0-8.0
HLB value: approx. 15

Surfactant Mixture:
HOE S 3755:

Compound for the production of dioxane free toiletries

Chemical composition:
Mixture of: 8 parts Sodium paraffin sulfonate (HOSTAPUR SAS)
2 parts Sodium methyl cocoyl taurate (HOSTAPON CT Paste)

Typical properties:
Appearance: clear liquid

Chemical data:
Active ingredient: approx. 30%
Inorganic salts: approx. 3%

Physical data:
Color (iodine color number 20C): max. 2
Viscosity (20C): 1200 cps
pH value (1% aqueous soln): 7.5±0.5
Cloud point: approx. -5C
Clear point: approx. +10C

HOECHST CELANESE CORP.: Products for Cosmetics(Continued):

HOSTACERIN CG:
Emulsifiable concentrate for the manufacture of semisolid creamy oil-in-water emulsions for cosmetics and pharmacy

Chemical composition:
mixture of ethoxylated oleic acid alkanolamides, ethoxylated phosphoric acid esters and secondary alkane sulphonate with fat alcohols (primarily natural cetyl alcohol)

Typical Properties:
Appearance: yellowish white, waxy flakes
Odor: weak

Chemical data:
Water (%): 1.0 max.
Acid value: approx. 4

Physical data:
Density (60C): 0.83 g/cm3
Solidification point: approx. 48C
Solidification range: 46-50C

HOSTACERIN DGO:
Emulsifier for the manufacture of water-in-oil emulsions for the cosmetics and pharmaceutical industries

Chemical composition: Oleic acid polyglycerol ester

Typical Properties:
Appearance: yellow liquid
Odor: weakly fatty odor

Chemical data:
Concentration: approx. 100%
Water: max. 1.0%
Acid number: max. 1
Saponification number: 165±15
Iodine value: 70±5

Physical data:
Iodine color number (20C): max. 10
Viscosity (20C): 600±200 cP (mPas)
Density (20C): approx. 0.97 g/cm3
pH (1% in 50% aqueous ethanol): 6.5±0.5
Flash point: approx. 250C
Setting point: approx. 0C

HOECHST CELANESE CORP.: Products for Cosmetics(Continued):

HOSTACERIN DGL:

Emulsifier for the manufacture of oil-in-water emulsions for the cosmetics and pharmaceutical industry

Chemical composition: ethoxylated fatty acid polyglycerol ester

CTFA designation: PEG-10 polyglyceryl-2 laurate

Physical Properties:
- Appearance (20C): yellowish, clear liquid
- Odor: weak

Chemical data:
- Active ingredient: approx. 100%
- Water: max. 1%
- Acid value: max. 3
- Saponification value: 73±3
- Iodine value: max. 1

Physical data:
- Iodine color number (20C): max. 2
- Viscosity (20C): approx. 200 mPas
- Density (20C): 1.05 g/cm3
- pH (1% in water, 20C): 6.0-8.0
- HLB value: approx. 15

HOSTACERIN DGS:

Emulsifier for the manufacture of oil-in-water emulsions for the cosmetics and pharmaceutical industry

Chemical composition: Ethoxylated fatty acid polyglycerol ester

CTFA designation: PEG-4 polyglycerol-2-stearates

Typical properties:
- Appearance (20C): white, waxy flakes
- Odor: virtually odorless

Chemical data:
- Active ingredient: approx. 100%
- Water: max. 1%
- Acid value: max. 2
- Saponification value: 125±10
- Iodine value: max. 1

Physical data:
- Iodine color number (50C): max. 2
- Viscosity (50C): approx. 80 mPas
- Density (50C): approx. 0.96 g/cm3
- pH (1% aqueous dispersion, 20C): 6-8
- Drop point: approx. 40C
- HLB Value: approx. 7

HOECHST CELANESE CORP.: Products for Cosmetics(Continued):

HOSTACERIN WO:
Emulsifier for the manufacture of water-in-oil emulsions for the cosmetics and pharmaceutical industries

Chemical Composition: Mixture of polyglycerol esters, stabilizers and waxes

Typical Properties:
Appearance at 20C: pale, soft patse
Odor: weak

Chemical data:
Concentration: approx 100%
Water: 2% max.
Acid number: approx. 25-35

Physical data:
Iodine color number at 70C: 5 max.
Density at 70C: approx. 0.9 g/cm3
pH (1% dispersion in water at 20C): 6.5-8.0
HLB value: 4-5

HOSTAPHAT KL 340 N:
Emulsifier for the manufacture of oil-in-water emulsions for the cosmetic and pharmaceutical industries

Chemical composition:
Mixture of mono-di- and tri-(alkyltetraglycol ether)-o-phosphoric acid esters alkyl = predominantly C12/C14

CTFA designation: Trilaureth-4 Phosphate

Typical Properties:
Appearance at 25C: Almost colorless, clear liquid
Odor: Slight

Chemical data:
Concentration: approx. 95%
Water: max. 5%
Acid number: max. 3

Physical data:
Color:
Iodine color number: max. 2
Gardner color number: max. 1.5
Viscosity at 25C: 200±30 mPas
Density at 25C: approx. 1.0 g/cm3
Refractive index at 25C: 1.455±0.002
pH (1% aqueous soln. @ 20C): 6.0-7.5
Setting point: max. 12C
Flash point: above 100C
Low temperature behavior: the product solidifies at low storage temperatures.

HOECHST CELANESE CORP.: Products for Cosmetics(Continued):

HOSTAPON KTW New:
Anionic surfactant for the cosmetics industry

Chemical composition:
Fatty acid tauride sodium salt

Typical Properties:
Appearance at 20C: white powder
Odor: weak, typical odor

Chemical data:
Active ingredient content: 50.0±2%
Unreacted fatty acid in the active ingredient: max. 4%
Organic salts: approx. 15-20%
Sodium chloride: 25±3%
Sodium sulfate: max. 1%
Water: max. 1%
Mean molecular weight: 346±6

Physical data:
pH of 1% aqueous active detergent solution at 20C: 7.5±0.5

HOSTAPON T Paste 33:
Oleoyl methyl tauride, sodium salt

Typical Properties:
Appearance: not darker than standard
pH (in 1% aqueous solution): 7.0-8.0
Active content (MW 423): 32-34%
Free fatty acid (MW 280): 2.5% max.
Sodium chloride content: 5-7%
Water content (Fischer): 50-54%
Reaction: practically neutral

Resistance:
Unaffected by hard water, diluted mineral acids and alkalis. Resistant to most metal salts under normal working conditions. In concentrations of more than 2 g/l, aluminum and iron salts cause precipitation.

Compatibility:
As anionic products, compatible with soaps, other anionic and nonionic substances.

Interfacial activity:
Good wetting action when warm; high foaming power, outstanding cleaning action, even in acid or alkali solutions and also in hard water; very good dispersing power for lime soap and pigment soil.

HOECHST CELANESE CORP.: Products for Cosmetics(Continued):

MEDIALAN KA Conc.:

Chemical composition:
Palm kernel/coconut fatty acid sarcoside sodium salt

CTFA designation: Sodium lauroyl sarcosinate

Typical Properties:
Appearance at 20C: yellowish, soft paste
Odor: weak, typical odor

Chemical data:
Active ingredient content: 49±2%
Unreacted fatty acid in the active ingredient: max. 13%
Water: 38±2%
Sodium chloride: max. 1%
Mean molecular weight: 310±5

Physical data:
pH of 1% aqueous SAA solution at 20C: 8.0±0.5

MEDIALAN LD:
Anionic surfactant for the cosmetics industry

Chemical Composition:
Lauroyl sarcoside sodium salt

Typical Properties:
Appearance at 20C: yellowish, clear liquid
Odor: weak, typical odor

Chemical data:
Active ingredient content: 30±1.5%
Sodium laurate: approx. 1.5%
Sodium chloride: 0.1% max.
Sodium sulphate: 0.1% max.
Water: 64±2%
Ethanol: 4±1%
Mean molecular weight: 293±2

Physical data:
Color:
Iodine color number: 2 max.
Gardner color number: 2 max.
Viscosity at 20C: max. 100 mPas
Density at 20C: approx. 1.065 g/cm3
pH of 1% aqueous AD solution at 20C: 8±1

HOECHST CELANESE CORP.: Surfactants:

GENAMIN CTAC:
Cationic surfactant for the cosmetics industry

CTFA designation: Cetrimonium Chloride

Physical Properties:
Appearance at 20C: Colorless to slightly yellowish, clear liquid
Odor: slight

Chemical data:
Active ingredient: 29±1%
Water: approx. 70%
Mean molecular weight: approx. 319

Physical data:
Solubility: soluble in water and lower alcohols in any ratio
Color:
Iodine color number: max. 2
Gardner color number: max. 1.5
pH (1% synthetic active agent in aqueous solution at 20C): 6.5±0.5
Low-temperature behavior: turbidity and flocculation can occur at low storage temperatures; heating to 30-40C and stirring renders the product fully reusable again.

GENAMIN KDM-F:
CTFA designation: Behentrimonium chloride
Physical Properties:
Appearance at 20C: white to yellowish waxy flakes
Odor: typical
Chemical Data:
Active ingredient: 80±3%
Water: max. 3%
Isopropanol: approx. 17%
Mean molecular weight: 400±5
Physical data:
Solubility: sparingly soluble in water; soluble in low concentrations in ethyl alcohol and isopropyl alcohol
Color of molten product (approx. 60C):
Iodine color number: max. 10
Gardner color number: max. 6
pH (1% active ingredient in 50% aqueous isopropanol): 6.0±1.0

HOECHST CELANESE CORP.: Surfactants(Continued):

GENAMIN KSE:
Cationic raw material for the formulation of after-shampoo hair-care preparations

Chemical composition:
Quaternary ammonium compound combined with nonionic emulsifiers and consistency modifiers
Typical properties:
Appearance: white, waxy flakes
Odor: weak (not unpleasant)
Chemical data:
Water content: max. 3%
Drying loss: max. 7%
Iodine value: max. 2
Physical data:
Iodine color number (60C): max. 4
pH of 1% aqueous emulsion: 3.8-5.0
Properties:
Solubility
Insoluble but readily dispersible in water; insoluble in isopropanol and ethanol.
Active ingredient content:
GENAMIN KSE contains 20% cationic active ingredient.

GENAMIN KSL
Cationic surfactant for the cosmetics industry
Chemical composition:
Alkyl trimethyl ammonium chloride
Typical properties:
Appearance: clear yellowish liquid
Chemical data:
Active ingredient: 30.0±1.0%
Water and ethylene glycol: 69.0±1.5%
Electrolyte (ash): max. 0.5%
Average molecular weight: 530
Physical data:
Solubility: soluble in water and low-molecular-weight alcohols
Color (20C): Iodine color number: max. 5
pH (of product as supplied): 6.0±1.0
General Properties:
Owing to its cationic charge, GENAMIN KSL has affinity for negatively charged surfaces and reacts with anionic compounds to form salts and complexes. In addition to these properties, it has a slight bactericidal and fungicidal effect in neutral, weakly acid and weakly alkaline solution.

HOECHST CELANESE CORP.: Surfactants(Continued):

GENAPOL AMS:
Anionisches Tensid fur die kosmetische Industrie

Aussehen (20C): gelbe, klare, niedrigviskose Flussigkeit

Geruch: schwach

Chemische Kenndaten:
Aktivsubstanz: ca. 40,0%
Trockengehalt: 43±1,0%
Wassergehalt: 57±1,0%

Physikalische Kenndaten:
Jod-Farbzahl: max. 10
Viskositat (20C): max. 100 mPAs (cP)
Dichte (20C): ca. 1,06 g-cm-3
pH-Wert (1% ige waBrige WAS-Losung, 20c): 5,0-7,0

GENAPOL LRO Liquid:
Anionic surfactant for the cosmetics industry

Chemical Composition:
Alkyl diglycol ether sulfate sodium salt based on natural fatty alcohols

Typical Properties:
Appearance: pale yellow, clear liquid
Odor: slight

Chemical data:
Alkylether sulphate (%): 27.0±0.5
Alcohol extract (%): approx. 28.0
Sodium sulphate (%): 1.0 max.
Sodium chloride (%): 0.5 max.
Water (%): 71.0±1.5
Average molecular weight: 382

Physical data:
Color
Iodine color number: max. 2.5
Gardner color number: max. 2.5
Lovibond color number (diameter 1 1/4"):
Red filter: max. 0.3 R
Yellow filter: max. 3.0Y
Viscosity (20C): max. 100 cP (mPas)
Density (20C): approx. 1.05 g/cm3
pH (1% aqueous synthetic active agent solution, 20C): 6.5-8.0

HOECHST CELANESE CORP.: Surfactants(Continued):

GENAPOL LRO Paste:
Anionic surfactant for the cosmetics industry

Chemical composition:
Alkyl diglycol ether sulphate salt based on natural fatty alcohols

Typical Properties:
Appearance at 20C: slightly yellowish, mobile paste
Odor: slight

Chemical data:
Alkyl ether sulphate: 69±1%
Alcohol extract: Approx. 70%
Sodium sulphate: max. 1.5%
Sodium chloride: max. 0.5%
Water: 27±1%
Average molecular weight: 382

Physical data:
Color (20C, 25% aqueous SAA solution)
Iodine color number: max. 2
Density (20C): approx. 1.0 g/cm3
pH (1% aqueous SAA solution (20C): 8.5±1.0

GENAPOL PGM Conc.:
Pearl-luster concentrate for the cosmetics industry

Chemical composition:
Dispersion of nonionic pearl-lustering substances in an alkyl ether sulfate sodium salt with viscosity-controlling and foam-stabilizing additive.

Typical Properties:
Appearance: White, homogeneous, fluid paste
Odor: Slight

Chemical data:
Active ingredient content (=alcohol extract): 40.0±1.0%
Alkyl ether sulfate: 21±1%
Electrolytes: max. 2.0%
Water: 58±2%

Physical data:
pH (1% synthetic active agent solution/20C): 7.2±0.8
Density (20C): approx. 1.0 g/cm3

HOECHST CELANESE CORP.: TYLOSE Water-Soluble Cellulose Ethers:

Hydroxyethylcellulose:

Trade Name:

TYLOSE H grades:
Chemical constitution: HEC, normally etherified
Form in which supplied: granules

TYLOSE H-p grades:
Chemical constitution: HEC, normally etherified
Form in which supplied: powders, readily soluble

TYLOSE H-yp grades:
Chemical constitution: HEC, normally etherified
Form in which supplied: powders, with retarded swelling properties

Moisture content: approx. 5% on filling
Degree of purity: a.i.# content approx 96%, dry weight basis
Viscosity stages in mPa s, 2% solutions, Hoppler viscosmeter, 20C:

TYLOSE H	10	TYLOSE H	6000 yp
TYLOSE H	20	TYLOSE H	15000 yp
TYLOSE H	300	TYLOSE H	30000 yp
TYLOSE H	4000	TYLOSE H	60000 yp
TYLOSE H	10000	TYLOSE H	100000 yp

TYLOSE H	20p
TYLOSE H	200p
TYLOSE H	300p
TYLOSE H	4000p
TYLOSE H	10000p

The TYLOSE H grades are of the nonionic cellulose ether group. They are readily soluble in water at any temperature, forming clear solutions. The solutions have particularly good compatibility with salts.

a.i.: active ingredient

HOECHST CELANESE CORP.: TYLOSE Water-Soluble Cellulose Ethers (Continued):

Methylhydroxyethylcellulose:

Trade Name:

TYLOSE MH grades:
Chemical constitution: MHEC, normally etherified
Form in which supplied: granules

TYLOSE MH-p grades:
Chemical constitution: MHEC, normally etherified
Form in which supplied: powders

TYLOSE MHB grades:
Chemical constitution: MHEC, highly etherified
Form in which supplied: granules

TYLOSE MHB-p grades:
Chemical constitution: MHEC, highly etherified
Form in which supplied: powders

TYLOSE MB grades:
Chemical constitution: MC, highly etherified
Form in which supplied: granules

Moisture content: approx. 5% on filling
Degree of purity: a.i. content at least 98.5%, dry weight basis
Viscosity stages in mPa s, 2% solutions, Hoppler viscometer, 20C:

TYLOSE MH	20	TYLOSE MHB	1000
TYLOSE MH	50	TYLOSE MHB	3000
TYLOSE MH	300	TYLOSE MHB	10000
TYLOSE MH	1000	TYLOSE MHB	30000
TYLOSE MH	2000		
TYLOSE MH	4000	TYLOSE MHB	3000p
		TYLOSE MHB	10000p
TYLOSE MH	300p	TYLOSE MHB	30000p
TYLOSE MH	1000p		
TYLOSE MH	2000p	TYLOSE MB	1500
TYLOSE MH	4000p		

The TYLOSE MH and MHB grades are of the nonionic cellulose ether group

TYLOSE MHB grades are soluble in water and also in mixtures of chlorinated hydrocarbons and alcohols, e.g. methylene chloride/methanol.

TYLOSE MB 1500 is a special grade for use in the tobacco and pharmaceutical industries.

HOECHST CELANESE CORP.: TYLOSE Water-Soluble Cellulose Ethers (Continued):

Methylhydroxyethylcellulose(Continued):

Trade name:

TYLOSE MH-K grades:
Chemical constitution: MHEC, normally etherified
Form in which supplied: granules, readily soluble

TYLOSE MH-xp grades:
Chemical constitution: MHEC, normally etherified
Form in which supplied: powders, with retarded swelling properties

TYLOSE MHB-y grades:
Chemical constitution: MHEC, highly etherified
Form in which supplied: fine granules with retarded swelling properties

TYLOSE MHB-yp grades:
Chemical constitution: MHEC, highly etherified
Form in which supplied: powders, with retarded swelling properties

Moisture content: approx. 5% on filling
Degee of purity: a.i. content at least 98.5%, dry weight basis
Viscosity stages in mPa s, 2% solutions, Hoppler viscometer, 20C:

TYLOSE MH	200K	TYLOSE MHB	3000y
TYLOSE MH	1500K	TYLOSE MHB	6000y
TYLOSE MH	2000K	TYLOSE MHB	15000y
TYLOSE MH	4000K	TYLOSE MHB	30000y
TYLOSE MH	6000K		
TYLOSE MH	10000K	TYLOSE MHB	10000yp
TYLOSE MH	30000K	TYLOSE MHB	30000yp
TYLOSE MH	200xp		
TYLOSE MH	2000xp		
TYLOSE MH	6000xp		

The TYLOSE MH-K grades contain additives which improve their dissolving and processing properties and increase the stability of their solutions. Hence the moisture content of these grades may be slightly higher on filling (approx. 7%).

TYLOSE MH-xp grades are powders with short-time retarded swelling.

TYLOSE MHB-y grades are particularly free-flowing, low-dusting fine granules with a special particle size distribution spectrum and longer retarded swelling.

TYLOSE MHB-y grades are powders with longer retarded swelling.

Swelling-retarded TYLOSE MHB grades are not intended for use in organic solvent mixtures.

HOECHST-CELANESE CORP.: TYLOSE Water-Soluble Cellulose Ethers (Continued):

Sodium carboxymethylcellulose:

Trade name:

TYLOSE C grades:
Chemical constitution: NaCMC, normally etherified
Form in which supplied: granules

TYLOSE C-p grades:
Chemical Constitution: NaCMC, normally etherified
Form in which supplied: powders

TYLOSE CB grades:
Chemical constitution: NaCMC, highly etherified
Form in which supplied: granules

TYLOSE CB-p grades:
Chemical constitution: NaCMC, highly etherified
Form in which supplied: powders

Moisture content: approx. 7% on filling
Degree of purity: a.i. content approx. 99%, dry weight basis
Viscosity stages in mPa s, 2% solutions, Hoppler viscosmeter, 20C:

TYLOSE C	30#	TYLOSE CB	200
TYLOSE C	100	TYLOSE CB	30000
TYLOSE C	300		
TYLOSE C	600	TYLOSE CB	30000p
TYLOSE C	1000		
TYLOSE C	6000		
TYLOSE C	10000		
TYLOSE C	300p		
TYLOSE C	1000p		
TYLOSE C	10000p		

The TYLOSE C and CB grades are anionic cellulose ethers

available with a slightly lower viscosity under the designation TYLOSE C30NV.

HOECHST CELANESE CORP.: TYLOSE Water-Soluble Cellulose Ethers (Continued):

Sodium Carboxymethylcellulose, technical grades:

Trade Name:

TYLOSE CR grades:
Chemical constitution: 65% NaCMC#, normally etherified
Form in which suppied: granules

TYLOSE CR-p grades:
Chemical constitution: 65% NaCMC#, normally etherified
Form in which supplied: powders

TYLOSE CR-w grades:
Chemical constitution: >65% NaCMC#, normally etherified
Form in which supplied: fine granules

TYLOSE CR-n grades:
Chemical constitution: 55% NaCMC#, normally etherified
Form in which supplied: granules

TYLOSE CBR Grades:
Chemical constitution: 60% NaCMC#, highly etherified
Form in which supplied: granules

TYLOSE CBR-np grades:
Chemical constitution: 60% NaCMC#, highly etherified
Form in which supplied: powders

Moisture content: approx. 7% on filling
Reaction of aqueous solution: weakly alkaline with TYLOSE CR, CR-p, CR-w, CBR
neutral with TYLOSE CR-n, CBR-np
Viscosity stages in mPa s, 5% solutions, Hoppler viscometer, 20C:

TYLOSE CR	1500	TYLOSE CBR	30
		TYLOSE CBR	4000
TYLOSE CR	1500p	TYLOSE CBR	6000
		TYLOSE CBR	10000##
TYLOSE CR	1500w##		
		TYLOSE CBR	5000np
TYLOSE CR	700n		

percentage content of the commercial grade
for the detergent industry: TYLOSE CR 1500w (supplied as fine granules)
TYLOSE CBR 10000 (pH below 10)

GEO. A. HORMEL & CO.: Collagen Family of Proteins:

Procollagen:

Collagen biosynthesis occurs in the fibroblast cells of the dermis. Production of collagen begins with the elaboration of a precursor peptide (115-140,000 mol. wt.), the Procollagen Chain. This peptide is generated by ribosomal synthesis, as are all natural proteins. Unlike many proteins, it undergoes extensive post-translational modification in the rough endoplasmic reticulum.

Collagen Monomer:

The first step towards maturation of collagen is the enzymatic modification of procollagen to a Collagen Monomer by the procollagen peptidase. This enzyme clips off the telopeptide forming the collagen monomer having a molecular weight of 300,000. Schematically, this has been represented as the removal of the three 'tails' from the far left side of the procollagen molecule. Collagen Monomer retains the rigid-rod helical structure of Procollagen and is also uncrosslinked, soluble, and highly water retentive. The collagen monomer remains in the extracellular space for crosslinking into mature collagen.

Soluble Collagen:

The two members of the collagen family previously discussed, Procollagen and Collagen Monomer, constitute the soluble fraction of skin collagen. This fraction decreases markedly during aging, and, as such, appears to correlate with other age dependent signs such as dryness, flaking, wrinkling, and loss of elasticity. The use of Soluble Collagen in skin care products is due to its large hydration potential, its ability to bind and retain many times its weight in water.

Mature Collagen:

Collagen Monomer is secreted by the cells into the extracellular space where it forms fibers and becomes crosslinked during maturation. Crosslinking is partially enzyme mediated by oxidation of lysine residues leading to both intra and intercellular covalent crosslinks. This process yields an extensive network of sturdy fibers which form the basic support structure of the skin. This Mature Collagen is no longer water soluble. The crosslinking process continues throughout the life of the animal such that the crosslink density increases with age and exposure to UV radiation. A major consequence of crosslink density is the dramatic reduction in water soluble collagen. As such, the process may be related to skin dryness and wrinkling associated with aged skin.

Mature Collagen is the starting material for the production of Gelatin. The process entails extraction of the collagen by a weak acid at elevated temperatures.

Hydrolyzed Animal Protein:

Hydrolyzed Animal Protein is produced by the enzyme hydrolysis of collagen rich raw materials. Hydrolysis can be accomplished by chemical, high temperature, or high pressure processes.

GEO. A. HORMEL & CO.: PEPTEIN CAA Collagen Amino Acids:

PEPTEIN CAA are enzymatically prepared natural collagen amino acids produced under closely controlled conditions from high collagen content raw materials.

Collagen is the principle structural protein comprising about 30% of the proteins in animals. An important feature of collagen is that it can be water extracted, purified, and then carefully modified in further chemical processes.

A natural humectant, collagen amino acids demonstrate strong moisture binding.

Substantivity: Hair Care Performance:
PEPTEIN CAA is ideally suited for applications in hair care. PEPTEIN CAA is highly substantive, able to penetrate the hair shaft, actually becoming part of the hair, keeping hair more soft and manageable. PEPTEIN CAA will repair damaged hair, restore elasticity and mend split ends.

Humectancy: Skin Care Performance:
PEPTEIN CAA demonstrates its efficacy when used in skin care treatment products by enhancing the moisture binding quality of the product and slowing transepidermal water loss. PEPTEIN CAA replenishes lost moisture and increases the skins' ability to bind moisture by bonding moisture in the stratum corneum.

Typical Analysis:
Physical Attributes:
Color (Gardner): 11 maximum
Specific Gravity: 1.15 minimum

Chemical Profile:
pH (diluted 2 to 1): 5.0-6.5
Molecular Weight: 1000 maximum
Total Solids: 50% minimum
Total Nitrogen: 8.5-9.5%
Ash: 1% maximum
Iron: 5 ppm maximum
Copper: 3 ppm maximum

Microbiological Profile:
APC: less than 1000/gm
Gram Neg. Rods: negative
Gram Pos. Cocci: negative for S. aureus

GEO. A. HORMEL & CO.: PEPTEIN 2000:

PEPTEIN 2000 is an enzymatically hydrolyzed animal protein that is produced under closely controlled conditions from high collagen content raw materials.

Performance:

PEPTEIN 2000, hydrolyzed animal protein, is especially beneficial to hair. PEPTEIN 2000 will penetrate through the cuticle layer of the hair into the cortex, actually becoming part of the hair shaft. PEPTEIN 2000 will repair damaged hair, increase tensile strength and elasticity, mend split ends, add body, shine and manageability, improving the total health of the hair.

Substantivity:

Hydroxyproline Absorption Technique

The right protein in hair care formulations can contribute a desirable conditioning effect, but only if the protein is substantive and becomes absorbed. The substantivity of PEPTEIN 2000 has been demonstrated by hydroxyproline absorption techniques.

Typical Analysis:

Physical Attributes:

Color (Gardner): 12 maximum
Viscosity: 100 cps maximum
Specific Gravity: 1.15 minimum
Odor: bland

Chemical Profile:

pH: 5.8-6.3
Molecular Weight: 1500-2500
Total Solids: 55% minimum
Total Nitrogen: 8-10%
Ash: 2% maximum
Iron: 5 ppm maximum
Copper: 3 ppm maximum

Microbiological Profile:

APC: less than 1000/gm
Gram Neg. Rods: negative
Gram Neg. Rods: negative for S. aureus

HULS: Raw Materials for Cosmetic Products:

Anionic Surfactants:
Field of application: Detergents for foam baths, shower gels, shower foams, hair shampoos and cosmetic cleansers.

Fatty alcohol ether sulphates:

MARLINAT 242/28:
Na salt of C12-C14 fatty alcohol ether sulphate

MARLINAT 242/70:
Na salt of C12-C14 fatty alcohol ether sulphate

SERDET DPK 30:
Na salt of C12-C15 fatty alcohol ether sulphate

SERDET DPK 3/70:
Na salt of C12-C15 fatty alcohol ether sulphate

SERDET DCK 30:
Na salt of C12-C14 fatty alcohol ether sulphate

SERDET DCK 3/70:
Na salt of C12-C14 fatty alcohol ether sulphate

Fatty alcohol sulphates:

SERDET DFK 30:
Sodium lauryl sulphate

SERDET DFL 40:
Triethanolammonium lauryl sulphate

SERDET DFN 30:
Ammonium lauryl sulphate

MARLINAT KT 50:
Sodium coconut/tallow fatty alcohol sulphate

Sulphonates:

MARLON PS:
Sodium alkanesulphonate

MARLON PF 40:
Alkanesulphonate/ether sulphate compound

MARLOPON AT 50:
Triethanolammonium alkylbenzene-sulphonate

HULS: Raw Materials for Cosmetic Products(Continued):

Anionic Surfactants(Continued):

Fatty alcohol carboxylates:

MARLINAT CM 100:
Alkyl polyglycol ether carboxylic acid

MARLINAT CM 105:
Sodium alkyl polyglycol ether carboxylate

Sulphosuccinates:

MARLINAT SRN 30:
Compound based on sulphosuccinates

MARLINAT DF8:
Na salt of diisooctyl sulphosuccinate

SERVOXYL VLB 1123:
Disodium fatty alcohol polyglycol ether sulphosuccinate

Phosphoric acid esters:

MARLOPHOR T 10 Na salt:
Na salt of C16-C18 fatty alcohol polyglycol ether phosphate

MARLOPHOR T 10 acid:
Partial ester of C16-C18 fatty alcohol polyglycol phosphoric acid

Ampholytes:

Field of application: Detergents for foam baths, shower gels, shower foams, hair shampoos and cosmetic cleansers. Particularly suitable for baby shampoos and specialized mild shampoos, and as the basis for cationics.

SERVO AMFOLYT JA 140:
Na salt of N-2-hydoxyethyl-N-2-carboxyethyl fatty acid amidoethylamine

SERVO AMFOLYT JB 130:
Cocoamidopropylbetaine

HULS: Raw Materials for Cosmetic Products(Continued):

3. Nonionic Surfactants:

Field of application: Detergents for foam baths, shower gels, shower foams, hair shampoos and cosmetic cleansers, emulsifiers for o/w emulsions, and solubilizers for active ingredients and oils.

Fatty alcohol polyglycol ethers:

MARLIPAL 1218/5:
C12-C18 fatty alcohol polyglycol ether

MARLIPAL 1218/10:
C12-C18 fatty alcohol polyglycol ether

MARLIPAL MG:
C12-C14 fatty alcohol polyglycol ether

MARLIPAL ML:
C12-C14 fatty alcohol polyglycol ether

MARLIPAL SU:
C16-C18 fatty alcohol polyglycol ether

MARLIPAL 24/100:
C12-C14 fatty alcohol polyglycol ether

Fatty alcohol polyalkylene glycol ethers:

MARLOX FK 64:
Alkyl polyalkylene glycol ether

MARLOX FK 69:
Alkyl polyalkylene glycol ether

4. Superfatting Agents and Thickeners:

Field of application: Shower gels, hair shampoos and foam baths

Fatty acid polyglycol esters:

MARLOSOL FS:
Fatty acid polyglycol ester

MARLOSOL BS:
Fatty acid polyglycol ester

HULS: Raw Materials for Cosmetic Products(Continued):

5. Superfatting Agents and Foam Stabilizers:

Field of application: Shower gels, hair shampoos and foam baths

Fatty acid alkanolamides:

MARLAMID DF 1218:
Coconut fatty acid diethanolamide

MARLAMID DF 1818:
Soya fatty acid diethanolamide

MARLAMID M 1218:
Coconut fatty acid monoethanolamide

MARLAMID M 1618:
Tallow fatty acid monoethanolamide

Fatty acid monoethanolamide polyglycol ethers:

DIONIL OC:
Oleic acid monoethanolamide polyglycol ether

DIONIL SH 100:
Oleic acid monoethanolamide polyglycol ether

6. Silk Lustre Agents/Pearlescent Agents:

Field of application: Foam baths, shower gels, hair shampoos and hair conditioners

Silk lustre components:

MARLAMID KL:
Fatty acid alkylolamide

MARLAMID KLP:
Ether sulphate/amide compound

Pearlescent components:

MARLAMID PG 20:
Fatty acid glycol/ester/fatty acid alkylolamide/fatty acid polyglycol ether compound

HULS: Raw Materials for Cosmetic Products(Continued):

7. Emulsifiers:

Field of application: Emulsifiers for o/w and w/o emulsions for skin creams, skin lotions, oil-based foam baths and perfume oils

MARLOWET 4603:
HLB: 5.5
Glycerol monooleate/dioleate nEO
MARLOWET NF:
HLB: 6.8
Castor oil nEO
MARLOWET LVS:
HLB: 6.9
Castor oil nEO dioleate
MARLOWET LVX:
Carboxylic acid polyglycol ester compound
MARLOWET SLS:
HLB: 8.9
Glycerol nEO dioleate
MARLOWET WOE:
HLB: 9.0
C18 fatty alcohol nEO
MARLOWET 5001:
HLB: 9.3
C18 fatty alcohol nEO/nPO
MARLOWET BL:
HLB: 10.0
C12-C14 fatty alcohol nEO
MARLOWET PW:
HLB: 10.3
C16-C18 fatty alcohol nEO
MARLOWET EF:
HLB: 13.1
Castor oil nEO
MARLOWET 4800:
HLB: 16.0
C16-C18 fatty alcohol nEO
MARLOWET FOX:
HLB: 16.4
C16-C18 fatty alcohol nEO
MARLOWET RNP:
Anionic/nonionic compound
MARLOWET SAF:
Alkyl polyglycol ether compound

HULS: Raw Materials for Cosmetic Products(Continued):

8. Polyethylene Glycols:

Field of application: As active ingredient bases in perfumes, face lotions and sticks containing active ingredients, for improving the spreadability in emulsions of all types, as water-soluble bases for creams and lipsticks, and for preventing tooth-pastes from drying out.

- PEG n: n = molecular weight 200, 300, 400, 600, 800, 1,000, 1,550, 2,000, 3,000, 4,000, 6,000, 12,000, 20,000 and 50,000

- PEG 550 MG: Compound PEG

9. Solvents:

Field of application: Solubilizers for bases and oils, humectants in emulsions

Solvent APV Spez.: Diethylene glycol monoethyl ether
2-Ethyl-1,3-hexanediol Phenoxyethanol: Monophenyl glycol
1,2-Propanediol: 1,2-propylene glycol
1,3-Butanediol: 1,3-butylene glycol
Ethanol
Isopropanol

10. Propellants:

Field of application: Propellants for sprays
Propellants based on isobutane, butane and propane
Pressurized as required

- DRIVOSOL

HULS: Raw Materials for Cosmetic Products(Continued):

11. Fragrances, Fragrance Precursors:

Field of application: Perfume compositions, synthesis of fragrances

Allylcyclohexyl propionate:
Allylhexahydrophenyl propionate
Odour: Sweet-fruity, reminiscent of pineapple
AROVA:
1,4-dioxacyclohexadecane-5,16-dione
Odour: Mild musk odour
2-tert.-butyl-cyclohexyl acetate:
Odour: Fruity, reminiscent of citrus
4-tert.-butyl-cyclohexyl acetate:
Odour: Powerful woody with a floral note
Capric aldehyde:
Odour: Strictly fatty, reminiscent of orange peel
Caprylic aldehyde:
Odour: Sharp: reminiscent of citrus when dilute
9-decen-1-ol:
Odour: Reminiscent of roses with waxy-aldehydic components when dilute
Phenoxyethanol O:
Odour: Mild, reminiscent of roses
Phenoxyethyl isobutyrate:
Odour: Sweet, floral-fruity
Phenoxyethyl propionate:
Odour: Warm, floral-fruity, not so sweet
Phenylmethylcarbinylacetate:
Odour: Fruity-green, reminiscent of gardenia
2-Phenylpropanal:
Hydratropaldehyde
Odour: Green, reminiscent of hyacinth
3-Phenylpropanal:
Hydrocinnamaldehyde
Odour: Balsamic-green, oily-earthy with a warm-spicy undertone
2-Phenyl-1-propanol Hydratope alcohol:
Odour: Sweet-floral, rather heavy, reminiscent of lily/hyacinth
3-Phenylpropanol:
Hydrocinnamic alcohol
Odour: Floral-balsamic, slightly reminiscent of hyacinths

HUMKO CHEMICAL DIVISION: HYSTRENE Behenic and Arachidic Acids

HYSTRENE 3022:

30% Behenic and Aracidic
CTFA: (Hydrogenated Menhaden Acid)
Titer C: 50-54
Iodine Value: 5
Acid Value: 193-201
Sap Value: 193-202
% Unsap Max: 1
Color: % Trans 440/550 nm, Min: 50/90
Other: 15.0Y-1.5R Lovibond
Saturated: C14: 8
C6: 27
C8: 29
C20/C22: 30
Unsat.: C18:1: 2

HYSTRENE 5522:

70% Behenic and Arachidic
CTFA: (Hydrogenated Menhaden Acid)
Titer C: 60-63
Iodine Value: 5
Acid Value: 178-185
Sap Value: 179-186
% Unsap Max: 2
Color: % Trans 440/550 nm, Min: 60/90
Other: 15.0Y-1.5R Lovibond
Saturated: C14: 1
C6: 14
C8: 30
C20/C22: 52
Unsat.: C18:1: 3

HYSTRENE 7022:

70% Behenic and Arachidic
CTFA: (Behenic Acid)
Titer C: 63-66
Iodine Value: 3.5
Acid Value: 170-180
Sap Value: 170-181
% Unsap Max: 2
Color: % Trans 440/550 nm, Min: 60/90
Other: 15.0Y-1.5R Lovibond
Saturated: C6/C8: 30
C20/C22: 70

HUMKO CHEMICAL DIVISION: HYSTRENE Behenic and Arachidic Acids (Continued):

HYSTRENE 9022:
90% Behenic and Aracidic
CTFA: (Behenic Acid)
Titer C: 67-71
Iodine Value: 3
Acid Value: 165-175
Sap Value: 165-176
% Unsap Max: 1.5
Color: % Trans 440/550 nm, Min: 60/90
Other: 15.0Y-1.5R Lovibond
Saturated: C8: 10
C20/C22: 88
Unsat.: C18:1: 2

HUMKO CHEMICAL DIVISION: HYSTRENE **Tallow Type Acids:**

HYSTRENE 1835:
Soap Acid Blend
CTFA: (Mixture Tallow/Coconut Acid)
Titer C: 40 Max
Iodine Value: 36-42
Acid Value: 214-222
Sap Value: 211-220
% Unsap Max: 0.5
Color: % Trans 440/550 nm, Min: 82/96
Other: 2.0Y-0.2R Lovibond
Saturated: C8: 1.5
C10: 2
C12: 10
C14: 5
C16: 20
C18: 18.5
Unsaturated: C14:1: 0.5
C16:1: 3
C18:1: 36
C18:2: 3.5

HUMKO CHEMICAL DIVISION: HYSTRENE/INDUSTRENE Stearic and Palmitic Acids:

HYSTRENE 9016:
90% Palmitic
CTFA: (Palmitic Acid)
Titer C: 58.0-60.5
Iodine Value: 0.5 Max
Acid Value: 216-220
Sap Value: 216-221
% Unsap Max: 0.2
Color: % Trans 440/550 nm, Min
Other: 1.0Y-0.1R Lovibond
Saturated: C14: 1
C16: 92
C18: 7

INDUSTRENE 4518:
Single Pressed Grade
CTFA: (Stearic Acid)
Titer C: 53-55
Iodine Value: 8-11
Acid Value: 204-211
Sap Value: 204-212
% Unsap Max: 1
Color: % Trans 440/550 nm, Min: 60/94
Other: 2.0Y-0.2R Lovibond
Saturated: C14: 4
C15: 0.5
C16: 50
C17: 2
C18: 37
Unsaturated: C18:1: 6.5

INDUSTRENE 5016:
Double Pressed Grade
CTFA: (Stearic Acid)
Titer C: 53-56
Iodine Value: 4-7
Acid Value: 207-210
Sap Value: 207-211
% Unsap Max: 0.5
Color: % Trans 440/550 nm, Min: 88/96
Other: 1.0Y-0.1R
Saturated: C14: 3
C15: 0.5
C16: 50
C17: 1.5
C18: 42
Unsat.: C18:1: 3

HUMKO CHEMICAL DIVISION: HYSTRENE Stearic and Palmitic Acids (Continued):

HYSTRENE 5016:

Triple Pressed Grade
CTFA: (Stearic Acid)
Titer C: 54.5-56.5
Iodine Value: 0.5 Max
Acid Value: 206-210
Sap Value: 206-211
% Unsap Max: 0.2
Color: % Trans 440/550 nm, Min: 92/98
Other: 1.0Y-0.1R Lovibond
Saturated: C14: 1
C15: 0.5
C16: 52
C17: 2.5
C18: 44

HYSTRENE 4516:

45% Palmitic
CTFA: Palmitic Acid
Titer C: 54.5-60
Iodine Value: 1.0 Max.
Acid Value: 203-209
Sap Value: 204-210
% Unsap Max: 1
Color: % Trans 440/550 nm, Min: 90/97
Other: 1.0Y-0.1R
Saturated: C14: 1
C15: .5
C16: 42
C17: 1.5
C18: 55

HYSTRENE 7018:

70% Stearic
CTFA: (Stearic Acid)
Titer C: 58-62.5
Iodine Value: 0.8 Max
Acid Value: 200-206
Sap Value: 200-207
% Unsap Max: 0.5
Color: % Trans 440/550 nm, Min: 92/98
Other: 1.0Y-0.1R Lovibond
Saturated: C14: 2
C15: 0.5
C16: 30
C17: 2.5
C18: 65

HUMKO CHEMICAL DIVISION: HYSTRENE Stearic and Palmitic Acids (Continued):

HYSTRENE 9718:
92% Stearic
CTFA: (Stearic Acid)
Titer C: 66.5-68.5
Iodine Value: 0.8 Max
Acid Value: 196-201
Sap Value: 196-202
% Unsap Max: 0.3
Color: % Trans 440/550 nm/Min: 82/96
Other: 2.0Y-0.2R Lovibond
Saturated: C16: 5
C18: 95

HUMKO CHEMICAL DIVISION: HYSTRENE Lauric and Myristic Acids

HYSTRENE 5012:
Hydrogenated Stripped Coconut
CTFA: (Hydrogenated Coconut Acid)
Titer C: 24-33
Iodine Value: 2 Max
Acid Value: 250-266
Sap Value: 250-266
% Unsap Max: 0.5
Color: % Trans 440/550 nm, Min: 74/96
Lovibond 5 1/4" cell: 4.0Y-0.4R
Saturated: C8: 1
C10: 1
C12: 55
C14: 24
C16: 12.5
C18: 6.5

HYSTRENE 9512:
95% Lauric
CTFA: (Lauric Acid)
Titer C: 41-43
Iodine Value: 0.5 Max
Acid Value: 276-281
Sap Value: 276-282
% Unsap Max: 0.25
Color: % Trans 440/550 nm, Min: 78/97
Lovibond 5 1/4" cell: 3.0Y-0.3R
Saturated: C10: 2
C12: 95
C14: 3

HUMKO CHEMICAL DIVISION: HYSTRENE Lauric and Myristic Acids (Continued):

HYSTRENE 9912:
99% Lauric
CTFA: (Lauric Acid)
Titer C: 43-44
Iodine Value: 0.2 Max
Acid Value: 277-281
Sap Value: 278-281
% Unsap Max: 0.25
Color: % Trans 440/550 nm, Min: 78/97
Lovibond 5 1/4" cell: 3.0Y-0.3R
Saturated: C12: 99
C14: 1

HYSTRENE 9014:
90% Myristic
CTFA: (Myristic Acid)
Titer C: 51-53
Iodine Value: 0.5 Max
Acid Value: 238-243
Sap Value: 238-246
% Unsap Max: 0.3
Color: % Trans 440/550 nm, Min: 78/97
Lovibond 5 1/4" cell: 3.0Y-0.3R
Saturated: C12: 4
C14: 90
C16: 6

HYSTRENE 9514:
95% Myristic
CTFA: (Myristic Acid)
Titer C: 51.9-53.5
Iodine Value: 0.5 Max
Acid Value: 243-246
Sap Value: 243-247
% Unsap Max: 0.3
Color: % Trans 440/550 nm, Min: 92/98
Lovibond 5 1/4" cell: 1.0Y-0.1R
Saturated: C12: 2
C14: 97
C16: 1

HUMKO CHEMICAL DIVISION: KEMESTER: Specialty Esters:

The growing line of KEMESTER fatty esters for personal care applications allows Humko Chemical to offer these products for such varied formulations as lotions, creams, stick products and liquid soaps. In addition to their function as emollients, certain KEMESTER esters also act as pearlescing and bodying agents.

High performance esters include adipate, dimer and monoesters which are characterized by improved viscosity, volatility and high- and low-temperature properties. Their controlled structures can be designed to provide optimal properties for many applications, such as products for textile, machinery and metal-working lubricant formulations.

KEMESTER BE:
CTFA Name: Behenyl Erucate
Color, Gardner (1963) Max: 1
Iodine Value (Wijs) Max: 50
Sap Value: 80-95
Acid Value Max: 2
Melt/Freeze C, Typical: 40-44

KEMESTER DMP:
CTFA Name: Dimethyl Phthalate
Color, Gardner (1963) Max: 10
Acid Value Max: 0.1
Melt/Freeze C, Typical: 1

KEMESTER EGDS:
CTFA Name: Glycol Distearate
Color, Gardner (1963) Max: 2
Iodine Value (Wijs) Max: 1
Sap Value: 190-200
Acid Value Max: 6
Melt/Freeze C, Typical: 60-63

KEMESTER EGMS:
CTFA Name: Glycol Stearate
Color, Gardner (1963) Max: 1
Iodine Value (Wijs) Max: 1
Sap Value: 179-195
Acid Value Max: 5
Melt/Freeze C, Typical: 54-60

HUMKO CHEMICAL DIVISION: KEMESTER: Specialty Esters(Continued):

KEMESTER THFO:
CTFA Name: Tetrahydrofurfuryl Oleate
Color, Gardner (1963) Max: 1
Iodine Value (Wijs) Max: ±68
Sap Value: ±156
Acid Value Max: 0.5
Melt/Freeze C, Typical: <-20

KEMESTER 1000:
CTFA Name: Triolein (Glycerol Trioleate)
Color, Gardner (1963) Max: 6
Iodine Value (Wijs) Max: 85-90
Sap Value: 190-198
Acid Value Max: 5
Melt/Freeze C, Typical: -8

KEMESTER 2000:
CTFA Name: Glyceryl Oleate
Color, Gardner (1963) Max: 6
Iodine Value (Wijs) Max: 73-83
Sap Value: 160-175
Acid Value Max: 3
Melt/Freeze C, Typical: 25

KEMESTER 3681:
CTFA Name: Di-Octyl Dimerate
Color, Gardner (1963) Max: 10
Acid Value Max: 4
Melt/Freeze C, Typical: -38

KEMESTER 4000:
CTFA Name: Butyl Oleate
Color, Gardner (1963) Max: 2
Iodine Value (Wijs) Max: 72-81
Sap Value: 164-172
Acid Value Max: 2
Melt/Freeze C, Typical: 2

KEMESTER 5221SE:
CTFA Name: PEG-2 Stearate SE
Color, Gardner (1963) Max: 3
Iodine Value (Wijs) Max: 3
Sap Value: 163-178
Acid Value Max: 95-105
Melt/Freeze C, Typical: 46-56

HUMKO CHEMICAL DIVISION: KEMESTER: Specialty Esters(Continued):

KEMESTER 5415:
CTFA Name: Isobutyl Stearate
Color, Gardner (1963) Max: <1
Iodine Value (Wijs) Max: 1
Sap Value: 170-176
Acid Value Max: 4
Melt/Freeze C, Typical: 21

KEMESTER 5500:
CTFA Name: Glyceryl Stearate
Color, Gardner (1963) Max: 3
Iodine Value: 0.5
Sap Value: 164-177
Acid Value Max: 3
Melt/Freeze C, Typical: 56-60

KEMESTER 5510:
CTFA: Butyl Stearate
Color, Gardner (1963) Max.: <1
Iodine Value (Wijs) Max: 1
Sap Value: 170-176
Acid Value Max: 2
Melt/Freeze C, Typical: 23

KEMESTER 5654:
CTFA Name: Ditridecyl Adipate
Color, Gardner (1963) Max: 2
Acid Value Max: 0.1
Melt/Freeze C, Typical: -50

KEMESTER 5721:
CTFA Name: Tridecyl Stearate
Color, Gardner (1963) Max: 1
Iodine Value (Wijs) Max: 4
Sap Value: 117-126
Acid Value Max: 2
Melt/Freeze C, Typical: 8

KEMESTER 5822:
CTFA Name: Isocetyl Stearate
Color, Gardner (1963) Max: 1
Iodine Value (Wijs) Max: 4
Sap Value: 102-114
Acid Value Max: 2
Melt/Freeze C, Typical: 0

HUMKO CHEMICAL DIVISION: KEMESTER: Specialty Esters(Continued):

KEMESTER 6000:
CTFA Name: Glyceryl Stearate
Color, Gardner (1963) Max: 3
Iodine Value (Wijs) Max: 2
Sap Value: 162-176
Acid Value Max: 3
Melt/Freeze C, Typical: 57-60

KEMESTER 6000SE:
CTFA Name: Glyceryl Stearate SE
Color, Gardner (1963) Max: 3
Iodine Value (Wijs) Max: 3
Sap Value: 140-156
Acid Value Max: 10
Melt/Freeze C, Typical: 56-61

HUMKO CHEMICAL DIVISION: KEMESTER: Homosalate:

Homomenthyl salicylate (Homosalate) is a safe, effcetive sunscreen agent. It is easily utilized in a variety of cosmetic skin preparations designed to reduce the adverse effects of solar radiation.

KEMESTER HMS:
Homomenthyl Salicylate
CTFA: Homosalate
Appearance: Clear liquid at ambient temperature
Free salicylic acid: Not more than faint purple
Specific gravity: 1.049-1.053
Refractive index: 1.516-1.519
Composition by GLC: 99% Min HMS

HUMKO CHEMICAL DIVISION: KEMSTRENE: Glycerine:

Four grades of USP glycerine and a high gravity glycerine are available. Glycerine is used as a humectant in cosmetics, cough syrups, tobacco, liquid soaps and in food products. It is also used extensively as an an intermediate in the production of polyester, polyurethane and alkyd resin formulations.

KEMSTRENE 99.7%:
99.7% USP
CTFA: Glycerine
Specific gravity 25/25 Min: 1.26124
Color: Not darker than FeCl3 standard
% Ash Max: 0.01
% Chloride Max: 0.001
% Sulfate Max: 0.002
% Arsenic Max: 0.00015
% Heavy metals Max: 0.0005
% Chlorinated compounds Max: 0.003
Fatty acids and esters: 1.0 ml of 0.5N NaOH Max

KEMSTRENE 99.5%:
99.5% USP
CTFA: Glycerine
Specific gravity 25/25C Min: 1.26073
Color: Not darker than FeCl3 standard
% Ash Max: 0.01
% Chloride Max: 0.001
% Sulfate Max: 0.002
% Arsenic Max: 0.00015
% Heavy metals Max: 0.0005
% Chlorinated compounds Max: 0.003
Fatty acids and esters: 1.0 ml of 0.5N NaOH Max

KEMSTRENE 99.0%:
99.0% USP
CTFA: Glycerine
Specific gravity 25/25C Min: 1.25945
Color: Not darker than FeCl3 standard
% Ash Max: 0.01
% Chloride Max: 0.001
% Sulfate Max: 0.002
% Arsenic Max: 0.00015
% Heavy metals Max: 0.0005
% Chlorinated compounds Max: 0.003
Fatty acids and esters: 1.0 ml of 0.5N NaOH Max

HUMKO CHEMICAL DIVISION: KEMSTRENE: Glycerine(Continued):

KEMSTRENE 96.0%:
96.0% USP
CTFA: Glycerine
Specific gravity 25/25 Min: 1.25165
Color: Not darker than FeCl3 standard
% Ash Max: 0.01
% Chloride Max: 0.001
% Sulfate Max: 0.002
% Arsenic Max: 0.00015
% Heavy metals Max: 0.0005
% Chlorinated compounds Max: 0.003
Fatty acids and esters: 1.0 ml of 0.5N NaOH Max

KEMSTRENE High Gravity:
High Gravity
CTFA: Glycerine
Specific gravity 25/25 Min: 1.2587
Color: Not darker than standard
% Ash Max: 0.10
% Chloride Max: 0.01
Sap Equivalent: 0.05% Max
Acidity/Alkalinity measures 0.6 ml of 0.5N HCl or 0.5 NaOH

HUMKO CHEMICAL DIVISION: Quaternary Ammonium Chlorides:

KEMAMINE: Benzyl and methyl chloride quaternaries:

The fatty quaternary ammonium chlorides are salts of ammonia formed by the alkylation of the corresponding tertiary amines. As such, they exhibit strong cationic characteristics. Unlike the amines, they are water-miscible or -dispersible. They are extensively used as antistats, textile-softening agents, dyeing aids, corrosion inhibitors and emulsifiers.

KEMAMINE Q-2802C:
Dimethyl Di-Behenyl
CTFA: Dibehenyl Dimonium Chloride
% Active Min: 75
% Amine Max: 2
% Amine HCl Max: 2
Color Gardner, 1963 Max: 2
Average molecular weight: 690
pH of 5% solution Max: 9
% Ash Max: 0.2
Saturated: C16: 2
C18: 5
C20: 10
C22: 80
C24: 3

KEMAMINE Q-9702C:
Dimethyl Di-Hydrogenated Tallow
CTFA: Quaternium-18
% Active Min: 75
% Amine Max: 1.5
% Amine HCl Max: 0.5
Color Gardner, 1963 Max: 2
Average Molecular weight: 575
pH of 5% solution Max: 9
% Ash Max: 0.2
Saturated: C14: 4
C16: 29
C18: 67

KEMAMINE Q-9743CHGW:
Trimethyl Monoalkyl
CTFA: Tallow Trimonium Chloride
% Active Min: 65
% Amine Max: 1.5
% Amine HCl Max: 0.5
Color Gardner, 1963 Max: 4
Average molecular weight: 335-355
pH of 5% solution Max: 9
% Ash Max: 0.5
Saturated: C14: 4
C16: 29
C18: 25
Unsaturated: C18:1: 38
C18:2: 4

HUMKO CHEMICAL DIVISION: Quaternary Ammonium Chlorides (Continued):

KEMAMINE Q-9743C:
Trimethyl Monoalkyl
CTFA: Tallow Trimonium Chloride
% Active Min: 65
% Amine Max: 1.5
% Amine HCl Max: 0.5
Color Gardner, 1963 Max: 4
Average molecular weight: 335-355
pH of 5% solution Max: 9
% Ash Max: 0.5
Saturated: C14: 4
C16: 29
C18: 25
Unsaturated: C18:1: 38
C18:2: 4

KEMAMINE BQ-9742C:
Dimethyl Tallow Benzyl
CTFA: Tallow Alkonium Chloride
% Active Min: 75
% Amine Max: 1.5
% Amine HCl Max: 0.5
Color Gardner, 1963 Max: 6
Average molecular weight: 420
pH of 5% solution Max: 9
% Ash Max: 0.2
Saturated: C14: 4
C16: 29
C18: 25
Unsaturated: C18:1: 38
C18:2: 4

HUMKO CHEMICAL DIVISION: Tertiary Amines:

KEMAMINE: Mono- and Dimethyl:

The tertiary amines are derivatives of ammonia having all three hydrogens substituted with hydrocarbon chains. In KEMAMINE tertiary amines either one or two of the alkyl groups are methyl. Tertiary amines find the widest application as intermediates for quaternary ammonium derivatives, which are used for various applications in the cosmetic and textile industries. Another application is as an acid scavenger in petroleum products.

KEMAMINE T-6502D:
CTFA: Distilled Dimethyl Cocc
% Tertiary Min: 95
Total amine value Min: 230
Color Gardner, 1963 Max: 1
% H2O Max.: 0.5
Saturated: C10: 2
C12: 61
C14: 22
C16: 8
C18: 3
Unsaturated: C18:1: 2

KEMAMINE T-9701:
CTFA: Methyl Di-Hydrogenated Tallow
% Tertiary Min: 95
Total amine value Min: 103
Color Gardner, 1963 Max.: 3
% H20 Max: 0.5
Saturated: C14: 4
C16: 29
C18: 67

KEMAMINE T-9892D:
CTFA: Distilled Dimethyl Oleyl
% Tertiary Min: 95
Total amine value Min: 180
Color Gardner, 1963 Max: 1
% H20 Max: 0.5
Saturated: C14: 4
C16: 14
C18: 10
Unsaturated: C18:1: 65
C18:2: 7

HUMKO CHEMICAL DIVISION: Tertiary Amines(Continued):

KEMAMINE T-9902D:
CTFA: Distilled Dimethyl Stearyl Dimethyl Stearamine
% Tertiary Min: 95
Total amine value Min: 180
Color Gardner, 1963 Max: 1
% H2O Max: 0.5
Saturated: C16: 10
C18: 90

KEMAMINE T-9992D:
CTFA: Distilled Dimethyl Oleic-Linolenic
% Tertiary Min: 95
Total amine value Min: 180
Color Gardner, 1963 Max.: 2
% H2O Max: 0.5
Saturated: C16: 15
C18: 7
Unsaturated: C18:1: 53
C18:2: 22
C18:3: 3

ICI SPECIALTY CHEMICALS: ATLAS Sorbitol Products:

SORBO Sorbitol Solution, USP:

Form: Clear, viscous liquid
Taste: Faint, sweet, cooling and pleasant. About 60% as sweet as sucrose
Sp. Gravity at 25C: Above 1.285
Refractive Index (25C): Approx. 1.458
Viscosity (at 25C): Approx. 110 cps
Flash Point: Above 300F
Fire Point: Above 300F
Boiling Point: Approx. 105C

Crystalline Sorbitol 1162, NF:

Form: White, crystalline solid
Taste: Faint, sweet, cooling and pleasant. About 60% as sweet as sucrose.
Flash Point: Above 300F
Fire Point: Above 300F

Crystalline Sorbitol 712, NF:

Form: White, crystalline solid
Taste: Faint, sweet, cooling and pleasant. About 60% as sweet as sucrose.
pH (25% in H2O): Approx. 5
Flash Point: Above 300F
Fire Point: Above 300F

Pure Sorbitol:

Form: White, crystalline solid
Mol. Wt.: 182
Taste: Faint, sweet, cooling and pleasant. About 60% as sweet as sucrose
Sp. Gravity at 25C: 1.49
Refractive Index (25C): 1.35 (in 10% W/W aqueous sol.)
Melting Point: 93C (metastable form)
97.7C (stable form)
Heat of Solution: -26.5 calories per gram, at 25C
Heat of Combustion: 3971 ± 5 calories per gram
Chemical Stability: Stable in dry state or in sterile aqueous solutions. In solutions, it is not attaacked by cold dilute acids or alkalies, or by atmospheric oxygen in the absence of catalysts.

ICI SPECIALTY CHEMICALS: ATLAS Sorbitol Products-Standard Specifications:

SORBO Sorbitol Solution, USP:
Polyhydric Alcohol, %: 70.0 min.
Total Sugars, %: 0.45 max.
Reducing Sugars, %: 0.10 max.
Normal Ash, %: 0.01 max.
Iron, ppm: 3 max.
Color: Water white
Taste: Pleasantly sweet
Odor: Essentially odor-free
Moisture, %: 30% or less
Conforms to Food Chemicals Codex and USP requirements

Crystalline Sorbitol 1162, NF:
Polyhydric Alcohol, %: 99.0 min.
Total Sugars, %: 0.75 max.
Reducing Sugars, %: 0.14 max.
Normal Ash, %: 0.02 max.
Iron, ppm: 4 max.
Color: Water white (25% in water)
Taste: Pleasantly sweet
Odor: Essentially odor-free
Moisture, %: 1.0 max.
Mesh Size: 0.5% max. on U.S. Sieve No. 16
10% max. on U.S. Sieve No. 20
5% max. through U.S. Sieve No. 325
Conforms to Food Chemicals Codex and NF requirements

Crystalline Sorbitol 712, NF:
Polyhydric Alcohol, %: 99.0 min.
Total Sugars, %: 0.75 max.
Reducing Sugars, %: 0.14 max.
Normal Ash, %: 0.02 max.
Iron, ppm: 4 max.
Color: Water white (25% in water)
Taste: Pleasantly sweet
Odor: Essentially odor-free
Moisture, %: 1.0 max.
Mesh Size: 0.5% max. on U.S. Sieve No. 30
5% max. on U.S. Sieve No. 40
30% max. on U.S. Sieve No. 200
Conforms to Food Chemicals Codex and NF requirements

ICI SPECIALTY CHEMICALS: Mannitol USP:

Three forms of Mannitol, USP are available: Granular, Powdered and Reagent Grade.

Appearance:
Mannitol is a white, odorless substance, available either in granular or powder form.

Taste:
Mannitol has a subtle taste--distinctively pleasant, slightly sweet and cool. It has a smooth mouthfeel that is not gritty, chalky or pasty. About half as sweet as sucrose, it is free from cloying sweetness.

General Characteristics:
- Form: White, crystalline granules
- Mol. Wt.: 182.17
- Taste: Faint, sweet, cooling
- Sp. Gravity at 20C: 1.52
- Refractive Index (20C): 1.3476 (in 10% W/W aqueous solution)
- Hygroscopicity: Nonhygroscopic
- Flash Point: Above 300F
- Fire Point: Above 300F
- Heat of Solution: -28.9 calories per gram at 25C
- Heat of Combustion: 3960 ± 5 calories per gram
- Chemical Stability: Stable in dry state or in sterile aqueous solutions. In solutions, it is not attacked by cold dilute acids or alkalies, nor by atmospheric oxygen in the absence of catalysts.

Standard Specifications:
- Moisture, %: 0.30 max.
- Reducing Sugars (USP test): Passes
- Total Sugars: Undetectable
- Normal Ash, %: 0.02 max.
- Water Insolubles, %: 0.02 max.
- Iron, ppm: 1 max.
- Chlorides, ppm: 50 max.
- Sulfates, as Na_2SO_4, ppm: 50 max.
- Arsenic, ppm: 1 max.
- Heavy Metals (as Pb, Codex test): Passes
- Acidity (USP test): 0.3 max.
- Melting Range, C: (USP test): 165-168
- Clarity and Color (20% in water @ 30C): Clear, colorless
- Odor: None
- Particle Size (Cenco-Meinzer Sieve Shaker):
  - Granular--Approx. 99% through 16 mesh
    Approx. 25% through 80 mesh
  - Powdered--Approx. 98% through 40 mesh
- Conforms to Food Chemicals Codex and USP requirements.

ICI SPECIALTY CHEMICALS: Nonionic Surfactants for Cosmetics and Pharmaceuticals:

ARLACEL 20:
Sorbitan monolaurate
CTFA: Sorbitan Laurate
Yellow liquid
HLB: 8.6
Viscosity at 25C: 6500 cps

ARLACEL 40:
Sorbitan monopalmitate
CTFA: Sorbitan Palmitate
Cream solid (granules)
HLB: 6.7
Pour Point: 48C

ARLACEL 60:
Sorbitan monostearate
CTFA: Sorbitan Stearate
Cream solid (granules)
HLB: 4.7
Pour Point: 53C

ARLACEL 80:
Sorbitan monooleate
CTFA: Sorbitan Oleate
Yellow liquid
HLB: 4.3
Viscosity at 25C: 1900 cps

ARLACEL 83:
Sorbitan sesquioleate
CTFA: Sorbitan Sesquioleate
Yellow liquid
HLB: 3.7
Viscosity at 25C: 1500 cps

ARLACEL C:
Sorbitan sesquioleate
CTFA: Sorbitan Sesquioleate
Amber liquid
HLB: 3.7
Viscosity at 25C: 1000 cps

ARLACEL 85:
Sorbitan trioleate
CTFA: Sorbitan Trioleate
Yellow liquid
HLB: 1.8
Viscosity at 25C: 250 cps

ICI SPECIALTY CHEMICALS: Nonionic Surfactants for Cosmetics and Pharmaceuticals(Continued):

ARLACEL 165:
Glycerol monostearate and polyoxyethylene stearate (acid-stable, self-emulsifying)
CTFA: Glyceryl Stearate (and) PEG-100 Stearate
White solid (granules)
HLB: 11.0
Pour Point: 54C

ARLACEL 186:
Glycerol monooleate diluted with propylene glycol and with 0.02% BHA and 0.01% citric acid added as preservatives
CTFA: Glyceryl Oleate (and) Propylene Glycol
Clear pale yellow liquid
HLB: 2.8
Viscosity at 25C: 150 cps

ARLAMOL E:
Polyoxypropylene 15 stearyl ether with 0.01% BHA and 0.005% citric acid added as preservatives
CTFA: PPG-15 Stearyl Ether
Colorless to light yellow liquid
Viscosity at 25C: 80 cps

ARLATONE G:
Polyoxyethylene 25 hydrogenated castor oil
CTFA: PEG-25 Hydrogenated Castor Oil
Yellow liquid WSS
HLB: 10.8
Viscosity at 25C: 1400 cps

ARLATONE T:
Polyoxyethylene 40 sorbitol septaoleate
CTFA: PEG-40 Sorbitan Peroleate
Yellow Liquid
HLB: 9
Viscosity at 25C: 175 cps

ATLAS G-1441:
Polyoxyethylene 40 sorbitol lanolin derivative
CTFA: PEG-40 Sorbitan Lanolate
Red brown solid
HLB: 14.0
Pour Point: 33C

ATLAS G-1471:
Polyoxyethylene 75 sorbitol lanolin derivative
CTFA: PEG-75 Sorbitan Lanolate
Red brown solid
HLB: 16.0
Pour Point: 38C

ICI SPECIALTY CHEMICALS: Nonionic Surfactants for Cosmetics and Pharmaceuticals(Continued):

ATLAS G-1702:
Polyoxyethylene 6 sorbitol beeswax derivative
CTFA: PEG-6 Sorbitan Beeswax
Tan solid
HLB: 5
Pour Point: 68C

ATLAS G-1726:
Polyoxyethylene 20 sorbitol beeswax derivative
CTFA: PEG-20 Sorbitan Beeswax
Tan solid
HLB: 5
Pour Point: 63C

ATLAS G-1790:
Polyoxyethylene 20 lanolin derivative
CTFA: PEG-20 Lanolate
Brown solid
HLB: 11.0
Pour Point: 33C

ATLAS G-1795:
Polyoxyethylene 50 lanolin derivative
CTFA: PEG-50 Lanolate
Light tan solid
HLB: 17.0
Pour Point: 43C

ATLAS G-2162:
Polyoxyethylene 25 propylene glycol stearate
CTFA: PEG-25 Propylene Glycol Stearate
Cream solid
HLB: 16.0
Pour Point: 23C

BRIJ 30:
Polyoxyethylene 4 lauryl ether
CTFA: Laureth-4
Colorless liquid
HLB: 9.7
Viscosity at 25C: 30 cps

BRIJ 30 SP:
Polyoxyethylene 4 lauryl ether with 0.01% BHA and 0.005% citric acid added as preservatives
CTFA: Laureth-4
Colorless liquid
HLB: 9.5
Viscosity at 25C: 30 cps

ICI SPECIALTY CHEMICALS: Nonionic Surfactants for Cosmetics and Pharmaceuticals(Continued):

BRIJ 35:
Polyoxyethylene 23 lauryl ether
CTFA: Laureth-23
White solid
HLB: 16.9
Pour Point: 33C

BRIJ 35 SP:
Polyoxyethylene 23 lauryl ether with 0.01% BHA and 0.005% citric acid added as preservatives
CTFA: Laureth-23
White solid
HLB: 16.9
Pour Point: 33C

BRIJ 52:
Polyoxyethylene 2 cetyl ether with 0.01% BHA and 0.005% citric acid added as preservatives
CTFA: Ceteth-2
White solid
HLB: 5.3
Pour Point: 33C

BRIJ 56:
Polyoxyethylene 10 cetyl ether with 0.01% BHA and 0.005% citric acid added as preservatives
CTFA: Ceteth-10
White solid
HLB: 12.9
Pour Point: 31C

BRIJ 58:
Polyoxyethylene 20 cetyl ether with 0.01% BHA and 0.005% citric acid added as preservatives
CTFA: Ceteth-20
White solid
HLB: 15.7
Pour Point: 38C

BRIJ 72:
Polyoxyethylene 2 stearyl ether with 0.01% BHA and 0.005% citric acid added as preservatives
CTFA: Steareth-2
White solid
HLB: 4.9
Pour Point: 43C

ICI SPECIALTY CHEMICALS: Nonionic Surfactants for Cosmetics and Pharmaceuticals(Continued):

BRIJ 76:
Polyoxyethylene 10 stearyl ether with 0.01% BHA and 0.005% citric acid added as preservatives
CTFA: Steareth-10
White solid
HLB: 12.4
Pour Point: 38C

BRIJ 78:
Polyoxyethylene 20 stearyl ether with 0.01% BHA and 0.005% citric acid added as preservatives
CTFA: Steareth-20
White solid
HLB: 15.3
Pour Point: 38C

BRIJ 700:
Polyoxyethylene 100 stearyl ether with 0.01% BHA and 0.005% citric acid added as preservatives
CTFA: Steareth-100
White solid
HLB: 18.8
Pour Point: 55C

BRIJ 721:
Polyoxyethylene 21 stearyl ether with 0.01% BHA and 0.005% citric acid added as preservatives
CTFA: Steareth-21
White solid
HLB: 15.5
Pour Point: 45C

BRIJ 92:
Polyoxyethylene 2 oleyl ether with 0.01% BHA and 0.005% citric acid added as preservatives
CTFA: Oleth-2
Yellow liquid
HLB: 4.9
Viscosity at 25C: 30 cps

BRIJ 93:
Polyoxyethylene 2 oleyl ether (low color and odor) with 0.01% BHA and 0.005% citric acid added as preservatives
CTFA: Oleth-2
Pale yellow liquid
HLB: 4.9
Viscosity at 25C: 30 cps

ICI SPECIALTY CHEMICALS: Nonionic Surfactants for Cosmetics and Pharmaceuticals(Continued):

BRIJ 96:
Polyoxyethylene 10 oleyl ether with 0.01% BHA and 0.005% citric acid added as preservatives
CTFA: Oleth-10
Yellow liquid WSS
HLB: 12.4
Viscosity at 25C: 100 cps

BRIJ 97:
Polyoxyethylene 10 oleyl ether (low color and odor) with 0.01% BHA and 0.005% citric acid added as preservatives
CTFA: Oleth-10
Pale yellow liquid WSS
HLB: 12.4
Viscosity at 25C: 100 cps

BRIJ 98:
Polyoxyethylene 20 oleyl ether with 0.01% BHA and 0.005% citric acid added as preservatives
CTFA: Oleth-20
Cream solid
HLB: 15.3
Pour Point: 30C

BRIJ 99:
Polyoxyethylene 20 oleyl ether (low color and odor) with 0.01% BHA and 0.005% citric acid added as preservatives
CTFA: Oleth-20
Cream solid
HLB: 15.3
Pour Point: 33C

MYRJ 45:
Polyoxyl 8 stearate (Polyoxyethylene 8 stearate)
CTFA: PEG-8 Stearate
Cream solid
HLB: 11.1
Pour Point: 28C

MYRJ 52:
Polyoxyl 40 stearate, USP (Polyoxyethylene 40 stearate)
CTFA: PEG-40 Stearate
Ivory solid
HLB: 16.9
Pour Point: 38C

ICI SPECIALTY CHEMICALS: Nonionic Surfactants for Cosmetics and Pharmaceuticals(Continued):

MYRJ 52S:
Polyoxyethylene 40 stearate
CTFA: PEG-40 Stearate
White solid (granules)
HLB: 16.9
Pour Point: 38C

MYRJ 53:
Polyoxyethylene 50 stearate
CTFA: PEG-50 Stearate
Cream solid
HLB: 17.9
Pour Point: 36C

MYRJ 59:
Polyoxyethylene 100 stearate
CTFA: PEG-100 Stearate
Off-white to light tan solid
HLB: 18.8
Pour Point: 46C

SPAN 20:
Sorbitan monolaurate
CTFA: Sorbitan Laurate
Amber liquid
HLB: 8.6
Viscosity at 25C: 4250 cps

SPAN 40:
Sorbitan monopalmitate
CTFA: Sorbitan Palmitate
Tan solid
HLB: 6.7
Pour Point: 48C

SPAN 60:
Sorbitan monostearate
CTFA: Sorbitan Stearate
Tan solid (granules)
HLB: 4.7
Pour Point: 53C

SPAN 65:
Sorbitan tristearate
CTFA: Sorbitan Tristearate
Tan solid (granules)
HLB: 2.1
Pour Point: 53C

ICI SPECIALTY CHEMICALS: Nonionic Surfactants for Cosmetics and Pharmaceuticals(Continued):

SPAN 80:
Sorbitan Monooleate, NF
CTFA: Sorbitan Oleate
Amber liquid
HLB: 4.3
Viscosity: 1000 cps

SPAN 85:
Sorbitan trioleate
CTFA: Sorbitan Trioleate
Amber liquid
HLB: 1.8
Viscosity: 210 cps

TWEEN 20:
Polysorbate 20 (Polyoxyethylene 20 sorbitan monolaurate)
CTFA: Polysorbate-20
Yellow liquid
HLB: 16.7
Viscosity: 400 cps

TWEEN 21:
Polyoxyethylene 4 sorbitan monolaurate
CTFA: Polysorbate-21
Yellow liquid
HLB: 13.3
Viscosity: 500 cps

TWEEN 40:
Polysorbate 40 (Polyoxyethylene 20 sorbitan monopalmitate)
CTFA: Polysorbate-40
Yellow liquid
HLB: 15.6
Viscosity at 25C: 500 cps

TWEEN 60:
Polysorbate 60 (Polyoxyethylene 20 sorbitan monostearate)
CTFA: Polysorbate-60
Yellow liquid
HLB: 14.9
Viscosity at 25C: 600 cps

TWEEN 61:
Polyoxyethylene 4 sorbitan monostearate
CTFA: Polysorbate-61
Tan solid
HLB: 9.6
Pour Point: 38C

ICI SPECIALTY CHEMICALS: Nonionic Surfactants for Cosmetics and Pharmaceuticals(Continued):

TWEEN 65:
Polysorbate 65 (Polyoxyethylene 20 sorbitan tristearate)
CTFA: Polysorbate-65
Tan solid
HLB: 10.5
Pour Point: 33C

TWEEN 80:
Polysorbate 80, USP (Polyoxyethylene 20 sorbitan monooleate)
CTFA: Polysorbate-80
Yellow liquid
HLB: 15.0
Viscosity at 25C: 425 cps

TWEEN 81:
Polyoxyethylene 5 sorbitan monooleate
CTFA: Polysorbate-81
Amber liquid
HLB: 10.0
Viscosity at 25C: 450 cps

TWEEN 85:
Polysorbate 85 (Polyoxyethylene 20 sorbitan trioleate)
CTFA: Polysorbate-85
Amber liquid
HLB: 11.0
Viscosity at 25C: 300 cps

INOLEX CHEMICAL CO.: Biocides:

BRONOPOL:
CTFA: 2-Bromo-2-Nitropropane-1,3-Diol
Crystal
Conc.: 100%
Powerful anti-bacterial agent with high water solubility and excellent stability in the presence of surfactants, including nonionics.

MYACIDE SP:
CTFA: 2,4-Dichlorobenzyl Alcohol
Crystal
Conc.: 100%
Antifungal agent with quick lethal action, stability at pH 3 to 10, and most useful where contamination by spoilage yeasts and molds is the problem.

LEXGARD M:
CTFA: Methylparaben
Powder
Conc.: 100%

LEXGARD P:
CTFA: Propylparaben
Powder
Conc.: 100%

LEXGARD B:
CTFA: Butylparaben
Powder
Conc.: 100%
The preservatives most often used by the cosmetic and pharmaceutical industries. Used singly or in combination, they inhibit the growth of bacteria, molds, and yeast extending product shelf life.

INOLEX CHEMICAL CO.: Concentrates & Blends:

LEXATE BPQ:

CTFA: Lauramidopropyl Betaine (and) TEA Coco hydrolyzed animal protein (and) Oleamidopropyl Dihydoxypropyl Dimonium Chloride

Liquid

Conc.: 35%

A complete blend of detergent, conditioner, and protein used as an economical base for shampoo, bath gel, liquid soaps, bubble baths, and cleansers.

LEXATE PX:

CTFA: Petrolatum (and) Lanolin (and) Ozokerite

Soft Solid

Conc.: 100%

Lanolin-rich base forms elegant w/o emulsions for a wide variety of cosmetic, treatment and topical creams and ointments.

LEXATE TA:

CTFA: Glyceryl Stearate (and) Isopropyl Myristate (and) Stearyl Stearate

Flakes

Conc.: 100%

Balanced emollient blend. Easily emulsified, non-greasy feel. For cosmetic creams and lotions.

LEXATE TL:

CTFA: Glyceryl Stearate (and) Butyl Stearate (and) Stearyl Stearate

Solid

Conc.: 100%

Easily emulsified emollient blend for superfatting fine toilet soaps and for incorporation in barrier cream and lotion formulations.

LEXATE CRC:

CTFA: Stearamidopropyl Dimethylamine (and) Glycol Stearate (and) Ceteth 2

Flake

Conc.: 100%

Cream rinse concentrate. Balanced blend of conditioners and emulsifiers providing all the functionality of a conditioner in a single easy-to-use product.

INOLEX CHEMICAL CO.: Emollients:

LEXOL GT 865:
CTFA: Caprylic/Capric Triglyceride
Liquid
Conc.: 100%

LEXOL GT 855:
CTFA: Caprylic/Capric Triglyceride
Liquid
Conc.: 100%

Naturally derived, highly emollient oils with a very elegant non-oily skin feel. Superb moisturizers. Odorless, tasteless, excellent oxidative stability. For creams, lotions, bath oils, lipsticks and make-up items. Outstanding solvents for perfume and flavor ingredients; vehicles for medicinals, antibiotics and vitamins. Alcohol soluble.

LEXOL PG 865:
CTFA: Propylene Glycol Dicaprylate/Dicaprate
Liquid
Conc.: 100%

LEXOL PG 855:
CTFA: Propylene Glycol Dicaprylate/Dicaprate
Liquid
Conc.: 100%

Luxuriant emollients, moisturizers with excellent lubricity and non-oily skin deposition for creams, lotions and make-up items. Low viscosity and unusually low freezing points for bath oils, pre-electric shave lotions, aerosol systems. Excellent hydro-alcohol solubility. Ideal vehicle for flavors, fragrances and pigmented cosmetics; also vitamins, antibiotics and medicinals.

LEXOL PG-800:
CTFA: Propylene Glycol Dioctanoate
Liquid
Conc.: 100%

Highly refined emollient oil designed specifically for cosmetic and pharmaceutical applications. Exhibits an elegant non-oily feel, excellent lubricity and unique skin absorbance. Outstanding solvency and oxidation stability, low viscosity and cloud point. Applications in a wide variety of cosmetic creams, lotions and topicals as an emollient and moisturizer. Colorless and odorless, it has applications in bath oils, aftershave lotions, aerosol systems, flavors and fragrences; also a vehicle for vitamins, antibiotics and medicinals.

INOLEX CHEMICAL CO.: Emollients(Continued):

LEXOL PG-900:
CTFA: Propylene Glycol Dipelargonate
Liquid
Conc.: 100%
Highly refined emollient oil possessing good penetration and elegant spreading properties without a residual oily feel. For use in luxurious emollient creams and lotions. Its light color and unusually low freezing point suggest applications in bath oils, preshave lotions, aerosol systems and other clear liquid products requiring high emollient concentrations. Applications in lipsticks, lip glosses and makeup bases.

LEXOL IPM:
CTFA: Isopropyl Myristate
Liquid
Conc.: 100%
Highly purified emollient oil for cosmetic applications. Colorless and essentially odorless; low viscosity, freezing point; and outstanding spreading properties and good solubility characteristics.

LEXOL IPP:
CTFA: Isopropyl Palmitate
Liquid
Conc.: 100%
Stable, odor-free, colorless emollient. Excellent spreading and solubilizing properties make it an effective carrier for many cosmetic and topical pharmaceutical ingredients.

LEXOL IPP-A:
CTFA: Isopropyl Palmitate
Liquid
Conc.: 100%
Same as above but with a higher level of Palmitic ester content.

LEXOL IPP-NF:
CTFA: Isopropyl Palmitate
Liquid
Conc.: 100%
Same as above. In addition, conforms to NF specs.

INOLEX CHEMICAL CO.: Emollients(Continued):

LEXOL 60:
CTFA: Isopropyl Palmitate (and) Isopropyl Myristate (and) Isopropyl Stearate
Liquid
Conc.: 100%
Essentially odorless and colorless emollient with emolliency and solvency similar to LEXOL IPM. Applications in bath oils, electric pre-shave lotions, aerosol antiperspirants, creams and lotions.

LEXOL 3975:
CTFA: Isopropyl Palmitate (and) Isopropyl Myristate (and) Isopropyl Stearate
Liquid
Conc.: 100%
Combines the emolliency and solvency of its components. Essentially odorless and colorless; oxidative resistant, for a longer product shelf life. For cosmetics and topicals.

LEXOL EHP:
CTFA: Octyl Palmitate
Liquid
Conc.: 100%
Essentially colorless and odorless emollient oil. Particularly recommended for non-occlusive creams and lotions; also bath oils, anti-perspirants and other cosmetic and topical formulations where a less occlusive film is desired.

LEXOL SS:
CTFA: Stearyl Stearate
Flakes
Conc.: 100%
A solid emollient/opacifier that provides a less oily skin feel than mineral oil.

INOLEX CHEMICAL CO.: Emulsifiers:

LEXEMUL AR:

CTFA: Glyceryl Stearate (and) Stearamidoethyl Diethylamine
Flakes
Conc.: 100%
HLB: 4.1

Acid-stable, cationic, self-emulsifying GMS for use as an emulsifier, stabilizer, opacifier and emollient in cationic systems, especially deodorant and anti-perspirant formulations.

LEXEMUL AS:

CTFA: Glyceryl Stearate (and) Sodium Lauryl Sulfate
Flakes
Conc.: 100%
HLB: 4.9

Acid-stable, anionic, self-emulsifying GMS for use as above in anionic systems, particularly low pH. Applications in all types of cosmetics, toiletries and topical pharmaceuticals.

LEXEMUL T:

CTFA: Glyceryl Stearate SE
Flakes
Conc.: 100%
HLB: 5.5

Self-emulsifying GMS for use as emulsifier, opacifier, stabilizer, and emollient in neutral to slightly alkaline anionic systems.

LEXEMUL 55G:

CTFA: Glyceryl Stearate
Flakes
Conc.: 100%
HLB: 3.6

High purity GMS. Non-self-emulsifying. For use in cosmetics and topical pharmaceuticals. Improves lubricity, gloss in creams and lotions.

LEXEMUL 503:

CTFA: Glyceryl Stearate
Flakes
Conc.: 100%
HLB: 3.2

Neutral, non-self-emulsifying GMS for use as emulsifier, stabilizer, thickener, opacifier in nonionic, anionic or cationic emulsions or surfactants systems.

INOLEX CHEMICAL CO.: Emulsifiers(Continued):

LEXEMUL 515:
CTFA: Glyceryl Stearate
Flakes
Conc.: 100%
HLB: 3.8
Non-self-emulsifying GMS. Emulsifier, stabilizer, thickener, opacifier. Produces rich smooth emulsions of unexcelled texture in conjunction with other surface active agents.

LEXEMUL 530:
CTFA: Glyceryl Stearate SE
Flakes
Conc.: 100%
HLB: 5.0
Highly stable self-emulsifying GMS. Low odor, light color. For cosmetic creams, lotions, hair dressings, shaving creams. Anionic emulsifier.

LEXEMUL 561:
CTFA: Glyceryl Stearate (and) PEG-100 Stearate
Flakes
Conc.: 100%
HLB: 10.0
Nonionic modified self-emulsifying GMS. Versatile, compatible, easily compounded. Provides attractive appearance and luxurious feel to cosmetic creams and lotions.

LEXEMUL P:
CTFA: Propylene Glycol Stearate SE
Flakes
Conc.: 100%
HLB: 4.0
Self-emulsifying PGMS for low viscosity emulsions. Produces lotions and light creams of excellent stability and rich soft texture.

LEXEMUL EGMS:
CTFA: Glycol Stearate
Flakes
Conc.: 100%
HLB: 2.3
Particularly recommended for use as opacifying and pearlizing agent. Applications in cream shampoos, rinses and conditioners; as emulsifier in creams, lotions and topicals.

INOLEX CHEMICAL CO.: Emulsifiers(Continued):

LEXEMUL EGDS:
CTFA: Glycol Distearate
Flakes
Conc.: 100%
HLB: 1.5
Particularly recommended for use as opacifying and pearlizing agent for cosmetic surfactant systems including cream shampoos, rinses and conditioners.

LEXEMUL CS-20:
CTFA: Cetearyl Alcohol (and) Ceteareth 20
Flakes
Conc.: 100%
HLB: 5.4
Specialty non-ionic emulsifier containing no ester linkages. Completely stable at temperature and pH conditions where GMS, PGMS, etc. cannot be used. Forms O/W emulsions for creams, lotions, cream shampoos, topical preparations.

LEXEMUL GDL:
CTFA: Glyceryl Dilaurate
Flakes
Conc.: 100%
HLB: 2.4
Nonionic water dispersible lipid for creams and lotions formulated to provide a dry skin feel. Melts just below body temperature and provides a smooth emolliency. By also acting as a coupling agent for more lipophilic materials, it also promotes emulsion stability.

LEXQUAT AMG-O:
CTFA: Oleamidopropyl Dihydroxypropyl
Liquid
Conc.: 25 min.
HLB: 102

LEXQUAT AMG-M:
CTFA: Lauramidopropyl Dihydroxypropyl
Liquid
Conc.: 10 min.
HLB: 14

LEXQUAT AMG-WC:
CTFA: Cocamidopropyl Dihydroxypropyl
Liquid
Conc.: 30 min.
HLB: 12

LEXQUAT AMG-BEO:
CTFA: Behenamidopropyl Dihydroxypropyl
Solid
Conc.: 25 min.
HLB: 8

LEXQUAT AMG-IS:
CTFA: Isostearylamidopropyl Dihydroxypropyl
Solid
Conc.: 30 min.
HLB: 9
Unique cationic primary emulsifiers. Delivers emollient feel. Compatible with anionic surfactants.

INOLEX CHEMICAL CO.: Proteins:

Hydrolysates:

LEXEIN X250:
CTFA: Hydrolyzed Animal Protein
Water Solution
Conc.: 55%
Most highly regarded collagen protein available to the cosmetic industry. This original substantive protein is recommended for high performance hair and skin formulations: the protein of choice where proven complete efficacy is required.

LEXEIN X350:
CTFA: Hydrolyzed Animal Protein
Water Solution
Conc.: 55%
Substantive collagen protein. A cost-efficient way to obtain high substantivity to hair and skin formulations.

LEXEIN X300:
CTFA: Hydrolyzed Animal Protein
Dry Powder
Conc.: 94%
Convenient and cost efficient form of LEXEIN X350. Advantages include bacterial stability, improved odor, ease of handling and storage.

POLYPEPTIDE SF:
CTFA: Hydrolyzed Animal Protein
Water Solution
Conc.: 57%
Premium high solids beef collagen protein with low ash content. This provides for greater emulsion stability. For use in products sensitive to electrolytes.

POLYPEPTIDE LSN-A:
CTFA: Hydrolyzed Animal Protein
Dry Powder
Conc.: 94%
Spray-dried version of POLYPEPTIDE LSN. White, free-flowing powder, miscible with water in all proportions. Applications where water is not desired, such as aerosols, soaps, syndet bars, anhydrous shampoos. Higher levels of ash than LEXEIN X300.

INOLEX CHEMICAL CO.: Proteins(Continued):

Hydrolysates(Continued):

POLYPEPTIDE 37:
CTFA: Hydrolyzed Animal Protein
Water Solution
Conc.: 49%
Beef collagen protein for products where cost is of utmost importance. For products not sensitive to electrolytes such as shampoos and conditioners.

Specialties:

LEXEIN QX3000:
CTFA: Quaternium-76
Hydrolyzed Animal Protein
Water Solution
Conc.: 30%
A breakthrough in protein technology, it is a fully reacted protein derivative which incorporates both a quaternized nitrogen and a lipophilic moiety. A strongly cationic molecule that provides optimum substantivity, outstanding body, greater ease in wet and dry combing. On the skin, it provides a renewable skin feel due to its unique structure.

LEXEIN CP125:
CTFA: Oleamidopropyl Dimethylamine Hydrolyzed Animal Protein
Water Solution
Conc.: 50%
A cationic salt of a collagen hydrolysate. In shampoo and cream rinse products, improves combing and bodying effects without buildup.

Condensates:

LEXEIN A200:
CTFA: Myristoyl Hydrolyzed Animal Protein
Freeze-dried Powder
Conc.: 100%
A film forming collagen protein derivative. The clear, flexible film and formulating versatility suggest uses in hair sprays, makeups, protective skin lotions and creams.

LEXEIN A210:
CTFA: Myristoyl Hydrolyzed Animal Protein
Form: Ethanol Solution
Conc.: 20%
Alcohol solution of LEXEIN A200. Forms water-insoluble, clear, flexible, coherent film from anhydrous alcohol and hydroalcohol products.

INOLEX CHEMICAL CO.: Proteins(Continued):

Condensates(Continued):

LEXEIN A220:
CTFA: TEA-Myristoyl Hydrolyzed Animal Protein
Water Solution
Conc.: 20%
Water solution of LEXEIN A200. Forms water-soluble, hygroscopic clear film and can be used to improve film forming in face makeups, to impart a protective film in skin creams and lotions.

LEXEIN A240:
CTFA: Isopropyl Myristate (and) Sorbitan Oleate (and) Myristoyl Hydrolyzed Animal Protein
Oil Solution
Conc.: 20%
Oil soluble form of LEXEIN A210. Recommended wherever a cosmetic protein is to be used in an oil-based formulation.

LEXEIN A520:
CTFA: TEA-Abietoyl Hydrolyzed Animal Protein
Water Solution
Conc.: 30%
Sebum control additive. Causes a significant delay in the re-fatting of the scalp when used in shampoo formulations.

MAYPON 4C:
CTFA: Potassium Coco-Hydrolyzed Animal Protein
Water Solution
Conc.: 35%
Low irritation, high foaming protein surfactant with anti-irritation effects. Ideally suited for baby shampoos, shower gels, bubble baths, facial cleansers, anywhere mildness is most important.

MAYPON 4CT:
CTFA: TEA-Coco-Hydrolyzed Animal Protein
Water Solution
Conc.: 39%
Properties similar to MAYPON 4C. Viscosity building and foam modifying effects will vary because of low inorganic salt content.

INOLEX CHEMICAL CO.: Proteins(Continued):

Condensates(Continued):

SUPERPRO 5A:
CTFA: TEA-Coco-Hydrolyzed Animal Protein and Sorbitol
Water Solution
Conc.: 69%
Specialty blend of a mild, high foaming cleansing agent and a humectant. The creamy, dense, stable lather, mildness and moisturizing properties make it an ideal surfactant for sensitive skin.

MAYPON UD:
CTFA: Potassium Undecylenoyl Hydrolyzed Animal Protein
Water Solution
Conc.: 35%
The beneficial anti-mycotic and anti-bacterial effects of the undecylenyl grouping is united with the substantive properties and skin conditioning action of the polypeptide. It possesses the foaming and cleansing properties typical of protein condensates. Excellent for dermatological preparations.

INOLEX CHEMICAL CO.: Specialties:

LEXEIN QX3000:
CTFA: Quaternium-76
Water Solution
Conc.: 30%
A breakthrough in protein technology, it is a fully reacted protein derivative which incorporates both a quaternized nitrogen and a lipophilic moiety. A strongly cationic molecule that provides optimum substantivity, outstanding body, greater ease in wet and dry combing. On the skin, it provides a renewable skin feel due to its unique structure.

LEXQUAT AMG-O:
CTFA: Oleamidopropyl Dihydroxypropyl Dimonium Chloride
Liquid
Conc.: 25%

LEXQUAT AMG-M:
CTFA: Lauramidopropyl Dihydroxypropyl Dimonium Chloride
Liquid
Conc.: 30%

LEXQUAT AMG-WC:
CTFA: Cocamidopropyl Dihydroxypropyl Dimonium Chloride
Liquid
Conc.: 30%

LEXQUAT AMG-IS:
CTFA: Isostearylamidopropyl Dihydroxypropyl Dimonium Chloride
Liquid
Conc.: 30%

LEXQUAT AMG-BEO:
CTFA: Behenamidopropyl Dihydroxypropyl Dimonium Chloride
Solid
Conc.: 30%
Unique multifunctional cationic conditioners, emulsifiers, emollients, completely compatible with all classes of surfactants. Will form clear gels. Broad pH stability. Very high salt tolerance. Will not adversely effect foaming of other surfactants. Extremely mild. Very substantive. Excellent conditioning.

LEXQUAT 2240:
CTFA: Polymethacrylamidopropyl Trimonium Chloride
Liquid
Conc.: 25%
Film forming cationic polymer. Polymeric conditioner.

INOLEX CHEMICAL CO.: Surfactants:

LEXAINE C:
CTFA: Cocamidopropyl Betaine
Liquid
Conc.: 35%
High performance mild amphoteric surfactant. Excellent foam boosting, viscosity building properties. For shampoos and other hair, bath and skin care products, liquid soaps.

LEXAINE CS:
CTFA: Cocamidopropyl Betaine
Liquid
Conc.: 35%
Mild amphoteric surfactant with excellent foam boosting and viscosity building properties. Applications in shampoos and other hair products. Also bath and skin care products, liquid soaps.

LEXAINE CG-30:
CTFA: Cocamidopropyl Betaine
Liquid
Conc.: 35%
Mild amphoteric surfactant, foam booster and thickener for shampoos, bubble baths, foaming conditioners. Stable in acid and alkaline systems.

LEXAINE LM:
CTFA: Lauramidopropyl Betaine
Liquid
Conc.: 35%
Mild amphoteric surfactant with excellent foam boosting properties due to the lauric fatty moiety incorporated into its structure. Excellent for mild, high foaming bath products, shampoos, liquid soaps, and dishwash liquids.

LEXAINE CSB-50:
CTFA: Cocomidopropyl Hydroxysultaine
Liquid
Conc.: 48%
Exhibits outstanding foaming, wetting and surface activity. Applications in low irritation and baby shampoos, bath baths, bubble baths, skin cleansers, and other hair and skin products.

INOLEX CHEMICAL CO.: Surfactants(Continued):

LEXAMINE C-13:
CTFA: Cocamidopropyl Dimethylamine
Soft Solid
Conc.: 100%

LEXAMINE-13:
CTFA: Lauramidopropyl Dimethylamine
Solid
Conc.: 100%

LEXAMINE O-13:
CTFA: Oleamidopropyl Dimethylamine
Liquid
Conc.: 100%

LEXAMINE P-13:
CTFA: Palmitamidopropyl Dimethylamine
Solid
Conc.: 100%

LEXAMINE S-13:
CTFA: Stearamidopropyl Dimethylamine
Flakes
Conc.: 100%

LEXAMINE 22:
CTFA: Stearamidoethyl Diethylamine
Solid
Conc.: 100%

Cationic Emulsifiers. The LEXAMINE series of amidotertiary amines offers the formulator great flexibility in preparing cationic emulsions. The LEXAMINE bases are readily soluble in acidic media to form powerful cationic o/w emulsifiers which produce stable emulsions in combination with suitable nonionic agents and emollients. The large selection of lipophilic ends provide a wide range of attainable physical properties.

The acidic cationic salts are highly substantive to proteins and cellulosic substrates to which they impart antistatic properties and enhanced protein substantivity.

Some of the advantages of LEXAMINE based hair conditioning systems:

Much lower irritation than normal quaternary based systems;

Enhanced hair conditioning effects without greasiness due to lower hair deposition;

Enhanced protein substantivity;

Reduced hygroscopicity under humid conditions.

Applications in cream rinse conditioners, protein conditioners, conditioning shampoos and as emulsifiers in acidic based cationic creams and lotions.

LEXAMINE S-13 Lactate:
Stearamidopropyl Dimethylamine Lactate
Liquid
Conc.: 23%

Easy, convenient to handle, its principal use is as a conditioner in both shampoos and conditioners. It does not leave hair limp or overconditioned, yet produces noticeable improvement in wet combing and static control.

JOJOBA GROWERS & PROCESSORS INC.: Jojoba Products:

HYDROBA-70:
Product Description: Crystalline Flakes
CTFA Nomenclature: Hydrogenated Jojoba Oil
Appearance: Free flowing hard wax flakes
Color: White to off-white
Odor: Slight fatty
Saponification No.: 90-95
Iodine Value: 3.0
Melting Point: 69C. (Fisher Johns)
Bacteriological: TPC less than 50/gm.
Bulk Density: 0.49 g/cc

JOJOBEADS 40/60 Mesh:
Product Description: Jojoba Wax Capsules
CTFA Nomenclature: Jojoba Wax
Appearance: Free flowing beads
Color: White to off-white
Odor: Slight fatty
Saponification No.: 94
Iodine Value: 2.5
Particle Size: 80% @ 40/60 Mesh
Melting Point: 69C (Fisher Johns)
Bacteriological: TPC = 40/gm.
Bulk Density: 0.49 g/cc

TRANSJOJOBA-30:
Product Description: Transesterified Jojoba Oil and Wax
CTFA Nomenclature: Jojoba Esters
Appearance: Soft white to off-white cream
Acid Value: 2.0
Saponification Number: 91
Bacteriological: TPC = 50/gm.
Iodine Value: 48
Color: 3.0 (Gardner)
Odor: Typical fatty
Melting Point: 29C. (Fisher Johns)

JOJOBA GROWERS & PROCESSORS INC.: Jojoba Products(Continued):

Pasteurized Jojoba Oils:

Pure Grade:
- Gardner color: Maximum - 9
- Odor: Typical fatty
- Acid value: Less than 1.00
- Saponification number: 90-95
- Iodine value: 80-85
- Total plate count: Less than 50/gram
- Melting point: 7 to 9C
- Flash point: 295C
- Fire point: 338C

SONORA Grade:
- Gardner color: Maximum - 6
- Odor: Slightly fatty
- Acid value: Less than 1.00
- Saponification number: 90-95
- Iodine value: 80-85
- Total plate count: Less than 50/gram
- Melting point: 7 to 9C
- Flash point: 295C
- Fire point: 338C

Refined Grade:
- Gardner color: Maximum - 2
- Odor: Odorless
- Acid value: Less than 1.00
- Saponification number: 90-95
- Iodine value: 80-85
- Total plate count: Less than 50/gram
- Melting point: 7 to 9C
- Flash point: 295C
- Fire point: 338C

Jojoba oil consists of straight chain esters, formed predominantly from combinations of C-20 and C-22 fatty acids and fatty alcohols. The average molecular weight of jojoba oil is 606 grams per mole.

JOJOBA GROWERS & PROCESSORS INC.: Jojoba Products(Continued):

Jojoba Derivative Products:

JOJOBUTTER 31:
Color: Light amber
Odor: Slightly fatty
Appearance: Semi-solid wax
Melting point: 28-31C
Iodine value: 80-85

JOJOBUTTER 51:
Color: Light amber
Odor: Slightly fatty
Appearance: Semi-solid wax
Melting point: 48-51C
Iodine value: 65-75

Hydrogenated Jojoba Wax:
Color: White
Odor: Odorless
Appearance: Crystalline
Melting point: 70-72C
Iodine value: less than 2

JOJOBUTTER 31 (Patent No. 4,329,298) is trans-isomerized jojoba oil.

JOJOBUTTER 51 (Patent No. 4,360,387) is an isomorphic mixture of trans-isomerized jojoba oil and hydrogenated jojoba wax.

LIPO CHEMICALS INC.: LIPO Glyceryl, Ethylene Glycol, and Propylene Glycol Esters:

General purpose emulsifiers and emulsion stabilizers. Widely used as emollients and viscosity builders in creams and lotions. Neutral and self-emulsifying grades for wider versatility.

Product:

LIPO GMS-450:
CTFA: Glyceryl Stearate
Appearance at 25C: White beads or flakes
Sap. Value: 165-182
Acid Value: 5 max
HLB±1: 3.6

LIPO GMS-470:
CTFA: Glyceryl Stearate SE
Appearance at 25C: White beads or flakes
Sap. Value: 138-152
Acid Value: 5 max
HLB±1: 5.8

LIPOMULSE 165:
CTFA: Glyceryl Stearate (and) PEG-100 Stearate
Appearance at 25C: White beads or flakes
Sap. Value: 90-100
Acid Value: 2 max
HLB±1: 11.0

LIPO EGMS:
CTFA: Glycol Stearate
Appearance at 25C: White/off-white beads or flakes
Sap. Value: 175-190
Acid Value: 6 max
HLB±1: 2.0

LIPO EGDS:
CTFA: Glycol Distearate
Appearance at 25C: White/off-white beads or flakes
Sap. Value: 190-205
Acid Value: 7 max
HLB±1: 1.0

LIPO DGLS:
CTFA: PEG-2 Laurate SE
Appearance at 25C: Yellow liquid
Sap. Value: 160-170
Acid Value: 4 max
HLB±1: 8.3

LIPO PGMS:
CTFA: Propylene Glycol Stearate
Appearance at 25C: White flakes
Sap. Value: 180-192
Acid Value: 6 max
HLB±1: 3.0

LIPO CHEMICALS INC.: LIPOCOL Polyoxyethylene Ethers:

A broad range of acid and alkaline stable nonionic surfactants particularly useful in antiperspirants, depilatories, creams, lotions and pigment dispersions. Also used as emulsifiers, defoamers, wetting agents, solubilizers and conditioning agents in shampoos, detergents, bleaches and dyes.

Product:

LIPOCOL L-4:
CTFA: Laureth-4
Appearance at 25C: Colorless liquid
Acid Value: 2 max
Hydroxyl Value: 145-160
HLB±1: 9.7

LIPOCOL L-12:
CTFA: Laureth-12
Appearance at 25C: White solid wax
Acid Value: 1 max
Hydroxyl Value: 72-87
HLB±1: 14.5

LIPOCOL L-23:
CTFA: Laureth-23
Appearance at 25C: White solid wax
Acid Value: 2 max
Hydroxyl Value: 42-52
HLB±1: 16.9

LIPOCOL C-2:
CTFA: Ceteth-2
Appearance at 25C: White solid wax
Acid Value: 1 max
Hydroxyl Value: 160-180
HLB±1: 5.3

LIPOCOL C-10:
CTFA: Ceteth-10
Appearance at 25C: White solid wax
Acid Value: 1 max
Hydroxyl Value: 75-90
HLB±1: 12.9

LIPOCOL C-20:
CTFA: Ceteth-20
Appearance at 25C: White solid wax
Acid Value: 2 max
Hydroxyl Value: 50-58
HLB±1: 15.7

LIPO CHEMICALS INC.: LIPOCOL Polyoxyethylene Ethers(Continued):

Product:

LIPOCOL S-2:
CTFA: Steareth-2
Appearance at 25C: White solid wax
Acid Value: 1 max
Hydroxyl Value: 155-165
HLB±1: 4.9

LIPOCOL S-10:
CTFA: Steareth-10
Appearance at 25C: White solid wax
Acid Value: 1 max
Hydroxyl Value: 75-90
HLB±1: 12.4

LIPOCOL S-20:
CTFA: Steareth-20
Appearance at 25C: White solid wax
Acid Value: 1 max
Hydroxyl Value: 45-60
HLB±1: 15.3

LIPOCOL O-2:
CTFA: Oleth-2
Appearance at 25C: Yellow liquid
Acid Value: 1 max
Hydroxyl Value: 160-180
HLB±1: 4.9

LIPOCOL O-10:
CTFA: Oleth-10
Appearance at 25C: Yellow liquid
Acid Value: 2 max
Hydroxyl Value: 74-84
HLB±1: 12.4

LIPOCOL O-20:
CTFA: Oleth-20
Appearance at 25C: White solid wax
Acid Value: 2 max
Hydroxyl Value: 45-65
HLB±1: 15.3

LIPOCOL SC-4:
CTFA: Ceteareth-4
Appearance at 25C: White solid wax
Acid Value: 1 max
Hydroxyl Value: 120-140
HLB: 8.0

LIPO CHEMICALS INC.: LIPOCOL Poloxyethylene Ethers(Continued):

Product:

LIPOCOL SC-15:
CTFA: Ceteareth-15
Appearance at 25C: White solid wax
Acid Value: 2 max
Hydroxyl Value: 50-65
HLB±1: 14.3

LIPOCOL SC-20:
CTFA: Ceteareth-20
Appearance at 25C: White solid wax
Acid Value: 1 max
Hydroxyl Value: 45-60
HLB±1: 15.5

LIPOCOL TD-12:
CTFA: Trideceth-12
Appearance at 25C: White paste
Acid Value: 1 max
Hydroxyl Value: 70-85
HLB±1: 14.6

LIPO CHEMICALS INC.: LIPONATE Emollient Esters:

A wide selection of emollients for adjusting rub-in and afterfeel of creams, lotions or bath preparations. Also used for thickening and viscosity control.

Product:

LIPONATE CL:
CTFA: Cetyl Lactate
Appearance at 25C: White soft solid/liquid
Sap. Value: 174-195
Acid Value: 3 max

LIPONATE GC:
CTFA: Caprylic/Capric Triglyceride
Appearance at 25C: Colorless liquid
Sap. Value: 325-355
Acid Value: 0.1 max

LIPONATE PC:
CTFA: Propylene Glycol Dicaprylate/Dicaprate
Appearance at 25C: Colorless liquid
Sap. Value: 315-335
Acid Value: 0.1 max

LIPONATE DPC-6:
CTFA: Dipentaerythrityl Hexacaprylate/Hexacaprate
Appearance at 25C: Colorless visc. liquid
Sap. Value: 320-340
Acid Value: 0.5 max

LIPONATE IPM:
CTFA: Isopropyl Myristate
Appearance at 25C: Colorless liquid
Sap. Value: 202-211
Acid Value: 2 max

LIPONATE IPP:
CTFA: Isopropyl Palmitate
Appearance at 25C: Colorless liquid
Sap. Value: 183-190
Acid Value: 2 max

LIPONATE MM:
CTFA: Myristyl Myristate
Appearance at 25C: White solid wax
Sap. Value: 120-135
Acid Value: 5 max

LIPO CHEMICALS INC.: LIPONATE Emollient Esters(Continued):

Product:

LIPONATE NPCG-2:
CTFA: Neopentylglycol Dicaprylate/Dicaprate
Appearance at 25C: Colorless liquid
Sap. Value: 292-312
Acid Value: 0.5 max

LIPONATE PB-4:
CTFA: Pentaerythrityl Tetrabehenate
Appearance at 25C: Off-white flakes
Sap. Value: 150-165
Acid Value: 10 max

LIPONATE PO-4:
CTFA: Pentaerythrityl Tetraoleate
Appearance at 25C: Yellow liquid
Sap. Value: 185-195
Acid Value: 10 max

LIPONATE PS-4:
CTFA: Pentaerythrityl Tetrastearate
Appearance at 25C: Off-white flakes
Sap. Value: 183-198
Acid Value: 10 max

LIPONATE SPS:
CTFA: Cetyl Esters
Appearance at 25C: White/cream flakes
Sap. Value: 109-120
Acid Value: 5 max

LIPONATE SS:
CTFA: Stearyl Stearate
Appearance at 25C: Off-white flakes
Sap. Value: 103-117
Acid Value: 5 max

LIPONATE TDS:
CTFA: Tridecyl Stearate
Appearance at 25C: Colorless liquid
Sap. Value: 110-130
Acid Value: 1.0 max

LIPONATE TDTM:
CTFA: Tridecyl Trimellitate
Appearance at 25C: Pale yellow viscous liquid
Sap. Value: 238-258
Acid Value: 0.5 max

LIPONATE 2-DH:
CTFA: (PEG-4 Diheptanoate)
Appearance at 25C: Colorless liquid
Sap. Value: 249-269
Acid Value: 0.5 max

LIPONATE PE-810:
CTFA: (Pentaerythrityl Tetracaprylate/Tetracaprate)
Appearance at 25C: Colorless liquid
Sap. Value: 315-335
Acid Value: 1.0 max

LIPO CHEMICALS INC.: LIPONIC Humectants:

Humectants control the moisture exchange between a product and the atmosphere thus helping to retard drying of the product in the package and in use. Humectants are incorporated into creams, lotions, antiperspirants, beauty masks, depilatories and wavesets. LIPO offers a wide selection of humectants including the novel ethoxylated glycerines--LIPONIC EG-1 and LIPONIC EG-7.

LIPAMIDE MEAA: (Acetamide MEA)

LIPONIC EG-1: (Glycereth-26)

LIPONIC EG-7: (Glycereth-7)

LIPONIC SO-20: (Sorbeth-20)

LIPONIC 70-NC: (Sorbitol)

LIPONIC 76-NC: (Sorbitol)

LIPO POLYOL NC: (Hydrogenated Starch Hydrolysate)

Sorbitol Solution, 70% USP:

LIPO CHEMICALS INC.: LIPOLAN: Lanolin Derivatives:

A select variety of lanolin derivatives that contribute perceivable benefits to finished cosmetics, toiletries or pharmaceuticals. They are used as emollients, emulsifiers, solubilizers, stabilizers, conditioners and moisturizers in skin care products, makeup, lipstick, shampoos and rinses, soap and bath specialties, shaving preparations, sun products, ointments, acne preparations and veterinary products

LIPOLAN: (Hydrogenated Lanolin)

LIPOLAN DISTILLED: (Distilled Hydrogenated Lanolin)

LIPOLAN R: (Lanolin Oil)

LIPOLAN 31: (PEG-24 Hydrogenated Lanolin)

LIPO CHEMICALS INC.: LIPOPEG Polyoxyethylene Glycol Esters:

A basic line of mild fatty esters for use in bath oils, creams and lotions for spreading, emulsification, dispersion and lubrication.

Product:

LIPOPEG 4-L:
CTFA: PEG-8 Laurate
Appearance at 25C: Yellow liquid
Sap. Value: 90-100
Acid Value: 5 max
HLB±1: 13.0

LIPOPEG 2-DL:
CTFA: PEG-4 Dilaurate
Appearance at 25C: Yellow liquid
Sap. Value: 170-185
Acid Value: 10 max
HLB±1: 6.0

LIPOPEG 4-DL:
CTFA: PEG-8 Dilaurate
Appearance at 25C: Yellow liquid
Sap. Value: 125-142
Acid Value: 10 max
HLB±1: 10.0

LIPOPEG 4-DO:
CTFA: PEG-8 Dioleate
Appearance at 25C: Amber liquid
Sap. Value: 113-128
Acid Value: 10 max
HLB±1: 7.2

LIPOPEG 4-S:
CTFA: PEG-8 Stearate
Appearance at 25C: Cream paste
Sap. Value: 80-90
Acid Value: 5 max
HLB±1: 11.2

LIPO CHEMICALS INC.: LIPOPEG Polyoxyethylene Glycol Esters (Continued):

Product:

LIPOPEG 10-S:
CTFA: PEG-20 Stearate
Appearance at 25C: White solid wax
Sap. Value: 39-49
Acid Value: 5 max
HLB±1: 15.2

LIPOPEG 39-S:
CTFA: PEG-40 Stearate
Appearance at 25C: White solid wax or flakes
Sap. Value: 23-35
Acid Value: 2 max
HLB±1: 16.9

LIPOPEG 100-S:
CTFA: PEG-100 Stearate
Appearance at 25C: Tan flakes or beads
Sap. Value: 9-20
Acid Value: 1 max
HLB±1: 18.8

LIPOPEG 4-DS:
CTFA: PEG-8 Distearate
Appearance at 25C: Cream soft wax
Sap. Value: 113-128
Acid Value: 10 max
HLB±1: 8.0

LIPOPEG 6000-DS:
CTFA: PEG-150 Distearate
Appearance at 25C: Off-white flakes
Sap. Value: 12-20
Acid Value: 10 max
HLB±1: 18.4

LIPO CHEMICALS INC.: LIPOSORB Polyoxyethylene Sorbitan Esters:

All-purpose nonionic hydrophilic surfactants used for solubizing oils and in conjunction with LIPOSORB sorbitan esters for their emulsification, lubrication and antistatic properties.

Product:

LIPOSORB L-10:
- CTFA: PEG-10 Sorbitan Laurate
- Appearance at 25C: Yellow liquid
- Sap. Value: 66-76
- Hydroxyl Value: 150-170
- HLB±1: 14.9

LIPOSORB L-20:
- CTFA: Polysorbate 20
- Appearance at 25C: Yellow liquid
- Sap. Value: 40-50
- Hydroxyl Value: 96-108
- HLB±1: 16.7

LIPOSORB P-20:
- CTFA: Polysorbate 40
- Appearance at 25C: Yellow liquid
- Sap. Value: 40-53
- Hydroxyl Value: 90-107
- HLB±1: 15.6

LIPOSORB S-20:
- CTFA: Polysorbate 60
- Appearance at 25C: Yellow paste
- Sap. Value: 45-55
- Hydroxyl Value: 81-96
- HLB±1: 14.9

LIPOSORB TS-20:
- CTFA: Polysorbate 65
- Appearance at 25C: Tan solid wax
- Sap. Value: 88-98
- Hydroxyl Value: 44-60
- HLB±1: 10.5

LIPOSORB O-20:
- CTFA: Polysorbate 80
- Appearance at 25C: Yellow liquid
- Sap. Value: 45-55
- Hydroxyl Value: 65-80
- HLB±1: 15.0

LIPOSORB TO-20:
- CTFA: Polysorbate 85
- Appearance at 25C: Yellow liquid
- Sap. Value: 82-95
- Hydroxyl Value: 39-52
- HLB±1: 11.0

LIPO CHEMICALS INC.: LIPOSORB Sorbitan Esters:

All-purpose nonionic lipophilic surfactants used primarily in conjunction with LIPOSORB polyoxyethylene sorbitan esters for emulsification, thickening, lubricating and antistatic effects.

Product:

LIPOSORB L:
- CTFA: Sodium Laurate
- Appearance at 25C: Amber liquid
- Sap. Value: 158-170
- Hydroxyl Value: 330-360
- HLB±1: 8.6

LIPOSORB P:
- CTFA: Sorbitan Palmitate
- Appearance at 25C: Tan beads or flakes
- Sap. Value: 139-151
- Hydroxyl Value: 272-306
- HLB±1: 6.7

LIPOSORB S:
- CTFA: Sorbitan Stearate
- Appearance at 25C: Cream beads or flakes
- Sap. Value: 147-157
- Hydroxyl Value: 235-260
- HLB±1: 4.7

LIPOSORB TS:
- CTFA: Sorbitan Tristearate
- Appearance at 25C: Cream beads or flakes
- Sap. Value: 175-190
- Hydroxyl Value: 65-80
- HLB±1: 2.1

LIPOSORB O:
- CTFA: Sorbitan Oleate
- Appearance at 25C: Yellow/amber liquid
- Sap. Value: 145-160
- Hydroxyl Value: 193-210
- HLB±1: 4.3

LIPOSORB SDQ:
- CTFA: Sorbitan Sesquioleate
- Appearance at 25C: Amber liquid
- Sap. Value: 145-160
- Hydroxyl Value: 185-215
- HLB±1: 3.7

LIPOSORB TO:
- CTFA: Sorbitan Trioleate
- Appearance at 25C: Amber liquid
- Sap. Value: 171-185
- Hydroxyl Value: 58-69
- HLB±1: 1.8

LIPO CHEMICALS INC.: LIPOVOL MOS Series#:

Blend of specialty esters which offer the formulating versatility of esters and exhibit the tactile properties of mineral oil.

Product:

LIPOVOL MOS-70:
CTFA: Tridecyl Stearate (and) Neopentylglycol Dicaprylate/Dicaprate (and) Tridecyl Trimellitate
Appearance at 25C: Colorless liquid
Sap. Value: 206-226
Acid Value: 0.5 max

LIPOVOL MOS-130:
CTFA: Tridecyl Stearate (and) Tridecyl Trimellitate (and) Dipentaerythrityl Hexacaprylate/Hexacaprate
Appearance at 25C: Colorless liquid
Sap. Value: 187-207
Acid Value: 0.5 max

LIPOVOL MOS-350:
CTFA: Dipentaerythrityl Hexacaprylate/Hexacaprate (and) Tridecyl Trimellitate (and) Tridecyl Stearate (and) Neopentylglycol Dicaprylate/Dicaprate
Appearance at 25C: Colorless to straw liquid
Sap. Value: 268-288
Acid Value: 0.5 max

Patent Number 4,659,573

LIPO CHEMICALS INC.: LIPOVOL Natural Oils:

Highest quality refined natural emollient oils for use as lubricants and conditioners in luxury skin products, hair-care products, makeups, fine soaps, bath oils and anhydrous systems.

Product:

LIPOVOL A:
CTFA: Avocado Oil
Appearance at 25C: Yellow to green oil
Sap. Value: 177-198
Acid Value: 3 max
Iodine Value: 65-95

LIPOVOL ALM:
CTFA: Sweet Almond Oil
Appearance at 25C: Yellow oil
Sap. Value: 185-200
Acid Value: 2 max
Iodine Value: 95-115

LIPOVOL G:
CTFA: Grape Seed Oil
Appearance at 25C: Yellow/amber oil
Sap. Value: 183-205
Acid Value: 5 max
Iodine Value: 132-152

LIPVOL HS:
CTFA: Hydrogenated Soybean Oil
Appearance at 25C: Yellow oil
Sap. Value: 186-197
Acid Value: 0.5 max
Iodine Value: 101-114

LIPOVOL J:
CTFA: Jojoba Oil
Appearance at 25C: Yellow oil
Sap. Value: 85-110
Acid Value: 5 max
Iodine Value: 75-95

LIPOVOL P:
CTFA: Apricot Kernel Oil
Appearance at 25C: Yellow oil
Sap. Value: 185-195
Acid Value: 1 max
Iodine Value: 90-115

LIPOVOL PAL:
CTFA: Palm Oil
Appearance at 25C: Off-white paste
Sap. Value: 195-205
Acid Value: 1 max
Iodine Value: 44-59

LIPO CHEMICALS INC.: LIPOVOL Natural Oils(Continued):

Product:

LIPOVOL SAF:
CTFA: Safflower Oil
Appearance at 25C: Yellow oil
Sap. Value: 182-202
Acid Value: 2 max
Iodine Value: 135-155

LIPOVOL SES:
CTFA: Sesame Oil
Appearance at 25C: Yellow oil
Sap. Value: 188-195
Acid Value: 0.2 max
Iodine Value: 103-116

LIPOVOL SO:
CTFA: Hybrid Safflower Oil
Appearance at 25C: Yellow oil
Sap. Value: 184-196
Acid Value: 1 max
Iodine Value: 90-105

LIPOVOL SOY:
CTFA: Soybean Oil
Appearance at 25C: Yellow oil
Sap. Value: 180-200
Acid Value: 1 max
Iodine Value: 120-145

LIPO SS:
CTFA: Hydrogenated Vegetable Oil
Appearance at 25C: White to tan wax
Sap. Value: 230-250
Acid Value: 0.1 max
Iodine Value: 5 max

LIPOVOL SUN:
CTFA: Sunflower Seed Oil
Appearance at 25C: Yellow oil
Sap. Value: 185-195
Acid Value: 2 max
Iodine Value: 120-140

LIPOVOL WGO:
CTFA: Wheat Germ Oil
Appearance at 25C: Yellow/brown oil
Sap. Value: 175-195
Acid Value: 5 max
Iodine Value: 120-140

LIPO CHEMICALS INC.: LIPOWAX Emulsifying Waxes:

Emulsifying waxes are formulated waxes that can serve as complete emulsion systems in themselves. Useful in oil-in-water emulsions such as creams, lotions and ointments, the completely nonionic waxes are stable in both acid and alkaline formulations and are compatible with cationic ingredients.

LIPOWAX D: (Cetearyl Alcohol (and) Ceteareth-20)

LIPOWAX G: (Stearyl Alcohol (and) Ceteareth-20)

LIPOWAX NI: (Cetearyl Alcohol (and) Ceteth-20)

LIPOWAX P: (Emulsifying Wax, NF)

LIPOWAX PR: (Cetearyl Alcohol (and) Polysorbate 60 (and) PEG-150 Stearate (and) Steareth-20)

LIPOWAX P-SPEC: (Cetearyl Alcohol (and) Polysorbate 60)

LIPOWAX P-31: (Emulsifying Wax)

LIPO CHEMICALS INC.: LIPO Natural Abrasives:

Natural meals and flours for use as abrasives in facial scrubs, abrasive body scrubs and foot products.

LIPO AMS: (Almond Meal)

LIPO APS 40/60: (Apricot Seed Powder)

LIPO LUFA 30/100: (Luffa)

LIPO PP 40/60: (Peach Pit Powder)

LIPO PP 60/100: (Peach Pit Powder)

LIPO WSF 35/60: (Walnut Shell Flour)

LIPO WSF 60/100: (Walnut Shell Flour)

LIPO CHEMICALS INC.: LIPO Miscellaneous Products:

LIPO offers a wide variety of specialty raw ingredients for cosmetics, pharmaceuticals and toiletries, including fatty alcohols, amides, polyols, as well as special order products to meet the customer's own requirements. Some examples of these are:

LIPAMINE SPA: (Stearamidopropyl Dimethylamine)

LIPOBEE 102: (Synthetic Beeswax)

LIPOCOL C: (Cetyl Alcohol)

LIPOCOL L: (Lauryl Alcohol)

LIPOCOL O: (Oleyl Alcohol)

LIPOQUAT R: (Ricinoleamidopropyl Ethyldimonium Ethosulfate)

ORGASOL 2002D Ex Nat. Cos.: (Nylon-12)

ORGASOL 2002 UD Nat. Cos.: (Nylon-12)

UNICIDE U-13: (Imidazolidinyl Urea)

UNIPERTAN P-24: (Hydrolyzed Animal Collagen (and) Tyrosine (and) Riboflavin)

UNIPERTAN P-242: (Hydrolyzed Animal Collagen (and) Tyrosine (and) Adenosine Triphosphate)

UNIPHEN P-23: (Phenoxyethanol (and) Methylparaben (and) Ethylparaben (and) Propylparaben (and) Butylparaben)

D-PANTHENYLTRIACETATE: (Panthenyltriacetate)

SAFESTER A-75: (Ethyl Linoleate)

UNIPABOL U-17: (PEG-25 PABA)

UNITRIENOL T-27: (Farnesyl Acetete (and) Farnesol (and) Panthenyltriacetate)

UVATONE 2-6: (Octyl Dimethyl PABA)

LONZA INC.: AMPHOTERGE K/AMPHOTERGE K-2:

Typical Properties:

AMPHOTERGE K:
Cocoamphopropionate#
NaCl (%): None
pH: 9-11
Water Solubility: Soluble in all proportions
Appearance: Amber fluid
Ionic character: Amphoteric
Stability: Stable to acids, alkali and electrolytes
Specific Gravity: 1.06 (8.8 lb./gal.)

AMPHOTERGE K-2:
Cocoamphodipropionate#
NaCl (%): None
pH: 9-11
Water Solubility: Soluble in all proportions
Appearance: Amber fluid
Ionic character: Amphoteric
Stability: Stable to acids, alkali and electrolytes
Specific Gravity: 1.07 (8.9 lb./gal.)

CTFA adopted name

Cosmetics and Toiletries: AMPHOTERGE K and AMPHOTERGE K-2 are recommended where salt-free products are required.

DR. MADIS LABORATORIES, INC.: Cosmetic Ingredients:

Agar	Gelidium or gracilaria species
Aloe Vera Gel	Purified aloe vera extract
Annatto	Bixa orellana
Benzoin	Styrax species
Bromelain	Ananas comosus
Buchu	Barosma species
Canada Balsam	Abies balsamea
Chamomile	Matricaria chamomilla
Copaiba Balsam	Copifera species
Deer's Tongue	Trilisa odoratissima
Elder Flowers	Sambucus canadensis
Ficin	Ficus species
Fir Oregon Balsam	Pseudotsuga taxifolia
Galbanum	Ferula galbaniflua
Grape skin	Vitis vinifera
Guar	Cyamopsis tetragonolobus
Labdanum	Cistus ladaniferus
Locust Bean	Ceratonia siliqua
Myrrh	Commiphora species
Opopanax	Opopanox chironium
Papain	Carica papaya
Peruvian Balsam	Myroxylon pereirae
Sage	Salvia officinalis
Storax	Liquidambar species
Tolu Balsam	Myroxylon balsamum
Tragacanth	Astragalus species
Tumeric	Curcuma longa
VERAGEL	Purified aloe vera extract
Witch Hazel	Hamamelis virginiana

DR. MADIS LABORATORIES, INC.: VERAGEL Aloe Vera Gel:

Powder:
Strength: 1:200
Two Hundred Fold
Description:
Powder:

VERAGEL 200 is a concentrated purified Aloe Vera Gel free flowing powder. One part of VERAGEL 200 (two hundred fold) is produced from 200 parts of fresh Aloe inner gel. VERAGEL 200 dissolves slowly in water forming a viscous colloidal solution which is soluble in up to 15% alcohol, in propylene glycol and glycerin; insoluble in chloroform, acetone, ether and other organic solvents. When stored at or below room temperature in tight containers protected from moisture and light, VERAGEL 200 is stable up to 5 years. Because of its stability to microbial degradation, no preservative is added. However, once VERAGEL 200 is solubilized in water, it is quite susceptible to microbial attack and a preservative must be added to the liquid.

Purity and Quality:
Organoleptic Tests:
- Consistency: Powder
- Color: Off-White
- Odor: Odorless
- Taste: Bland

Borntrager Reaction: Negative
Emodin: Negative
Aloe Leaf Fibers: None
Total Microbial Count: Less than 1,000/g
Pathogenic Bacteria: None
pH (0.5% Solution): 4.5-7.5
Acid Insoluble Ash: Max: 0.6%
Heavy Metals: Max: 0.001%
Water Insoluble Matter: Max. 0.050%
Dispersion Rate: Max: 1½ hrs.
Infrared Spectrum: Conforms to Standard

DR. MADIS LABORATORIES, INC.: VERAGEL Aloe Vera Gel(Continued):

Water Soluble VERAGEL Aloe Vera Gel:

Liquid
Strength: 1:1
Single Fold
Description:
Liquid:

Liquid VERAGEL is a clear to very slightly opalescent, viscous liquid. One part of VERAGEL Liquid 1:1 is produced from one part of fresh Aloe inner gel. One part of VERAGEL Liquid 1:10 (tenfold) is produced from 10 parts of fresh Aloe inner gel. VERAGEL Liquid is soluble in water, propylene glycol, glycerin; insoluble in alcohol over 20% (higher alcohol content produces white flocculent precipitate), chloroform, acetone, ether and other organic solvents.

Purity and Quality:
Organoleptic Tests:
Consistency: Slightly Viscous
Color: Colorless
Odor: Odorless
Taste: Almost Tasteless
Borntrager Reaction: Negative
Emodin: Negative
Aloe Leaf Fibers: None
Specific Gravity: 1.002-1.015
Total Microbial Content: Less than 500/g
Pathogenic Bacteria: None
pH (0.5% Solution): 4.5-7.5
Heavy Metals: Max: 0.001%
Water Insoluble Matter: Max: 0.001%
Infrared Spectrum: Conforms to Standard

Liquid:
Strength: 1:10
Ten Fold
Purity and Quality:
Organoleptic Tests:
Consistency: Very Viscous
Color: Off-White
Odor: Odorless
Taste: Slightly Bland
Brontrager Reaction: Negative
Emodin: Negative
Aloe Leaf Fibers: None
Specific Gravity: 1.020-1.170
Total Microbial Count: Less than 500/g
Pathogenic Bacteria: None
pH (0.5% Solution): 4.5-7.5
Heavy Metals: Max: 0.001%
Water Insoluble Matter: Max: 0.010%
Infrared Spectrum: Conforms to Standard

DR. MADIS LABORATORIES, INC.: VERAGEL Lipoid: Aloe Vera Leaf Extract:

Strength: 1:1
Single Fold

Description:

VERAGEL Lipoid is clear to very slightly opalescent oily liquid. One part of VERAGEL LIPOID 1:1 is produced from one part of the fresh Aloe Vera leaf. One part of VERAGEL Lipoid 1:10 has a tenfold strength and is produced from ten parts of the fresh Aloe Vera leaf.

VERAGEL Lipoid is completely soluble in vegetable, animal and fixed oils such as mineral oil. It is insoluble in water or alcohol (upon special request an alcohol soluble VERAGEL Lipoid could also be supplied).

The VERAGEL Lipoids have been developed for specific applications in cosmetic formulations where selected oil soluble ingredients of the Aloe leaf are preferred. Suggested use levels ranging from 1-2% of 1:1 strength or .1 to .5% of the tenfold strength in suntan and after-tan oils, creams, ointments, lipsticks, bath oils, etc. are recommended.

Purity and Quality:
Organoleptic Tests:
Consistency: Oily Liquid
Color: Pale Yellow
Odor: Slight
Taste: Slightly Oily
Borntrager Reaction: Negative
Emodin: Negative
Aloe Leaf Fibers: None
Pathogenic Bacteria: None
Total Microbial Count: Max: 200/g
Specific Gravity: 0.840-0.910
Heavy Metals: Max: 0.001%
Moisture: Less than 1%
Acid Value: Less than .1
Saponification Number: 25-35
Preservatives: Not Needed
Anti-Oxidants: Not Needed
Infrared Spectrum: Conforms to Standard

DR. MADIS LABORATORIES, INC.: VERAGEL Lipoid: Aloe Vera Leaf Extract(Continued):

Strength: 1:10
Ten Fold

Description:

VERAGEL Lipoid is a clear to very slightly opalescent oily liquid. One part of VERAGEL Lipoid 1:1 is produced from one part of the fresh Aloe Vera leaf. One part of VERAGEL Lipoid 1:10 has a tenfold strength and is produced from ten parts of the fresh Aloe Vera leaf.

VERAGEL Lipoid is completely soluble in vegetable, animal and fixed oils such as mineral oil. It is insoluble in water or alcohol (upon special request an alcohol soluble VERAGEL Lipoid could also be supplied).

The VERAGEL Lipoids have been developed for specific applications in cosmetic formulations where selected oil soluble ingredients of the Aloe leaf are preferred. Suggested use levels ranging from 1-2% of 1:1 strength or .1 to .5% of the tenfold strength in suntan and after-tan oils, creams, ointments, lipsticks, bath oils, etc. are recommended.

Purity and Quality:
Organoleptic Tests:
Consistency: Oily Liquid
Color: Yellow
Odor: Slight
Taste: Slightly Oily
Borntrager Reaction: Negative
Emodin: Negative
Aloe Leaf Fibers: None
Pathogenic Bacteria: None
Total Microbial Count: Max.: 200/g
Specific Gravity: 0.880-0.950
Heavy Metals: Max. 0.001%
Moisture: Less than 1%
Acid Value: Less than .1
Saponification Number: 25-35
Preservatives: Not Needed
Anti-Oxidants: Not Needed
Infrared Spectrum: Conforms to Standard

MCINTYRE GROUP LTD.: MACKAM/Amphoterics:

MACKAM 2CY:
Capryloamphodiacetate
Form: Liquid
Conc.: 50
pH: 11.0

MACKAM 1C:
Cocoamphoacetate
Form: Liquid
Conc.: 45
pH: 11.0

MACKAM 2C:
Cocoamphodiacetate
Form: Liquid
Conc.: 50
pH: 8.5

MACKAM 2C-75:
Cocoamphodiacetate
Form: Liquid
Conc.: 37
pH: 8.0

MACKAM 1L
Lauroamphoacetate
Form: Liquid
Conc.: 44
pH: 10.0

MACKAM 2L:
Lauroamphodiacetate
Form: Liquid
Conc.: 50
pH: 9.0

MACKAM 1W:
Wheat Germamphoacetate
Form: Liquid
Conc.: 35
pH: 9.5

MACKAM 2W:
Wheat Germamphodiacetate
Form: Liquid
Conc.: 35
pH: 9.5

MCINTYRE GROUP LTD.: MACKAM/Amphoterics(Continued):

MACKAM MEJ:
Mixed Alkylamphocarboxylate
Form: Liquid
Conc.: 34
pH: 10.0

MACKAM 2CYSF:
Caprylamphodipropionate
Form: Liquid
Conc.: 50
pH: 9.8

MACKAM CSF:
Cocoamphopropionate
Form: Liquid
Conc.: 39
pH: 10.0

MACKAM 2CSF:
Cocoamphodipropionate
Form: Liquid
Conc.: 39
pH: 10.0

MACKAM 2LSF:
Lauroamphodipropionate
Form: Liquid
Conc.: 39
pH: 10.0

MACKAM 2CT:
Cocoamphodiacetate (and) Sodium Trideceth Sulfate (and) Hexylene Glycol
Form: Liquid
Conc.: 50
pH: 8.5

MACKAM MLT:
Lauroamphoacetate (and) Sodium Trideceth Sulfate
Form: Liquid
Conc.: 35
pH: 10.0

MACKAM 151L:
Lauraminopropionic Acid
Form: Liquid
Conc.: 40
pH: 5.0

MCINTYRE GROUP LTD.: MACKAM/Amphoterics(Continued):

MACKAM 151C:
Cocaminopropionic Acid
Form: Liquid
Conc.: 40
pH: 5.0

MACKAM 160C:
Sodium Lauriminodipropionate
Form: Liquid
Conc.: 38
pH: 7.0

MACKAM TM:
Dihydroxyethyl Tallow Glycinate
Form: Liquid
Conc.: 40
pH: 5.0

MCINTYRE GROUP LTD.: MACKAM/Betaines:

MACKAM 35:
Cocamidopropyl Betaine (Via Glyceride)
Form: Liquid
Conc.: 35
pH: 6.0

MACKAM 35HP:
Cocamidopropyl Betaine
Form: Liquid
Conc.: 35
pH: 6.0

MACKAM J:
Cocamidopropyl Betaine
Form: Liquid
Conc.: 35
pH: 6.0

MACKAM LMB:
Lauramidopropyl Betaine
Form: Liquid
Conc.: 35
pH: 6.0

MACKAM LMB-LS:
Lauramidopropyl Betaine - Low Salt
Form: Liquid
Conc.: 32
pH: 6.0

MCINTYRE GROUP LTD.: MACKAM/Betaines(Continued):

MACKAM HV:
Oleamidopropyl Betaine
Form: Liquid
Conc.: 35
pH: 6.5

MACKAM ISA:
Isostearamidopropyl Betaine
Form: Liquid
Conc.: 33
pH: 7.5

MACKAM BA:
Behenamidopropyl Betaine
Form: Liquid
Conc.: 25
pH: 6.5

MACKAM WGB:
Wheat Germamidopropyl Betaine
Form: Liquid
Conc.: 34
pH: 6.5

MACKAM RA:
Ricinoleamidopropyl Betaine
Form: Liquid
Conc.: 35
pH: 6.5

MACKAM CB-35:
Coco Betaine
Form: Liquid
Conc.: 35
pH: 8.0

MACKAM CB-LS:
Coco Betaine - Low Salt
Form: Liquid
Conc.: 33
pH: 7.5

MACKAM OB-30:
Oleyl Betaine
Form: Liquid
Conc.: 30
pH: 7.0

MCINTYRE GROUP LTD.: MACKAM/Betaines(Continued):

MACKAM CAP:
Cocamidopropyl Dimethylaminopropionate
Form: Liquid
Conc.: 30
pH: 5.0

MACKAM LAP:
Lauramidopropyl Dimethylaminopropionate
Form: Liquid
Conc.: 30
pH: 5.0

MACKAM ISP:
Isostearamidopropyl Dimethylaminopropionate
Form: Liquid
Conc.: 30
pH: 5.0

MACKAM NLP:
Oleamidopropyl Dimethylaminopropionate (and) Palmitamidopropyl Dimethylaminopropionate (and) Palmitoleamidopropyl Dimethylaminopropionate
Form: Liquid
Conc.: 30
pH: 6.0

MCINTYRE GROUP LTD.: MACKAMIDE/Alkanolamides:

MACKAMIDES are nonionic surfactants made from the condensation of mono or diethanolamine and a fatty acid or ester.

They have a wide variety of uses both for personal care and industrial products. MACKAMIDES basically function as thickeners, foam boosters and stabilizers when blended with anionic surfactants such as lauryl sulfates, ether sulfates, olefin sulfonates or alkylbenzene sulfonates.

When back titrated with a fatty acid they have many applications for industrial products. Some applications are: detergent for high alkaline heavy duty cleaners, solubilizer for inorganic builders, corrosion inhibitors, cutting oil lubricants and chain lubricants.

Monoethanol amides are solid at room temperature. They are used for solid applications such as a slow release detergent block, impregnated soap pads, or bar soap additive.

MACKAMIDE/Alkanolamides:

MACKAMIDE C:
Cocamide DEA (1:1)
Form: Liquid
Conc.: 100
pH: 10.0

MACKAMIDE CD:
Cocamide DEA (2:1)
Form: Liquid
Conc.: 100
pH: 10.0

MACKAMIDE CS:
Cocamide DEA (1:1)
Form: Liquid
Conc.: 100
pH: 10.0

MACKAMIDE MC:
Cocamide DEA (1:1)
Form: Liquid
Conc.: 100
pH: 10.0

MACKAMIDE PK:
Palmkernelamide DEA
Form: Liquid
Conc.: 100
pH: 10.0

MCINTYRE GROUP LTD.: MACKAMIDE/Alkanolamides(Continued):

MACKAMIDE L95:
Lauramide DEA (95% lauric)
Form: Solid
Conc.: 100
pH: 10.0

MACKAMIDE LMD:
Lauramide DEA (70% lauric)
Form: Solid
Conc.: 100
pH: 10.0

MACKAMIDE LLM:
Lauramide DEA
Form: Liquid
Conc.: 100
pH: 10.0

MACKAMIDE L-10:
Lauramide DEA
Form: Liquid
Conc.: 100
pH: 10.0

MACKAMIDE CD-10:
Capramide DEA
Form: Liquid
Conc.: 100
pH: 10.0

MACKAMIDE MO:
Oleamide DEA (1:1)
Form: Liquid
Conc.: 100
pH: 10.0

MACKAMIDE NOA:
Oleamide DEA (1:1)
Form: Liquid
Conc.: 100
pH: 10.0

MACKAMIDE O:
Oleamide DEA (2:1)
Form: Liquid
Conc.: 100
pH: 10.0

MCINTYRE GROUP LTD.: MACKAMIDE/Alkanolamides(Continued):

MACKAMIDE S:
Soyamide DEA (1:1)
Form: Liquid
Conc.: 100
pH: 10.0

MACKAMIDE SD:
Soyamide DEA (2:1)
Form: Liquid
Conc.: 100
pH: 10.0

MACKAMIDE LOL:
Linoleamide DEA
Form: Liquid
Conc.: 100
pH: 10.0

MACKAMIDE ISA:
Isostearamide DEA
Form: Liquid
Conc.: 100
pH: 10.0

MACKAMIDE R:
Ricinoleamide DEA
Form: Liquid
Conc.: 100
pH: 10.0

MACKAMIDE AME-75:
Acetamide MEA
Form: Liquid
Conc.: 75
pH: 7.0

MACKAMIDE AME-100:
Acetamide MEA
Form: Liquid
Conc.: 100
pH: 7.0

MACKAMIDE LME:
Lactamide MEA
Form: Liquid
Conc.: 100
pH: 5.0

MCINTYRE GROUP LTD.: MACKAMIDE/Alkanolamides(Continued):

MACKAMIDE CMA:
Cocamide MEA
Form: Flake
Conc.: 100
pH: 10.0

MACKAMIDE PKM:
Palmkernelamide MEA
Form: Flake
Conc.: 100
pH: 10.0

MACKAMIDE LMM:
Lauramide MEA
Form: Flake
Conc.: 100
pH: 10.0

MACKAMIDE SMA:
Stearamide MEA
Form: Flake
Conc.: 100
pH: 10.0

MACKAMIDE OP:
Oleamide MIPA
Form: Paste
Conc.: 100
pH: 10.0

MCINTYRE GROUP LTD.: MACKAMIDE/Modified:

MACKAMIDE CDM:
Cocamide DEA (and) DEA Oleate
Form: Liquid
Conc.: 100
pH: 9.0

MACKAMIDE CDT:
Cocamide DEA (and) Tall Oil Soap
Form: Liquid
Conc.: 100
pH: 9.0

MACKAMIDE CD-25:
Cocamide DEA (and) Tall Oil Soap
Form: Liquid
Conc.: 100
pH: 9.0

MCINTYRE GROUP LTD.: MACKAMIDE/Modified(Continued):

MACKAMIDE CD-8:
Cocamide DEA (and) Mixed Soaps
Form: Liquid
Conc.: 100
pH: 9.0

MACKAMIDE CD-6:
Cocamide DEA (and) DEA Caprate
Form: Liquid
Conc.: 100
pH: 9.0

MACKAMIDE CDS-80:
Cocamide DEA (and) DEA Dodecylbenzene Sulfonate
Form: Liquid
Conc.: 80
pH: 9.0

MACKAMIDE CDC:
Cocamide DEA (and) DEA Coconate
Form: Liquid
Conc.: 100
pH: 9.0

MACKAMIDE CCDM:
Cocamide (and) Mixed Soaps
Form: Liquid
Conc.: 100
pH: 9.0

MACKAMIDE PG:
Cocamide DEA (and) Mixed Soaps
Form: Liquid
Conc.: 100
pH: 9.0

MACKAMIDE ODM:
Oleamide DEA (and) DEA Oleate
Form: Gel
Conc.: 100
pH: 9.0

MACKAMIDE BE-29:
Cocamide DEA (and) Mixed Soaps
Form: Liquid
Conc.: 100
pH: 9.0

MACKAMIDE BE-30:
Cocamide DEA (and) Mixed Soaps
Form: Liquid
Conc.: 100
pH: 9.0

MCINTYRE GROUP LTD.: MACKANATE/Sulfosuccinates:

MACKANATES are sulfosuccinate surfactants from both mono and diesters of maleic acid.

Personal Care:
The monoester type is primarily used in personal care because it is very mild to both skin and eyes. They also reduce the irritation properties of high foaming anionic surfactants and are low cost compared to other mild specialty surfactants. Ethoxylation improves mildness even further and the low irritation properties are unaffected by concentration.

Other Applications:
The diesters, especially the dioctyl, have a variety of applications. They are powerful wetting agents and one of the most effective interfacial tension depressants commercially available.

MACKANATE/Monoesters:

MACKANATE CM:
Disodium Cocamido MEA Sulfosuccinate
Form: Liquid
Conc.: 40
pH: 6.0

MACKANATE CM-100:
Disodium Cocamido MEA Sulfosuccinate
Form: Powder
Conc.: 100
pH: 6.0

MACKANATE CP:
Disodium Cocamido MIPA Sulfosuccinate
Form: Liquid
Conc.: 40
pH: 6.0

MACKANATE LM-40:
Disodium Lauramido MEA Sulfosuccinate
Form: Liquid
Conc.: 40
pH: 6.0

MACKANATE OD-35:
Disodium Oleamido PEG-2 Sulfosuccinate
Form: Liquid
Conc.: 35
pH: 6.0

MCINTYRE GROUP LTD.: MACKANATE/Sulfosuccinates(Continued):

MACKANATE/Monoesters(Continued):

MACKANATE OP:
Disodium Oleamido MIPA Sulfosuccinate
Form: Liquid
Conc.: 38
pH: 6.0

MACKANATE OM:
Disodium Oleamido MEA Sulfosuccinate
Form: Liquid
Conc.: 35
pH: 6.0

MACKANATE RM:
Disodium Ricinoleamido MEA Sulfosuccinate
Form: Liquid
Conc.: 40
pH: 6.0

MACKANATE LO:
Disodium Lauryl Sulfosuccinate
Form: Paste
Conc.: 40
pH: 6.0

MACKANATE LO-Special:
Disodium Lauryl Sulfosuccinate
Form: Paste
Conc.: 40
pH: 6.0

MACKANATE LO-100:
Disodium Lauryl Sulfosuccinate
Form: Powder
Conc.: 100
pH: 6.0

MACKANATE LA:
Diammonium Lauryl Sulfosuccinate
Form: Liquid
Conc.: 40
pH: 6.0

MACKANATE L-1:
Disodium Laureth Sulfosuccinate
Form: Liquid
Conc.: 40
pH: 6.0

MCINTYRE GROUP LTD.: MACKANATE/Sulfosuccinates(Continued):

MACKANATE/Monoesters(Continued):

MACKANATE L-2:
Disodium Laureth Sulfosuccinate
Form: Liquid
Conc.: 40
pH: 6.0

MACKANATE EL:
Disodium Laureth Sulfosuccinate
Form: Liquid
Conc.: 40
pH: 6.0

MACKANATE TDS:
Disodium Tridecyl Sulfosuccinate
Form: Liquid
Conc.: 40
pH: 6.0

MACKANATE UM:
Disodium Undecylenamido MEA Sulfosuccinate
Form: Liquid
Conc.: 45
pH: 6.0

MACKANATE WGD:
Disodium Wheatgermamido PEG-2 Sulfosuccinate
Form: Liquid
Conc.: 35
pH: 6.0

MACKANATE NLD:
Disodium Oleamido PEG-2 Sulfosuccinate (and) Disodium Palmamido PEG-2 Sulfosuccinate (and) Disodium Palmitoleamido PEG-2 Sulfosuccinate
Form: Liquid
Conc.: 35
pH: 6.0

MACKANATE A-102:
Disodium Deceth-6 Sulfosuccinate
Form: Liquid
Conc.: 30
pH: 6.0

MACKANATE A-103:
Disodium Nonoxynol-10 Sulfosuccinate
Form: Liquid
Conc.: 35
pH: 6.0

MCINTYRE GROUP LTD.: MACKANATE/Sulfosuccinates(Continued):

MACKANATE/Monoesters(Continued):

MACKANATE DC-30:
Disodium Dimethicone Copolyol Sulfosuccinate
Form: Liquid
Conc.: 30
pH: 5.0

MACKANATE DC-30A:
Diammonium Dimethicone Copolyol Sulfosuccinate
Form: Liquid
Conc.: 30
pH: 5.0

MACKANATE/Diesters:

MACKANATE DOS-40:
Dioctyl Sodium Sulfosuccinate
Form: Liquid
Conc.: 40
pH: 6.0

MACKANATE DOS-70:
Dioctyl Sodium Sulfosuccinate
Form: Liquid
Conc.: 70
pH: 6.0

MACKANATE DOS-75:
Dioctyl Sodium Sulfosuccinate
Form: Liquid
Conc.: 75
pH: 6.0

MACKANATE DOS-70PG:
Dioctyl Sodium Sulfosuccinate (and) Propylene Glycol
Form: Liquid
Conc.: 70
pH: 6.0

MACKANATE DOS-70MS:
Dioctyl Sodium Sulfosuccinate (and) Mineral Spirits
Form: Liquid
Conc.: 70
pH: 6.0

MACKANATE DOS-70BC:
Dioctyl Sodium Sulfosuccinate (and) Butyl Carbitol
Form: Liquid
Conc.: 70
pH: 6.0

MACKANATE 85P:
Dioctyl Sodium Sulfosuccinate (and) Sodium Benzoate
Form: Powder
Conc.: 85
pH: 6.0

MACKANATE AY-65TD:
Diamyl Sodium Sulfosuccinate (and) Tridecyl Alcohol
Form: Liquid
Conc.: 65
pH: 6.0

MCINTYRE GROUP LTD.: MACKADET/Blends:

MACKADETS are completely formulated products in a concentrate form. They are a combination of components in which the ratios are maximized to provide optimum performance at minimum cost. Blends provide a significant cost savings to the manufacturer for the following reasons:

- They reduce manufacturing steps.
- They eliminate the need to inventory and handle several individual components.

MACKADET SBC-8:
Applications: Multipurpose shampoo, hand soap, bubble bath concentrate
Form: Liquid
Conc.: 46.0
pH: 6.5

MACKADET CA:
Applications: High foaming shampoo concentrate that provides high viscosity at low concentration
Form: Liquid
Conc.: 42.0
pH: 7.0

MACKADET BSC:
Very mild baby shampoo concentrate
Form: Liquid
Conc.: 45.0
pH: 7.0

MACKADET BBC:
Very mild childrens bubble bath concentrate
Form: Liquid
Conc.: 35.0
pH: 6.5

MACKADET CBC:
Conditioner concentrate for viscous cream consistency
Form: Flakes
Conc.: 100.0
pH: 4.0

MACKADET LCB:
Liquid conditioner concentrate that can be cold blended
Form: Liquid
Conc.: 30.0
pH: 3.0

MACKADET INC:
Leave on conditioner concentrate
Form: Liquid
Conc.: 16.5
pH: 4.5

MCINTYRE GROUP LTD: MACKADET/Blends(Continued):

MACKADET RS:
Rug shampoo concentrate that leaves dry residue
Form: Liquid
Conc.: 35.0
pH: 7.0

MACKADET WHC:
Waterless hand cleaner concentrate
Form: Liquid
Conc.: 100.0
pH: 8.0

MACKADET 40K:
Potassium coconate for liquid hand cleansers
Form: Liquid
Conc.: 38.0
pH: 9.0

MCINTYRE GROUP LTD.: MACKINE/Amidoamines :

MACKINES are amidoamines which become cationic surfactants when neutralized. These salts provide excellent hair conditioning properties. MACKINES are unusual in that they are compatible with anionic surfactants and will not depress foam properties of a shampoo.

MACKINES are also excellent corrosion inhibitors and the salts produce emulsifying properties for acid systems.

They are used primaily as precursers in the manufacture of betaines, amine oxides, quaternary ammonium compounds and cationic surfactants. The salts of the morpholine derivatives are extremely mild to skin and eyes. They are excellent conditioners for baby products and mild skin cleansers.

MACKINE 101:
Cocamidopropyl Dimethylamine
Form: Liquid
Conc.: 100

MACKINE 201:
Ricinoleamidopropyl Dimethylamine
Form: Liquid
Conc.: 100

MACKINE 301:
Stearamidopropyl Dimethylamine
Form: Flake
Conc.: 100

MCINTYRE GROUP LTD.: MACKINE/Amidoamines(Continued):

MACKINE 321:
Stearamidopropyl Morpholine
Form: Flake
Conc.: 100

MACKINE 401:
Isostearamidopropyl Dimethylamine
Form: Liquid
Conc.: 100

MACKINE 421:
Isostearamidopropyl Morpholine
Form: Liquid
Conc.: 100

MACKINE 501:
Oleamidopropyl Dimethylamine
Form: Liquid
Conc.: 100

MACKINE 601:
Behenamidopropyl Dimethylamine
Form: Flake
Conc.: 100

MACKINE 701:
Wheat Germamidopropyl Dimethylamine
Form: Paste
Conc.: 100

MACKINE 801:
Lauramidopropyl Dimethylamine
Form: Solid
Conc.: 100

MACKINE 901:
Soyamidopropyl Dimethylamine
Form: Paste
Conc.: 100

MCINTYRE GROUP LTD.: MACKALENE/Cationic Conditioners:

They are excellent conditioners for both skin and hair care products. The MACKALENES are unique in that they are compatible with anionic surfactants and therefore can be used as conditioners for clear shampoos. Compared to a quaternary ammonium compound, MACKALENES do not depress the foam properties of anionic surfactants. Also the morpholine types are non-irritating and are recommended for applications which require extra mildness, such as baby shampoos and leave on conditioners.

MACKALENE 116:
Cocamidopropyl Dimethylamine Lactate
Form: Liquid
Conc.: 25
pH: 5.0

MACKALENE 117:
Cocamidopropyl Dimethylamine Propionate
Form: Liquid
Conc.: 40
pH: 6.5

MACKALENE 216:
Ricinoleamidopropyl Dimethylamine Lactate
Form: Liquid
Conc.: 95
pH: 6.0

MACKALENE 316:
Stearamidopropyl Dimethylamine Lactate
Form: Liquid
Conc.: 25
pH: 4.5

MACKALENE 326:
Stearamidopropyl Morpholine Lactate
Form: Liquid
Conc.: 25
pH: 4.5

MACKALENE 416:
Isostearamidopropyl Dimethylamine Lactate
Form: Liquid
Conc.: 25
pH: 6.0

MACKALENE 426:
Isostearamidopropyl Morpholine Lactate
Form: Liquid
Conc.: 25
pH: 4.0

MCINTYRE GROUP LTD.: MACKALENE/Cationic Conditioners(Continued):

MACKALENE 616:
Behenamidopropyl Dimethylamine Lactate
Form: Liquid
Conc.: 25
pH: 4.5

MACKALENE 716:
Wheat Germamidopropyl Dimethylamine Lactate
Form: Paste
Conc.: 95
pH: 6.0

MACKALENE NLC:
Oleamidopropyl Dimethylamine Lactate (and) Palmitamidopropyl Dimethylamine Lactate (and) Palmitoleamidopropyl Dimethylamine Lactate
Form: Paste
Conc.: 95
pH: 5.5

MCINTYRE GROUP LTD.: **MACKAMINE/Amine Oxides:**

MACKAMINES are alkyl amine oxides which are mildly cationic in acid media.

Their major applications in personal care are as hair conditioners, viscosity builders and foam boosters. They have an advantage over alkanolamides in that the amine oxide will not cause pH drift in acid systems, thus, providing greater stability to formulations.

For highly alkaline heavy duty cleaners, MACKAMINES provide excellent detergent and wetting properties.

MACKAMINE CAO:
Cocamidopropylamine Oxide
Form: Liquid
Conc.: 30
pH: 7.0

MACKAMINE LAO:
Lauramidopropylamine Oxide
Form: Liquid
Conc.: 30
pH: 7.0

MACKAMINE OAO:
Oleamidopropylamine Oxide
Form: Gel
Conc.: 50
pH: 7.0

MCINTYRE GROUP LTD.: MACKAMINE/Amine Oxides(Continued):

MACKAMINE WGO:
Wheat Germamidopropylamine Oxide
Form: Gel
Conc.: 30
pH: 7.0

MACKAMINE IAO:
Isostearamidopropylamine Oxide
Form: Gel
Conc.: 30
pH: 7.0

MACKAMINE SAO:
Stearamidopropylamine Oxide
Form: Paste
Conc.: 25
pH: 7.0

MACKAMINE BAO:
Behenamidopropylamine Oxide
Form: Paste
Conc.: 25
pH: 7.0

MACKAMINE CO:
Cocamine Oxide
Form: Liquid
Conc.: 30
pH: 7.0

MACKAMINE LO:
Lauramine Oxide
Form: Liquid
Conc.: 30
pH: 7.0

MACKAMINE O2:
Oleamine Oxide
Form: Liquid
Conc.: 35
pH: 7.5

MACKAMINE SO:
Stearamine Oxide
Form: Paste
Conc.: 25
pH: 7.0

MACKAMINE ISMO:
Isostearamidopropylmorpholine Oxide
Form: Liquid
Conc.: 30
pH: 7.0

MCINTYRE GROUP LTD.: MACKERNIUM/Quaternary Ammonium Compounds:

MACKERNIUMS are quaternary ammonium compounds which provide excellent anti-static and conditioning properties.

They are highly substantive to hair and have a greater affinity to the most damaged portion.

MACKERNIUM SDC-25:
Stearalkonium Chloride
Form: Paste
Conc.: 25
pH: 4.0

MACKERNIUM SDC-85:
Stearalkonium Chloride
Form: Flake
Conc.: 100
pH: 6.0

MACKERNIUM NLE:
Oleyl/Palmitoyl/Palmitoleamidopropyl/Epoxy Dimonium Chloride
Form: Liquid
Conc.: 100
pH: 7.0

MACKERNIUM KP:
Olealkonium Chloride
Form: Liquid
Conc.: 50
pH: 5.0

MCINTYRE GROUP LTD.: MACKESTER/Organic Esters:

MACKESTERS are condensates of an alcohol and organic acid.

They have a variety of applications from cosmetic emollients, pearl agents, and emulsifiers to high pressure lubricants. Applications are found in metal working, textile lubricants, plastics, and paper industries. The MACKESTERS provide lubricity, emulsification, static control and defoaming characteristics.

MACKESTER TD-88:
Triethylene Glycol Dioctoate
Form: Liquid
Conc.: 100

MACKESTER IDO:
Isodecyl Oleate
Form: Liquid
Conc.: 100

MACKESTER TDCC:
Triethylene Glycol Dicaprylate/Dicaprate
Form: Liquid
Conc.: 100

MACKESTER SP:
Glycol Stearate Modified
Form: Flake
Conc.: 100

MACKESTER EGMS:
Glycol Stearate
Form: Flake
Conc.: 100

MACKESTER JJO:
Jojoba Oil Substitute
Form: Liquid
Conc.: 100

MCINTYRE GROUP LTD.: MACKPRO/Quaternized Proteins:

MACKPROS are cosmetic grade protein derivatives.

They are quaternized proteins in which a lipophile is grafted to a protein backbone. In the quaternized form the MACKPROS become highly attracted to both hair and skin forming a protective natural conditioning effect.

MACKPRO NLP:
(Natural Lipid Protein)
Quaternium-79 Hydrolyzed Animal Protein
Form: Liquid
Conc.: 40
pH: 6.5

MACKPRO NLP-Special:
(Natural Lipid Protein)
Quaternium-79 Hydrolyzed Animal Protein
Form: Liquid
Conc.: 40
pH: 5.0

MACKPRO NSP:
Oleyl/Palmityl/Palmitoleamidopropyl/Silkhydroxypropyl Dimonium Chloride
Form: Liquid
Conc.: 33
pH: 5.0

MACKPRO KLP:
Oleyl/Palmityl/Palmitoleyl/Keratin Hydroxypropyl/Dimonium Chloride/Lactate
Form: Liquid
Conc.: 35
pH: 5.0

MCINTYRE GROUP LTD.: MACKSTAT DM/DMDM Hydantoin:

MACKSTAT DM is a cosmetic grade preservative DMDM Hydantoin.

It is an excellent broad spectrum cosmetic preservative which is a very effective for shampoos, skin cleansers, bath products, lotions and creams. MACKSTAT DM disperses readily in cold systems and is very economical.

MACKSTAT DM:
DMDM Hydantoin
Form: Liquid
Conc.: 55
pH: 7.0
MACKSTAT DM100:
DMDM Hydantoin
Form: Powder
Conc.: 100
pH: 7.0

MCINTYRE GROUP LTD.: **MACKOL/Fatty Alcohols:**

MACKOLS are fatty alcohol compounds.
R--OH
They are used as thickeners and co-emulsifiers in many cosmetic formulations, such as, hair conditioners, lotions, skin cleansers and shampoos.

MACKOL 16:
Cetyl Alcohol
Form: Flake
Conc.: 100
MACKOL 1618:
Cetearyl Alcohol
Form: Flake
Conc.: 100
MACKOL 18:
Stearyl Alcohol
Form: Flake
Conc.: 100

MCINTYRE GROUP LTD.: **MACKPEARL/Liquid : Pearling Agents:**

MACKPEARLS are liquid pearling agents that readily disperse in a cold surfactant blend to provide an elegant pearl effect. MACKPEARLS eliminate the need to heat the product to a high temperature saving both time and energy.
MACKPEARL LV:
Proprietary Blend
Form: Liquid
Conc.: 15
pH: 8.5
MACKPEARL LVD:
Proprietary Blend
Form: Liquid
Conc.: 45
pH: 7.0

MEARL CORP.: BIJU TX and BIJU ULTRA TX Concentrates:

BIJU TX Concentrates contain a special brilliant nacreous bismuth oxychloride. The resulting nail enamels are extremely close in appearance to those containing natural pearl essence. BiOCl is widely used for frosted nail enamels in place of natural pearl essence plates, for reasons of economy and availability.

BIJU ULTRA TX Concentrates provide nail enamels even more brilliant than those made with natural pearl essence plates. At lower pigment concentrations, BIJU ULTRA imparts an opulent pearl luster. At higher concentrations it creates silvery, more metallic effect.

Frost (Pearlescent) Qualities:

Product:
BIJU:

TX-SAL-E-6:
Color: Bright Pearl
Pigment Type: BiOCl
TX-SAL-D-6:
Color: Bright Pearl
Pigment Type: BiOCl
TX-LAL-E-6:
Color: Bright Pearl
Pigment Type: BiOCl
TX-LAL-B-6:
Color: Bright Pearl
Pigment Type: BiOCl

BIJU ULTRA:

TX-SCL-E-6:
Color: Very Bright Pearl
Pigment Type: BiOCl
TX-SCL-D-6:
Color: Very Bright Pearl
Pigment Type: BiOCl
TX-LCL-E-6:
Color: Very Bright Pearl
Pigment Type: BiOCl
TX-LCL-A-6:
Color: Very Bright Pearl
Pigment Type: BiOCl

MEARLMAID:

TX-EPM-8-6:
Color: Bright Pearl
Pigment Type: Natural Pearl

MEARL CORP.: BIJU TX and BIJU ULTRA TX Concentrates(Continued):

Frost (Pearlescent) Qualities(Continued:

FLAMENCO VELVET:

TX-LVF-H-6:
Color: Lustrous Satin Pearl
Pigment Type: Mica-TiO2
Dilution Ratio: 1:5

TX-LVF-L-6:
Color: Lustrous Satin Pearl
Pigment Type: Mica-TiO2
Dilution Ratio: 1:5

Metallic and Iridescent Qualities:

FLAMENCO:
Blue TX-LBF-L-6:
Color: Blue
Pigment Type: Mica-TiO2
Dilution Ratio: 1:5

Red TX-LRF-L-6:
Color: Red
Pigment Type: Mica-TiO2
Dilution Ratio: 1:5

Gold TX-LYF-L-6:
Color: Gold
Pigment Type: Mica-TiO2
Dilution Ratio: 1:5

CLOISONNÉ:
Gold TX-LYC-6:
Color: Gold
Pigment Type: Mica-TiO2-Fe2O3
Dilution Ratio: 1:5

Bronze TX-LZC-6:
Color: Bronze
Pigment Type: Mica-TiO2-Fe2O3
Dilution Ratio: 1:5

Copper TX-LCC-6:
Color: Copper
Pigment Type: Mica-TiO2-Fe2O3
Dilution Ratio: 1:5

Rouge TX-LRC-6:
Color: Red
Pigment Type: Mica-TiO2-Fe2O3
Dilution Ratio: 1:5

MEARL CORP.: CLOISONNE' Color Pigments:

The CLOISONNE' Colors are highly lustrous pigments consisting primarily of titanium dioxide-coated mica. They derive color both from light interference and light absorption. The combination of these two color-producing mechanisms results in colors of unique leveliness and impact.

<u>Color:</u>

Blue:
Composition: TiO_2-Mica + Iron Blue
Bulk Density Approx. lb./ft^3: 17

Blue-Green:
Composition: TiO_2-Mica + Cr_2O_3
Bulk Density Approx. lb./ft^3: 18

Green:
Composition: TiO_2-Mica + Cr_2O_3
Bulk Density Approx. lb./ft^3: 19

Supergreen:
Composition: TiO_2-Mica + Fe_2O_3 + Iron Blue
Bulk Density Approx. lb./ft^3: 21

Orange:
Composition: TiO_2-Mica + Fe_2O_3
Bulk Density Approx. lb./ft^3: 13

Red:
Composition: TiO_2-Mica + Carmine
Bulk Density Approx. lb./ft^3: 13

Rouge Flambe':
Composition: TiO_2-Mica + Fe_2O_3
Bulk Density Approx. lb/ft^3: 11

Copper:
Composition: TiO_2-Mica + Fe_2O_3
Bulk Density Approx. lb/ft^3: 12

Gold:
Composition: TiO_2-Mica + Fe_2O_3
Bulk Density Approx. lb/ft^3: 11

MEARL CORP.: CLOISONNE' Color Pigments(Continued):

Color:

Golden Bronze:
Composition: TiO2-Mica + Fe2O3
Bulk Density Approx. lb/ft3: 10

NU-ANTIQUE Red:
Composition: TiO2-Mica + Carmine + Iron Oxides
Bulk Density Approx. lb/ft3: 15

NU-ANTIQUE Rouge Flambe':
Composition: TiO2-Mica + Iron Oxides
Bulk Density Approx. lb/ft3: 14

NU-ANTIQUE Green:
Composition: TiO2-Mica + Cr2O3 + Iron Oxides
Bulk Density Approx. lb/ft3: 19

NU-ANTIQUE Supergreen:
Composition: TiO2-Mica + Iron Blue + Iron Oxides
Bulk Density Approx. lb/ft3: 19

Violet:
Composition: TiO2-Mica + Carmine + Iron Blue
Bulk Density Approx. lb/ft3: 16

NU-ANTIQUE Blue:
Composition: TiO2-Mica + Iron Blue + Iron Oxides
Bulk Density Approx. lbs/ft3: 17

NU-ANTIQUE Gold:
Composition: TiO2-Mica + Iron Oxides
Bulk Density Approx. lb/ft3: 14

NU-ANTIQUE Bronze:
Composition: TiO2-Mica + Iron Oxides
Bulk Density Approx. lbs/ft3: 13

NU-ANTIQUE Copper:
Composition: TiO2-Mica + Iron Oxides
Bulk Density Approx. lbs/ft3: 14

Additional Properties of CLOISONNE' Colors:
Particle Size Average(range): 25 microns
Specific Gravity: Approx. 3.0
Refractive Index: TiO2: 2.3
Mica: 1.59
Trace Impurities: Lead--Maximum 15 ppm
Arsenic--Maximum 2 ppm

MEARL CORP.: DUOCHROME Iridescent Colors:

The DUOCHROME Iridescent Colors are a new series of lustrous pigments, available as powders, which produce a two-color effect. They derive color from both light interference and light absorption. The combination of these two color-producing mechanisms results in a unique double color.

In each DUOCHROME Color, the pigment particle consists of titanium dioxide-coated mica platelet which also contains a thin layer of colored pigment. Light interference from the layered structure produces the reflection color, which is seen as a colored highlight. The added colored component imparts a second color by light absoprtion; this is the background color. Applied to a curved surface, such as an eyelid or fingernail, the colored highlight contrasts with the differently colored background.

DUOCHROME:

BR(Blue/Red):
Absorption Color(Background): Blue
Reflection Color(Highlight): Red
Colored Component: Iron Blue
Bulk Density: 15 Approx.

BG(Blue/Green):
Absorption Color: Blue
Reflection Color: Green
Colored Component: Iron Blue
Bulk Density: 18 Approx.

BY(Blue/Gold):
Absorption Color: Blue
Reflection Color: Gold
Colored Component: Iron Blue
Bulk Density: 16 Approx.

BV(Blue/Violet):
Absorption Color: Blue
Reflection Color: Violet
Colored Component: Iron Blue
Bulk Density: 19 Approx.

RB(Red/Blue):
Absorption Color: Reddish-Blue
Reflection Color: Blue
Colored Component: Carmine
Bulk Density: 14 Approx.

RG(Red/Green):
Absorption Color: Reddish-Blue
Reflection Color: (yellowish) Green
Colored Component: Carmine
Bulk Density: 15 Approx.

MEARL CORP.: DUOCHROME Iridescent Colors(Continued):

DUOCHROME:

RY(Red/Gold):
Absorption Color: Reddish-blue
Reflection Color: Gold
Colored Component: Carmine
Bulk Density: 12 Approx.

RV(Red/Violet):
Absorption Color: Red
Reflection Color: Violet
Colored Component: Carmine
Bulk Density: 14 Approx.

RO(Red/Orange):
Absorption Color: Red
Reflection Color: Orange
Colored Component: Carmine
Bulk Density: 12 Approx.

YG(Gold/Green):
Absorption Color: Gold
Reflection Color: Green
Colored Component: Iron Oxide
Bulk Density: 16 Approx.

YB(Gold/Blue):
Absorption Color: Gold
Reflection Color: Blue
Colored Component: Iron Oxide
Bulk Density: 13 Approx.

YR(Gold/Red):
Absorption Color: Gold
Reflection Color: Red
Colored Component: Iron Oxide
Bulk Density: 14 Approx.

GY(Green/Gold):
Absorption Color: Greenish-Gold
Reflection Color: Gold
Colored Component: Iron Oxide Iron Blue
Bulk Density: 13 Approx.

GR(Green/Red):
Absorption Color: Greenish-Gold
Reflection Color: Red
Colored Component: Iron Oxide Iron Blue
Bulk Density: 19 Approx.

GB(Green/Blue):
Absorption Color: Greenish-Gold
Reflection Color: Blue
Colored Component: Iron Oxide Iron Blue
Bulk Density: 15 Approx.

Other Properties:
Form: Fine lustrous colored powder with contrasting highlight.
Particle Size: Most platelets 6-50 um in longest dimension.
Specific Gravity: Approx. 3.0
Trace Impurities: Lead--Less than 15 ppm/Arsenic--Less than 2 ppm.

MEARL CORP.: FLAMENCO Pearl and Color Pigments:

The FLAMENCO products are nacreous pigments of great brilliance and versatility, available in four qualities of white and a range of iridescent pearl colors. They consist of titanium dioxide-coated mica platelets, and are thus similar in composition to the TIMICA Luster Pigments. However, the FLAMENCO Pigments have significantly higher brilliance and greater smoothness. Additionally, FLAMENCO Superpearl is exceptionally transparent and FLAMENCO Velvet has very good coverage.

FLAMENCO Pigments:

FLAMENCO Velvet:
Reflection Color: White
Transmission Color: Very Dense White
Bulk density lbs/ft3: 10

FLAMENCO Satina:
Reflection Color: White
Transmission Color: Dense White
Bulk Density lbs/ft3: 10

FLAMENCO Pearl:
Reflection Color: White
Transmission Color: Translucent White
Bulk Density lbs/ft3: 10

FLAMENCO Superpearl:
Reflection Color: White
Transmission Color: Transparent White
Bulk Density lbs/ft3: 10

FLAMENCO BLUE:
Reflection Color: Blue
Transmission Color: Yellow
Bulk Density lbs/ft3: 15

FLAMENCO Green:
Reflection Color: Green
Transmission Color: Red
Bulk Density lbs/ft3: 16

FLAMENCO Gold:
Reflection Color: Gold
Transmission Color: Blue
Bulk Density lbs/ft3: 12

FLAMENCO Red:
Reflection Color: Red
Transmission Color: Green
Bulk Density lbs/ft3: 13

FLAMENCO Violet:
Reflection Color: Violet
Transmission Color: Yellow-Green
Bulk Density lbs/ft3: 15

FLAMENCO Orange:
Reflection Color: Orange
Transmission Color: Blue-Green
Bulk Density lbs/ft3: 13

MEARL CORP.: GEMTONE Color Pigments:

The GEMTONE Color Pigments are jewel-like lustrous pigments that offer a wide range of new color possibilities for distinctive cosmetics. They are rich, lustrous powders based primarily on titanium dioxide-coated mica. Like certain precious gems, the GEMTONE Color Pigments derive color both from light interference and light absorption. This combination of color-producing mechanisms results in colors of unique liveliness and intensity. Details of the range of colors, composition and properties are supplied in the following tables:

<u>GEMTONE:</u>

Aquamarine:
Composition: Mica, TiO2, Iron Oxide, Iron Blue
Color Description: Blue-Green
Bulk Density: 18

Azurite:
Composition: Mica, TiO2, Iron Blue
Color Description: Silvery Blue
Bulk Density: 17

Copperstone:
Composition: Mica, TiO2, Iron Oxide
Color Description: Red-Brown
Bulk Density: 14

Goldstone:
Composition: Mica, TiO2, Iron Oxide Iron Blue
Color Description: Medium Gold
Bulk Density: 13

Mauve Quartz:
Composition: Mica, TiO2, Carmine Iron Blue
Color Description: Vibrant Violet
Bulk Density: 15

Purple Agate:
Composition: Mica, TiO2, Iron Blue, Carmine
Color Description: Golden Lavender
Bulk Density: 13

Sunstone:
Composition: Mica, TiO2, Iron Oxide, Carmine
Color Description: Orange-Gold
Bulk Density: 13

Garnet:
Composition: Mica, TiO2, Iron Oxide, Carmine
Color Description: Pink Rose
Bulk Density: 12

MEARL CORP.: GEMTONE Color Pigments(Continued):

<u>GEMTONE:</u>

Amethyst:
Composition: Mica, TiO_2, Iron Blue, Carmine
Color Description: Plum Purple
Bulk Density: 14

Sapphire:
Composition: Mica, TiO_2, Iron Blue, Carmine
Color Description: Deep Sky Blue
Bulk Density: 16

Turquoise:
Composition: Mica, TiO_2, Iron Blue, Iron Oxide
Color Description: Greenish-Blue
Bulk Density: 15

Emerald:
Composition: Mica, TiO_2, Chromium Oxide, Iron Blue
Color Description: Sea Green
Bulk Density: 16

Jade:
Composition: Mica, TiO_2, Chromium Oxide, Iron Oxide
Color Description: Apple Green
Bulk Density: 15

Topaz:
Composition: Mica, TiO_2, Iron Oxide, Iron Blue
Color Description: Chartreuse
Bulk Density: 13

Ruby:
Composition: Mica, TiO_2, Iron Oxide, Carmine
Color Description: Red Wine
Bulk Density: 12

Amber:
Composition: Mica, TiO_2, Iron Oxide
Color Description: Burnished Gold
Bulk Density: 11

Moonstone:
Composition: Mica, TiO_2, Black Iron Oxide
Color Description: Shimmering Silver
Bulk Density: 12

Tan Opal:
Composition: Mica, Iron Oxide, TiO_2
Color Description: Champagne
Bulk Density: 9

Additional Properties of GEMTONE Colors:
Particle Size Average (range): 25 microns (6-50)
Specific Gravity: Approx. 3.0
Refractive Index: TiO_2: 2.3
Mica: 1.59
Trace Impurities: Lead--Maximum 15 ppm
Arsenic--Maximum 2 ppm

MEARL CORP.: MEARLMAID Natural Pearl Essence:

MEARLMAID Natural Pearl Essence is a suspension or paste of the natural guanine crystals which occur in fish, and give them their typical pearly lustrous appearance.

There are two main categories of natural pearl essence; a quality of high brilliance based on broad-surfaced crystals called plates; and a quality of satin luster based on thin, elongated crystals called needles.

Properties of Natural Pearl Essence Crystals:
- Refractive Index: 1.85
- Specific Gravity: 1.6
- Average Crystal Dimensions: Plates: 30x6 microns
Needles: 30x1 microns
- Light Stability: Excellent
- Heat Stability: Excellent
- Trace Imputities: Lead--Max. 15 ppm
Arsenic--Max. 2 ppm

Standard MEARLMAID Pastes for Lotions and Makeups:

<u>Grade:</u>

MEARLMAID AA:
- Type of Crystal: Plates
- Crystal Concentration: 20%
- Vehicle: Aqueous methyl cellulose solution
- Lotions and eye makeups based on water systems.

MEARLMAID OL:
- Type of Crystal: Plates
- Crystal Concentration: 20%
- Vehicle: Isopropanol
- Lotions, nail enamel removers and liquids based on alcohol water mixtures

MEARLMAID FL:
- Type of Crystal: Plates
- Crystal Concentration: 20%
- Vehicle: Castor Oil
- Lipsticks. Also suitable for use in transparent soaps.

MEARLMAID TR:
- Type of Crystal: Plates
- Crystal Concentration: 20%
- Vehicle: Aqueous lauryl sulfate isopropanol solution
- Shampoos and lotions.

MEARL CORP.: MEARLMAID Natural Pearl Essence(Continued):

Standard MEARLMAID Pastes for Lotions and Makeups(Continued):

Grade:

MEARLMAID PLN:
Type of Crystal: Plates
Crystal Concentration: 0.3%
Vehicle: Aqueous gel
Lotions, gels, shave balms

Standard MEARLMAID Pastes for Nail Enamels:

MEARLMAID CP:
Type of Crystal: Plates
Crystal Conventration: 11%
Vehicle: Nitrocellulose lacquer
Nail enamels of highest brilliance.

MEARLMAID CKD:
Types of Crystal: Plates
Crystal Concentration: 22%
Vehicle: Nitrocellulose lacquer
Nail enamels of highest brilliance.

MEARLMAID KN:
Type of Crystal: Needles
Crystal Concentration: 11%
Vehicle: Nitrocellulose lacquer
Nail enamels with soft satin sheen.

MEARLMAID KND:
Type of Crystal: Needles
Crystal Concentration: 22%
Vehicle: Nitrocellulose lacquer
Nail enamels with soft satin sheen.

MEARL CORP.: TIMICA Luster Pigments:

The TIMICA products are nacreous pigments consisting primarily of mica platelets coated with titanium dioxide. All nacreous pigments contain transparent plates of high refractive index; the TIMICA pigments acquire the platelet shape from the mica and the high refractive index from the titanium dioxide coating.

The White Pearlescents are examples of this TiO_2--mica combination in a variety of lusters and particle sizes. The Metallic and Earth Colors incorporate Fe_2O_3 for deeper color effects and greater opacity.

The TIMICA pigments are offered as sparkling, free-flowing powders as follows:

White Pearlescents:

Grade:

SILKWHITE:
Particle Size: Fine (4-75 microns)
Opacity: Excellent
Bulk Density: 9 Approx.
Smooth, satin luster; excellent coverage.

PEARLWHITE:
Particle Size: Medium (6-90 microns)
Opacity: Very Good
Bulk Density: 9 Approx.
Good luster and coverage, most widely used and economical quality.

SPARKLE:
Particle Size: Large (10-110 microns)
Opacity: Fair
Bulk Density: 13 Approx.
Glittery, silver luster; quite transparent.

Extra Large SPARKLE:
Particle Size: Extra Large (10-150 microns)
Opacity: Low
Bulk Density: 18 Approx.
Low proportions provide a high gloss "wet look" finish.

MEARL CORP.: TIMICA Luster Pigments(Continued):

Metallic and Earth Colors:

Grade:

Brilliant Gold:
Particle Size: Medium (5-80 microns)
Opacity: Very Good
Bulk Density: 12 Approx.
Deep gold color by reflection, yellow transmission. Good luster, good color intensity.

Gold SPARKLE:
Particle Size: Large (10-110 microns)
Opacity: Fair
Bulk Density: 16 Approx.
Similar to Brilliant Gold in a larger particle-size range. For sparkling gold highlights.

Golden Bronze:
Particle Size: Medium (6-90 microns)
Opacity: Very Good
Bulk Density: 11 Approx.
Bright, bronze color, can be shifted toward yellow with Brilliant Gold, and toward red with Copper.

Copper:
Particle Size: Medium (6-90 microns)
Opacity: Very Good
Bulk Density: 14 Approx.
Deep copper color, can be blended with Golden Bronze for intermediate shades.

NU-ANTIQUE Silver:
Particle Size: Medium (6-90 microns)
Opacity: Very Good
Bulk Density: 10 Approx.
Dark gunmetal color; can be blended with White TIMICAS to pewter or to bright silvers.

NU-ANTIQUE Gold:
Particle SIze: Medium (5-80 microns)
Opacity: Very Good
Bulk Density: 14 Approx.
Dark metallic gold color; can be blended to medium gold shades with Brilliant Gold.

NU-ANTIQUE Bronze:
Particle Size: Medium (6-90 microns)
Opacity: Very Good
Bulk Density: 13 Approx.
Very dark bronze color, designed to be blended with Golden Bronze for antique effects.

NU-ANTIQUE Copper:
Particle Size: Medium (6-90 microns)
Opacity: Very Good
Bulk Density: 14 approx.
Very dark copper color, designed to be blended with TIMICA Copper for antique effects.

MEARL CORP.: VELVET Color Pigments:

VELVET Colors consist of a new series of nacreous pigment powders offering high color intensity with subdued, soft luster effects. They are composed primarily of titanium dioxide--coated mica which derive color from either light interference or a combination of light interference and absorption. All grades are based on ultra-fine platelets which result in greater opacity, a smoother feel, and easier pressability when compared with other conventional pigments of this type.

Each VELVET Sheer grade is an interference color that exhibits a twin-color iridescent color play. These colors do not arise from light absorption, as is the case with ordinary pigments, but are optical effects produced only by light interference.

In the VELVET Deep grades, light absorbing components are present within the pigment platelets, so that each platelet is colored both by interference and light absorption. In this case, the two-color play, characteristic of the VELVET Sheer grades, is eliminated in favor of a more intense single color.

VELVET Sterling Silver is similar to FLAMENCO VELVET, a white pearlescent pigment of the same particle size, but it has incorporated in it a black iron oxide for a deeper, more silvery effect and greater opacity.

<u>Pigment:</u>

Sheer Red:
Composition: TiO_2 and Mica
Bulk Density: 17

Sheer Blue:
Composition: TiO_2 and Mica
Bulk Density: 18

Sheer Green:
Composition: TiO_2 and Mica
Bulk Density: 20

Sheer Gold:
Composition: TiO_2 and Mica
Bulk Density: 10

Sterling Silver:
Composition: Mica, TiO_2, Iron Oxide
Bulk Density: 10

Deep Red:
Composition: TiO_2, Mica, Carmine
Bulk Density: 15

Deep Blue:
Composition: TiO_2, Mica, Iron Blue
Bulk Density: 18

Deep Green:
Composition: TiO_2, Mica, Chromium Oxide
Bulk Density: 20

Deep Gold:
Composition: TiO_2, Mica, Iron Oxide
Bulk Density: 15

M. MICHEL AND CO., INC.: CACHALOT Fatty Alcohols:

CACHALOT Code:

O-3:
oleyl
Grade: nf
Boiling Range C: 310-350

O-8:
oleyl
Grade: Ctfa
Boiling Range C: 300-350

O-15:
oleyl
Grade: cosm
Boiling Range C: 300-350

O-27:
oleyl
Grade: tech
Boiling Range C: 290-350

S-56:
stearyl
Grade: usp/ctfa
Boiling Range C: 330-350

S-54:
stearyl
Grade: usp
Boiling Range C: 330-350

S-53:
stearyl
Grade: tech
Boiling Range C: 315-350

C-52:
cetyl
Grade: nf/ctfa
Boiling Range C: 310-330

C-50:
cetyl
Grade: nf/ctfa
Boiling Range C: 310-330

C-51:
cetyl
Grade: nf
Boiling Range C: 310-340

M-43:
myristyl
Grade: cosm
Boiling Range C: 280-295

L-90:
lauryl
Grade: cp
Boiling Range C: 250-265

L-50:
lauryl
Grade: tech
Boiling Range C: 260-340

M. MICHEL AND CO., INC.: MICHEL XO-150 Series Guerbet-Reaction Branched-Chain Alcohols:

MICHEL XO-150-12:
iso-dodecanol
iso-lauryl alcohol
CAS number: 3913-02-8
dominant chain: C=12 branched
peak % (approx): 95%
boiling point C (approx): 243

MICHEL XO-150-16:
iso-hexadecanol
iso-cetyl alcohol
CAS number: 36311-34-9
dominant chain: C=16 branched
peak % (approx): 95%
boiling point C (approx): 294

MICHEL XO-150-18:
iso-octadecanol
iso-stearyl alcohol
CAS number: 27458-93-1
dominant chain: C=18 branched
peak % (approx): 95%
boiling point C (approx): 300

MICHEL XO-150-20:
iso-eicosanol
iso-eicosyl alcohol
CAS number: 5333-42-6
dominant chain: C=20 branched
peak % (approx): 95%
boiling point C (approx): 324

MICHEL XO-150-1620:
mixed branched alcohols
C16-C20 alcohols
CAS number: 70693-04-8
boiling point C (approx): 290

MIRANOL INC.: MIRANOL Amphoteric Anionic Salts:

These comprise a class of materials which are mild to the skin, nonirritating to the eyes, as well as non eye-stinging. Care must be taken in formulating with this series of aqueous products so as not to impair these properties. Thus, excessive amounts of anionic surfactants should be avoided and only fragrances used which do not affect these properties.

MIRANOL 2MCA MODIFIED:
CTFA Adopted Name: Cocoamphodiacetate (and) Sodium Lauryl Sulfate (and) Hexylene Glycol

Typical Properties:
Appearance: clear, thin, yellow liquid
Solids (including diol): 46.5%
Chloride (as NaCl): 6.9%
pH (20%, 25C): 8.2

Special Characteristics:
- Lauryl sulfate salt of an amphoteric surfactant derived from coconut imidazoline
- Base for the formulation of nonirritating and, in particular, baby shampoos

MIRANOL 2MCAS MODIFIED:
CTFA Adopted Name: Cocoamphodiacetate (and) Sodium Lauryl Sulfate (and) Sodium Laureth Sulfate (and) Propylene Glycol

Typical Properties:
Appearance: clear, thin, light yellow liquid
Solids (including diol): 47.0%
Chloride (as NaCl): 6.3%
pH (30C): 7.8

Special Characteristics:
- Mixed lauryl sulfate and laureth-3 sulfate salts of an amphoteric surfactant derived from coconut imidazoline
- Base for the preparation of nonirritating shampoos

MIRANOL 2MCA-ESF:
CTFA Adopted Name: Cocoamphodipropionate (and) Sodium Lauryl Sulfate

Typical Properties:
Appearance: viscous, amber liquid
Solids: 34.0%
Chloride (as NaCl): 0.1%
pH: 7.7

Special Characteristics:
- Lauryl sulfate salt of an amphoteric surfactant derived from coconut imidazoline
- Similar to MIRANOL 2MCA MODIFIED, but has very low salt content

MIRANOL INC.: MIRANOL Amphoteric Anionic Salts(Continued):

MIRANOL 2MCT MODIFIED:
CTFA Adopted Name: Cocoamphodiacetate (and) Sodium Trideceth Sulfate (and) Hexylene Glycol

Typical Properties:
Appearance: clear, thin, yellow liquid
Solids (including diol): 50.0%
Chloride (as NaCl): 5.8%
pH (30C): 7.7

Special Characteristics:
- Trideceth-3 sulfate salt of an amphoteric surfactant derived from coconut imidazoline
- Particularly recommended for nonirritating and non eye-stinging cleansing and makeup removal products
- Can be used in clear liquid or lotionized products

MIRANOL 2MHT MODIFIED:
CTFA Adopted Name: Lauroamphodiacetate (and) Sodium Trideceth Sulfate (and) Hexylene Glycol

Typical Properties:
Appearance: clear, thin yellow liquid
Solids (including diol): 49.0%
Chloride (as NaCl): 6.5%
pH (20%, 30C): 8.2

Special Characteristics:
- Trideceth-3 sulfate salt of an amphoteric surfactant derived from lauric imidazoline
- Possesses ultimate mildness characteristics
- Recommended for nonirritating and non-eye stinging cleansing and make-up removal products
- Can be used in clear liquids or lotionized products

MIRANOL MHT:
CTFA Adopted Name: Lauroamphoacetate (and) Sodium Trideceth Sulfate

Typical Properties:
Appearance: clear, viscous, yellow liquid
Solids: 34.5%
Chloride (as NaCl): 3.4%
pH (25C): 9.8

Special Characteristics:
- Trideceth-3 sulfate salt of an amphoteric surfactant derived from lauric imidazoline
- Recommended as the base surfactant in skin cleansers and in baby shampoos
- Nonirritating

MIRANOL INC.: MIRANOL Amphoteric Anionic Salts(Ccntinued):

MIRANOL BT:
CTFA Adopted Name: Lauroamphodiacetate (and) Sodium Trideceth Sulfate

Typical Properties:
Appearance: clear, slightly viscous, amber liquid
Solids: 37.5%
Chloride (as NaCl): 4.0%
pH (25C): 9.2

Special Characteristics:
- Trideceth-3 sulfate salt of an amphoteric surfactant derived from lauric/myristic imidazoline
- Used in applications which require a high foaming surfactant
- Recommended as the base surfactant in skin cleansers and nonirritating shampoos

MIRANOL INC.: MIRANOL Amphoteric Surfactants (Carboxylates):

This series of surface active agents comprises aqueous solutions of mono- and di-carboxymethyl derivatives as well as salt-free mono- and di-carboxyethyl derivatives of 1-hydroxyethyl-2-alkylimidazolines. As a class, these materials are mild to the skin and eyes, nontoxic and biodegradable. They are compatible with all types of surfactants, tolerate high levels of electrolytes, and perform well in the presence of greasy soil.

MIRANOL CM CONC. N.P.:
CTFA Adopted Name: Cocoamphoacetate
CAS No. 68608-65-1

Typical Properties:
Appearance: clear, viscous light amber liquid
Solids: 44.0%
Chloride (as NaCl): 7.0%
pH (30C): 9.2
Solubility:
water: soluble
alcohol: insoluble
nonpolar solvents: insoluble

Special Characteristics:
- Derived form coconut imidazoline
- Mild foaming and cleansing agent
- Can reduce the overall irritation of products in which it is included

MIRANOL CM-SF CONC.:
CTFA Adopted Name: Cocoamphopropionate
CAS No. 68919-41-5

Typical Properties:
Appearance: clear, thin, light amber liquid
Solids: 37.0%
Chloride (as NaCl): 0.02%
pH (30C): 10.0
Solubility:
water: soluble
alcohol: soluble
nonpolar solvents: insoluble

Special Characteristics:
- Essentially free of sodium chloride
- Derived from coconut imidazoline
- Mild foaming and cleansing agent
- Tolerates high levels of electrolytes

MIRANOL INC.: MIRANOL Amphoteric Surfactants(Carboxylates) (Continued):

MIRANOL C2M Conc.:
CTFA Adopted Name: Cocoamphodiacetate
CAS No. 68647-53-0

Typical Properties:
Appearance: viscous, yellow liquid
Solids: 50.0%
Chloride (as NaCl): 11.5%
pH (20%, 30C): 8.2
Solubility:
water: soluble
alcohol: insoluble
nonpolar solvents: insoluble

Special Characteristics:
- Derived from coconut imidazoline
- Available in two grades:
  Old Process (O.P.)--may require heating before use.
  New Process (N.P.)--remains clear when stored at temperatures above 15C
- Mild fomaing and cleansing agent
- Acts as emulsifier, solubilizer and emulsion stabilizer
- Reduces irritant effects of other ingredients (e.g. alcohol sulfates and alcohol ether sulfates)
- Recommended for nonirritating shampoos, skin cleansers, make-up removers and pharmaceuticals

Note: Also available with propylene glycol present for ease of handling.

MIRANOL C2M-SF CONC. and MIRANOL C2M-SF 70%:
CTFA Adopted Name: Cocoamphodipropionate
CAS No. 68910-41-5
Typical Properties:

	MIRANOL C2M-SF Conc.:	MIRANOL C2M-SF 70%:
Appearance:	clear, thin amber liquid	viscous, amber paste
Solids:	39.0%	71.0%
pH (25C):	9.6 (as is)	9.5 (10%)
Solubility:		
water:	soluble	soluble
alcohol:	soluble	soluble
nonpolar solvents:	insoluble	insoluble

Special Characteristics:
- Essentially free of sodium chloride
- Derived from coconut imidazoline
- Mild foaming and cleansing agent
- Acts as a coupler for many cationic conditioning agents
- Deposits some cationic conditioners onto hair more effectively than other surfactants

Note: MIRANOL C2M-SF 70% has the same general properties as MIRANOL C2M-SF CONC., but, due to its low water content, is also recommended for use in bar soaps.

MIRANOL INC.: MIRANOL Amphoteric Surfactants(Carboxylates) (Continued):

MIRANOL HM CONC.:
CTFA Adopted Name: Lauroamphoacetate
CAS No. 68608-66-2

Typical Properties:
Appearance: clear, viscous, light amber liquid
Solids: 43.5%
Chloride (as NaCl): 7.3%
pH (30C): 9.2
Solubility:
water: soluble
alcohol: insoluble
nonpolar solvents: insoluble

Special Characteristics:
- Derived from lauric imidazoline prepared from a 99% pure grade of lauric acid
- Mild foaming and cleansing agent
- Similar to MIRANOL CM CONC. N.P. but is a slightly better foaming agent

MIRANOL H2M CONC.:
CTFA Adopted Name: Lauroamphodiacetate
CAS No. 68608-66-2

Typical Properties:
Appearance: clear, viscous, yellow liquid
Solids: 50.0%
Chloride (as NaCl): 11.7%
pH (20%, 30C): 8.2
Solubility:
water: soluble
alcohol: insoluble
nonpolar solvents: insoluble

Special Characteristics:
- Derived from lauric imidazoline prepared from a 99% pure grade of lauric acid
- Mild foaming and cleansing agent
- Similar to MIRANOL C2M CONC. but is a slightly better foaming agent
- Foams especially well in the presence of greasy soil

MIRANOL INC.: MIRANOL Amphoteric Surfactants(Carboxylates) (Continued):

MIRANOL H2M-SF CONC. and MIRANOL H2M-SF 70%:
CTFA Adopted Name: Lauroamphodipropionate
CAS No. 68610-43-5

Typical Properties:
MIRANOL H2M-SF CONC.:
Appearance: clear, thin, yellow liquid
Solids: 39.0%
pH: 9.5 (as is, 30C)
Solubility:
water: soluble
alcohol: soluble
nonpolar solvents: insoluble

MIRANOL H2M-SF 70%:
Appearance: viscous, amber paste
Solids: 71.0%
pH: 9.5 (10%, 25C)
Solubility:
water: soluble
alcohol: soluble
nonpolar solvents: insoluble

Special Characteristics:
- Essentially free of sodium chloride
- Derived from lauric imidazoline prepared from a 99% pure grade of lauric acid
- Mild foaming and cleansing agent
- Tolerates high levels of electrolytes
- Similar in its applications to MIRANOL C2M-SF CONC.
- Suitable for particularly sensitive cosmetic formulations where a pure lauric derivative is desired

Note: MIRANOL H2M-SF 70% has the same general properties as MIRANOL H2M-SF CONC., but, in addition, permits the formulation of high-active products such as concentrated shampoos, foam baths, and soap bars, where low water content is essential.

MIRANOL INC.: MIRANOL Amphoteric Surfactants(Carboxylates) (Continued):

MIRANOL BM CONC.:
CTFA Adopted Name: Lauroamphodiacetate

Typical Properties:
Appearance: clear, viscous, light amber liquid
Solids: 38.0%
Chloride (as NaCl): 7.5%
pH (25C): 9.0
Solubility:
water: soluble
alcohol: insoluble
nonpolar solvents: insoluble

Special Characteristics:
- Derived from a fatty imidazoline which has been prepared from a blend of lauric and myristic acids
- Mild foaming and cleansing agent
- Designed specifically as a base for the formulation of nonirritant shampoos

MIRANOL S2M CONC.:
CTFA Adopted Name: Caproamphodiacetate
CAS No. 70750-05-9

Typical Properties:
Appearance: clear, thin, yellow liquid
Solids: 50.5%
Chloride (as NaCl): 12.5%
pH (20%, 30C): 8.2

Special Characteristics:
- Derived from capric imidazoline which has been prepared from a 99% pure grade of capric acid
- Completely compatible with fatty acid soap
- Mild surfactant producing a high but open foam
- Can be used in shampoos in combination with MIRANOL C2M CONC. to improve the flash foam
- Improves hard water resistance of products containing soap

MIRANOL INC.: MIRANOL Amphoteric Surfactants(Carboxylates) (Continued):

MIRANOL SM CONC.:
CTFA Adopted Name: Caproamphoacetate
CAS No. 68608-61-7

Typical Properties:
Appearance: clear, thin, yellow liquid
Solids: 43.0%
Chloride (as NaCl): 8.0%
pH (30C): 9.0

Special Characteristics:
- Derived from capric imidazoline prepared from a 99% pure grade of capric acid
- High foaming, mild surfactant
- Completely compatible with soap
- Prevents the deposition of a dull film on the hair
- In liquid soaps which are partly based on fatty acid soap, it enhances the foaming ability and does not interfere with the superior lubricity of the soap

MIRANOL S2M-SF CONC.:
CTFA Adopted Name: Caproamphodipropionate
CAS No. 68815-45-2

Typical Properties:
Appearance: clear, thin, yellow liquid
Solids: 39.0%
pH (30C): 9.0

Special Characteristics:
- Essentially free of sodium chloride
- Derived from capric imidazoline prepared from a 99% pure grade of capric acid
- Completely compatible with fatty acid soap
- Mild surfactant producing a high but open foam
- Used in shampoos to improve the flash foam
- In formulations containing soap it improves hard water resistance

Note: Formulations containing MIRANOL SM CONC. are also applicable to MIRANOL S2M-SF CONC.

MIRANOL INC.: MIRANOL Amphoteric Surfactants(Carboxylates) (Continued):

MIRANOL DM and MIRANOL DM CONC. 45%:
CTFA Adopted Name: Stearoamphoacetate
CAS No.: 68608-63-9

Typical Properties:
MIRANOL DM:
Appearance: thick, white paste
Solids: 26.0%
Chloride (as NaCl): 5.3%
pH: 5.6 (as is, 65C)

MIRANOL DM CONC. 45%:
Appearance: thick, white paste
Solids: 45.0%
Chloride (as NaCl): 9.5%
pH: 5.5 (50%, 65C)

Special Characteristics:
- Derived from stearic imidazoline
- Supplied in the form of a creamy white paste which is readily pourable at 60C
- Used as a conditioning agent in formulations requiring a material which is not irritating to skin or eyes
- Has antistatic properties and produces outstanding softening effects on the skin and hair
- Suggested for the formulation of creme rinses

MIRANOL INC.: MIRANOL Amphoteric Surfactants(Sulfonates):

The MIRANOL Amphoteric Sulfonates are analogous to the MIRANOL Amphoteric Carboxylates, not only in their method of preparation, but also in the fact that they too possess most of the outstanding properties of the carboxylates. Thus, these materials are mild to the skin and eyes, are nontoxic and biodegradable. They are compatible with all types of surfactants, tolerate high levels of electrolytes and are stable over a wide pH range. In addition, these products offer excellent wetting, greater solubilizing power than the carboxylates, as well as lime soap dispersing properties.

MIRANOL CS CONC.:
CTFA Adopted Name: Cocoamphohydroxypropylsulfonate
CAS No. 68604-73-9

Typical Properties:
Appearance: thin, yellow liquid
Solids: 45.0%
Chloride (as NaCl): 7.2%
pH (30C): 8.0

Special Characteristics:
- Derived from coconut imidazoline
- Mild foaming and cleansing agent
- Has excellent solubility, solubilizing and lime soap dispersing properties
- Recommended for use in high-foaming shampoos and foam baths
- Reduces the irritation of other, more irritating surfactants

MIRANOL OS-D:
CTFA Adopted Name: Oleoamphohydroxypropylsulfonate
CAS No. 68610-38-8

Typical Properties:
Appearance: slightly hazy, viscous, yellow liquid
Solids: 25.0%
Chloride (as NaCl): 3.7%
pH (30C): 9.2

Special Characteristics:
- Derived from oleic imidazoline
- Mild foaming and cleansing agent
- Has excellent solubility and solubilizing properties
- Viscosity builder and foam enhancer in formulations employing other surfactants (e.g. MIRANOL C2M CONC. and MIRANOL CS CONC.)
- Especially useful for foam baths and shampoos

MIRANOL INC.: MIRAPOL Cationic Polymers:

The MIRAPOLS are a group of polymeric polyquaternary ammonium chlorides designated for use as conditioning agents in skin and hair care products. Due to their cationic nature and polymeric structure, these materials are particularly substantive. At the same time they are readily water soluble, which allows for ease of formulation and prevents build-up, thus overcoming two major problems of most other products of this type.

MIRAPOL A-15:
CTFA Adopted Name: Polyquaternium-2
CAS No. 68555-36-2

Typical Properties:
Appearance: viscous, yellow liquid
Solids: 64.0%
Water: 36.0%
Chloride (as Cl-): 12.1%
pH (25C): 8.5
Color (Gardner 1933): 8
av. mol. wt.: 2260

Special Characteristics:
- Readily dissolves in water in all proportions
- In shampoos and hair conditioners if confers good wet-combing, dry-combing, and produces a soft natural feel
- Reduces static "fly-a-way"
- Provides an emollient effect and has substantivity to skin surfaces

Note: In formulating with MIRAPOL A-15, it is necessary to dissolve the MIRAPOL A-15 in about 75% of the required water. The amphoteric surfactant is then added, followed by the anionic and the remaining ingredients. Use levels of MIRAPOL A-15 should be in the range of 0.6-2.5%.

MIRAPOL AD-1 (Patent Pending):
CTFA Adopted Name: Polyquaternium-17
CAS No.: 90624-75-2

Typical Properties:
Appearance: viscous, amber liquid
Solids: 62.0%
Chloride (as Cl-): 9.5%
Water: 38.0%
Color (Gardner 1933): 6.0
pH (10%, 25C): 8.0
av. mol. wt.: 50,000

Special Characteristics:
- Provides superior conditioning to cosmetic products intended for use on the hair and skin
- Enables the achievement of de-tangling and anti-static effects
- Does not build up with repeated use
- Can be used without filtration in all types of anionic or amphoteric systems or blends
- Does not require special formulating techniques
- Compatible with electrolytes and is unaffected by the pH of the formulation.

MIRANOL INC.: MIRAPOL Cationic Polymers(Continued):

MIRAPOL AZ-1 (Patent Pending):
CTFA Adopted Name: Polyquaternium-18
CAS No. 90624-76-3

Typical Properties:
Appearance: viscous, amber liquid
Solids: 62.0%
Chloride (as Cl-): 8.5%
Water: 38.0%
Color (Gardner 1933): 6.0
pH (10%, 25C): 8.0
av. mol. wt.: 50,000

Note: MIRAPOL AZ-1 is closely-related to MIRAPOL AD-1 and, in many instances, the two may be used interchangeably. It has the unique property of remaining clear when anionic systems containing it are diluted with water.

MIRAPOL 9, MIRAPOL 95, MIRAPOL 175 (Patent Pending):
CTFA Adopted Name: Polyquaternium-27

These products are all polycationic block copolymers containing the structural elements of MIRAPOL AD-1 and MIRAPOL A-15.

Typical Properties:
MIRAPOL 9:
Active: 62.0%
Chloride (as Cl-): 9.8%
Water: 38.0%
pH (10%, 25C): 8.0
av. mol. wt.: 20,000

MIRAPOL 95:
Active: 62.0%
Chloride (as Cl-): 9.6%
Water: 38.0%
pH (10%, 25C): 8.0
av. mol. wt.: 20,000

MIRAPOL 175:
Active: 62.0%
Chloride (as Cl-): 10.1%
Water: 38.0%
pH (10%, 25C): 8.0
av. mol. wt.: 20,000

Special Characteristics:
- Intermediate in substantivity between MIRAPOL AD-1 and MIRAPOL A-15
- Compatible with anionic and nonionic detergents
- Compatible with electrolytes
- Require formulating procedure described for MIRAPOL A-15

MIRANOL INC.: MIRATAINE Aminopropionates:

The MIRATAINE aminopropionates are a series of aqueous solutions of amphoteric surfactants which are anionic at alkaline pH, cationic at acid pH and zwitterionic in their isoelectric range. They are compatible with all surfactants.

Although materials of this type are not frequently used in cosmetics and personal care products, these are included because there are some applications in which they confer desirable results.

MIRATAINE H2C:
CTFA Adopted Name: Disodium Lauriminodipropionate
CAS No. 3655-00-3
Typical Properties:
Appearance: thin, yellow liquid
Solids: 30.0%
pH (25C): 10.5

MIRATAINE H2C-HA:
CTFA Adopted Name: Sodium Lauriminodipropionate
CAS No. 3546-96-1
Typical Properties:
Appearance: thin, yellow liquid
Solids: 30.0%
pH (25C): 7.0
Special Characteristics:
- Partial salt, with one carboxylic acid group available for neutralization
- Mild surfactant which is substantive to skin and hair
- Recommended for conditioning shampoos, liquid soaps and waterless hand cleaners

MIRATAINE T2C:
CTFA Adopted Name: Disodium Tallowiminodipropionate
CAS No. 61791-56-8
Typical Properties:
Appearance: thin, yellow liquid
Solids: 30.0%
pH (25C): 11.5

MIRATAINE XL:
CTFA Adopted Name: DEA-Lauryl Sulfate (and) DEA-Lauraminopropionate (and) Sodium Lauraminopropionate (and) Propylene Glycol
Typical Properties:
Appearance: clear, thin, yellow liquid
Solids: 39.0%
Chloride (as NaCl): 0.5%
pH (25C): 8.0
Special Characteristics:
- High foaming surfactant which is not defoamed by the presence of oil or greasy soil
- Substantive to hair and provides a pleasant feel without a "drying-out" effect
- Used in shampoos as the main surfactant

MIRANOL INC.: MIRATAINE Betaines and Sultaines:

This group of products consists of aqueous solutions of alkyl- and amidopropylbetaines and hydroxypropyl sultaines. These materials are mild and are compatible with most other surfactants. With the exception of the tallow-based products, they exhibit excellent foaming properties over a wide pH range.

MIRATAINE CB:
CTFA Adopted Name: Cocamidopropyl Betaine
CAS No. 70851-07-9

Typical Properties:
Appearance: clear, thin, yellow liquid
Solids: 35.0%
Chloride (as NaCl): 5.0%
pH (25C): 8.5

Special Characteristics:
- Derived from hydrogenated coconut fatty acids
- Has excellent foaming properties alone, or in combination with such surfactants as sodium lauryl sulfate and sodium lauryl ether sulfate
- Compatible with many cationic hair conditioning agents, maintaining clear solutions in their presence
- Used as the base for many shampoos

MIRATAINE CBR:
CTFA Adopted Name: Cocamidopropyl Betaine
CAS No. 70851-07-9

Typical Properties:
Appearance: clear, thin, yellow liquid
Solids: 35.0%
Chloride (as NaCl): 5.0%
pH (25C): 5.0

Special Characteristics:
- Derived from hydrogenated coconut fatty acid
- Excellent foaming agent throughout the entire pH range
- Recommended for the formulation of shampoos, bubble baths and skin cleansers

MIRANOL INC.: MIRATAINE Betaines and Sultaines(Continued):

MIRATAINE CBC:
CTFA Adopted Name: Cocamidopropyl Betaine
CAS No. 70851-07-9

Typical Properties:
Appearance: clear, thin, yellow liquid
Solids: 35.0%
Chloride (as NaCl): 5.0%
pH (25C): 6.0

Special Characteristics:
- Derived from coconut oil
- Excellent foaming agent throughout the entire pH range
- Has lime soap dispersing properties and low irritancy
- Recommended for the formulation of shampoos and liquid soaps

MIRATAINE BB:
CTFA Adopted Name: Lauramidopropyl Betaine
CAS No. 86438-78-0

Typical Properties:
Appearance: clear, thin, yellow liquid
Solids: 35.0%
Chloride (as NaCl): 5.0%
pH: 8.5

Special Characteristics:
- Derived from mixed lauric and myristic acids
- Outstanding foaming agent throughout the entire pH range
- Recommended for use in shampoo formulations

MIRATAINE COB (U.S. 4,490,355):
CTFA Adopted Name: Coco/Oleamidopropyl Betaine
CAS No. 86438-79-1

Typical Properties:
Appearance: thin, yellow liquid
Solids: 34.0%
Chloride (as NaCl): 4.5%
pH (25C): 7.0

Special Characteristics:
- Derived from mixed coconut and oleic acids
- Outstanding foaming agent throughout the entire pH range
- Extremely effective foam enhancer and viscosity builder
- Offers conditioning properties and may be used to impart "after-feel" to shampoos and other personal care products
- Excellent solubilizer for natural oils, such as jojoba, enabling clear products to be obtained

MIRANOL INC.: MIRATAINE Betaines and Sultaines(Continued):

MIRATAINE ODMB-35:
CTFA Adopted Name: Oleyl Betaine
CAS No. 871-37-4
Typical Properties:
Appearance: Slightly hazy, viscous, yellow gel
Solids: 35.0%
Chloride (as NaCl): 5.0%
pH (10%, 25C): 7.0
Special Characteristics:
- Derived from an oleyl fatty amine
- Excellent foaming agent over a wide pH range in both hard and soft water
- Contributes a unique "after-feel" when used in shampoos and other personal care products
- Recommended for skin cleansers, liquid soaps and particularly for conditioning shampoos

MIRATAINE CBS:
CTFA Adopted Name: Cocamidopropyl Hydroxysultaine
CAS No. 70851-08-0
Typical Properties:
Appearance: clear, thin, yellow liquid
Solids: 50.0%
Chloride (as NaCl): 6.5%
pH (25C): 8.2
Special Characteristics:
- Derived from hydrogenated coconut fatty acids
- Foams profusely over a wide pH range in both hard and soft water
- Base for mild shampoos, high foaming bath products and liquid soaps
- Particularly suited for use with alkyl sulfates and alkyl ether sulfates
- Compatible with many cationic hair conditioning agents, maintaining clear solutions in their presence

MIRATAINE TM:
CTFA Adopted Name: Dihydroxy Ethyl Tallow Glycinate
CAS No. 61791-25-1
Typical Properties:
Appearance: clear to hazy viscous amber liquid
Solids: 40.0%
Chloride (as NaCl): 5.0%
pH (25C): 5.0
Special Characteristics:
- Derived from a tallow fatty amine
- Used where a mild conditioning agent is desired
- Compatible with anionic surfactants

Note: The compatibility of MIRATAINE TM with anionic surfactants was illustrated by the fact that solutions containing 1% active MIRATAINE TM and 6% active of the respective anionic, adjusted to pH 7.0 with hydrochloric acid, remained clear.

MIRANOL INC.: Miscellaneous Products:

In addition to the more traditional line of amphoterics and cationic polymers, MIRANOL offers a number of other products as well as several special blends for the cosmetics industry.

MIRANATE LEC:
CTFA Adopted Name: Sodium Laureth-13 Carboxylate
CAS No. 70632-06-3

Typical Properties:
Appearance: clear to slightly hazy gel
Solids: 70%
pH (10% solution): 8.0

Special Characteristics:
- Anionic surfactant possessing some non-ionic properties
- Compatible with cationic materials
- Provides improved lime soap dispersion
- Recommended for use in shampoos and other personal care products, particularly those containing a conditioning agent

MIRANATE LSS:
CTFA Adopted Name: Disodium Lauryl Sulfosuccinate
CAS No. 36409-57-1

Typical Properties:
Appearance: Creamy white solid
Solids: 38.0%
Sodium Sulfate: 0.5%
pH (10% solution): 7.0

Special Characteristics:
- Anionic surfactant noted for its high foaming and low irritation properties
- Suggested for use in shampoos, hand soaps and other personal care products
- Compatible with amphoteric surfactants

MIRANOL INC.: Miscellaneous Products(Continued):

COMPOUND MS-1 and COMPOUND MS-2:
CTFA Adopted Name: See Composition

Typical Properties:
MS-1:
Appearance: clear, yellow liquid
Solids: 40.0%
Chloride (as NaCl): 1.1%
Color (Gardner 1933): 2
pH: 8.5

MS-2:
Appearance: clear, yellow liquid
Solids: 40.0%
Chloride (as NaCl): 1.5%
Color (Gardner 1933): 2
pH: 8.5

Composition:
MS-1:
PEG-80 Sorbitan Laurate: 19.4
Sodium Trideceth Sulfate (70%): 17.2
PEG-150 Distearate: 5.0
Cocamidopropyl Hydroxysultaine: 5.2
Lauroamphodiacetate: 10.6
Sodium Laureth-13 Carboxylate: 2.0
Quaternium 15: 0.1
Water: 40.5

MS-2:
PEG-80 Sorbitan Laurate: 17.0
Sodium Trideceth Sulfate (70%): 15.0
PEG-150 Distearate: 6.5
Cocamidopropyl Hydroxysultaine: 11.6
Lauroamphodiacetate: 10.0
Sodium Laureth-13 Carboxylate: 2.0
Quaternium 15: 0.1
Water: 37.8

COMPOUND SBC:
CTFA Adopted Name: See Composition

Typical Properties:
Appearance: Thin yellow liquid
Solids: 34.0%
Chloride (as NaCl): 3.0%
Color (Gardner 1933): 2.0
pH (25C): 7.2

Special Characteristics:
- Concentrated
- For formulating mild shampoo
- Contains conditioners
- Simple formulating procedure

MIRANOL INC.: Miscellaneous Products(Continued):

CEDAPAL Products:

CEDEPAL TD407MF and CEDEPAL TD404M are Tridecylpolyoxyethylene (3.0) Sodium Sulfate (CAS No. 25446-78-0) with the CTFA designation of Sodium Trideceth Sulfate. They vary in their active content but in most formulas may be used interchangeably providing that compensation is made for differences in active content.

Typical Properties:

CEDEPAL TD 407MF:

Active: 74.0%
Sodium Sulfate: 0.9%
pH (25C): 9.0

CEDEPAL TD 404M:

Active: 34.0%
Sodium Sulfate: 0.4%
pH (25C): 8.0

Special Characteristics:

- High foaming anionic surfactants
- Recommended for use in mild baby shampoos and baby baths
- Compatible with amphoterics
- Suggested for use in liquid hand soaps, bubble baths and personal care products

CEDEPON LS30PM:

CEDEPON LS30PM is an aqueous solution of Sodium Lauryl Sulfate (CAS No. 151-21-3) especially low in free fatty alcohol and salt content. As an anionic high foaming material, CEDEPON LS30PM is suggested for use in shampoos, hand soaps, bubble baths and other cosmetic specialties.

Typical Properties:

Appearance: Colorless to pale straw liquid
Active: 30.0%
Inorganic Sulfate: 0.6%
pH: 8.2
Unsulfated: 0.8%

MIRANOL ESTER PO-LM4(Patent Pending):

CTFA Adopted Name: Polypentaerythrityl Tetralaurate

Typical Properties:

Appearance: Off-white viscous liquid to semi-solid
Solids: 100%
Acid Value: 2
Saponification value: 190

Special Characteristics:

- Functions as emollient, moisturizer and secondary emulsifier
- Recommended for use in skin and hair care products
- Leaves skin smooth and supple with no greasy or sticky "after-feel"
- Noncomedogenic
- Can replace wholly, or in part, occlusive emollients

MONA INDUSTRIES, INC.: MONAMATES:

The MONAMATE series of sulfosuccinate surfactants has been designed specifically for the personal care and household areas. The unusually low eye and skin irritation, excellent lathering properties, and controlled cleaning action enable these products to offer benefits unobtainable using traditional anionic surfactants.

General Functional Characteristics:

- Low skin and eye irritation properties.
- Excellent foaming properties ranging from high flash foam with open bubble structure to rich, long lasting dense lather.
- Soap-like feel at low pH.
- Talc-like after-feel.
- Outstanding cleaning properties on skin and hair without overreduction of natural epidermal lipids.

<u>Product Name:</u>

MONAMATE LNT-40:
CTFA Designation: Ammonium Lauryl Sulfosuccinate
% Active: 40
Physical Form: Liquid

MONAMATE OPA-30:
CTFA Designation: Disodium Oleamido PEG-2 Sulfosuccinate
% Active: 30
Physical Form: Liquid

MONAMATE C-1142:
CTFA Designation: Disodium Cocamido MIPA Sulfosuccinate
% Active: 40
Physical Form: Liquid

MONAMATE CPA-40:
CTFA Designation: Disodium Cocamido MIPA Sulfosuccinate
% Active: 40
Physical Form: Liquid

MONAMATE LA-100:
CTFA Designation: Disodium Lauryl Sulfosuccinate
% Active: 100
Physical Form: Powder

MONAMATE OPA-100:
CTFA Designation: Disodium Oleamido PEG-2 Sulfosuccinate
% Active: 100
Physical Form: Powder

MONAMATE CPA-100:
CTFA Designation: Disodium Cocamido MIPA Sulfosuccinate
% Active: 100
Physical Form: Powder

MONA INDUSTRIES, INC.: MONAMATES (Continued):

MONAMATE CPA-40%:

CTFA Designation: Disodium Cocamido MIPA Sulfosuccinate
Chemical Description: Sulfosuccinate half ester of an alkanolamide, di-sodium salt.

Use Characteristics:

1. Low order of eye and skin irritation at typical use levels.
2. High foamer, producing a dense, rich lather.
3. Develops a broad range of viscosities with alcohol sulfates and alkanolamides for use in mild shampoos, bubble baths and skin cleansers.
4. Imparts a soft and silky feel to the skin and hair.
5. It dries to a crisp friable white powder which makes it an ideal ingredient for rug shampoos and spray dried bubble bath or detergent formulations.
6. Being a liquid it can be easily handled and formulated into a broad range of products.

Typical Properties:
Chemical Description: Sulfosuccinate half ester of an alkanolamide, disodium salt
Color: 6 Max.
Appearance: Clear to Hazy
pH (as is): 6
% Moisture (Karl Fischer): 60
% Activity (By difference): 40

MONAMATE LA-100:
MONAMATE LA-100 is a highly active, powdered anionic surfactant. Its high foaming, non-irritating, soap-like feel, and talc-like after feel properties make it an excellent choice for personal care formulations such as bubble baths, hand and body liquid soaps, shampoos, and shave creams. Being a powder, it is especially useful in soap or detergent bars and powdered formulations, and its formulation dry-down properties coupled with high detergency create excellent rug shampoos.

CTFA Designation: Disodium Lauryl Sulfosuccinate

Tentative Specifications:
Physical Appearance: Fine White Powder
Ionic Nature: Anionic
Activity: 98.0% Min.
% Moisture: 2.0% Max.
Acid Number: 0-13
pH (10% Solution): 6.0-7.0

MONA INDUSTRIES, INC.: MONAMATES (Continued):

MONAMATE LNT-40:
Chemical & CTFA Description: Ammonium Lauryl Sulfosuccinate

Use Characteristics:
1) Produces extremely high flash foam and copious long-lasting lather.
2) Very low eye and skin irritation.
3) Rinses readily, leaving a soft talc-like feel.
4) Very light color and little or no odor.

Toxicity: MONAMATE LNT-40, when tested at 10.0% active, is not a primary skin irritant and is only minimally irritating to the eye.

Typical Properties:
Physical Appearance: Light colored liquid
CTFA Description: Ammonium Lauryl Sulfosuccinate
Ionic Nature: Anionic
Color (GVCS): 1-2 Max.
Total Solids: 40
pH (as is): 6.5
% Karl Fischer: 60
Specific Gravity: 1.07
Weight/Gallon: 8.88 lbs.

MONAMATE OPA-30:
Chemical Description: Sulfosuccinate half ester of an alkanolamide, di-sodium salt.

Use Characteristics:
1. Completely non-irritating to the eyes and skin when tested at 15.0% active, according to independent toxicological tests.
2. Produces excellent flash foam and a rich, dense lather.
3. Excellent rinsing properties.
4. Permits control of viscosity over a broad range with alcohol sulfates, amphoterics and alkanolamides.

Typical Chemical and Physical Properties:
Chemical Description: Sulfosuccinate half ester of alkanolamide, di-sodium salt.
CTFA Description: Disodium Oleamido PEG-2 Sulfosuccinate
Physical Appearance: Light Yellow Liquid
Ionic Nature: Anionic
% Solids: 30.0%
pH (as is): 5.6
Specific Gravity: 1.06
Pounds/Gallon: 8.85

MONA INDUSTRIES, INC.: MONAMINES & MONAMIDS:

The MONAMINES and MONAMIDS, a series of surfactants with many functional properties, find use in a wide variety of consumer and industrial products, such as:

- Cosmetics
- Bubble Baths
- Shampoos
- Dry Cleaning Detergents
- Dishwashing Detergents
- Metal Cleaners
- Hand Soaps
- Waterless Hand Cleaners
- Textile Detergents
- Fiber & Hair Conditioners
- Leather and Fur Preparations
- Lubricants
- Rust Inhibitors
- Rug Shampoos
- Metal Cutting Fluids
- Dustless Soap Powders
- Metal Polishes
- Emulsifiable Waxes
- Pigment Dispersions
- Agricultural Sprays
- Fuel Oil Additives, etc.

Functional Properties:

The following functional characteristics of the MONAMINES and MONAMIDS make them outstanding:

- Foam boosters and stabilizers
- Emulsifiers
- Detergents
- Wetting agents
- Corrosion inhibitors
- Viscosity Builders
- Lubricants
- Dispersants

Basic Chemistry:

The MONAMINES and MONAMIDS are basically fatty acid-alkanolamides derived from alkanolamines, such as monoethanolamine, monoispropanolamine and diethanolamine.

Their surface active properties originate from the balance between the lipophilic fatty acid radicals and the hydrophilic hydroxyalkyl groups on the substituted nitrogen atom in the molecules.

MONAMINES:

MONAMINES are essentially derived from a 2:1 molar reaction of diethanolamine and various chain-length fatty acids.

MONAMINES: 1:2 FA-Diethanolamides:

Product:

AA-100:

Fatty Acid: Distilled Coconut
Physical State @ 20C: Liquid
Ionic Nature: N-A
Lbs./Gal.: 8.30
Sp. Gr. @ 20C: 1.00
pH 10% Sol.: 9.5-10.5
% Active: 100
Max. Color: Gvcs-33: 7
Acid No.: 28-32
Alkali No.: 165-185

MONA INDUSTRIES, INC.: MONAMINES & MONAMIDS(Continued):

MONAMINES(Continued):

Product:

AC-100:
Fatty Acid: Mixed
Physical State @ 20C: Liquid
Ionic Nature: N-A
Sp.Gr. @ 20C: 1.00
Lbs./Gal.: 8.30
pH 10% Sol.: 9.5-10.5
% Active: 100
Max. Color Gvcs-33: 13
Acid No.: 22-32
Alkali No.: 170-190

ACO-100:
Fatty Acid: Lauric
Physical State @ 20C: Paste
Ionic Nature: N-A
Sp. Gr. @ 20C: 1.01
Lbs./Gal.: 8.40
pH 10% Sol.: 9.5-10.5
% Active: 100
Max. Color Gvcs-33: 5
Acid No.: 10-14
Alkali No.: 180-200

AD-100:
Fatty Acid: Coconut
Physical State @ 20C: Liquid
Ionic Nature: N
Sp. Gr. @ 20C: 1.02
Lbs./Gal.: 8.50
pH 10% Sol.: 9.5-10.5
% Active: 100
Max. Color Gvcs.-33: 11
Acid No.: 2-8
Alkali No.: 105-125

ADD-100:
Fatty Acid: Coconut
Physical State @ 20C: Liquid
Ionic Nature: N
Sp. Gr. @ 20C: 1.02
Lbs./Gal.: 8.50
pH 10% Sol.: 9.5-10.5
% Active: 100
Max. Color Gvcs-33: 3
Acid No.: 2-6
Alkali No.: 105-125

MONA INDUSTRIES, INC.: MONAMINES and MONAMIDS(Continued):

MONAMINES(Continued):

Product:

ADS-100:
Fatty Acid: Mixed
Physical State @ 20C: Liquid
Ionic Nature: N-A
Sp. Gr. @ 20C: 1.00
Lbs./Gal.: 8.30
pH 10% Sol.: 9.0-10.0
% Active: 100
Max. Color Gvcs-33: 11
Acid No.: 48-52
Alkali Number: 110-125

ADY-100:
Fatty Acid: Mixed
Physical State @ 20C: Liquid
Ionic Nature: N
Sp. Gr. @ 20C: 0.98
Lbs./Gal.: 8.20
pH 10% Sol.: 10.5-11.5
% Active: 100
Max. Color Gvcs-33: 9
Acid No.: 0-2
Alkali No.: 110-130

ALX-80SS:
Fatty Acid: Modified Coconut
Physical State @ 20C: Liquid
Ionic Nature: N-A
Sp. Gr. @ 20C: 1.05
Lbs./Gal.: 8.75
pH 10% Sol.: 8.5-9.5
% Active: 80
Max. Color Gvcs-33: 6
Acid No.: 52-60
Alkali No.: 65-75

ALX-100S:
Fatty Acid: Modified Coconut
Physical State @ 20C: Liquid
Ionic Nature: N-A
Sp. Gr. @ 20C: 1.06
Lbs./Gal.: 8.85
pH 10% Sol.: 8.5-9.5
% Active: 100
Max. Color Gvcs.-33: 8
Acid No.: 62-70
Alkali No.: 70-90

MONA INDUSTRIES, INC.: MONAMINES and MONAMIMIDS(Continued):

MONAMINES:

Product:

CF-100M:
Fatty Acid: Mixed
Physical State @ 20C: Liquid
Ionic Nature: N-A
Sp. Gr. @ 20C: 1.01
Lbs/Gal.: 8.40
pH 10% Sol.: 8.5-9.5
% Active: 100
Max. Color Gvcs-33: 11
Acid No.: 56-64
Alkali No.: 110-120

LM-100:
Fatty Acid: 70/30 Lauric Myristic
Physical State @ 20C: Liquid
Ionic Nature: N-A
Sp. Gr. @ 20C: 1.02
Lbs./Gal.: 8.50
pH 10% Sol.: 9.5-10.5
% Active: 100
Max. Color Gvcs-33: 5
Acid No.: 18-23
Alkali No.: 160-175

T-100:
Fatty Acid: Tall Oil
Physical State @ 20C: Liquid
Ionic Nature: N-A
Sp. Gr. @ 20C: 0.97
Lbs./Gal.: 8.10
pH 10% Sol.: 10.0-11.0
% Active: 100
Max. Color Gvcs-33: 10
Acid No.: 10-16
Alkali No.: 100-120

I-76:
Fatty Acid: Coconut
Physical State @ 20C: Liquid
Ionic Nature: N-A
Sp. Gr. @ 20C: 1.00
Lbs./Gal.: 8.30
pH 10% Sol.: 8.5-9.5
% Active : 100
Max. Color Gvcs-33: 12
Acid No.: 45-55
Alkali No.: 80-100

MONA INDUSTRIES, INC.: MONAMINES and MONAMIDS(Continued):

MONAMINES(Continued):

Product:

R8-26:
Fatty Acid: Mixed
Physical State @ 20C: Liquid
Ionic Nature: N-A
Sp. Gr. @ 20C: 0.99
Lbs./Gal.: 8.25
pH 10% Sol.: 9.0-10.0
% Active: 100
Max. Color Gvcs-33: 11
Acid No.: 75-85
Alkali No.: 177-187

MONAMIDS:
1:1 FA-Diethanolamides:
150-AD:
Fatty Acid: Coconut
Physical State @ 20C: Liquid
Ionic Nature: N
Sp. Gr. @ 20C: 0.99
Lbs./Gal.: 8.25
pH 10% Sol.: 9.8-10.8
% Active: 100
Max. Color Gvcs-33: 11
Acid No.: 0-3
Alkali No.: 55-70

150-ADD:
Fatty Acid: Coconut
Physical State @ 20C: Liquid
Ionic Nature: N
Sp. Gr. @ 20C: 1.00
Lbs./Gal.: 8.30
pH 10% Sol.: 10-11
% Active: 100
Max. Color Gvcs-33: 4
Acid No.: 0-3
Alkali No.: 58-68

150-ADY:
Fatty Acid: Mixed
Physical State @ 20C: Liquid
Ionic Nature: N
Sp. Gr. @ 20C: 0.97
Lbs./Gal.: 8.10
pH 10% Sol.: 10-11
% Active: 100
Max. Color Gvcs-33: 10
Acid No.: 0-1
Alkali No.: 30-45

MONA INDUSTRIES, INC.: MONAMINES and MONAMIDS(Continued):

MONAMIDS(Continued):

Product:

150-DR:
Fatty Acid: Coconut
Physical State @ 20C: Liquid
Ionic Nature: N
Sp. Gr. @ 20C: 0.98
Lbs./Gal.: 8.20
pH 10% Sol.: 9.0-10
% Active: 100
Max. Color Gvcs-33: 5
Acid No.: 0-5
Alkali No.: 10-25

150-LMW-C:
Fatty Acid: 70/30 Lauric Myristic
Physical State @ 20C: Solid
Ionic Nature: N
Sp. Gr. @ 20C: 0.98 @ 40C
Lbs/Gal.: 8.20
pH 10% Sol.: 10.2-11.2
% Active: 100
Max. Color Gvcs-33: 3
Acid No.: 0-1
Alkali No.: 30-45

150-LW:
Fatty Acid: Lauric
Physical State @ 20C: Solid
Ionic Nature: N
Sp. Gr. @ 20C: 0.98 @ 40C
Lbs./Gal.: 8.20
pH 10% Sol.: 10-11
% Active: 100
Max. Color Gvcs-33: 3
Acid No.: 0-1
Alkali No.: 30-45

150-LWA:
Fatty Acid: Lauric
Physical State @ 20C: Solid
Ionic Nature: N
Sp. Gr. @ 20C: 0.98 @ 40C
Lbs./Gal.: 8.20
pH 10% Sol.: 9.5-10.5
% Active: 100
Max. Color Gvcs-33: 3
Acid No.: 0-1
Alkali No.: 10-25

MONA INDUSTRIES, INC.: MONAMINES and MONAMIDS(Continued):

MONAMIDS(Continued):

Product:

150-MW:
Fatty Acid: Myristic
Physical State @ 20C: Solid
Ionic Nature: N
Sp. Gr. @ 20C: 0.98 @ 45C
Lbs./Gal.: 8.20
pH 10% Sol.: 9.5-10.5
% Active: 100
Max. Color Gvcs-33: 2
Acid No.: 0-3
Alkali No.: 35-50

150-IS:
Fatty Acid: Iso-Stearic
Physical State @ 20C: Liquid
Ionic Nature: N
Sp. Gr. @ 20C: 0.96
Lbs./Gal.: 8.00
pH 10% Sol.: 8.8-9.8
% Active: 100
Max. Color Gvcs-33: 6
Acid No.: 5-10
Alkali No.: 30-60

150-CW:
Fatty Acid: Capric
Physical State @ 20C: Liquid
Ionic Nature: N
Sp. Gr. @ 20C: 0.99
Lbs./Gal.: 8.25
pH 10% Sol.: 10.3-11.3
% Active: 100
Max. Color Gvcs-33: 7
Acid No.: 0-2
Alkali No.: 40-55

716:
Fatty Acid: Modified Lauric
Physical State @ 20C: Liquid
Ionic Nature: N
Sp. Gr. @ 20C: 0.98
Lbs./Gal.: 8.20
pH 10% Sol.: 10.0-11.0
% Active: 100
Max. Color Gvcs-33: 5
Acid No.: 0-3
Alkali No.: 45-60

MONA INDUSTRIES, INC.: MONAMINES and MONAMIDS(Continued):

MONAMIDS(Continued):

Product:

718:
Fatty Acid: Stearic
Physical State @ 20C: Solid
Ionic Nature: N
Sp.Gr. @ 20C: 0.96 @ 45C
Lbs./Gal.: 8.00
pH 10% Sol.: 9.3-10.3
% Active: 100
Max. Color Gvcs-33: 4
Acid No.: 21±3
Alkali No.: 45-65

770:
Fatty Acid: Modified Coconut
Physical State @ 20C: Liquid
Ionic Nature: N
Sp.Gr. @ 20C: 0.99
Lbs./Gal.: 8.25
pH 10% Sol.: 9.2-10.2
% Active: 85
Max. Color Gvcs-33: 4
Acid No.: 0-1
Alkali No.: 35-45

7-100:
Fatty Acid: Coconut
Physical State @ 20C: Liquid
Ionic Nature: N
Sp.Gr. @ 20C: 0.96
Lbs./Gal.: 8.00
pH 10% Sol: 8-9
% Active: 100
Max. Color Gvcs.-33: 6
Acid No.: 0-2
Alkali No.: 5-20

7-153CS:
Fatty Acid: Modified Coconut
Physical State @ 20C: Liquid
Ionic Nature: N-A
Sp.Gr. @ 20C: 1.05
Lbs./Gal.: 8.75
pH 10% Sol.: 8.5-9.5
% Active: 100
Max. Color Gvcs.-33: 3
Acid No.: 0-2
Alkali No.: 4-10

MONA INDUSTRIES, INC.: MONAMINES and MONAMIDS(Continued):

MONAMIDS(Continued):

Product:

15-70W:
Fatty Acid: Linoleic
Physical State @ 20C: Liquid
Ionic Nature: N
Sp.Gr. @ 20C: 0.96
Lbs./Gal.: 8.00
pH 10% Sol.: 10-11
% Active: 100
Max. Color Gvcs-33: 11
Acid No.: 0-1
Alkali No.: 25-40

R31-42:
Fatty Acid: 70/30 Lauric Myristic
Physical State @ 20C: Liquid
Ionic Nature: N
Sp.Gr. @ 20C: 0.99
Lbs./Gal.: 8.25
pH 10% Sol.: 10-11
% Active: 80
Max. Color Gvcs-33: 2
Acid No.: 0-1
Alkali No.: 25-35

150GLT:
Fatty Acid: Coconut
Physical State @ 20C: Liquid
Ionic Nature: N
Sp.Gr. @ 20C: 0.96
Lbs./Gal.: 8.00
pH 10% Sol.: 10.3-11.3
% Active: 100
Max. Color Gvcs-33: 6
Acid No.: 0-1
Alkali No.: 30-45

1:1 FA-Monoethanolamides:
CMA:
Fatty Acid: Coconut
Physical State @ 20C: Granular
Ionic Nature: N
pH 10% Sol.: 9.4-10.8
% Active: 100
Max. Color Gvcs-33: Tan
Acid No.: 0-1
Alkali No.: 6-12
Solid. Pt. C.: 87±2

MONA INDUSTRIES, INC.: MONAMINES and MONAMIDS(Continued):

1:1 FA-Monoethanolamides(Continued):

LMA:
Fatty Acid: Lauric
Physical State @ 20C: Granular
Ionic Nature: N
pH 10% Sol.: 10-11
% Active: 100
Max. Color Gvcs-33: Tan
Acid No.: 0-1
Alkali No.: 5-12
Solid Pt. C: 80±2

LMMA:
Fatty Acid: Lauric Myristic
Physical State @ 20C: Granular
Ionic Nature: N
pH 10% Sol.: 9.7-10.7
% Active: 100
Max. Color Gvcs-33: Tan
Acid No.: 0-1
Alkali No.: 5-12
Solid Pt. C: 80±2

S:
Fatty Acid: Stearic
Physical State @ 20C: Granular
Ionic Nature: N
pH 10% Sol.: 9.5-11.0
% Active: 100
Max. Color Gvcs-33: Tan
Acid No.: 0-1
Alkali No.: 5-18
Solid Pt. C: 87±2

1:1 FA-Monoisopropanolamides:

LIPA:
Fatty Acid: Lauric
Physical State @ 20C: Granular
Ionic Nature: N
pH 10% Sol.: 10.3-11.3
% Active: 100
Max. Color Gvcs-33: Tan
Acid No.: 0-1
Alkali No.: 12-22
Solid Pt. C: 55±3

MONA INDUSTRIES, INC.: MONAMINES and MONAMIDS(Continued):

MONAMID 150-CW:

MONAMID 150-CW is a "superamide" specifically developed to increase flash foam, at concentrations of 1-3%, in a variety of cosmetic products such as shampoos, bubble baths, bath oils, etc. MONAMID 150-CW also augments the foam boosting and foam stabilization properties of other "super-amides" without significantly affecting viscosities.

MONAMID 150-CW is an effective emulsifier and solubilizing agent in cosmetic creams, lotions and other toiletry products. As a result of these functions, it can also be used as a viscosity control agent in these applications.

Typical Properties:

Chemical Composition: 1:1 Capric Diethanolamide
Appearance: Amber Liquid--Paste
Color (GVCS-1933): 3
pH (10% Solution): 11.0
Activity: 100%
Acid Number: 0
Alkali Number: 50
Ionic Nature: Nonionic
Specific Gravity @ 25C: 0.99

MONAMID 150-GLT:

MONAMID 150-GLT is a 100% active, modified lauric diethanolamide designed as an outstanding thickener and foam stabilizer for use in cosmetic and toiletry products. This product is liquid at room temperature and is therefore much easier to handle and formulate than the commonly used solid amides.

The foam and viscosity building properties are equal or superior to most other alkanolamides, and MONAMID 150-GLT is specifically recommended as a replacement for the solid lauric and/or the lauric/myristic alkanolamides.

Tentative Specifications:

Chemical Composition:: Modified Lauric Diethanolamide
Physical Appearance: Clear Amber Liquid
Acid #: 1 Max.
Alkali #: 30-45
pH (1% solution): 9.0-10.5
Color (GVCS-1933): 6 Max.
Ionic Nature: Nonionic
Activity: 100%

MONA INDUSTRIES, INC.: MONAMINES and MONAMIDS(Continued):

MONAMID 150-IS:

MONAMID 150-IS is a high activity, liquid isostearic alkanolamide which can impart excellent lubricity and mildness to a broad range of products.

In cosmetic 1-3% MONAMID 150-IS will substantially improve slip and comb-out properties to hair care products and lubricity to creams and lotions. This type of functionality also suggests its use as a lubricant, emulsifier and corrosion inhibitor in synthetic coolants, fiber lubricants and similar product areas.

Specifications:

- Chemical Composition: 1:1 Isostearic Diethanolamide
- Appearance: Light Amber Liquid
- Color (GVCS. 1933) Molten: 5 Max.
- Activity: 100%
- Acid #: 5-10
- Alkali #: 30-60
- Ionic Nature: Nonionic

Typical Properties:

- pH (10% Dispersion): 9.5±0.5
- Cloud Point (C): <-5C

MONAMID 150-MW:

MONAMID 150-MW is a myristic alkanolamide which is non-irritating to the eyes and recommended as an outstanding thickener, foam stabilizer and emulsifier in aqueous or nonaqueous cosmetic and toiletry products.

Thickening Properties:

The addition of 5% MONAMID 150-MW, to a formulation containing 10% active sodium lauryl sulfate or sodium lauryl ether sulfate, develops approximately 3 times the viscosity achieved with a 70/30% lauric/myristic alkanolamide and approximately 5-10 times the viscosity normally obtained with a lauric alkanolamide. This increased thickening power is even more pronounced at the 7% level where it produces stiff, clear gels. The same concentrations of the lauric/myristic or lauric diethanolamide develop, at best, only pourable gels. This viscosity increase is much less pronounced when MONAMID 150-MW is combined with TEA lauryl sulfate.

Typical Properties:

- Chemical Composition: 1:1 Myristic Diethanolamide
- Appearance: White Solid Wax
- Color (GVCS 1933) Molten: 3
- pH (10% Dispersion): 11.0
- Specific Gravity @ 45C: Approx. 0.98
- Activity: 100%
- Acid #: 1
- Alkali #: 42
- Ionic Nature: Nonionic
- Melting Point (C): Approx. 50

MONA INDUSTRIES, INC.: MONAMINES and MONAMIDS(Continued):

MONAMID 716:
CTFA Designation: Lauramide DEA

MONAMID 716 is a 100% active, liquid "super" amide offering improved flash foam, along with excellent foam boosting, foam stabilizing and viscosity building properties as illustrated. It is an outstanding solubilizer and coupler for essential oils and other additives, and offers excellent stability in systems ranging from 4 to 12.

MONAMID 716 is suggested for all types of shampoos, bubble baths and liquid detergent formulations where quality, performance, mildness and maximum ease of handling are required. It often serves as a direct replacement for solid lauric amides. In aerosol formulations it is frequently used for its coupling, foam stabilizing and wetting properties.

Typical Properties:
CTFA Name: Lauramide DEA
Appearance: Clear Light Amber Liquid
Color (GVCS 1933): 3
pH (10% Solution): 10.7
Acid #: 1.0
Alkali #: 52
Activity: 100%
Cloud Point: <1C
Specific Gravity @ 25C: 0.98

MONAMID 716:
MONAMID 716 is a 100% active, liquid "super" amide highly recommended for use in all types of shampoos, cleansers, bubble baths and liquid detergent formulations where quality performance, extreme mildness and maximum ease of handling are required. It is an outstanding solubilizer and coupler for essential oils and other additives, and offers excellent stability in systems ranging in pH from pH 4 to 11.

Features:
- Excellent detergent
- Light color/odor
- Superior flash foamer
- Non-irritating
- Outstanding solubilizer
- 100% active liquid

Applications:
- Facial cleansers
- Shower gels
- Household liquid detergents
- Shampoos and bubble baths
- Body cleansers

Typical Chemical and Physical Properties:
Chemical Composition: Lauric Diethanolamide
Appearance: Clear Light Amber Liquid
Color (GVCS 1933): 5
Activity: 100%
Alkali #: 55
pH (10% Solution): 10.8
- Acid #: 2
- Ionic Nature: Nonionic
- Specific Gravity: 0.98

MONA INDUSTRIES, INC.: MONAMINES and MONAMIDS(Continued):

MONAMID 759:

MONAMID 759 is a 100% active proprietary liquid super amide, with extremely low eye irritation, created to give excellent foam boosting, foam stabilizing, viscosity building and general conditioning properties to shampoos, bubble baths, body cleansers and related toiletry products.

The extreme mildness of MONAMID 759 is verified by independent Draize Tests which show it to be completely non-irritating at a concentration of 5%.

Typical Properties:

Chemical Nature: 1:1 Mixed Fatty Acid Diethanolamide
Physical Appearance: Amber Liquid at Room Temperature
Color (GVCS-1933): 6
pH (10% solution): 11.0
Acid Number: 1
Alkali Number: 35
Cloud Point: 0C
Clear Point: 33C

MONAMID 1007:

MONAMID 1007 is a 1:1 mixed fatty acid diethanolamide offering a combination of excellent foam boosting, foam stabilizing and viscosity building properties with liquidity at room temperature for optimum ease of handling.

It is specifically designed to give the same high foam and viscosity properties obtained with the classic 70/30 lauric/myristic diethanolamide, but without the difficulties and expense associated with handling a product which is solid at room temperature.

MONAMID 1007 is recommended at concentrations of 4-6% for use in most shampoos, bubble baths and skin cleaners, as well as household and industrial liquid detergents.

Specifications:

CTFA Description: Lauramide DEA (and) Linoleamide DEA
Appearance: Clear Amber Liquid
Color, GVCS 33': 8 Max.
pH (10% Aqueous): 10.0-11.5
Acid #: 1 Max.
Alkali #: 25-45
Iodine Value: 35-45

MONA INDUSTRIES, INC.: MONAMINES and MONAMIDS(Continued):

MONAMID LMIPA:

MONAMID LMIPA is a 1:1 lauric/myristic monoisopropanolamide which produces rich, dense, stable foam and excellent viscosities in typical shampoo, bubble bath and detergent systems.

The addition of 5% MONAMID LMIPA to a formulation containing 15% active ammonium lauryl sulfate or sodium lauryl sulfate develops up to 3 times the viscosity normally obtained with a lauric or lauric/myristic diethanolamide.

Typical Properties:

Physical appearance: Off-white solid
Acid number: 1 Max.
Alkali number: 10-25
Color (GVCS): 3 Max.
Activity: 100%
Ionic nature: Nonionic
pH (10% aqueous dispersion): 10.5±0.5
Melting point: 61±2C

MONAMINE 779:

MONAMINE 779 is a unique 100% active liquid primary surfactant with the following functional properties:

1. Low degree of eye and skin irritation.
2. Produces high volume stable foam at low concentrations.
3. A wide range of viscosities can be obtained by simple water dilution or the addition of alkanolamides and other surfactants.
4. Good detergency and soil suspending power.
5. Excellent hard water and soap tolerance.
6. Good compatibility and stability with alkali.
7. Excellent compatibility with many other surfactants.
8. Good wetting and penetrating properties.
9. Effective surface tension reduction at low concentrations.
10. Good solubility in a wide range of solvents.
11. Excellent solubilizer for essential oils and colors.

Superior foaming ability and hard water tolerance coupled with low irritation properties make MONAMINE 779 particularly effective as a primary surfactant in shampoos, bubble baths, foaming bath oils and various body and skin cleansers. As a totally anhydrous material it has value in aerosol and enscapulated products and can be adsorbed on powdered detergents.

CTFA Designation: Cocamide DEA (and) DEA Laureth Sulfate.

Specifications:

Chemical Composition: Amido Ether Sulfate Complex
Appearance: Clear to hazy viscous yellow liquid
Activity: 100%
Alkali #: 65±10
Color (Gardner 1933): 6 Max.

MONA INDUSTRIES, INC.: MONAQUAT:

MONAQUAT P Series:

The MONAQUAT P Series is a new line of tri-quaternary phosphate esters. Initially, Mona is offering the following grades in developmental quantities for evaluation:

MONAQUAT P-TC
MONAQUAT P-TD
MONAQUAT P-TL
MONAQUAT P-TZ

Typical Properties:

MONAQUAT P-TC:
Chemical & CTFA Description: Cocamidopropyl PG-Dimonium Chloride Phosphate
Appearance: Clear Amber Liquid
% Active: 40.0
pH (10% Solution): 7.0
Wetting (Draves) (1.0% Active @ 25C): 1'4"
Surface Tension Dynes/cm (1.0% Active @ 25C): 37.7
Foam mm 1 min./5 min. (1.0% As Is): 91/84
Specific Gravity @ 25C: 1.10
Lbs. per Gallon: 9.1

MONAQUAT P-TD:
Chemical & CTFA Description: Lauramidopropyl PG-Dimonium Chloride Phosphate
Appearance: Clear Amber Liquid
% Active: 34.0
pH (10% Solution): 7.5
Wetting (Draves) (1.0% Active @ 25C): 1'31"
Surface Tension Dynes/cm (1.0% Active @ 25C): 43.1
Foam mm 1 min./5 min. (1.0% As Is): 96/88
Specific Gravity @ 25C: 1.05
Lbs. per Gallon: 8.7

MONAQUAT P-TL:
Chemical & CTFA Description: Lauroampho PG-Glycinate Phosphate
Appearance: Clear-Opaque Viscous Liquid
% Active: 30.0
pH (10% Solution): 7.5
Wetting (Draves) (1.0% Active @ 25C): 4"
Surface Tension Dynes/cm (1.0% Active @ 25C): 33.7
Foam mm 1 min./5 min. (1.0% As Is): 61/48
Specific Gravity @ 25C: 1.10
Lbs. per Gallon: 9.1

MONA INDUSTRIES, INC.: MONAQUAT(Continued):

MONAQUAT P Series(Continued):

Typical Properties(Continued):

MONAQUAT P-TZ:
Chemical & CTFA Description: Cocohydroxyethyl PG-Imidazolinium Chloride Phosphate
Appearance: Clear Amber Liquid
% Active: 30.0
pH (10% Solution): 7.0
Wetting (Draves) (1.0% Active @ 25C): 10"
Surface Tension Dynes/cm (1.0% Active @ 25C): 37.6
Foam mm 1 min./5 min. (1.0% As Is): 67/59
Specific Gravity @ 25C: 1.07
Lbs. per Gallon: 8.9

Comparative Functional Properties:

MONAQUAT P-TC:
Very mild, excellent foamer, easy to formulate, frequently shows best substantivity as measured by the Rubine Dye Test, excellent bacteriocidal activity, good hydrotrope and co-emulsifier, good viscosity builder, wetter and surface tension reducer.

MONAQUAT P-TD:
Excellent hydrotrope and co-emulsifier, highest foamer, easy to handle and formulate, good wetter but only moderate surface tension reducer, very mild to the skin but somewhat irritating to the eyes.

MONAQUAT P-TL:
The mildest to both skin and eyes, good foamer, excellent hydrotrope, best wetter and surface tension reducer.

MONAQUAT P-TZ:
Very mild to the eyes but slightly irritating to the skin, good foamer, good wetter and surface tension reducer.

Functional Characteristics:
The unique feature of these products is that they combine in one molecule exceptional:
Mildness: to the skin and eyes
Substantivity: conditioning, softening and antistatic properties
Surface Activity: detergency, foaming, emulsifying, solubiklizing, dispersing, thickening, surface tension reduction and wetting properties
Bacteriocidal Activity

MONA INDUSTRIES, INC.: MONAQUAT (Continued):

MONAQUAT P-TS:
CTFA Adopted Designation: Synthetic Phospholipid
Skin Benefits:
- Provides long lasting skin smoothing properties
- No adverse epidermal cellular effects
- Does not accelerate transepidermal water loss
- Not occlusive
- Rapid skin pH recovery after application

Functional Properties:
- Efficient Primary Emulsifier
- Requires only 0.6% active for lotions and only 0.9% active for cremes
- Forms stable low pH emulsions without adjustment
- Produces emulsions with stable viscosity on long term storage
- Compatible with emollients, oils and waxes
- Effective skin and hair conditioning
- Highly substantive
- Eliminates the tacky feeling of high glycerin formulations

Typical Chemical and Physical Properties:
CTFA Adopted Designation: Stearamidopropyl PG-Dimonium Chloride Phosphate
Appearance: Light yellow paste
Total Solids: 35%
Chloride (as Cl-): 2.3%
Activity: 30%
pH: (10% in 50/50 IPA/H2O): 7.6
Melting Point: 40-50C
Specific Gravity @ 25C: 1.01
Wt. Per Gallon: 8.4 lbs.

MONAQUAT ISIES:
MONAQUAT ISIES is a 100% active, liquid quaternary compound offering excellent antistatic, lubricating, fiber softening and corrosion inhibiting properties in aqueous and non-aqueous systems.

Cosmetics:
MONAQUAT ISIES is suggested as a base for cream rinses @ 0.5-3.0% where it contributes outstanding manageability, lubricity, softness and "sheen" to the hair. The combination of functional and solubility characteristics also suggest its use as an additive to hair sprays and hair grooms.

Typical Properties:
Chemical Composition: Quaternized Heptadecyl Imidazoline
Activity: 100%
Physical Form: Amber liquid
pH of 10% Solution: 6.9
Specific Gravity @ 25C: 1.03

MONA INDUSTRIES, INC.: MONAQUAT(Continued):

MONAQUAT TG:

MONAQUAT TG is a new quaternary designed to give exceptional performance in hair and skin products. MONAQUAT TG has been found, through extensive testing, to be an excellent hair rinse conditioner. It is a superior product when compared to other quaternaries such as Quaternium 16, stearyl dimethyl benzyl ammonium chloride, and polymers based on PVP and cellulosics.

In addition, combinations of MONAQUAT TG and the polymeric quaternaries have been found to be an excellent low cost conditioner for hair rinses and shampoos. MONAQUAT TG is compatible with all commonly used rinse and shampoo ingredients and exhibits no build-up characteristic. Its cationic antistatic and high foaming properties suggest its use in fine fabric laundering and softening products, as well as for personal care preparations.

Typical Properties:
- Appearance: Clear to slightly hazy liquid
- Activity: 30%
- Ionic Nature: Cationic
- Color (Gardner 1933): 2
- % Hexylene Glycol: 10
- % NaCl: 5
- pH @ 10%: 4
- Specific Gravity: 1.011
- Weight/Gallon: 8.39 lbs.
- Cloud Point: +3C
- Solid Point: -10 to -12C

MONA INDUSTRIES, INC.: MONATERIC:

MONATERIC 951A:

Chemical & CTFA Description: Lauroamphocarboxyglycinate

Use Characteristics:

1) Exceptional flash foaming
2) Very light color
3) Compatible with lauryl sulfates, alkanolamides, alpha olefin sulfonates, sulfosuccinate half esters, and betaines.
4) Viscosities of MONATERIC 951A systems can be readily built through the addition of betaines and alkanolamides.

MONATERIC 951A is 100% biodegradable.

Typical Properties:

Chemical Description: Lauroamphocarboxyglycinate
Appearance: Clear pale yellow liquid
pH as is: 8.8
Total Solids: 30
% Active: 24
% NaCl: 5.7
Specific Gravity: 1.08
Weight/Gallon: 8.96 lb.

MONATERIC 1202:

CTFA Designation: Dihydroxyethyl Tallow Glycinate

Features:

- High Degree of Effective Conditioning
- Selective Substantivity
- No Build-up
- Non-Oily
- Compatible with All Other Surfactant Types
- Easily Soluble in Water

MONATERIC 1202 is a multi-functional tallow based betaine especially useful in hair care products. The conditioning effects it provides coupled with its surfactant properties are especially suited for creme rinses, conditioning shampoos, and conditioning/styling mousse formulations.

Typical Properties:

Physical Appearance: Clear Viscous Liquid
Total Solids: 40.0%
Chloride (as NaCl): 5.0%
Activity: 35.0%
pH: 5.5

MONA INDUSTRIES, INC.: MONATERIC(Continued):

MONATERIC 1203:
CTFA Designation: Sodium Hydrogenated Tallow Dimethyl Glycinate

Features:
- High Degree of Effective Conditioning
- Selective Substantivity
- No Build-up
- Non-Oily
- Compatible with all Other Surfactant Types

MONATERIC 1203 is a tallow based amphoteric surfactant designed to provide a high degree of oil-free conditioning effects to hair. Everyday shampoos formulated with MONATERIC 1203 replenish the daily requirements of hair resulting in a clean, fresh and healthy look.

Typical Properties:
Physical Appearance: Opaque off-white paste
Total Solids: 35.0%
Chloride (as NaCl): 5.0%
Activity: 30.0%
pH (5% Solution): 6.5

MONATERIC CA-35%:
MONATERIC CA-35% is an amphoteric surfactant based on coconut Imidazoline

Use Characteristics:
This amphoteric surfactant is supplied as a 35% active aqueous solution in the form of its Zwitterion or inner salt at a pH in the range of 4.8-6.5. Adjustment of the pH to 4-4.5 results in a material which behaves as a cationic agent while increasing the pH to 8.5 or above produces an anionic surface active agent.

MONATERIC CA-35% is an excellent detergent, wetting, emulsifying and dispersing agent.

MONATERIC CA-35% produces good foam or lather either by itself or in combination with other surfactants.

The excellent chemical stability, solubility and surface active properties make this compound a valuable aid in formulating of a wide range of cosmetic, household and industral products.

Specifications:
Appearance: Amber Liquid
Activity: 35±1
Moisture (K.F.): 65±1
Acid #: 50±5
Alkali #: 55±5

MONA INDUSTRIES, INC.: MONATERIC Surfactants:

Group I: Imidazoline Derived:
A) Monocarboxylic
1. with salt (Glycinates)

MONATERIC Surfactant:

CM-36S:
CTFA: Cocoamphoglycinate
Counter Ion: Na+
Appearance: Clear Amber Liquid
pH @ 10%: 11.9
% Total Solids: 42
% NaCl: 6
% Active: 36
Wetting 1% Active: 5 sec.
Specific Gravity (25C): 1.10
Lbs/Gal: 9.2

LMM-30:
CTFA: Lauroampoglycinate
Counter Ion: Na+
Appearance: Viscous Amber Liquid
pH @ 10%: 9.2
% Total Solids: 36
% NaCl: 6
% Active: 30
Wetting 1% Active: 5 sec.
Specific Gravity (25C): 1.09
Lbs/Gal: 9.1

2. Salt free (Propionates):

CA-35:
CTFA: Cocoamphopropionate
Counter Ion: H+
Appearance: Clear Amber Liquid
pH @ 10%: 5.7
% Total Solids: 35
% Active: 35
Wetting 1% Active: 11 sec.
Specific Gravity (25C): 1.02
Lbs/Gal: 8.5

CAM-40:
CTFA: Cocoamphopropionate
Counter Ion: Na+
Appearance: Clear Amber Liquid
pH @ 10%: 9.3
% Total Solids: 40
% Active: 40
Wetting 1% Active: 40 sec.
Specific Gravity (25C): 1.05
Lbs/Gal: 8.8

MONA INDUSTRIES, INC.: MONATERIC Surfactants(Continued):

Group I: Imidazoline Derived(Continued):
2. Salt free (Propionates)(Continued):

<u>MONATERIC</u> <u>Surfactant:</u>

ISA-35:
CTFA: Isostearoamphopropionate
Counter Ion: H+
Appearance: Clear Amber Liquid
pH @ 10%: 5.4
% Total Solids: 35
% Active: 35
Wetting 1% Active: 10 min.
Specific Gravity (25C): 1.01
Lbs/Gal: 8.4

Cy NA-50:
CTFA: Capryloamphopropionate
Counter Ion: Na+
Appearance: Dark Amber Liquid
pH @ 10%: 10.6
% Total Solids: 50
% Active: 50
Wetting 1% Active: 29 sec.
Specific Gravity (25C): 1.10
Lbs/Gal.: 9.2

LF Na-50:
CTFA: Not assigned--mixed short chain
Counter Ion: Na+
Appearance: Clear Brown Liquid
pH @ 10%: 11.5
% Total Solids: 50
% Active: 50
Wetting 1% Active: 10 min.
Specific Gravity (25C): 1.09
Lbs/Gal.: 9.1

LF-100:
CTFA: Not assigned--mixed short chain
Appearance: Clear Brown Liquid
pH @ 10%: 11.7
% Total Solids: 100
% Active: 100
Wetting 1% Active: 25 sec.
Specific Gravity (25C): 1.04
Lbs/Gal.: 8.7

MONA INDUSTRIES, INC.: MONATERIC Surfactants(Continued):

Group I: Imidazoline Derived(Continued):
2. Salt free (Propionates)(Continued):

MONATERIC Surfactant:

810-A-50:
CTFA: Not assigned--Caprylic/Capric
Counter Ion: H+
Appearance: Clear Brown Liquid
pH @ 10%: 4.4
% Total Solids: 50
% Active: 50
Wetting 1% Active: 5 sec.
Specific Gravity (25C): 1.07
Lbs/Gal.: 8.9

TA-35:
CTFA: Not assigned--Tall Oil
Counter Ion: H+
Appearance: Dark Brown Gel
pH @ 10%: 5.2
% Total Solids: 35
% Active: 35
Wetting 1% Active: 7 min.
Specific Gravity (25C): 1.02
Lbs/Gal.: 8.5

CNa-40:
CTFA: Not assigned--Coconut
Counter Ion: Na+
Appearance: Clear Amber Liquid
pH @ 10%: 10.9
% Total Solids: 40
% Active: 40
Wetting 1% Active: 6 sec.
Specific Gravity (25C): 1.09
Lbs/Gal.: 9.1

B. Dicarboxylic:
1. with salt (Glycinates):

CDX-38:
CTFA: Cocoamphocarboxyglycinate
Counter Ion: Na+
Appearance: Viscous Yellow Liquid
pH @ 10%: 8.5
% Total Solids: 50
% NaCl: 11
% Active: 39
Wetting 1% Active: 18 sec.
Specific Gravity (25C): 1.18
Lbs/Gal.: 9.8

MONA INDUSTRIES, INC.: MONATERIC Surfactants(Continued):

Group I: Imidazoline Derived(Continued):
B. Dicarboxylic(Continued):
1. with salt (Glycinates)(Continued):

MONATERIC Surfactant:

CDX-38 Mod:
CTFA: Cocoamphocarboxyglycinate
Counter Ion: Na+
Appearance: Clear Yellow Liquid
pH @ 10%: 8.8
% Total Solids: 50
% NaCl: 11
% Active: 39
Wetting 1% Active: 20 sec.
Specific Gravity (25C): 1.18
Lbs/Gal.: 9.8

CSH-32:
CTFA: Cocoamphocarboxyglycinate
Counter Ion: Na+
Appearance: Clear Yellow Liquid
pH @ 10%: 8.4
% Total Solids: 40
% NaCl: 8
% Active: 32
Wetting 1% Active: 20 sec.
Specific Gravity (25C): 1.13
Lbs/Gal.: 9.4

2. Salt free (Propionates):

CEM-38:
CTFA: Cocoamphocarboxypropionate
Counter Ion: Na+
Appearance: Clear to Hazy Amber Liquid
pH @ 10%: 8.6
% Total Solids: 39
% Active: 39
Wetting 1% Active: 4.5 min.
Specific Gravity (25C): 1.05
Lbs/Gal.: 8.8

CEM-38CG:
CTFA: Cocoamphocarboxypropionate
Counter Ion: Na+
Appearance: Clear to Hazy Amber Liquid
pH @ 10%: 9.8
% Total Solids: 38
% Active: 38
Wetting 1% Active: 10 min.
Specific Gravity (25C): 1.07
Lbs/Gal.: 8.9

MONA INDUSTRIES, INC.: MONATERIC Surfactants(Continued):

Group I: Imidazoline Derived(Continued):
B. Dicarboxylic(Continued):
2. Salt free (Propionates)(Continued):

MONATERIC Surfactant:

811:
CTFA Designation: Not assigned--Caprylic
Counter Ion: Na+
Appearance: Clear to Hazy Amber Liquid
pH @ 10%: 11.4
% Total Solids: 50
% Active: 50
Wetting 1% Active: Instant
Specific Gravity (25C): 1.04
Lbs/Gal.: 8.7

1000:
CTFA: Caprylamphopropionate
Counter Ion: Na+
Appearance: Clear to Hazy Amber Liquid
pH @ 10%: 11.8
% Total Solids: 50
% Active: 50
Wetting 1% Active: Instant
Specific Gravity (25C): 1.05
Lbs/Gal.: 8.8

CyA-50:
CTFA: Not assigned--Caprylic
Counter Ion: H+
Appearance: Clear Dark Brown Liquid
pH @ 10%: 5.6
% Total Solids: 50
% Active: 50
Wetting 1% Active: 9 sec.
Specific Gravity (25C): 1.07
Lbs/Gal: 8.9

CyMM-40:
CTFA: Not assigned--Caprylic
Counter Ion: Na+
Appearance: Clear Amber Liquid
pH @ 10%: 9.8
% Total Solids: 40
% Active: 40
Wetting 1% Active: 4.5 min.
Specific Gravity (25C): 1.10
Lbs/Gal.: 9.2

MONA INDUSTRIES, INC.: MONATERIC Surfactants(Continued):

Group II: Betaines:

MONATERIC Surfactant:

CAB:

CTFA: Cocamidopropyl Betaine
Counter Ion: Inner Salts
Appearance: Clear Light Yellow Liquid
pH @ 10%: 7.1
% Total Solids: 35
% NaCl: 5
% Active: 30
Wetting 1% Active: 18 sec.
Specific Gravity (25C): 1.04
Lbs/Gal.: 8.7

MCB:

CTFA: Cocamidopropyl Betaine
Counter Ion: Inner Salts
Apppearance: Clear Light Yellow Liquid
pH @ 10%: 4.8
% Total Solids: 33
% NaCl: 3
% Active: 30
Wetting 1% Active: 18 sec.
Specific Gravity (25C): 1.02
Lbs/Gal.: 8.5

COAB:

CTFA: Cocamidopropyl Betaine
Counter Ion: Inner Salts
Appearance: Clear Yellow Liquid
pH @ 10%: 7.9
% Total Solids: 37
% NaCl: 5
% Active: 32
Wetting 1% Active: 10 sec.
Specific Gravity (25C): 1.04
Lbs/Gal.: 8.7

ADA:

CTFA: Cocamidopropyl Betaine
Counter Ion: Inner Salts
Appearance: Clear Amber Liquid
pH @ 10%: 7.7
% Total Solids: 38
% NaCl: 5
% Active: 33
Wetting 1% Active: 13 sec.
Specific Gravity (25C): 1.05
Lbs/Gal.: 8.8

MONA INDUSTRIES, INC.: MONATERIC Surfactants(Continued):

Group II : Betaines(Continued):

<u>MONATERIC</u> <u>Surfactant:</u>

LMAB:
CTFA: Lauramidopropyl Betaine
Counter Ion: Inner Salts
Appearance: Clear Light Yellow Liquid
pH @ 10%: 8.3
% Total Solids: 35
% NaCl: 5
% Active: 30
Wetting 1% Active: 9 sec.
Specific Gravity (25C): 1.04
Lbs/Gal.: 8.7

Group III: Blends:

985A:
CTFA: Lauroamphoglycinate/Sodium Trideceth Sulfate
Appearance: Clear Amber Liquid
pH @ 10%: 9.3
% Total Solids: 39
% NaCl: 3
% Active: 36
Wetting 1% Active: Instant
Specific Gravity (25C): 1.07
Lbs/Gal.: 8.9

805:
CTFA: Cocoamphocarboxyglycinate/Cocamido MIPA-SS
Counter Ion: Na2
Appearance: Clear Amber Liquid
pH @ 10%: 7.7
% Total Solids: 42
% NaCl: 2
% Active: 40
Wetting 1% Active: 5 sec.
Specific Gravity (25C): 1.10
Lbs/Gal.: 9.2

CDL:
CTFA: Cocoamphocarboxyglycinate/SLES SLS
Appearance: Clear Yellow Liquid
pH @ 10%: 8.5
% Total Solids: 37
% NaCl: 6
% Active: 31
Wetting 1% Active: 4 sec.
Specific Gravity (25C): 1.11
Lbs/Gal.: 9.2

MONA INDUSTRIES, INC.: MONATERIC Surfactants(Continued):

Group III: Blends(Continued):

CDTD:
CTFA: Cocoamphocarboxyglycinate/Sodium Trideceth Sulfate
Appearance: Clear Yellow Liquid
pH @ 10%: 8.3
% Total Solids: 50
% NaCl: 6
% Active: 44
Wetting 1% Active: Instant
Specific Gravity (25C): 1.11
Lbs/Gal.: 9.2

CDS:
CTFA: Cocoamphocarboxyglycinate/Sodium Lauryl Sulfate
Appearance: Clear Yellow Liquid
pH @ 10%: 8.5
% Total Solids: 37
% NaCl: 6
% Active: 31
Wetting 1% Active: 3 sec.
Specific Gravity (25C): 1.09
Lbs/Gal.: 9.1

Group IV: Proprietary Compounds:

ADFA:
CTFA: Not assigned
Appearance: Clear Amber Liquid
pH @ 10%: 7.8
% Total Solids: 34
% NaCl: 3
% Active: 31
Wetting 1% Active: 10 sec.
Specific Gravity (25C): 1.03
Lbs/Gal.: 8.6

MONA INDUSTRIES, INC.: MONAWET Surfactants:

Chemical and Physical Descriptions:

Covers only those products chemically classified as sodium dialkylsulfosuccinates. The alkyl (R) groups in their general structure represent isobutyl, hexyl, octyl or tridecyl chain lengths.

Their surface active and solubility characteristics vary with the individual alkyl groups. The short chain isobutyl derivatives exhibit the greatest solubility in water while the long chain tridecyl products show the highest oil solubility.

For broad application coverage the MONAWET grades are manufactured in a variety of solvents. All of the MONAWETS are liquid products.

MONAWET:

MB-45:
% Activity: 45
Dialkyl Groups: Isobutyl
Solvent System: Water
Appearance @ 25C: Clear Liquid

MM-80:
% Activity: 80
Dialkyl Groups: Hexyl
Solvent System: Water/Isopropanol
Appearance @ 25C: Clear Liquid

MO-70:
% Activity: 70
Dialkyl Groups: Octyl
Solvent System: Water/Butyl CARBITOL
Appearance @ 25C: Clear Liquid

MO-70E:
% Activity: 70
Dialkyl Groups: Octyl
Solvent System: Water/Ethyl Alcohol
Appearance @ 25C: Clear Liquid

MO-70R:
% Activity: 70
Dialkyl Groups: Octyl
Solvent System: Water/Propylene Glycol
Appearance @ 25C: Clear Liquid

MONA INDUSTRIES, INC.: MONAWET Surfactants(Continued):

MONAWET:

MO-75E:
% Activity: 75
Dialkyl Groups: Octyl
Solvent System: Water/Ethyl Alcohol
Appearance @ 25C: Clear Liquid

MO-84R2W#:
% Activity: 84
Dialkyl Groups: Octyl
Solvent System: Propylene Glycol
Appearance @ 25C: Viscous Liquid

MT-70:
% Activity: 70
Dialkyl Groups: Tri-Decyl
Solvent System: Water/Hexylene Glycol
Appearance: Clear Liquid

MT-70E:
% Activity: 70
Dialkyl Groups: Tri-Decyl
Solvent System: Water/Ethyl Alcohol/Butyl CARBITOL
Appearance: Clear Liquid

MT-80H2W#:
% Activity: 80
Dialkyl Groups: Tri-Decyl
Solvent System: Hexylene Glycol
Appearance: Viscous Liquid

Anhydrous Grades

Functional Properties:
The MONAWETS as a group of powerful anionic surfactants gain their importance from their ability to perform the following functions:
Reduce surface and interfacial tension
Wetting
Dispersing
Emulsifying
Penetrating
Solubilizing

MONA INDUSTRIES, INC.: MONAWET Surfactants(Continued):

MONAWET SNO-35:

MONAWET SNO-35 is an anionic surfactant with a wide variety of functional properties. Its versatility makes it suitable for use in a broad range of applications such as industrial detergents, agricultural, cosmetic and textile products and specialty emulsions or dispersions for polymerization systems.

Functional Properties:

1. High Alkaline Stability.
2. Excellent Solubility in High Electrolyte Salt Solutions.
3. Excellent Wetting Agent Above 40C.
4. Strong Solubilizing Properties.
5. High Calcium Tolerance.
6. Good Emulsifier and Dispersant.
7. Viscosity Depressant.
8. Mild Detergent.

Physical/Chemical Properties:

Chemical Name: Tetrasodium N-(1,2-Dicarboxyethyl) N-Alkyl (C18) Sulfosuccinamate
Average Molecular Weight: 653
Ionic Nature: Anionic
Appearance: Clear light amber liquid
Color-Gardner: 8 max.
Total Solids %: 35±1.5
Water %: 65±1.5
Specific Gravity @ 25C: 1.12-1.16 (9.5 lb/gal)
Acid Number: 2.0 max.
Iodine Number: 0.5 max.
Viscosity @ 25C--#2 Zahn Cup: 16-18 seconds
pH (as is): 7-8
Freezing Pt. F: 45±5
Critical Micelle Conc. @ 100% Activity: 0.06

Solubilities: Soluble in water but insoluble in most organic solvents.

Cosmetics:

Its mild detergency and low wetting properties suggest the use of MONAWET SNO-35 in shampoos, bubble baths, skin cleaners and hand soaps. It is also recommended as an emulsifier in creams and lotions to produce smoother textures and better spreading properties.

MONA INDUSTRIES, INC.: MONAWET Surfactants(Continued):

MONAWET MO-85P:

Chemical and Physical Properties:
Chemical Nature: Sodium dioctyl sulfosuccinate
Appearance: Fine White Powder
Activity: 85.0%
Diluent: Sodium Benzoate, 15%
Solubility in Water @ Room Temperature: 0.85%
pH (1.0%): 6.5
Moisture: 1% maximum
Surface Tension (1% in Water): 26 dynes/cm
CMC (approx.): .06%

MONAWET TD-30:

MONAWET TD-30 is the half ester of sulfosuccinic acid based on an ethoxylated fatty alcohol.

The general physical, chemical and surface active properties of MONAWET TD-30 suggest its use in a variety of end uses.

Of particular interest is its use in emulsion polymerization.

Specifications:
Appearance: Light yellow liquid
% Activity (Total Solids): 30±1
Acid #: 6 Max.
Alkali #: 20-30

Typical Properties:
pH (as is): 5-6
Weight/gallon: 9 lbs.

The combination of MONAWET TD-30's foam and emulsifying properties, suggests its use in a broad range of cosmetic and fine fabric detergent formulations. MONAWET TD-30 is also recommended in textile wet processing for its good wetting and low surface tension properties.

MONA INDUSTRIES, INC.: Miscellaneous:

AVAMID 150:

CTFA Designation: Avocadamide DEA (and) Avocado Oil

Features:

- Self-solubilizing in aqueous cleaning systems--forms clear micro-emulsions
- Imparts smooth, silky feel to skin and hair
- Provides foam stabilization, viscosity building, and lubricity
- Ideally suited for:
  - Clear Conditioning Shampoos
  - Hair Rinses
  - Mousses
  - Creams and Lotions

Typical Properties:

Appearance: Clear amber liquid
Activity: 100%
Color (GVCS-33): 8
pH (10%): 10.5
Free Amine: 2.0%

NAPP CHEMICALS INC.: Pharmaceutical and Cosmetic Products:

Preservatives:
- Methyl Paraben
- Propyl Paraben
- Butyl Paraben
- Ethyl Paraben
- 8-Hydroxyquinoline
- 8-Hydroxyquinoline Benzoate
- 8-Hydroxyquinoline Sulfate
- Sodium Caprylate

Generics:
- Acetazolamide
- Benzocaine
- Bisacodyl
- Chlorpheniramine Maleate
- Cyclandelate
- Dicyclomine Hydrochloride
- Diethylcarbamazine Citrate
- Diiodohyroxyquin
- Diphenyhydramine Hydrochloride
- Ephedrine Hydrochloride
- Furazolidone
- Hydralazine Hydrochloride
- Imipramine Hydrochloride
- Iodochlorhydroxyquin
- Isosorbide Dinitrate
- Lidocaine
- Meclizine Hydrochloride
- Methocarbamol
- Neomycin Sulfate
- Nitrofurantoin
- Nitrofurazone
- Nystatin
- Orphenadrine Citrate
- Pheniramine Maleate
- Phenyltoloxamine Dihydrogen Citrate
- Povidone Iodine
- Probenecid
- Promethazine Hydrochloride
- Propantheline Bromide
- Propoxyphene Hydrochloride
- Resorcinol
- Tetrahydrozoline Hydrochloride
- Trihexyphenidyl Hydrochloride
- Vitamin B-12 with Intrinisic Factor

NAPP CHEMICALS, INC.: Pharmaceutical and Cosmetic Products (Continued):

Sulfonamides:
- Sulfacetamide
- Sulfacetamide Sodium
- Sulfadiazine
- Sulfadimethoxine
- Sulfamerazine
- Sulfamethazine
- Sulfamethazine Sodium
- Sulfanilamide
- Sulfaquinoxaline
- Sulfapyridine
- Sulfathizole
- Sulfoxazole
- Phthalylsulfacetamide

Steroids:
- Corticosterone
- Cortisone and Acetate
- Desoxycorticosterone
- Dexamethasone and Salts
- Hydrocortisone and Salts
- Prednisoline and Salts
- Prednisone and Salts
- Triamcinolone

Hormones:
- Aldosterone
- Chorionic Gonadotropin
- Estradiol and Salts
- Estrogenic Substance
- Estrone
- Ethinyl Estradiol
- Ethisterone
- Heparin, Sodium
- Methylandrostenediol
- Methyltestosterone
- Oxytocin Solution

Intermediates/Miscellaneous:
- Benzil
- Benzilic Acid
- Benzoin
- Methyl Benzilate
- Calcium Succinate
- Resorcinol

All products sold to conform to applicable U.S.P. or N.F. Specifications

PENRECO: Petrolatums:

White Petrolatum USP:

PENRECO ULTIMA:
Melting Point, F: 130/140
Viscosity, SUS @ 210F: 60/70
Maximum Lovibond Color 2" Cell: .5Y
Consistency @ 77F: 155/190
Typical Congealing Point F: 130

PENRECO SUPER:
Melting Point, F: 122/135
Viscosity, SUS @ 210F: 60/75
Maximum Lovibond Color 2" Cell: .5Y
Consistency @ 77F: 170/205
Typical Congealing Point F: 125

PENRECO SNOW:
Melting Point, F: 122/135
Viscosity, SUS @ 210F: 64/75
Maximum Lovibond Color 2" Cell: 2Y
Consistency @ 77F: 170/205
Typical Congealing Point F: 123

PENRECO REGENT:
Melting Point, F: 118/130
Viscosity, SUS @ 210F: 57/70
Maximum Lovibond Color 2" Cell: 2Y
Consistency @ 77F: 210/240
Typical Congealing Point F: 120

PENRECO LILY:
Melting Point, F: 122/135
Viscosity, SUS @ 210F: 64/75
Maximum Lovibond Color 2" Cell: 8Y .5R
Consistency @ 77F: 170/205
Typical Congealing Point F: 124

PENRECO CREAM:
Melting Point, F: 122/135
Viscosity, SUS @ 210F: 64/75
Maximum Lovibond Color 2" Color: 18Y .5R
Consistency @ 77F: 175/205
Typical Congealing Point F: 125

PENRECO: Petrolatums(Continued):

Petrolatum (USP):

PENRECO ROYAL:
Melting Point F: 118/130
Viscosity, SUS @ 210F: 57/70
Maximum Lovibond Color 2" Cell: 35Y 3R
Consistency @ 77F: 210/240
Typical Congealing Point F: 118

PENRECO BLOND:
Melting Point F: 122/135
Viscosity, SUS @ 210F: 68/82
Maximum Lovibond Color 2" Cell: 35Y 3R
Consistency @ 77F: 175/205
Typical Congealing Point F: 123

PENRECO AMBER:
Melting Point F: 122/135
Viscosity, SUS @ 210F: 68/82
Maximum Lovibond Color 2" Cell: 35Y 7R
Consistency @ 77F: 175/205
Typical Congealing Point F: 123

Technical Petrolatum:

PENRECO RED:
Melting Point F: 120/135
Viscosity, SUS @ 210F: 70/82
Maximum Lovibond Color 2" Cell: Red
Consistency @ 77F: 175/205

PENRECO 1520:
Melting Point F: 115/135
Viscosity, SUS @ 210F: 70/115
Maximum Lovibond Color 2" Cell: Dark Green
Consistency @ 77F: 170/260

PENRECO 3070:
Melting Point F: 125/140
Viscosity, SUS @ 210F: 70/95
Maximum Lovibond Color 2" Cell: Dark Green
Consistency @ 77F: 130/175

PENRECO: Petrolatums(Continued):

Ointment Based White Petrolatum, USP:

Ointment Base No. 3:
Melting Point F: 118/125
Viscosity, SUS @ 210F: 55/65
Maximum Lovibond Color 2" Cell: 1.5Y
Consistency @ 77F: 220/250
Congealing Point F: 104/115

Ointment Base No. 4:
Melting Point F: 118/125
Viscosity, SUS @ 210F: 60/70
Maximum Lovibond Color 2" Cell: 1.5Y
Consistency @ 77F: 250/285
Congealing Point F: 109/119

Ointment Base No. 6:
Melting Point F: 122/133
Viscosity, SUS @ 210F: 60/70
Maximum Lovibond Color 2" Cell: 1.5Y
Consistency @ 77F: 195/230
Congealing Point F: 120/130

PENRECO: Petrolatums(Continued):

Mineral Jelly:

Mineral Jelly No. 5:
Viscosity, SUS @ 210F: 38/43
Consistency @ 77F: Over 350
Pour Point F: 75/85
Maximum Lovibond Color 2" Cell: 2Y

Mineral Jelly No. 10:
Saybolt Melting Point, F: 97/105
Viscosity, SUS @ 210F: 40/43
Consistency @ 77F: 270/330
Pour Point F: 95/105
Maximum Lovibond Color 2" Cell: 1Y

Mineral Jelly No. 15:
Saybolt Melting Point, F: 97/108
Viscosity, SUS @ 210F: 40/44
Consistency @ 77F: 170/220
Pour Point, F: 95/105
Maximum Lovibond Color 2" Cell: .5Y

Mineral Jelly No. 20:
Saybolt Melting Point, F: 111/116
Viscosity, SUS @ 210F: 37/40
Consistency @ 77F: 170/220
Pour Point, F: 110/120
Maximum Lovibond Color 2" Cell: .5Y

Mineral Jelly No. 25:
Saybolt Melting Point, F: 103/108
Viscosity, SUS @ 210F: 38/40
Consistency @ 77F: 160/175
Pour Point, F: 100/110
Maximum Lovibond Color 2" Cell: .5Y

PENRECO: White Mineral Oils:

Mineral Oil USP:

DRAKEOL 35:
Viscosity: SUS @ 100F: 340/365
CST @ 40C: 65.8/71.0
API @ 60F: 28.0/31.1
Specific Gravity @ 60C: .870/.887
@ 77F: .864/.881
Flash Point: F: 420
C: 216
Pour Point: F: 5
C: -15

DRAKEOL 34:
Viscosity: SUS @ 100F: 370/410
CST @ 40C: 72.0/79.5
API @ 60F: 29.7/32.3
Specific Gravity @ 60F: .864/.878
@ 77F: .858/.872
Flash Point: F: 460
C: 238
Pour Point: F: 15
C: -9

DRAKEOL 32:
Viscosity: SUS @ 100F: 312/330
CST @ 40C: 60.0/63.3
API @ 60F: 28.9/32.5
Specific Gravity @ 60F: .863/.882
@ 77F: .856/.876
Flash Point: F: 415
C: 213
Pour Point: F: 10
C: -12

DRAKEOL 21:
Viscosity: SUS @ 100F: 200/215
CST @ 40C: 38.4/41.5
API @ 60F: 28.9/33.2
Specific Gravity @ 60F: .859/.882
@ 77F: .853/.876
Flash Point: F: 380
C: 193
Pour Point: F: 10
C: -12

PENRECO: White Mineral Oils(Continued):

Mineral Oil USP(Continued):

DRAKEOL 19:
Viscosity: SUS @ 100F: 180/190
CST @ 40C: 34.9/37.3
API @ 60F: 28.9/33.4
Specific Gravity @ 60F: .858/.882
@ 77F: .852/.876
Flash Point: F: 370
C: 188
Pour Point: F: 10
C: -12

Light Mineral Oil NF:

DRAKEOL 15:
Viscosity: SUS @ 100F: 145/155
CST @ 40C: 28.1/30.3
API @ 60F: 28.9/33.8
Specific Gravity: @ 60F: .856/.882
@ 77F: .850/.873
Flash Point: F: 370
C: 188
Pour Point: F: 10
C: -12

DRAKEOL 13:
Viscosity: SUS @ 100F: 125/135
CST @ 40C: 24.2/26.3
API @ 60F: 30.6/34.2
Specific Gravity: @ 60F: .854/.873
@ 77F: .848/.867
Flash Point: F: 365
C: 185
Pour Point: F: 15
C: -9

DRAKEOL 10:
Viscosity: SUS @100F: 95/105
CST @ 40C: 17.7/20.2
API @ 60F: 31.1
36.6
Specific Gravity: @ 60F: .842/.870
@ 77F: .838/.864
Flash Point: F: 360
C: 182
Pour Point: F: 15
C: -9

PENRECO: White Mineral Oils(Continued):

Light Mineral Oil NF(Continued):

DRAKEOL 10B:
```
   Viscosity: SUS @ 100F: 95/105
              CST @  40C: 17.7/20.2
   API @ 60F: 28.5/30.5
   Specific Gravity @ 60F: .874/.884
                    @ 77F: .867/.878
   Flash Point: F: 320
                C: 160
   Pour Point: F: -40
               C: -40
```

DRAKEOL 9:
```
   Viscosity: SUS @ 100F: 80/90
              CST @  40C: 14.2/17.0
   API @ 60F: 33.0/36.0
   Specific Gravity @ 60F: .845/.860
                    @ 77F: .838/.854
   Flash Point: F: 355
                C: 179
   Pour Point: F: 15
               C: -9
```

DRAKEOL 7:
```
   Viscosity: SUS @ 100F: 65/75
              CST @  40C: 10.8/13.6
   API @ 60F: 35.0/38.2
   Specific Gravity @ 60F: .834/.850
                    @ 77F: .828/.843
   Flash Point: F: 350
                C: 177
   Pour Point: F: 15
               C: -9
```

DRAKEOL 5:
```
   Viscosity: SUS @ 100F: 52/57
              CST @  40C: 7.6/8.7
   API @ 60F: 36.9/38.7
   Specific Gravity @ 60F: .831/.842
                    @ 77F: .825/.835
   Flash Point: F: 310
                C: 154
   Pour Point: F: 15
               C: -9
```

PENRECO: White Mineral Oils(Continued):

Light Mineral Oil NF(Continued):

DRAKETEX 50:
Viscosity: SUS @ 100F: 48/53
CST @ 40C: 6.5/7.8
API @ 60F: 37.4/40.2
Specific Gravity: @ 60F: .824/.838
@ 77F: .817/.832
Flash Point: F: 305
C: 152
Pour Point: F: 15
C: -9

Technical Mineral Oil:

PENETECK:
Viscosity: SUS @ 100F: 38/42
CST @ 40C: 3.4/4.7
API @ 60F: 41.5/43.6
Specific Gravity: @ 60F: .808/.818
@ 77F: .802/.811
Flash Point: F: 265
C: 129
Pour Point: F: 30
C: -1

PAROL 100:
Viscosity: SUS @ 100F: 95/105
CST @ 40C: 17.7/20.2
API @ 60F: 31.1/36.6
Specific Gravity: @ 60F: .842/.870
@ 77F: .838/.864
Flash Point: F: 360
C: 182
Pour Point: F: 15
C: -9

PAROL 80:
Viscosity: SUS @ 100F: 75/90
CST @ 40C: 13.2/17.0
API @ 60F: 32.5/37.6
Specific Gravity: @ 60F: .837/.863
@ 77F: .830/.857
Flash Point: F: 355
C: 179
Pour Point: F: 15
C: -9

PENRECO: White Mineral Oils(Continued):

Technical Mineral Oil(Continued):

PAROL 70:

Viscosity: SUS @ 100F: 65/75
CST @ 40C: 10.8/13.6
API @ 60F: 35.0/38.2
Specific Gravity: @ 60F: .834/.855
@ 77F: .828/.849
Flash Point: F: 350
C: 177
Pour Point: F: 15
C: -9

4463 Oil:

Viscosity: SUS @ 100F: 42/59
API @ 60F: 38.0/42.0
Specific Gravity: @ 60F: .815/.834
Flash Point: F: 310
C: 154
Pour Point: F: 50
C: 10

6970 Oil:

Viscosity: SUS @ 100F: 55/65
API @ 60F: 34.0/38.0
Specific Gravity @ 60F: .835/.855
Flash Point: F: 315
C: 157
Pour Point: F: 15
C: -9

Government Regulations:

- The products listed above are included in the EPA's TSCA (Toxic Substances Control Act) Inventory.
- The CAS Number for all products listed above is 8042-47-5

PETROLITE SPECIALTY POLYMERS GROUP: Microcrystalline Waxes and Synthetic Polymers:

Microcrystalline Waxes:
Petrolite supplies both soft and hard microcrystalline waxes to the cosmetics industry.

VYBAR Polymers:
VYBAR polymers are ethylene derived hydrocarbon products.

SILTEK Polymers:
SILTEK polyethylenes are used in a variety of applications.

UNILIN Alcohols and UNITHOX Ethoxylates:
UNILIN alcohols offer many formulating possibilities. They are long chain linear alcohols.

PETRONAUBA C:
PETRONAUBA C wax is a synthetic substitute for carnauba wax in water-based formulations.

Product:

VICTORY White:
Viscosity @ 210F, SUS: 85
Melt Point F: 174

VICTORY Amber:
Viscosity @ 210F, SUS: 85
Melt Point F: 174

BE SQUARE 175 Amber:
Viscosity @ 210F, SUS: 87
Melt Point F: 182

BE SQUARE 195 White:
Viscosity @ 210F, SUS: 95
Melt Point F: 198

BE SQUARE 195 Amber
Viscosity @ 210F, SUS: 97
Melt Point F: 198

PETROLITE SPECIALTY POLYMERS GROUP: Microcrystalline Waxes and Synthetic Polymers(Continued):

Product:

SILTEX M:
Viscosity @ 300F, SUS: 47
Melt Point F: 211

SILTEX L:
Viscosity @ 210F, SUS: 70
Melt Point F: 202

SILTEX M SUPER:
Viscosity @ 300F, SUS: 47
Melt Point F: 211

SILTEX PL:
Viscosity @ 210F, SUS: 54
Melt Point F: 190

VYBAR 825:
Viscosity @ 210F, SUS: 320

VYBAR 5013:
Viscosity @ 210F, SUS: 45
Melt Point F: 135

UNILIN 425:
Viscosity @ 300F, SUS: 39
Melt Point F: 197

UNILIN 550:
Viscosity @ 300F, SUS: 48
Melt Point F: 210

UNILIN 700:
Viscosity @ 300F, SUS: 60
Melt Point F: 222

UNITHOX 450:
Viscosity @ 210F: 102
Melt Point F: 194

UNITHOX 550:
Viscosity @ 300F: 71
Melt Point F: 209

UNITHOX 750:
Melt Point F: 221

PETRONAUBA C:
Viscosity @ 210F: 160
Melt Point F: 199

PFIZER CHEMICAL DIVISION: Pfizer Citric Acid: Pharmaceutical and Cosmetic Applications:

Add effervescence

Internal preparations

External uses

Oral dosage forms

Blood anticoagulants

Six ways citric aids cosmetics and toiletries:

The buffering, solubilizing and chelating properties of citric acid indicate usefulness in antioxidant systems, acid hair rinses and permanent wave neutralizer solutions. The improved luster and resilience of hair washed with low-pH (4.0-6.0) shampoos has resulted in a new generation of products containing citric acid. Citric's low toxicity and compatibility with a wide range of ingredients make it the product of choice in topicals and specialty items such as feminine hygiene products, lotions, creams and toothpastes.

Other citrates:

Potassium citrate:

Potassium citrate, a monohydrate salt in the form of odorless white granules, is readily soluble in water and extremely hygroscopic.

Potassium citrate can be used for most of the applications of sodium citrate. However, price, flavor characteristics and hygroscopicity, limit commercial uses.

Ferric ammonium citrate:

Ferric ammonium citrate is a dietary supplement, particularly useful in liquid products, including milk, as a palatable iron source.

Calcium citrate:

Calcium citrate is used as a firming agent for fruits and vegetables.

Ammonium citrate dibasic:

Where transportation of liquid citrate is inconvenient, ammonium citrate dibasic salt can be dissolved on site for use in rust and scale removal, and as a buffer for plating baths.

PFIZER CHEMICAL DIVISION: Pfizer Fragrance and Flavor Modifiers in Pharmaceutical and Cosmetic Applications:

VELTOL-Plus:
(ethyl maltol, FCC, Pfizer)

VELTOL:
(maltol, FCC, Pfizer)

Summary:
Pfizer modifiers (VELTOL, VELTOL-Plus) or combinations (VELTOL/VELTOL-Plus) create favorable results in pharmaceutical and cosmetic products:

- Masking bitter notes of vitamins and cough syrup
- Increasing fragrance impact in candles and talc
- Creating distinctive fragrance in shampoos
- Generating definitive flavor preference in toothpaste

Favorable Results:
Chewable Multivitamin Tablets
Cherry Cough Syrup with d-methophan
Compressed Mint Sorbitol Troche with d-methorphan
Blueberry-Scented Candles
Mild Spice-Scented Candles
Baby Talc
Hyacinth-Scented Shampoo
Wintergreen-Flavored Toothpaste

Food Additive Status:
The Food and Drug Administration (FDA) includes ethyl maltol (VELTOL-Plus) and maltol (VELTOL) in its list of "Synthetic falvoring substances and adjuvants" (21CFR 172.515) as safe for use in foods.

Physical and Chemical Properties:
Pfizer VELTOL-Plus is a white crystalline powder with a pleasant aroma.
Empirical Formula: $C_7H_8O_3$
Molecular Weight: 140.14
Melting Range: About 90C
3-Hydroxy-2-ethyl-4-pyrone

Pfizer VELTOL is a white crystalline powder with a pleasant aroma.
Formula: $C_6H_6O_3$
Molecular Weight: 126.11
Melting Range: 160-164C
3-Hydroxy-2-methyl-4-pyrone

PFIZER CHEMICAL DIVISION: Pfizer Products for the Pharmaceutical and Cosmetic Industries:

Acidulants:
- Citric Acid USP, FCC
- Fumaric Acid NF, FCC
- Tartaric Acid NF, FCC
- Potassium Acid Tartrate FCC
- Glucono-Delta-Lactone

Antibiotics:
- Dihydrostrepomycin Sulfate Veterinary
- Potassium Penicillin G Crude
- Potassium Penicillin G USP
- Potassium Phenoxymethyl Penicillin USP (VK)
- Procaine Penicillin G USP
- Polymyin B Sulfate USP
- Streptomycin Sulfate Technical

Buffering Agents--Salts:
- Potassium Citrate USP, FCC
- Sodium Citrate USP, FCC
- Sodium Gluconate FCC

Mineral Products:
- Calcium Citrate FCC
- Ferric Ammonium Citrate FCC

Preservatives:
- Potassium Benzoate--Food Grade
- Sodium Benzoate NF, FCC
- SORBISTAT (Sorbic Acid NF, FCC)
- SORBISTAT-K (Potassium Sorbate NF, FCC)

Vitamins:
- STABLETS (1% Cyanocobalamin USP, FCC Adsorbed on Resin)
- Ascorbic Acid USP, FCC
- Sodium Ascorbate USP, FCC

Other Pharmaceutical Products:
- Caffeine USP, FCC
- Sorbitol NF, FCC Crystalline
- Sorbitol Solution USP, FCC
- Polyol-P Noncrystallizing Polyol Solution Technical Grade
- VELTOL (Maltol FCC)
- VELTOL-Plus (Ethyl Maltol FCC)

PFIZER CHEMICAL DIVISION: Sorbitol In Pharmaceuticals and Cosmetics:

Sorbitol NF, FCC Crystalline (Pfizer):

Sorbitol Solution USP, FCC (Pfizer):

Non-Crystallizing Sorbitol Solution (NCS), Manufacturing Grade (Pfizer):

Summary:

Sorbitol NF, FCC Crystalline is a white hygroscopic granule or powder. Special Pfizer granulations (Coarse Powder, Tablet Type and 60 Mesh Powder) are designed for direct compression of chewable, non-chewable tablets and troches. Granular and powder also are available. Pfizer crystalline sorbitol is a pure, stable, sweet, cool-tasting polyol which is metabolized as a carbohydrate but at a slower rate. It is slowly fermented by oral microorganisms at rates far below sugars and many other polyols.

Sorbitol Solution USP, FCC is a clear, colorless, odorless, syrupy aqueous solution containing approximately 70% (w/w) of D-sorbitol.

Non-Crystallizing Sorbitol Solution (NCS), Manufacturing Grade is a clear colorless, odorless, syrupy aqueous 70% solution containing approximately 50% D-sorbitol plus smaller amounts of hydrogenated oligosaccharides.

Sorbitol solutions are superior humectants for moisture conditioning of wide range or products. They contribute body and flavor to vehicles for pharmaceuticals, cosmetics and toiletries. Non-crystallizing sorbitol is recommended for use in toothpastes, mouthwashes and other dental hygiene products.

Description:

Sorbitol NF, FCC Crystalline (Pfizer) is a white hygroscopic powder or granular material. It is about 60% as sweet as sucrose, and cooling to the taste owing to its negative heat of solution.

Sorbitol Crystalline

Empirical Formula: $C_6H_{14}O_6$

Solubility:

Very soluble in water (up to 83% w/w)

Quite soluble in glycerin, propylene glycol and hot ethanol

Heat of Solution: -26 calories per gram

Refractive Index: n^{25}/D = 1.3477 (10% aqueous solution)

Description:

Sorbitol Solution USP, FCC (Pfizer) is a water solution which contains 70 to 71% total solids.

Empirical Formula: $C_6H_{14}O_6$

Molecular Weight: 182.17

Solubility:

Miscible with water, glycerin and propylene glycol

Soluble in alcohol and practically insoluble in other common organic solvents

Specific Gravity (25C/25C): Minimum 1.285

Refractive Index (20C): 1.455-1.465

PILOT CHEMICAL CO.: CALFOAM Alcohol & Ether Sulfates:

CALFOAM ES-30:
Sodium Lauryl Ether Sulfate 30%
Liquid
Uses: A high quality flash foamer for shampoos, bubble baths, wool washing.

CALFOAM SEL-60:
Sodium Lauryl Ether Sulfate 60%
Liquid
Uses: For use in bubble baths, shampoos, car washing, liquid detergents, general all-purpose cleaning and wetting.

CALFOAM NEL-60:
Ammonium Lauryl Ether Sulfate 60%
Liquid
Uses: For use in bubble baths, shampoos, car washing, liquid detergents, general all-purpose cleaning and wetting.

CALFOAM SLS-30:
Sodium Lauryl Sulfate 30%
Liquid
Uses: Extremely mild detergent for shampoos, bubble bath, rug cleaner formulations, cosmetic emulsification.

CALFOAM AAL:
Concentrated blend of Anionic Detergents and Amide, 45%
Liquid
Uses: High performance detergent for dishwashing and for all purpose cleaning.

CALFOAM LLD:
Blend of Anionic Detergents and Amide 42%
Liquid
Uses: Light duty liquid for washing dishes and fine hand washables.

POKONOBE INDUSTRIES, INC.: Unrefined, Refined, and USP/NF Specialty Vegetable Oils for the Food, Cosmetic, and Pharmaceutical Industries:

Almond Oil (Sweet)
Aloe Vera Products
Apricot Kernel Oil
Avocado Oil
Camellia Oil
Canola Oil
Cocoa Butter
Coconut Oil
Corn Oil
Cottonseed Oil
Grapeseed Oil
Hazelnut Oil
Jojoba Oil
Kukui Nut Oil
Linseed Oil
Macadamia Nut Oil
Olive Oil
Passion Fruit Oil
Peanut Oil
Pecan Oil
Pumpkinseed Oil
Rice Bran Oil
Safflower Oil
Sesame Oil
Soybean Oil
Sunflower Oil
Walnut Oil
Wheat Germ Oil

Other oils may be available on special request.

POLYESTHER CORP.: Product List:

Fatty Acids:
A general range of drying, semi-drying and non-drying fatty acids, including: Castor, Coconut, Linseed, Marine Oil, Palm Kernel, Rapeseed, Soya, Safflower and Sunflower.

Fractionated Fatty Acids:
Natural: Caproic, Caprylic, Capric, Lauric, Myristic, Palmitic, Oleic, Stearic, Triple Based Stearic, Arachidic, Behenic, Erucic and various Blends
Synthetic: Caproic and Short Chain Fatty Acid Blends.
Food Grade/Kosher: Caproic, Caprylic, Capric, Lauric, Myristic, Palmitic, Stearic, Triple Pressed Stearic, Oleic, Palm Kernel and other Blends.

Specialty Fatty Acids: Ricinoleic, 12-Hydroxy Stearic, Iso-Stearic, Dehydrated Castor, Conjugated Fatty Acids and other Unique Fatty Acids.

Fatty Alcohols: Natural--Octyl, Decyl, Lauryl, Myristyl, Cetyl, Stearyl, Oleyl, Behenyl, Isostearyl and various Blends.

Glycerine: 99.5% and 96% Grades: U.S.P., Natural, Food Grade Kosher.

Fatty Acid Esters: Isopropyl Myristate (IPM), Isopropyl Palmitate (IPP), Medium Chain Triglycerides and various Stearates and Oleates.

Miscellaneous Oils: Sesame Oil N.F., Sunflowerseed Oil, Grapeseed Oil, Castor Oil, Rapeseed Oil (Canola Oil), Squalane, Squalene and various Cosmetic Oils.

Polysynlane: An Established Cosmetic Oil Base considered to be the First Worthy substitute for Squalane (Perhydrosqualene).

Pure Silk Powder: 100% Pure, Natural Silk Powder, Sterilized.

PPG/MAZER CHEMICALS: AVANEL S-150: Sodium C12-15 Pareth-15 Sulfonate (CTFA):

- Counter-Irritant Anionic Surfactant
- pH-Stable Nonirritating Emulsifier

AVANEL S-150 is Sodium C12-15 Pareth-15 Sulfonate. It is one of a family of specialty anionic surfactants. Its structure is unique among commercially available surfactants, with a hydrophobic portion consisting of a linear alkyl chain, and a hydrophilic portion consisting of 15 moles of ethoxylation terminated with a sulfonate anionic group. This results in physical and performance properties which are different from those of the common sulfonate- and sulfate-type surfactants. In some cases these properties are quite surprising. The remarkable mildness and counter-irritancy behavior of AVANEL S-150, for example. Also, the lower-than-expected HLB of 15.4 and the excellent chemical stability in extremely high pH or low pH systems.

Physical Properties:
Appearance: Clear liquid
Color, APHA: 150 max.
pH, 10% Aq: 6.0-8.0
Viscosity: 250 cps max.
Specific Gravity @ 25C: 1.07 typical
Solids: 33.5%-36.5%
Iron: less than 5 ppm
Avg. Molecular Weight: 950
CMC, Solids basis: 0.008%
HLB: 15.4

Performance Properties:
Surface Tension @ CMC: 41.5 dynes/cm
Foaming, Ross-Miles, 0.1% solids: Initial 160 mm/
5 min.: 130 mm
Chemical Stability, 48 hrs @ 80C:
in 10% NaOH: No change
in 10% HNO3: No change

Features:
Excellent hydrolytic stability in all pH ranges
Low CMC
Good foaming ability
Very low color and odor
Totally biodegradable
Good emulsifying characteristics

Mildness:
Primary skin and eye irritation, and oral toxicity data verify that AVANEL S-150 is among the mildest surfactants available.

PPG/MAZER CHEMICALS: JORDAPON CI Sodium Cocoyl Isethionate:

JORDAPON CI features the following attributes:

- Superior Mildness
- Excellent Detergency, Foaming, and Lime Soap Dispersancy
- Excellent Emolliency and feel on the skin
- Dense Foam
- Rich, Creamy Lather
- Highly Surface Active
- Mild Odor

Typical Performance:
Foam Height-Ross Miles(0.2% solution @ 50C): Initial 220 mm
5 Min. 220 mm
Lime-Soap Dispersion Index: 15-20
Draves Wetting: 0.04%
(Required % for wetting in 25 sec. @ 70C)
Surface Tension, dynes/cm @ 25C:
0.01% solution: 33
0.1% solution: 27
Interfacial Tension vs. Mineral Oil, dynes/cm @ 25C:
0.01% solution: 18
0.1% solution: 4
Solubility in Water (gms/100 ml):
@ 25C: 0.01
@ 70C: >50

JORDAPON CI Powder: Sodium Cocoyl Isethionate:

The specialized processing of JORDAPON CI Powder provides material of consistently high activity; essentially no sodium chloride; low free fatty matter; and very low moisture. These properties, along with low color and odor profiles, have made JORDAPON CI Powder the preferred ingredient in the formulation of syndet bars and soap/syndet combo bars.

Typical Properties:
Activity: 82%
Moisture: 0.5%
pH, 5% solution: 6.0
APHA Color, 5% solution: 15
Free Fatty Matter: 8%
NaCl: 0.05%

PPG/MAZER CHEMICALS: JORDAPON CI Flake: Sodium Cocoyl Isethionate (and) Stearic Acid:

This new form of JORDAPON CI, developed by PPG/Mazer, is a blend of Sodium Cocoyl Isethionate and stearic acid. JORDAPON CI Flake is flowable, nondusting, and offers the mildness and skin feel that has made JORDAPON CI the leading syndet used today for super mild cleansing bars. Originally designed to simplify the commercial production of syndet and soap/syndet bars. JORDAPON CI Flake also functions as an effective emulsifier for creams and lotions.

Typical Properties:
- Activity as Sodium Cocoyl Isethionate: 48%
- Fatty Acid: 45%
- Moisture: 0.5%
- pH, 5% solution: 5.8

JORDAPON CI Dispersion: Sodium Cocoyl Isethionate:

JORDAPON CI Dispersion is a new aqueous form containing 50% active Sodium Cocoyl Isethionate. It is specifically designed for use in liquids and emulsions. In liquids, it is readily solubilized by other surfactants, permitting clear products to be easily formulated. In these systems, JORDAPON CI Dispersion often exhibits viscosity building or potentiating properties. In emulsions, it serves as an emulsifier, opacifier, and foaming agent. All product forms benefit from the mildness, hard water tolerance, and skin after-feel which JORDAPON CI Dispersion can impart.

Typical Properties:
- Activity: 50%
- Moisture: 33%
- pH, 10% solution: 6.0

PPG/MAZER CHEMICALS: MASIL 556: Phenyl Trimethicone (CTFA):

MASIL 556 is a phenyl-substituted silicone fluid which is similar to dimethicone in its emolliency, water repellency, and lubricity, but with the important advantage of compatibility with organic emollients. The CTFA lists MASIL 556 as Phenyl Trimethicone. Chemically, it is a mixture of oligomers. In hair and skin products, MASIL 556 offers the following benefits:

- Shine, gloss
- Water barrier properties
- Reduced tackiness
- Improved lubricity
- Emolliency, moisturization
- Organic emollient compatibility

Physical Properties:
Appearance: Clear, water-white fluid
Viscosity: <50 cps
Specific Gravity: 0.980
Refractive Index: 1.460
Melting Point: <-40F/-40C
Boiling Point: >400F/200C
Flash Point: 250F/121C, closed cup

Solubility:
Soluble in: Mineral Oil
Dimethicone
Cyclomethicone
Corn Oil
Isopropyl Myristate
Ethanol (190 proof)
Dispersible in: Butylene Glycol
Propylene Glycol
Insoluble in: Water
Glycerine

PPG/MAZER CHEMICALS: MASILWAX 135 Silicone Wax:

A new, high purity silicone wax offering these features to the formulator of Personal Care Products:

- Emolliency (non-tacky, non-greasy)
- Excellent Slip
- Easily Emulsified
- Compatibility with non-polar organics
- Bland Taste
- Melting Point just below body temperature
- Non-Irritating

MASILWAX 135 is a new organosilicone wax consisting of a short dimethicone chain with pendant stearoxy groups. Physically, it is a soft white wax which melts slightly below body temperature. MASILWAX 135 performs as an emollient, a slip aid, and contibutes to hydrophobicity in skin care and makeup formulations. It is recommended for evaluation in antiperspirant sticks, lipstick and other makeup products as well as creams and lotions.

Typical Properties:
Physical Form: Soft wax
Melting Point, C: Approximately 33
Color, Gardner: 2
Acid Number: <0.5
Flash Point: Above 150C (300F)
Approximate MW: 1000
Free Stearyl Alcohol: <5%

Solubility (at 5%, w/w):
Insoluble in: Water
Ethanol
Propylene Glycol
Cyclomethicone
Dispersible in: PPG 14 Butyl Ether
Castor Oil
Dimethicone (1000 cps)
Soluble in: Isopropyl Myristate
Capric/Caprylic Triglyceride
Octyl Palmitate
Lanolin
Mineral Oil

Applications:
Antiperspirant Sticks
Lotions
Lipstick Base

PPG/MAZER CHEMICALS: M-QUAT Hair Conditioners:

M-QUAT JS-25:
Appearance: White paste (at 25C)
Activity (%): 17-19
CTFA: Stearalkonium Chloride

M-QUAT JO-50:
Appearance: Clear, viscous liquid
Activity (%): 50 min.
CTFA: Oleaklonium Chloride

M-QUAT 522:
Appearance: Clear, viscous liquid
Activity (%): 85 min.
CTFA: Isostearamidopropyl Ethyldimonium Ethosulfate

M-QUAT 1033:
Appearance: Clear liquid
Activity (%): 54-56
CTFA: Soya Ethyldimonium Ethosulfate

M-QUAT DIMER 18:
Appearance: Paste (at 25C)
Activity (%): 44-46
CTFA: Hydroxypropyl Bisstearyldimonium Chloride

Performance Highlights:
M-QUAT JS-25:
M-QUAT JS-25 is the traditional choice for after shampoo conditioners. The hair is left soft, manageable and easier to comb, either wet or dry.
M-QUAT JO-50:
Because of its water solubility, JO-50 is an excellent choice for clear hair rinses and conditioners. It provides excellent conditioning and antistatic properties. It is also recommended for evaluation in hair sprays and styling lotions and gels.
M-QUAT 522 and M-QUAT 1033:
These products provide excellent conditioning and are unique in that they are anionic compatible, that is, they form clear, stable solutions with anionic surfactants like Alkyl Sulfates while retaining their conditioning properties. These characteristics suggest their utility in conditioning shampoos. In addition they aid viscosity building in alpha-olefin sulfonate (AOS) systems. They are also adaptable to clear formulations.
M-QUAT DIMER 18:
M-QUAT DIMER 18 is PPG-Mazer's patented dimer quat which provides the following performance features:
- exceptional mildness (low skin and eye irritation) compared to stearalkonium chloride
- lower levels of build-up in multiple use compared to polymeric quats
- excellent conditioning characteristics

PROCTER & GAMBLE: Fatty Alcohols:

A line of natural fatty alcohol fractions ranging in chain lengths from C8-C18. Applications for fatty alcohols include: alkyl sulfates and ethoxylates, cosmetic ingredients, alkyl halides and esters.

CO-810 Octyl/Decyl:
Hydroxyl Value: 400
Acid Value: 0
Saponification Value: 0.2
Iodine Value: 0
Moisture, (%, KF): 0.04
Color--APHA: 3
Specific Gravity: 25C: 0.823
Melting Point: -15C
Appearance: water white mobile liquid
Composition (GC%):
C6: 0.8
C8: 58
C10: 38
C12: 1.7
Chemical Abstract Number: 68603-15-6

CO-1218 Broad Range:
Hydroxyl Value: 262
Acid Value: 0.1
Saponification Value: 0.8
Iodine Value: 0.3
Moisture, (%, KF): 0.1
Color--APHA: 10-20
Specific Gravity 25C: 0.826
Melting Point: 28C
Appearance: white semi-solid
Composition (GC%):
C8: 0
C10: 0.2
C12: 46
C14: 20
C16: 13
C18: 22
C20: 0.2
Chemical Abstract Number: 67762-25-8

PROCTER & GAMBLE: **Fatty Alcohols(Continued):**

CO-1214 Lauryl:

Hydroxyl Value: 285
Acid Value: 0.01
Saponification Value: 0.2
Iodine Value: 0
Moisture, (%, KF): 0.04
P&G Acid Heat Stability: 97
Color--APHA: 3
Specific Gravity: 25C: 0.823
Melting Point: 22C
Appearance: water white mobile liquid
Composition (GC%):
C8: 0
C10: 0.3
C12: 67
C14: 26
C16: 5.8
C18: 0.2
Chemical Abstract Number: 67762-41-8

CO-1695 Cetyl:

Hydroxyl Value: 229
Acid Value: 0
Saponification Value: 0.4
Iodine Value: 0.6
Moisture, (%, KF): 0.04
Color--APHA: 6-10
Specific Gravity: 25C: 0.814
Melting Point: 49C
Appearance: waxy white solid
Composition (GC%):
C14: 0.4
C16: 96.0
C18: 2.3
Chemical Abstract Number: 36653-82-4

CO-1895 Stearyl:

Hydroxyl Value: 206
Acid Value: 0
Saponification Value: 0.5
Iodine Value: 0.7
Moisture, (%, KF): 0.03
Color--APHA: 6-15
Specific Gravity: 65C: 0.811
Melting Point: 58C
Appearance: waxy white solid
Composition (GC%):
C16: 1.8
C18: 96.5
C20: 0.8
Chemical Abstract Number: 112-92-5

PROCTER & GAMBLE: **Fatty Alcohols(Continued):**

CO-1897 Stearyl:
Hydroxyl Value: 206
Acid Value: 0
Saponification Value: 0.3
Iodine Value: 0.6
Moisture, (%, KF): 0.04
Color--APHA: 6-15
Specific Gravity: 25C: 0.811
Melting Point: 58C
Appearance: waxy white solid
Composition (GC%):
C16: 0.6
C18: 97.9
C20: 0.8
Chemical Abstract Number: 112-92-5

TA-1618 Tallow Type:
Hydroxyl Value: 208
Acid Value: 0
Saponification Value: 1.8
Iodine Value: 0.5
Moisture, (%, KF): 0.03
Color--APHA: 10-20
Specific Gravity: 65C: 0.810
Melting Point: 53C
Appearance: waxy white solid
Composition (GC%):
C12: 0.1
C14: 1.7
C15: 0.1
C16: 26
C17: 0.9
C18: 67
C20: 0.9
Chemical Abstract Number: 67762-30-5

PROCTER & GAMBLE: Glycerine:

Glycerine is a very versatile chemical and is used in such diverse products as pharmaceuticals and toilet goods, tobacco, alkyds, food products, explosives, cellophane, urethane foam and a wide variety of other industries.

SUPEROL Glycerine-U.S.P. Food Grade:
Glycerol (Bosart & Snoddy tables): 99.9
Specific Gravity at 25/25C (77/77F): 1.2618
Color, APHA Pt-Co (Hazen) scale: 10 maximum (7)
Residue on ignition: 0.007% or 70 ppm max
Chlorides (as chlorine): 0.0006% or 6 ppm max
Sulfates: 0.002% or 20 ppm max
Arsenic (as As1): 0.00015% or 1.5 ppm max
Heavy Metals (as Pb): 0.0005% or 5 ppm max
Chlorinated Compounds (as Cl): 0.003% or 30 ppm max
Fatty Acids and Esters: Not more than 0.3 ml. N/2 NaOH is absorbed by 50 g of glycerine, which is equivalent to 0.009% as Na2O, maximum.

STAR Glycerine-U.S.P. Food Grade:
Glycerol (Bosart & Snoddy tables): 96.3
Specific Gravity at 25/25C (77/77F): 1.2525
Color, APHA Pt-Co (Hazen) scale: 10 maximum (6)
Residue on ignition: 0.007% or 70 ppm max
Chlorides (as chlorine): 0.0006% or 6 ppm max
Sulfates: 0.002% or 20 ppm max
Arsenic (As As1): 0.00015% or 1.5 ppm max
Heavy Metals (as Pb): 0.0005% or 5 ppm max
Chlorinated Compounds (as Cl): 0.003% or 30 ppm max
Fatty Acids and Esters: Not more than 0.3 ml. N/2 NaOH is absorbed by 50 g of glycerine, which is equivalent to 0.009% as Na2O, maximum.

Chemical Abstract Number: 58-81-5, for both brands

PROTAMEEN CHEMICALS INC.: Amine Condensate 1:2 FA-Diethanolamides--Monoethanolamides:

PROTAMIDE CKD:
Coconut Amine Condensate
Cocamide DEA

PROTAMIDE X-45-B:
Coconut Amine Condensate
Cocamide DEA

PROTAMIDE DCAW:
Coconut Amine Condensate
Cocamide DEA

PROTAMIDE AR:
Coconut Amine Condensate
Cocamide DEA

PROTAMIDE OFO:
Oleic Amine Condensate
Oleamide DEA

PROTAMIDE L-5560:
Lauric Amine Condensate
Lauramide DEA

PROTAMIDE LM-5560:
Lauric/Myristic Amine Cond.
Lauramide DEA

PROTAMIDE DB:
Lauric Amine Condensate
Lauramide DEA

PROTAMIDE N-1918:
Stearic Amine Condensate
Stearamide DEA

PROTAMIDE L-80M:
Lauric Amine Condensate
Lauramide DEA

PROTAMIDE CME-CO:
Coconut Amine Condensate
Cocamide DEA

PROTAMIDE LNO:
Linoleic Amide
Linoleamide DEA

PROTAMEEN CHEMICALS, INC.: Amphoterics:

PROTACHEM JS:
N-Cocamido Propyl-N, N-Dimethyl N-2 Hydroxy Propyl Sulfo Betaine
Cocamidopropyl Hydroxysultaine

PROTACHEM CB-32:
Cocamidopropyl Betaine
Cocamidopropyl Betaine

PROTERIC CAB:
Cocamidopropyl Betaine
Cocamidopropyl Betaine

PROTERIC CDX-38:
Cocoamphocarboxyglycinate
Cocoamphocarboxyglycinate

PROTERIC CDL:
Cocoamphocarboxyglycinate/SLES SLS

PROTERIC CDTD:
Cocoamphocarboxyglycinate/Sodium Trideceth Sulfate

PROTAMEEN CHEMICALS, INC.: Hydrolyzed Proteins:

PROCHEM 100CG Powder:
Hydrolyzed Collagen Protein (MW 1000)
Hydrolyzed Collagen Protein

PROCHEM SPA:
Hydrolyzed Animal Protein (MW 2000)
Hydrolyzed Collagen Protein

PROTAMEEN CHEMICALS, INC.: Vitamins:

Vitamin E USP:
DL Alpha Tocopheryl Acetate
Vitamin E Synthetic

PROTAMEEN CHEMICALS, INC.: Ethoxylated Aliphatic Amines:

PROTOX C-2:
Coco Amine POE-2
PEG 2 Cocamine

PROTOX C-5:
Coco Amine POE-5
PEG 5 Cocamine

PROTOX C-10:
Coco Amine POE-10
PEG 10 Cocamine

PROTOX C-15:
Coco Amine POE-15
PEG 15 Cocamine

PROTOX O-2:
Oleyl Amine POE-2
PEG 2 Oleamine

PROTOX O-5:
Oleyl Amine POE-5
PEG 5 Oleamine

PROTOX O-15:
Oleyl Amine POE-15
PEG 15 Oleamine

PROTOX S-2:
Soya Amine POE-2
PEG 2 Soyamine

PROTOX S-5:
Soya Amine POE-5
PEG 5 Soyamine

PROTOX S-10:
Soya Amine POE-10
PEG 10 Soyamine

PROTOX S-15:
Soya Amine POE-15
PEG 15 Soyamine

PROTOX T-2:
Tallow Amine POE-2
PEG 2 Tallowamine

PROTOX T-5:
Tallow Amine POE-5
PEG 5 Tallowamine

PROTAMEEN CHEMICALS, INC.: Ethoxylated Aliphatic Amines (Continued):

PROTOX T-15:
Tallow Amine POE-15
PEG 15 Tallowamine

PROTOX T-40:
Tallow Amine POE-40
PEG-40 Tallowamine

PROTOX T-50:
Tallow Amine POE-50
PEG 50 Tallowamine

PROTOX HTA-2:
Stearyl Amine POE-2
PEG 2 Stearamine

PROTOX HTA-5:
Stearyl Amine POE-5
PEG 5 Stearamine

PROTOX HTA-15:
Stearyl Amine POE-15
PEG 15 Stearamine

PROTOX HTA-50:
Stearyl Amine POE-50
PEG 50 Stearamine

PROTAQUAT 2HT-75:
Di-Methyl Distearyl Ammonium Chloride 75% Active

PROTAQUAT S-13:
Stearamido Propyl Dimethylamine

PROTAMEEN CHEMICALS, INC.: Fatty Esters and Glyceryl Esters:

PROTACHEM SDM:
Stearamidopropyl Dimethylamine Lactate

PROTACHEM MLD:
Glyceryl Monolaurate
Glyceryl Laurate

PROTACHEM EGMS:
Ethylene Glycol Monostearate
Glycol Stearate

PROTACHEM EGDS:
Ethylene Glycol Distearate
Glycol Distearate

PROTACHEM G-556G:
Glyceryl Monostearate POE-3
Polyglyceryl 3 Stearate

PROTACHEM G-5509:
Glyceryl Monostearate POE-9
Polyglyceryl 9 Stearate

PROTACHEM GMS-165:
Glyceryl Monosteaate Ethoxylate
Glyceryl Stearate & PEG 100 Stearate

PROTACHEM GMS-450:
Glyceryl Monostearate Pure
Glyceryl Stearate

PROTACHEM GMS-20:
Glyceryl Monostearate POE-20
Polyglyceryl 20 Stearate

PROTACHEM GMS-78:
Glyceryl Monostearate POE-78
Polyglyceryl 78 Stearate

PROTACHEM HMS:
Glyceryl Monostearate
Glyceryl Stearate

PROTACHEM DGS:
Diglycol Monostearate
Diglycol Stearate

PROTAMEEN CHEMICALS, INC.: Fatty Esters and Glyceryl Esters (Continued):

PROTACHEM DGS-C:
Diglycol Monostearate
Diglycol Stearate

PROTACHEM CER:
Cetyl Ricinoleate
Cetyl Ricinoleate

PROTACHEM IPM:
Isopropyl Myristate
Isopropyl Myristate

PROTACHEM IPP:
Isopropyl Palmitate
Isopropyl Palmitate

PROTACHEM CTG:
Caprylic Capric Triglyceride
Caprylic Capric Triglyceride

PROTACHEM MST:
Synthetic Spermaceti
Myristal Stearate

PROTACHEM GL-7:
Glycerine POE-7
Glycereth 7

PROTACHEM GL-26:
Glycerine POE-26
Glycereth 26

PROTAMEEN CHEMICALS, INC.: Lanolin and Lanolin Derivatives:

Anhydrous Lanolin USP:
Anhydrous Lanolin USP
Anhydrous Lanolin USP

PROTOLAN M-16:
Lanolin Absorption Base
Mineral Oil/Lanolin Alcohol

PROTALAN M-26:
Conc. Lanolin Absorption Base
Mineral Oil/Lanolin Alcohol

PROTALAN L-30:
Lanolin POE-30
PEG 30 Lanolin

PROTALAN L-60:
Lanolin POE-60
PEG 60 Lanolin

PROTALAN L-75/50:
Lanolin POE-75 50% Active
PEG 75 Lanolin

PROTALAN L-75:
Lanolin POE-75
PEG 75 Lanolin

PROTALAN 98:
Polysorbate 80 & Cetyl Acetate & Acetylated Lanolin Alcohol

PROTALAN SS-100:
Petrolatum and Lanolin and Lanolin Alcohol

PROTALAN AC:
Acetylated Cetyl Alcohol & Acetylated Lanolin Alcohol
Cetyl Acetate & Acetylated Lanolin Alcohol

PROTALAN MOD:
Acetylated Lanolin
Acetylated Lanolin

PROTALAN WAX:
Lanolin Wax
Lanolin Wax

PROTALAN OIL:
Lanolin Oil
Lanolin Oil

PROTALAN S-16:
Laneth 16 & Ceteth 16 & Oleth 16 & Steareth 16

PROTALAN H:
Hydroxylated Lanolin
Hydroxylated Lanolin

PROTAMEEN CHEMICALS, INC.: Polyethylene Glycol Esters:

PROTAMATE 200-OC:
PEG 200 Monooleate
PEG-4 Oleate

PROTAMATE 300-OC:
PEG 300 Monooleate
PEG-6 Oleate

PROTAMATE 400-OC:
PEG 400 Monooleate
PEG-8 Oleate

PROTAMATE 600-OC:
PEG 600 Monooleate
PEG-10 Oleate

PROTAMATE 1000-OC:
PEG 1000 Monooleate
PEG-20 Oleate

PROTAMATE 200-DPS:
PEG 200 Monostearate
PEG-4 Stearate

PROTAMATE 300-DPS:
PEG 300 Monostearate
PEG-6 Stearate

PROTAMATE 400-DPS:
PEG 400 Monostearate
PEG-8 Stearate

PROTAMATE 600-DPS:
PEG 600 Monostearate
PEG-12 Stearate

PROTAMATE 1000-DPS:
PEG 1000 Monostearate
PEG-20 Stearate

PROTAMATE 1500-DPS:
PEG 1500 Monostearate
PEG-6-32 Stearate

PROTAMATE 1540-DPS:
PEG 1540 Monostearate
PEG-32 Stearate

PROTAMATE 2000-DPS:
PEG-2000 Monostearate
PEG-40 Stearate

PROTAMEEN CHEMICALS, INC.: Polyethylene Glycol Esters (Continued):

PROTAMATE 4000-DPS:
PEG-4000 Monostearate
PEG-75 Stearate

PROTAMATE 4400-DPS:
PEG 4400 Monostearate
PEG-100 Stearate

PROTAMATE 200-ML:
PEG-200 Monolaurate
PEG-4 Laurate

PROTAMATE 400-ML:
PEG-400 Monolaurate
PEG-8 Laurate

PROTAMATE 600-ML:
PEG-600 Monolaurate
PEG-15 Laurate

PROTAMATE 1000-ML:
PEG-1000 Monolaurate
PEG-22 Laurate

PROTAMATE 200-T:
PEG-200 Monotallate
PEG-4 Tallate

PROTAMATE 400-T:
PEG-400 Monotallate
PEG-8 Tallate

PROTAMATE 600-T:
PEG-600 Monotallate
PEG-12 Tallate

PROTAMATE 1000-T:
PEG-1000 Monotallate
PEG-20 Tallate

PROTAMATE 400-DO:
PEG-400 Dioleate
PEG-8 Dioleate

PROTAMATE 600-DO:
PEG-600 Dioleate
PEG-15 Dioleate

PROTAMEEN CHEMICALS, INC.: Polyethylene Glycol Esters (Continued):

PROTAMATE 400-DS:
PEG-400 Distearate
PEG-8 Distearate

PROTAMATE 600-DS:
PEG-600 Distearate
PEG-12 Distearate

PROTAMATE 6000-DS:
PEG-6000 Distearate
PEG-150 Distearate

PROTAMATE 200-DL:
PEG-200 Dilaurate
PEG-5 Dilaurate

PROTAMATE 400-DL:
PEG-400 Dilaurate
PEG-8 Dilaurate

PROTAMATE 600-DL:
PEG-600 Dilaurate
PEG-12 Dilaurate

PROTAMATE PPG-2000 MO:
PPG-2000 Monooleate
PPG-2000 Monooleate

PROTAMEEN CHEMICALS, INC.: Polyethylene Sorbitan Esters:

PROTASORB P-20:
Sorbitan Monopalmitate POE-20
Polysorbate 40

PROTASORB L-5:
Sorbitan Monolaurate POE-5
Polysorbate 21

PROTASORB L-10-M:
Sorbitan Monolaurate POE-10
Polysorbate 10

PROTASORB L-20:
Sorbitan Monolaurate POE-20
Polysorbate 20

PROTASORB S-4:
Sorbitan Monostearate POE-4
Polysorbate 61

PROTASORB S-20:
Sorbitan Monostearate POE-20
Polysorbate 60

PROTASORB O-5:
Sorbitan Monooleate POE-5
Polysorbate 81

PROTASORB O-20:
Sorbitan Monooleate POE-20
Polysorbate 80

PROTASORB TO-20:
Sorbitan Trioleate POE-20
Polysorbate 85

PROTASORB STS-20:
Sorbitan Tristearate POE-20
Polysorbate 65

Polyethylene Sorbitan Esters--Kosher Grade:

PROTASORB O-20-K:
Sorbitan Monooleate POE-20
Polysorbate 80 K

PROTASORB L-20-K:
Sorbitan Monolaurate POE-20
Polysorbate 20 K

PROTASORB S-20-K:
Sorbitan Monostearate POE-20
Polysorbate 60 K

PROTAMEEN CHEMICALS, INC.: Polyoxyethylene Castor Oil Derivatives:

PROTACHEM CA-9:
Castor Oil POE-9
PEG-9 Castor Oil

PROTACHEM CA-25:
Castor Oil POE-25
PEG-25 Castor Oil

PROTACHEM CA-30:
Castor Oil POE-30
PEG-30 Castor Oil

PROTACHEM CA-40:
Castor Oil POE-40
PEG-40 Castor Oil

PROTACHEM CA-60:
Castor Oil POE-60
PEG-60 Castor Oil

PROTACHEM CA-100:
Castor Oil POE-100
PEG-100 Castor Oil

PROTACHEM CA-200:
Castor Oil POE-200
PEG-200 Castor Oil

PROTACHEM CAH-25:
Hydrogenated Castor Oil POE-25
PEG-25 Hydrogenated Castor Oil

PROTACHEM CAH-40:
Hydrogenated Castor Oil POE-40
PEG-40 Hydrogenated Castor Oil

PROTACHEM CAH-60:
Hydrogenated Castor Oil POE-60
PEG-60 Hydrogenated Castor Oil

PROTACHEM CAH-100:
Hydrogenated Castor Oil POE-100
PEG-100 Hydrogenated Castor Oil

PROTACHEM CAH-200:
Hydrogenated Castor Oil POE-200
PEG-200 Hydrogenated Castor Oil

PROTAMEEN CHEMICALS, INC.: Polyoxyethylene Ethers:

PROCOL LA-3:
Lauryl Alcohol POE-3
Laureth 3

PROCOL LA-4:
Lauryl Alcohol POE-4
Laureth 4

PROCOL LA-7:
Lauryl Alcohol POE-7
Laureth 7

PROCOL LA-10:
Lauryl Alcohol POE-10
Laureth 10

PROCOL LA-12:
Lauryl Alcohol POE-12
Laureth 12

PROCOL LA-23:
Lauryl Alcohol POE-23
Laureth 23

PROCOL LA-30:
Lauryl Alcohol POE-30
Laureth 30

PROCOL LA-40:
Lauryl Alcohol POE-40
Laureth 40

PROCOL MA-4:
Myristal Alcohol POE-4
Myreth 4

PROCOL OA-2:
Oleyl Alcohol POE-2
Oleth 2

PROCOL OA-4:
Oleyl Alcohol POE-4
Oleth 4

PROCOL OA-10:
Oleyl Alcohol POE-10
Oleth 10

PROCOL OA-20:
Oleyl Alcohol POE-20
Oleth 20

PROTAMEEN CHEMICALS, INC.: Polyoxyethylene Ethers(Continued):

PROCOL OA-23:
Oleyl Alcohol POE-23
Oleth 23

PROCOL SA-2:
Stearyl Alcohol POE-2
Steareth 2

PROCOL SA-4:
Stearyl Alcohol POE-4
Steareth 4

PROCOL SA-10:
Stearyl Alcohol POE-10
Steareth 10

PROCOL SA-20:
Stearyl Alcohol POE-20
Steareth 20

PROCOL SA-100:
Stearyl Alcohol POE-100
Steareth 100

PROCOL ST-20G:
Stearyl Alcohol and Ceteareth 20

PROCOL CS-4:
Cetyl Stearyl Alcohol POE-4
Ceteareth 4

PROCOL CS-5:
Cetyl Stearyl Alcohol POE-5
Ceteareth 5

PROCOL CS-10:
Cetyl Stearyl Alcohol POE-10
Ceteareth 10

PROCOL CS-20:
Cetyl Stearyl Alcohol POE-20
Ceteareth 20

PROCOL CS-20D:
Cetearyl Alcohol and Ceteareth 20

PROCOL TDA-3:
Tridecyl Alcohol POE-3
Trideth 3

PROCOL TDA-6:
Tridecyl Alcohol POE-6
Trideth 6

PROTAMEEN CHEMICALS, INC.: Polyoxyethylene Ethers(Continued):

PROCOL TDA-12:
Tridecyl Alcohol POE-12
Trideth 12

PROCOL TDA-15:
Tridecyl Alcohol POE-15
Trideth 15

PROCOL P:
Cetearyl Alcohol & Polysorbate 60 & PEG-150 Stearate & Steareth 20

PROCOL P-31:
Cetearyl Alcohol & Polysorbate 60 & PEG-150 Stearate & Steareth 20

PROCOL CA-1:
Cetyl Alcohol POE-1
Ceteth 1

PROCOL CA-2:
Cetyl Alcohol POE-2
Ceteth 2

PROCOL CA-5:
Cetyl Alcohol POE-5
Ceteth 5

PROCOL CA-10:
Cetyl Alcohol POE-10
Ceteth 10

PROCOL CA-20:
Cetyl Alcohol POE-20
Ceteth 20

PROCOL PSA-11:
PPG-11 Stearyl Ether
PPG-11 Stearyl Ether

PROCOL PSA-15:
PPG-15 Stearyl Ether
PPG-15 Stearyl Ether

PROCOL PMA-3:
PPG-3 Myristal Ether
PPG-3 Myristal Ether

PROTAMEEN CHEMICALS, INC.: Sorbitan Esters:

PROTACHEM SML:
Sorbitan Monolaurate
Sorbitan Laurate

PROTACHEM SMP:
Sorbitan Monopalmitate
Sorbitan Palmitate

PROTACHEM SMS:
Sorbitan Monostearate
Sorbitan Stearate

PROTACHEM SMO:
Sorbitan Monooleate
Sorbitan Oleate

PROTACHEM STO:
Sorbitan Trioleate
Sorbitan Trioleate

PROTACHEM STS:
Sorbitan Tristearate
Sorbitan Tristearate

PROTACHEM SOC:
Sorbitan Sesquioleate
Sorbitan Sesquioleate

PROTAMEEN CHEMICALS, INC.: Preservatives:

Methyl Paraben NF:
NF Grade
Methyl Paraben

Propyl Paraben NF:
NF Grade
Propyl Paraben

Butyl Paraben NF:
NF Grade
Butyl Paraben

Ethyl Paraben NF:
NF Grade
Ethyl Paraben

PROTASTAT P-211:
Blended Preservative

PROTAMEEN CHEMICALS, INC.: Superamides 1:1 FA-Diethanolamides--Monoethanolamides:

PROTAMIDE HCA-RC:
Coconut Oil Amide
Cocamide DEA

PROTAMIDE LM-73:
70/30 Lauric Myristic Amide
Lauramide DEA

PROTAMIDE LM-73L:
70/30 Lauric Myristic Amide
Lauramide DEA

PROTAMIDE LM-73PG:
70/30 Lauric Myristic Amide
Lauramide DEA

PROTAMIDE L-90:
Lauric Amide
Lauramide DEA

PROTAMIDE L-90A:
Lauric Amide
Lauramide DEA

PROTAMIDE LMAV:
Lauric Amide
Lauramide DEA

PROTAMIDE LMFAB:
Lauric Amide
Lauramide DEA

PROTAMIDE 1224:
Lauric Amide
Lauramide DEA

PROTAMIDE MRCA:
Myristic Amide
Myristamide DEA

PROTAMIDE CA:
Ricinoleate Amide
Ricoleamide DEA

PROTAMIDE 15-W:
Linoleic Amide
Linoleamide DEA

PROTAMIDE CME:
Coconut Amide
Cocamide MEA

PROTAMEEN CHEMICALS, INC.: Alkyl Phenol Ethoxylates:

PROTACHEM NP-4:
POE-4 Nonyl Phenol
Nonoxynol 4

PROTACHEM NP-6:
POE-6 Nonyl Phenol
Nonoxynol 6

PROTACHEM NP-9:
POE-9 Nonyl Phenol
Nonoxynol 9

PROTACHEM NP-11:
POE-11 Nonyl Phenol
Nonoxynol 11

PROTACHEM OP-9:
POE-9 Octyl Phenol
Octoxynol 9

PROTAMEEN CHEMICALS, INC.: Polyoxyethylene Phosphate Esters:

PROTAPHOS 400-A:
Oleyl Alcohol POE-4 Phosphate Ester
Oleth 4 Phosphate

PROTAPHOS SDA:
Lauryl Alcohol POE-8 Phosphate Ester
Laureth 8 Phosphate

PROTAPHOS P-610:
Nonyl Phenol POE-9 Phosphate Ester
Nonoxynol 9 Phosphate

PROTAMEEN CHEMICALS, INC.: Methyl Taurate Esters:

PROTAPON 33:
Sodium Methyl Oleyltaurate
Sodium Methyloleyltaurate

PROTAPON 30-A:
Sodium Methyl Cocoyl Taurate 30% Active
Sodium Coco Methyltaurate

PROTAPON AC-78:
Sodium Cocoyl Isethionate
Sodium Cocoyl Isethionate

PROTAPON Powder:
Sodium Methyl Cocoyl Taurate Powder Form 95% Active

QUANTUM CHEMICAL CORP.: EMERY CP/USP Glycerines:

EMERY glycerines are clear, odorless products of the highest quality commercially available. These viscous, hygroscopic liquids possess the high boiling points and excellent solvent properties characteristic of high purity glycerines.

EMERY 912 CP/USP Glycerine:

Glycerol, % min.: 96.0
Specific gravity:
 25/25C, min.: 1.2517
Color, APHA, max.: 20
Residue on ignition, ppm, max.: 100
Chloride, ppm, max.: 10
Sulfate, ppm, max.: 20
Arsenic, ppm, max.: 1.5
Heavy metals, ppm, max.: 5
Readily carbonizable substances, max.: matching fluid H
Chlorinated compounds, ppm, max.: 30
Fatty acids and esters, max.: 1.0

EMERY 916 CP/USP Glycerine:

Glycerol, % min.: 99.7
Specific gravity:
 25/25C, min.: 1.2607
Color, APHA, max.: 10
Residue on ignition, ppm, max.: 100
Chloride, ppm, max.: 10
Sulfate, ppm, max.: 20
Arsenic, ppm, max.: 1.5
Heavy Metals, ppm, max.: 5
Readily carbonizable substances, max.: matching fluid H
Chlorinated compounds, ppm, max.: 30
Fatty acids and esters, max.: 1.0

EMERY 918 CP/USP Ultra Glycerine:

Glycerol, % min.: 99.8
Specific gravity:
 25/25C, min.: 1.2612
 15.5/15.5C, min.: 1.2645
Color, APHA, max.: 100
Residue on ignition, ppm, max.: 100
Chloride, ppm, max.: 5
Sulfate, ppm, max.: 20
Arsenic, ppm, max.: 1.5
Heavy Metals, ppm, max.: 5
Readily carbonizable substances, max.: matching fluid H
Chlorinated compounds, ppm, max.: 30
Fatty acids and esters, max.: 0.18
Saponification value, max.: 0.1

QUANTUM CHEMICAL CORP.: EMERY Short-Chain and Coconut Fatty Acids:

Even-Carbon Acids:

EMERY 657 Caprylic Acid (C8):

QUANTUM produces a commercially pure caprylic acid. This product is a water-white liquid, containing approximately 99% caprylic acid and 1% capric acid.

(CAS #124-07-2)

EMERY 658 Caprylic-Capric Acid (C8-C10):

EMERY 658 is a blend of caprylic and capric acids which exhibits a titer considerably lower than either of the two pure acids alone. With a maximum iodine value of 0.3 and an intial color of 88/99 (% transmission, 450/550 nm), EMERY 658 reflects the same high quality apparent in each of Quantum's even-carbon acids.

(CAS #68937-75-7)

EMERY 659 Capric Acid (C10):

This product contains approximately 97% capric acid, 2% lauric acid and 1% caprylic acid. As can be noted from the titer specification, EMERY 659 is semi-solid at room temperature.

(CAS #334-48-5)

EMERY 650, 651 and 652 Lauric Acids (C12):

EMERY 650 typically contains 71% lauric and 28% myristic acids, while EMERY 651 and EMERY 652 contain 96% and 99% lauric acid, respectively. These acids exhibit light initial color, excellent color stability and low iodine values.

(CAS #143-07-7)

EMERY 655 Myristic Acid (C14):

EMERY 655 Myristic Acid contains no short-chain acids below C12. It typically contains 96% myristic acid, 3% palmitic acid and 1% lauric acid. This white, hard crystalline solid has an iodine value specification of 0.5 maximum.

(CAS #544-63-8)

QUANTUM CHEMICAL CORP.: EMERY Short-Chain and Coconut Fatty Acids(Continued):

Even-Carbon Acids(Continued):

EMERY 657 Caprylic Acid:
Titer, C: 14-16
Iodine value (Wijs), max.: 0.2
Color, % trans. 450/550 nm, min.: 88/99
Acid value: 385-390

EMERY 658 Caprylic-Capric Acid:
Titer, C: 1-6
Iodine value (Wijs), max.: 0.3
Color, % trans. 450/550 nm, min.: 88/99
Acid value: 356-366

EMERY 659 Capric Acid:
Titer, C: 28-31
Iodine value (Wijs), max.: 0.5
Color, % trans. 440/550 nm, min.: 88/99
Acid value: 322-326

EMERY 650 Lauric Acid:
Titer, C: 33-35
Iodine value (Wijs), max.: 0.4
Color, % trans. 440/550 nm, min.: 85/97
Acid value: 268-272

EMERY 651 Lauric Acid:
Titer, C: 41-43
Iodine value (Wijs), max.: 0.2
Color, % trans. 440/550 nm, min.: 90/98
Acid value: 276-282

EMERY 652 Lauric Acid:
Titer, C: 43 min.
Iodine value (Wijs), max.: 0.2
Color, % trans. 440/550 nm, min.: 90/98
Acid value: 277-281

EMERY 655 Myristic Acid:
Titer, C: 52.0-53.5
Iodine value (Wijs), max.: 0.5
Color, % trans. 440/550 nm, min.: 90/99
Acid value: 243-246

QUANTUM CHEMICAL CORP.: EMERY Short-Chain and Coconut Fatty Acids(Continued):

Odd-Carbon Acids:

EMERY 1202 Pelargonic Acid:

EMERY 1202 Pelargonic Acid is a clear, saturated aliphatic monobasic acid which typically contains 94% pelargonic acid (C9). EMERY 1202 possesses light initial color, excellent color stability and low iodine value.

(CAS #112-05-0)

EMERY 878 Special Acid:

EMERY 878 (formerly Emery 1205) is a mixture of low molecular weight short-chain fatty acids. It is a clear, light straw-colored liquid with a mild characteristic fatty odor which exhibits excellent oxidation and color stability. It is soluble in water and is readily soluble in ethyl alcohol and most organic solvents.

EMERY 1210 Low Molecular Weight Acid:

EMERY 1210 LMW Acid is a low titer short-chain acid mixture averaging 7 carbon atoms in chain length. Because of its unique chain length mix, EMERY 1210 has applications in a variety of synthetic lubricants, such as transformer fluids, transmission lubricants and aircraft lubricants.

(CAS #68603-84-9)

EMERY 1202 Pelargonic Acid:

- Titer, C: 8-11
- Iodine Value (Wijs), max.: 0.5
- Color, % trans. 440/550 nm, min.: 90/99
- Acid value: 345-355
- Composition:
  - Caprylic: 4
  - Pelargonic: 94
  - Capric: 2

EMERY 878 Special Acid:

- Titer, C: 20
- Iodine Value (Wijs), max.: 1
- Color, % trans. 440/550 nm, min.: 70/95
- Acid value: 295-315

EMERY 1210 LMW Acid:

- Titer, C: 4
- Iodine Value (Wijs), max.: 1.5
- Color, % trans. 440/550 nm, min.: 80/96
- Acid value: 400-430
- Composition:
  - Valeric: 4
  - Caproic: 25
  - Enanthic: 32
  - Caprylic: 8
  - Pelargonic: 29
  - Capric: 2

QUANTUM CHEMICAL CORP.: EMERY Short-Chain and Coconut Fatty Acids(Continued):

Coconut Fatty Acids:

EMERY 621 Coconut Fatty Acid:
EMERY 621 is designed for low cost applications requiring short-chain fatty acids. This product is recommended for applications where the light initial color and superior color stability of EMERY 622, 625 and 626 Coconut Fatty Acids are not required.
(CAS #61788-47-4)

EMERY 622 Coconut Fatty Acid:
EMERY 622 has been specially refined to offer excellent initial color. This product is manufactured for applications requiring a lower-cost, short-chain fatty acid.
(CAS #61788-47-4)

EMERY 626 Ultra and EMERY 625 Coconut Fatty Acids:
These products exhibit greatly improved initial color and color stability as a direct result of their lower iodine values. For the most critical applications requiring coconut fatty acids of lightest color and excellent color stability, EMERY 625 and EMERY 626 Coconut Fatty Acids are recommended.
A Kosher grade of EMERY 626 is marketed under the name EMERY 7026 Kosher Coconut Fatty Acid.
(CAS #68938-15-8).

EMERY 627 Low I.V., Stripped, Ultra and EMERY 629 Stripped Coconut Fatty Acids:

EMERY 627 and EMERY 629 have been specially distilled to reduce the low molecular weight C8 to C10 portion of the acid mixture found in "normal" fatty acids from approximately 13% to 1%.
EMERY 627 has been further refined to substantially reduce unsaturated acid content, thus promoting greater thermal stability and combining the best properties of a low iodine value acid and a stripped coconut fatty acid.
(CAS #68397-85-9)

QUANTUM CHEMICAL CORP.: EMERY Short-Chain and Coconut Fatty Acids(Continued):

Coconut Fatty Acids(Continued):

EMERY 621 Coconut Fatty Acid:
Titer, C: 23-27
Iodine value (Wijs): 5-16
Color, % trans. 440/550 nm, min.: 30/80
Color, Gardner 1963, max.: 5
Acid value: 258-268

EMERY 622 Coconut Fatty Acid:
Titer, C: 22-26
Iodine value (Wijs): 5-10
Color, % trans. 440/550 nm, min.: 65-96
Color, Gardner 1963, max.: 2
Acid value: 268-276

EMERY 625 Partially Hydrogenated Coconut Fatty Acid:
Titer, C: 23-25
Iodine value (Wijs): 5 max.
Color, % trans. 440/550 nm, min.: 85/98
Color, Gardner 1963, max.: 1
Acid value: 269-273

EMERY 626 Low IV Ultra Coconut Fatty Acid:
Titer, C: 23-26
Iodine value (Wijs): 1 max.
Color, % trans. 440/550 nm, min.: 85/99
Color, Gardner 1963, max.: 1
Acid value: 270-276

EMERY 627 Low IV, Stripped, Ultra Coconut Fatty Acid:
Titer, C: 28-32
Iodine value (Wijs): 1 max.
Color, % trans. 440/550 nm, min.: 90/99
Color, Gardner 1963, max.: 1
Acid value: 252-258

EMERY 629 Stripped Coconut Fatty Acid:
Titer, C: 27-30
Iodine value (Wijs): 6-10
Color, % trans. 440/550 nm, min.: 88/98
Color, Gardner 1963, max.: 1
Acid value: 253-259

QUANTUM CHEMICAL CORP.: QUANTUM Unsaturated Fatty Acids:

Monounsaturated Acids:
Monounsaturated acids are predominantly oleic acid.

EMERSOL 233 LL Oleic Acid:
The superior performance of EMERSOL 233 LL Oleic is primarily a result of its low-linoleic acid content. Its extremely high oxidation stability and resistance to rancidity give finished products maximum protection against breakdown during both processing and storage before use.

EMERSOL 6333 USP/NF LL Oleic Acid, Food Grade:
EMERSOL 6333 USP/NF LL Oleic is a light-colored, low linoleic content acid which has a low polyunsaturated acid content.

EMERSOL 221 USP/NF Low-titer White Oleic Acid:
EMERSOL 221 is a versatile oleic acid for use where a light color is required. This double-distilled oleic acid offers excellent color stability as well as light initial color. Its low solid acid content, and subsequent low titer, results in low melting derivatives which are generally more soluble than derivatives of products having a higher titer.

EMERSOL 223 USP/NF Ultra Oleic Acid:
EMERSOL 223 is a versatile, double-distilled monounsaturated fatty acid.

EMERSOL 6321 USP/NF Low-titer White Oleic Acid, Food Grade:
EMERSOL 6321 has properties similar to those of EMERSOL 221.

EMERSOL 213 USP/NF Low-titer Oleic Acid:
A rather light-colored, single-distilled oleic acid, EMERSOL 213 is a multi-purpose grade for use where extremely light color is not essential.

EMERSOL 6313 USP/NF Low-titer Oleic Acid, Food Grade:
EMERSOL 6313 possesses properties similar to those of EMERSOL 213.

EMERSOL 210 Oleic Acid:
EMERSOL 210 is a single-distilled acid.

Polyunsaturated Acids:
Linoleic acid is the chief component in each of Quantum's polysaturated acids.

EMERSOL 315 Linoleic Acid:
EMERSOL 315 contains about 76% polyunsaturated acids, the highest in Quantum's line, and has a very low saturated acid content(approximately 5%).

EMERY 618 Soya Fatty Acid:
The polyunsaturated acids content of EMERY 618 is approximately 57%.

EMERY 610 Soya Fatty Acid:
The polyunsaturated acid content of EMERY 610 is approximately 53% and its saturated acid content is nearly 21%.

QUANTUM CHEMICAL CORP.: QUANTUM Unsaturated Fatty Acids (Continued):

Monounsaturated Acids:

EMERSOL 210 Oleic Acid:
Titer C: 7-12
Iodine Value: 87-95
Color % Trans. 440/550 nm., min.: 2/30
Acid Value: 197-204

EMERSOL 213 USP/NF Low-titer Oleic Acid:
Titer C: 5 max.
Iodine Value: 88-95
Color % Trans. 440/550 nm., min.: 50/86
Acid Value: 199-204

EMERSOL 221 USP/NF Low-titer White Oleic Acid:
Titer C: 5 max.
Iodine Value: 88-95
Color % Trans. 440/550 nm., min.: 77/98
Acid Value: 199-204

EMERSOL 223 USP/NF Ultra Oleic Acid:
Titer C: 5 max.
Iodine Value: 88-95
Color % Trans. 440/550 nm., min.: 85/99
Acid Value: 199/204

EMERSOL 233 LL Oleic Acid:
Titer C: 6 max.
Iodine Value: 86-90
Color % Trans. 440/550 nm., min.: 78/99
Acid Value: 200-204

EMERSOL 6313 USP/NF Low-titer Oleic Acid:
Titer C: 6 max.
Iodine Value: 88-93
Color % Trans. 440/550 nm., min.: 75/98
Acid Value: 201-204

QUANTUM CHEMICAL CORP.: QUANTUM Unsaturated Fatty Acids (Continued):

Monounsaturated Acids(Continued):

EMERSOL 6321 USP/NF Low-titer White Oleic Acid:
Titer C: 6 max.
Iodine Value: 87-92
Color % Trans. 440/550 nm., min.: 85-99
Acid Value: 201-204

EMERSOL 6333 USP/NF LL Oleic Acid:
Titer C: 8-10
Iodine Value: 86-91
Color % Trans. 440/550 nm., min.: 85/99
Acid Value: 200-204

Polyunsaturated Acids:

EMERSOL 315 Linoleic Acid:
Titer C: 5 max.
Iodine Value: 145-160
Color % Trans. 440/550 nm., min.: 72/96
Acid Value: 195-202

EMERY 610 Soya Fatty Acid:
Titer C: 15-25
Iodine Value: 125-138
Color % Trans. 440/550 nm., min.: 60/90
Acid Value: 195-205

EMERY 618 Soya Fatty Acid:
Titer C: 15-23
Iodine Value: 138-145
Color % Trans. 440/550 nm., min.: 72/96
Acid Value: 197-203

QUEST INTERNATIONAL: Fragrance Ingredients:

AGRUDOR AB 600:
Identity: Compounded specialty
Odour: A fresh citrus peel odour, with a powerful lemon character.

ALLYL HEPTANOATE:
Identity: Propenyl heptanoate (Unsaturated aliphatic ester).
Odour: A fresh, fruity, apple, somewhat pungent, vinous banana-like odour.

AMBERLYN:
Identity: Compounded specialty.
Odour: A very powerful amber note with a distinct ambergris character and a rich woody quality analogous to patchouli and vetiver.

AMBER-SEC AB 652A JAP:
Identity: Compounded specialty.
Odour: The odour has a dry, amber note which is supported by warm musky undertones.

AMBRASOL AB 455:
Identity: Compounded specialty.
Odour: An extremely long lasting amber odour with an unobtrusive animal character.

AMBRETTILYN AB 1526:
Identity: Compounded specialty
Odour Description: A sweet yet warm musky aroma with a subtle pervasive floral softness and remotest hint of spice which is the hallmark of musk ambrette.

ANTHER:
Identity: Phenylethyl iso-pentyl ether. (Aryl substituted aliphatic ether).
Odour: Powerful concentrated odour of spring flowers-notably hyacinth. Penetrating and cool this material possesses a floral greenness. ANTHER gives considerable lift to a creation.

APPLINAL:
Identity: Ethyl-2-methyl-1,3-dioxolane-2-acetate. (Alicyclic ketal).
Odour: A diffusive, sweet pectin-fruit odour which is supported by a warm balsamic background.

QUEST INTERNATIONAL: Fragrance Ingredients(Continued):

AQUANAL:
Identity: 3(3,4-methylenedioxyphenyl)-2-methylpropanal. (Aryl substitituted aliphatic aldehyde).
Odour: A fresh muguet-watery odour with heliotropine-anisic aspects.

ARBOROMA:
Identity: 5 and 6-carbomethoxy-1-methyl-4-isopropyl-bicyclo-(2.2.2)-oct-2-ene. (Unsaturated alicyclic ester).
Odour: A rich woody odour reminiscent of patchouli, with a suggestion of the more floral quality of vetyver.

AURANTION:
Identity: Hydroxycitronellal-methylanthranilate (A Schiff's base).
Odour: A traditional material prepared from hydroxycitronellal and methyl anthranilate and possessing a strong orange-blossom odour.

AVALONE:
Identity: A reaction product whose major components are dodecanal, tridecanal and the corrsesonding unreacted alcohols. (Aliphatic aldehyde/aliphatic alcohol).
Odour: Characteristic powerful fresh note of the higher aliphatic aldehydes.

BANGALOL:
Identity: 2-ethyl-4-(2,2,3-trimethyl-3-cyclopenten-1-yl)-2-buten-1-ol (Unsaturated alicyclic alcohol)
Odour: A very powerful sandalwood note with a suggestion of a soft floral character.

BEAUVERTATE:
Identity: Methyl-2-nonenoate (Unsaturated aliphatic ester).
Odour: A very powerful green note with complex vegetable, aldehydic, earthy and floral nuances.

BENZALDEHYDE FCC:
Identity: Benzaldehyde (Aromatic aldehyde).
Odour: A powerful sweet odour reminiscent of freshly crushed bitter almonds.

QUEST INTERNATIONAL: Fragrance Ingredients(Continued):

BENZYL ACETATE FCC:
Identity: Benzyl acetate.
(Aromatic ester).
Odour: A pleasant, sweet, light, floral odour with a jasminic character.

BENZYL ALCOHOL FCC AROMA GRADE:
Identity: Benzyl alcohol.
(Aromatic alcohol).
Odour: A faint, non-descript, sweetish odour.

BENZYL BENZOATE FCC:
Identity: Benzyl benzoate.
(Aromatic ester).
Odour: A balsamic odour.

BENZYL PROPIONATE:
Identity: Benzyl propionate.
(Aromatic ester).
Odour: A sweet fruity odour with a jasminic undertone.

BOURGEONAL:
Identity: 3-(4-tert butylphenyl)-propanal
(Aryl substituted aliphatic aldehyde).
Odour: An extremely strong floral muguet note.

CALYXOL:
Identity: Ethyl-2-methyl-6-pentyl-4-oxo cyclohex-2-ene-carboxylate.
(Unsaturated alicyclic ester).
Odour: A subtle floral odour with a sweet jasmin character but also possessing peach and honeysuckle notes.

L-CARVONE FCC:
Identity: L-1-methyl-4-isopropenyl-6-cyclohexen-2-one
(Unsaturated alicyclic ketone).
Odour: Warm herbaceous, penetrating, diffuse odour and somewhat spicy overall reminiscent of spearmint oil.

CASSIS BASE AB 937:
Identity: Compounded specialty.
Odour: This material is evocative of blackcurrant juice and the bitter green of the fruit.

CERVOLIDE:
Identity: 1,2-oxahexadecanolide
(Macrocyclic lactone).
Odour: Sweet, tenacious and intensely musky. Even when used at a moderate level, CERVOLIDE shows a fine exalting effect.

QUEST INTERNATIONAL: Fragrance Ingredients(Continued):

CHANDILYN AB 1513:
Identity: Compounded specialty
Odour: A sandalwood, balsamic odour with a sweet floral connotation and a hint of musk.

CHRYSANTHAL:
Identity: 3-propyl-bicyclo(2.2.1)-hept-5-ene-2-carbaldehyde. (Alicyclic aldehyde).
Odour: A very powerful, green floral (chrysanthemum), herbaceous, fresh note.

CINNAMIC ALDEHYDE:
Identity: 3-phenylpropenal (Aryl substituted unsaturated aliphatic aldehyde).
Odour: A powerful warm, spicy, cinnamon-like aroma.

CISTULATE:
Identity: Methyl-2,2-dimethylbicylo-(2.2.1.)-heptane-3-carboxylate (Alicyclic ester).
Odour: A natural pine, with fruity, fresh, rosemary-sage like aspects.

CITRATHAL:
Identity: A reaction product whose major component is 1,2-diethoxy-3,7-dimethyl-2,6-octadiene. (Unsaturated aliphatic acetal).
Odour: A very powerful complex of citrus odours from which a lemon/lime character emerges distinctly. CITRATHAL combines high odour impact with persistency. This specialty is based on an essential oil and some minor changes in the odour consistent with natural variation may be detected.

CITROFURAN R-098.883:
Identity: Compounded specialty.
Odour: A strong citrus odour with green rhubarb and lavandacious notes.

CITRONELLYL NITRILE:
Identity: 3,7-dimethyl-6-octen-1-yl-nitrile. (Unsaturated aliphatic nitrile).
Odour: A fresh lemon odour with a greenish accent.

CITRONELLYL NITRILE FORTE:
Identity: 3,7-dimethyl-6-octen-1-yl-nitrile. (Unsaturated aliphatic nitrile).
Odour: A fresh lemon odour with a greenish accent. The forte quality possesses a slight flowery and aldehydic note.

QUEST INTERNATIONAL: Fragrance Ingredients(Continued):

CIVET AB 394A:
Identity: Compounded specialty
Odour: This specialty is a fine reproduction of Abyssinian civet absolute. It possesses the same odour, power and excellent fixative properties of the original material.

COSTUS OIL SYNTHETIC P-91538:
Identity: Compounded specialty
Odour: Soft but very tenacious and strong odour, with aspects of precious wood, orris, fatty and vetivert; often compared with the smell of human hair.

CUIRONAL AB 875A:
Identity: Compounded specialty
Odour: This material constitutes the powerful and complex heart of a man's fragrance. Its aromatic top-note develops into an ambery, mossy, woody, leather, chypre body.

CUMIN NITRILE:
Identity: 4-isopropylbenzonitrile.
(Aromatic nitrile).
Odour: A powerful spicy, cumin odour, less pungent than cumin aldehyde, but definitely stronger and more tenacious.

9-DECENAL:
Identity: 9-decanal.
(Unsaturated aliphatic aldehyde).
Odour: A novel aldehydic top note with a fresh, fruity and rose character.

DECUMAL AB 772:
Identity: Compounded specialty.
Odour: A powerful floral odour with jasmin notes predominating. On drying out nuances associated with both peach and honeysuckle develop.

DEWBERRY R-132.377:
Identity: Compounded specialty.
Odour: A complex fruity, berry-like and slightly green herbal odour. Reminiscent of lichees and raspberries.

DIHYDROJASMONE:
Identity: 2-pentyl-3-methyl-2-cyclopenten-1-one.
(Unsaturated alicyclic ketone).
Odour: An intense fresh, floral, jasmin raw material with celery and myrrh-like aspects.

QUEST INTERNATIONAL: Fragrance Ingredients(Continued):

DIHYDROMYRCENOL:
Identity: 2,6-dimethyl-7-octen-2-ol.
(Unsaturated aliphatic alcohol).
Odour: Fresh-citrus raw material with floral-lavender aspects.

DIHYDROMYRCENYL ACETATE:
Identity: 2,6-dimethyl-7-octen-2-yl-acetate.
(Unsaturated aliphatic ester).
Odour: Fresh, bergamot, citrus like odour.

DUPICAL:
Identity: 4-(tricyclo (5.2.1.0 2,6)-decylidene-8)-butanal.
(Unsaturated alicyclic aldehyde).
Odour: An intensive floral muguet with aldehydic aspects.

EFETAL:
Identity: 1-ethoxy-1-phenylethoxyethane.
(Aryl substituted aliphatic acetal).
Odour: Modern, fresh, green, floral note, muguet and hyacinth aspects.

ELINTAAL:
Identity: 1-ethoxy-1-(3,7-dimethyl-1,6-octadien-3-yloxy)-ethane.
(Unsaturated aliphatic acetal).
Odour: Fresh, floral, natural muguet with herbal aspects.

ELINTAAL FORTE:
Identity: 1-ethoxy-1-(3,7-dimethyl-1,6-octadien-3-yloxy)-ethane.
(Unsaturated aliphatic acetal).
Odour: Fresh, floral, natural muguet with herbal aspects.

EMPETAL:
Identity: 3 and 4-(4-methyl-3-penten-1-yl)-3-cyclohexene-1-carbaldehyde
(Unsaturated alicyclic aldehyde).
Odour: Powerful, fresh, aldehydic, citrus odour also muguet-like.

EPITONE:
Identity: 7 and 8-acetyl-5-isopropyl-2-methylbicyclo (2.2.2) -oct-2-ene.
(Unsaturated alicyclic ketone).
Odour: Woody, spicy and ambery odour.

QUEST INTERNATIONAL: Fragrance Ingredients(Continued):

EQUINOL:
Identity: 4-tertiary butyl methoxy benzene.
(Aromatic ether).
Odour: A vibrant, leathery note with sweet, woody, animalic nuances.

ETHYL SAFRANATE:
Identity: Ethyl dehydrocyclogeranate.
(Unsaturated alicyclic ester).
Odour: Intensive and diffusive, natural, fresh, damascon, rose, apple, with woody aspects.

EVERNIA AB 454B:
Identity: Compounded specialty
Odour: A novel moss base with sophisticated floral undertones and a subtle green character.

FELVINONE:
Identity: 7 and 8-acetyl-5-isopropyl-2-methylbicyclo (2.2.2)-oct-2-ene.
(Unsaturated alicyclic ketone).
Odour: Woody, spicy, balsamic and ambery notes.

FENYRANE:
Identity: 2,4-dimethyl-6-phenyldihydropyran.
(Aryl substituted unsaturated alicyclic ether).
Odour: Intensive rose-geranium, rose oxide-like odour.

FIORIVERT:
Identity: 1-ethoxy-1-phenylethoxyethane.
(Aryl substituted aliphatic acetal).
Odour: Modern floral green note, with a hint of hyacinth and lilac.

FLORANE:
Identity: 2-heptyltetrahydrofuran.
(Alicyclic ether).
Odour: Mild, somewhat fatty natural green odour 'like wet green rose-petals'.

FLOROCYCLENE:
Identity: 3a,4,5,6,7,7a-hexahydro-4,7-methano-1(3) H-inden-6-yl-propanoate.
(Unsaturated alicyclic ester).
Odour: An unusual sweet floral odour with considerable impact.

QUEST INTERNATIONAL: Fragrance Ingredients(Continued):

FRESCILE:
Identity: 3-methyldodecanonitrile.
(Aliphatic nitrile).
Odour: A strong orange odour which is suffused by a light, green, sea-fresh quality

FRUITLISE AB 876:
Identity: Compounded specialty.
Odour: A fresh, fruity, green accord, reminiscent of grapefruit, peach, mango, pineapple and other exotic tropical fruits. Floral and blackcurrant notes give a rounded background.

FRUTONILE:
Identity: 2-methyl-decanonitrile.
(Aliphatic nitrile)
Odour: A fine, light, jasmin/floral odour with a soft, peach undertone

GARDAMIDE:
Identity: N-methyl-N-phenyl-2-methyl butyramide.
(Aromatic amide).
Odour: Long lasting odour reminescent of styralyl acetate. Also rhubarb with aspects of vetiveryl acetate; in the tail a touch of noot katone is perceptible.

GARDOCYCLENE:
Identity: 3a,4,5,6,7,7a,hexahydro-4,7-methano-1(3)H-inden-6-yl-butanoate.
(Unsaturated alicyclic ester).
Odour: A sweet powerful complex floral odour with predominantly jasmin, tuberose and gardenia notes.

GYRANE:
Identity: 2-butyl-4,6-dimethyldihydropyran.
(Unsaturated alicyclic ether).
Odour: A radiant fresh-green floral with a touch of geranium and rose.

HEPTAVERT V-06253:
Identity: Compounded specialty.
Odour: A strong natural green character with a marked woody note in the background.

HERBOXANE:
Identity: 2-butyl-4,4,6-trimethyl-1,3-dioxane.
(Alicyclic ketal).
Odour: A herbal, spicy, camomile odour.

QUEST INTERNATIONAL: Fragrance Ingredients(Continued):

HEXYL BENZOATE:
Identity: Hexyl benzoate.
(Aromatic ester).
Odour: A fresh fruity top note whilst the body and dry-out are more sweet balsamic.

HYDROXYCITRONELLAL Replacement A-720.010 (AB 2040):
Identity: Compounded specialty.
Odour: Natural, fresh odour with a muguet character. Odour strength quickly increases and tenacity is good.

INONYL ACETATE and INONYL ACETATE EXTRA:
Identity: 3,5,5-trimethylhexylacetate
Odour: The odour lies between linalyl and terpinyl acetates with a vinous fruity shading. The INONYL ACETATE EXTRA possesses a slightly fruiter top note.

IONONE:
Identity: 4-(2,6,6-trimethyl-2-cyclohexen-1-yl)-3-buten-2-one.
(Unsaturated alicyclic ketone).
Odour: A sweet, warm, woody, balsamic, floral odour.

ISOJASMONE PURE:
Identity: 2-hexyl-2-cyclopentenone.
(Unsaturated alicyclic ketone).
Odour: Powerful jasmin with lactonic, celery-like aspects.

ISOLONGIFOLANONE:
Identity: Isomeric sesquiterpenoid ketones.
(Unsaturated alicyclic ketone).
Odour: Fresh woody raw material, longlasting with an amber nuance.

JACINTHFLOR V-06285:
Identity: Compounded specialty.
Odour: Floral, green, pungent, reminiscent of many spring flowers, especially hyacinth and the green aspect in lilac and gardenia.

JASILYN:
Identity: 4-acetoxy-3-pentyltetrahydropyran
(Ester substituted alicyclic ether).
Odour: A very rich, sweet, jasmin complex with a slightly waxy background.

QUEST INTERNATIONAL: Fragrance Ingredients(Continued):

JASMACYCLENE:

Identity: 3a,4,5,6,7,7a-hexahydro-4,7-methano-1(3)H-inden-6-yl-acetate. (Unsaturated alicyclic ester)

Odour: A powerful top-note material which continues to lend character to the body of the perfume. Fresh, green and floral.

JASMATONE:

Identity: 2-n-hexylcyclopentanone. (Alicyclic ketone).

Odour: A diffusive, warm, spicy jasmin with a hint of fruitiness.

JASMIN AB 410A:

Identity: Compounded specialty.

Odour: A jasmin base with character and depth. A soft jasmin, which preserves the long lasting end note of jasmin absolute.

JASMOPYRANE:

Identity: 4-acetoxy-3-pentyltetrahydropyran (Ester substituted alicyclic ether)

Odour: Powerful, floral-jasmin note with a herbaceous undertone.

JASMOPYRANE FORTE:

Identity: 4-acetoxy-3-pentyltetrahydropyran (Ester substituted alicyclic ether).

Odour: A less prevailing flowery complex and an added sparkle of freshness distinguishes the odour of the forte quality.

JESSATE:

Identity: Ethyl-2-hexylaceto-acetate (Aliphatic keto-ester)

Odour: Lasting, strong, fruity, jasmin odour characteristic for the undertone in a jasmin complex.

LEATHER BASE R-136.340:

Identity: Compounded specialty.

Odour: This material has a characteristic note of fine leather goods with woody and animalic under tones.

LIGANTRAAL:

Identity: Methyl-N-(2,4-dimethylcyclohexen-3-ylidenemethyl)-anthranilate. (Unsaturated-N-substituted aromatic ester).

Odour: Green, floral, with citrus/orange flower aspects.

QUEST INTERNATIONAL: Fragrance Ingredients(Continued):

LIGUSTRAL:
Identity: 2,4-dimethyl cyclohexene-3-carbaldehyde.
(Unsaturated alicyclic aldehyde).
Odour: Powerful, fresh, green, reminiscent of privet.

LIGUVERT V-06261:
Identity: Compounded specialty.
Odour: Powerful, leafy green, flowery note, reminiscent of pine-forest.

LIME BASE AB 651A JAP:
Identity: Compounded specialty
Odour: A sharp, refreshing lime top-note supported by a rich green citrus-peel background.

LIMETTAL V-06257:
Identity: Compounded specialty
Odour: A very intriguing sprightly fragrance combining modern aldehyde and floral notes with the tingling freshness of lime.

LIXETONE:
Identity: A reaction product whose major component is methyl cedryl ketone.
(Alicyclic ketone).
Odour: A long-lasting woody odour, very rich and sweet with a warm amber/musk background note. It has outstanding fixative properties, but is strong enough to impress its character on the top note.

MACEAL:
Identity: 7-formyl-5-isopropyl-2-methyl bicyclo (2.2.2)-oct-2-ene.
(Unsaturated alicyclic aldehyde).
Odour: Very intensive, fresh, green, spicy with nutmeg-like aspects.

MANZANATE:
Identity: Ethyl-2-methylpentanoate.
(Aliphatic ester).
Odour: A high impact fresh, fruity-green note, evocative of apple skins.

MEFRANAL:
Identity: 3-methyl-5-phenyl-1-pentanal.
(Aryl substituted aliphatic aldehyde).
Odour: Long lasting green aldehydic odor reminsiscent of Lily of the Valley.

QUEST INTERNATIONAL: Fragrance Ingredients(Continued):

MELONIS AB 441B JAP:
Identity: Compounded specialty.
Odour: A modern, green, fruity odour, reminiscent of melon and pears with a fresh floral background.

METHYL CEDRYLONE:
Identity: A reaction product whose largest component is methyl cedryl ketone.
(Alicyclic ketone).
Odour: A long-lasting woody odour, very rich and sweet with a sophisticated cedar body. It gives excellent fixative properties to a composition yet its power impresses itself on the top note.

METHYL DIHYDROJASMONATE:
Identity: Methyl-(2-pentyl-3-oxocyclopentan-1-yl)-acetate.
(Alicyclic kero ester).
Odour: A soft, warm, radiant floral fragrance with a distinct jasminic character.

METHYL IONONE:
Identity: 5-(2,6,6-trimethyl-2-cyclohexen-1-yl)-4-penten-3-one.
(Unsaturated alicyclic ketone)
Odour: A light, warm, woody floral odour.

METHYL IONONE ALPHA ISO:
Identity: 4-(2,6,6-trimethyl-2-cyclohexen-1-yl)-3-methyl-3-buten-2-one
(Unsaturated alicyclic ketone).
Odour: Possesses a somewhat more woody and violet odour than Methyl Ionone with a creamy nuance.

MEVANTRAAL:
Identity: Methyl-2-methylpentylidene anthranilate
(Unsaturated N-substituted aromatic ester).
Odour: A sweetish, floral, fruity, green odour.

MOSS BASE AB 311D:
Identity: Compounded specialty.
Odour: A powerful decolourised moss base possessing considerable impact and substantivity.

MOUSSE DE MER R-099.199:
Identity: Compounded specialty.
Odour: A complex, fresh and mossy green odour.

QUEST INTERNATIONAL: Fragrance Ingredients(Continued):

MUSK R-1:
Identity: 11-oxahexadecanolide
(Macrocyclic lactone)
Odour: Intense, diffusive, musk odour.

MUSK XYLENE SUBSTITUTE P-91580:
Identity: Compounded specialty.
Odour: Reminiscent of musk xylene.

NEOBERGAMATE:
Identity: 2-acetoxy-2-methyl-6-methylene-7-octene.
(Unsaturated aliphatic ester)
Odour: Powerful, fresh, herbaceous, bergamot-like odour.

NEOBERGAMATE FORTE:
Identity: 2-acetoxy-2-methyl-6-methylene-7-octene.
(Unsaturated aliphatic ester)
Odour: Powerful, fresh, herbaceous, bergamot-like odour.

NEOBERGAMATE SPECIAL:
Identity: 2-acetoxy-2-methyl-6-methylene-7-octene.
(Unsaturated aliphatic ester)
Odour: Powerful, fresh, herbaceous, bergamot-like odour.

NEOLAVANDATE:
Identity: 1-methyl-4-isopropyl-1-cyclohexen-8-isobutyrate
(Unsaturated alicyclic ester)
Odour: Fruity, floral, herbaceous, refreshing and moderately tenacious odour.

ORTHOLATE:
Identity: 2-tertiary-butylcyclohexyl acetate
(Alicyclic ester)
Odour: A fresh fruity odour with a slightly citrus-green character and a subtle woody background.

PARA-CRESYL METHYL ETHER:
Identity: 4-cresylmethylether
(Aromatic ether)
Odour: A powerful spice, animalic, floral odour suggesting ylang ylang and on dilution wallflowers.

PEACH AB 650:
Identity: Compounded specialty.
Odour: A moist, fruity odour of fresh peaches with rounded natural green undertones.

QUEST INTERNATIONAL: Fragrance Ingredients(Continued):

PELARGENE:
Identity: 2,4-dimethyl-6-phenyldihydropyran
(Aryl substituted unsaturated alicyclic ether)
Odour: A powerful geranium odour supported by a linden-type freshness and an interesting spice undertone.

PETIOLE:
Identity: Phenylethyl isopropyl ether.
(Aryl substituted aliphatic ether)
Odour: An intense green odour of nasturtium leaves with in addition a watercress character.

PETITGRAIN OIL SYNTHETIC P-91332:
Identity: Compounded specialty.
Odour: Similiar odour to the natural petitgrain oil Paraguay.

PHENOXYETHYL ISOBUTYRATE BETA:
Identity: 2-phenoxyethyl-2-methylpropanoate.
(Esters of aromatic ether)
Odour: A floral odour with rose predominating but also sweet, green, fruity notes reminiscent of pineapple and grapefruit.

PHENYLETHYL ACETATE FCC:
Identity: 2-phenylethyl acetate
(Aryl substituted aliphatic ester)
Odour: A rosy-fruity sweet odour, the fruity nuance has a leafy green tonality.

PHENYLETHYL N-BUTYL ETHER:
Identity: Phenylethyl n-butyl ether
(Aryl substituted aliphatic ether)
Odour: A refreshing spice green note of cress.

PIVALOXYCYCLENE:
Identity: 3a,4,5,6,7,7a-hexahydro-4,7 methano-1(3)H-inden-6-yl-2,2-dimethyl-propanoate.
(Unsaturated alicyclic ester)
Odour: Sweet, powdery floral odour with strong peach nuance.

PIVAROSE:
Identity: 2-phenylethylpivalate
(Aryl substituted aliphatic ester)
Odour: A floral rose odour reminiscent of phenylethyl isobutyrate but more tea-like and the odour is much more stable in difficult media.

QUEST INTERNATIONAL: Fragrance Ingredients(Continued):

PRENYL ACETATE:
Identity: 3-methylbuten-2-yl-acetate
(Unsaturated aliphatic ester)
Odour: Natural, fresh, fruity, banana and pear-like odour; diffusive, volatile, reminiscent of ylang top, more sophisticated than amyl acetate.

PTBCHA:
Identity: 4-tertiary-butyl-cyclohexyl-acetate
(Alicyclic ester)
Odour: Warm, woody note with a sharp and characteristically fruity top note. Sustained odour tenacity.

PTBCHA HIGH-CIS:
Identity: 4-tertiary-butyl-cyclohexyl-acetate
(Alicyclic ester)
Odour: A fine quality very sweet, woody, jasmin odour with a clean penetrating top note.

RHUBACITRIL R-098.882:
Identity: Compounded specialty.
Odour: A very powerful, green, rhubarb odour with grapefruit connotations. On dry-out the citrus notes become more dominant.

RHUBAFURAN:
Identity: 2,4-dimethyl-4-phenyl tetrahydrofuran.
(Aryl substituted alicyclic ether)
Odour and Usage: A very powerful and diffusive green, rhubarb and grapefruit-like odour.

ROSANIA V-06290:
Identity: Compounded specialty.
Odour: A strong, diffusive, rose/geranium odour with fresh and natural green aspects.

ROSE AB 380:
Identity: Compounded specialty.
Odour: A soft, rose-absolute type odour with a hint of herbal and animal undertones.

ROSE AB 404C:
Identity: Compounded specialty.
Odour: A warm, sweet, rose odour reminiscent of precious rose oils.

QUEST INTERNATIONAL: Fragrance Ingredients(Continued):

SAUTANE:
Identity: 3,3,5-trimethylcyclohexyl acetate. (Alicyclic ester)
Odour: A sweet, fruity odour, reminiscent of light wine. It also possesses a background odour of crushed leaves.

2-SEC-BUTYL-3-METHOXY-PYRAZINE:
Identity: 2-sec-butyl-3-methoxy-pyrazine (Substituted pyrazine)
Odour: An intensely powerful and highly diffusive odour which is reminiscent of ivy leaves and galbanum. It occurs naturally in galbanum and green peas.

SINOCITRIL NB 132 A-125.000:
Identity: Compounded specialty.
Odour: A powerful odour with a very cheerful, fresh, citrusy character with accents of tangerine and an appealing green, floral undertone.

TANGENIL V-06265:
Identity: Compounded specialty.
Odour: It has a fresh citrus character reminiscent of tangerine with floral undertones.

TETRAHYDROCONVALOL V-06286:
Identity: Compounded specialty.
Odour: Fresh, floral odour with bergamot/citrus aspects.

TOP ROSE P-91616:
Identity: Compounded specialty.
Odour: The top note of Bulgarian rose oil.

TRASEOLIDE:
Identity: 5-acetyl-3-isopropyl-1,1,2,6-tetramethylindane. (Alicyclic substituted aromatic ketone).
Odour: A clean and versatile musk odour.

TRASEOLIDE "70":
Identity: 5-acetyl-3-isopropyl-1,1,2,6-tetra methyl indane. (Alicyclic substituted aromatic ketone).
Odour: A clean versatile musk odour.

TRIDECENE-2-NITRILE:
Identity: Tridecene-2-nitrile. (Unsaturated aliphatic nitrile).
Odour: Very powerful, floral aldehydic with citrus (lime) aspects.

QUEST INTERNATIONAL: Fragrance Ingredients(Continued):

TUBEROSE AB 473:
Identity: Compounded specialty.
Odour: An outstanding reproduction of the natural oil of tuberose--tenacious and powerful.

VERDALIA A:
Identity: 3a,4,5,6,7,7a-hexahydro-4,7-methano-5-methoxy-1(3)H-indene.
(Unsaturated alicyclic ether).
Odour: Powerful, floral, green character analogous to hyacinth with a slightly sweet fruity quality suggesting gardenia.

VERDILYN:
Identity: 1-ethoxy-1-phenyl ethoxy ethane.
(Aryl substituted aliphatic acetal).
Odour: Cool, green, fresh, floral odour with rose overtones.

VERDINAL:
Identity: 3,5,5-trimethyl-hexanal.
(Aliphatic aldehyde).
Odour: A powerful and diffusive complex of linden green, aldehydic and woody notes.

VERDORACINE:
Identity: 4-isopropyl-1-methyl-2-propenylbenzene.
(Aryl substituted alkene).
Odour: Green, carrot-like somewhat earthy, reminiscent of vetivert, galbanum and tomato leaf.

VERTELON:
Identity: 4-methyl-2-(1-phenylethyl)-1,3-dioxolane.
(Aryl substituted aliphatic acetal).
Odour: An intense green odour which possesses the fruitiness of melons superimposed on the character of mushrooms.

VETIVERYL ACETATE SUBSTITUTE P-91455:
Identity: Compounded specialty.
Odour: Reminiscent of vetiveryl acetate.

VIGOROSE V-06255:
Identity: Compounded specialty.
Odour: Radiant, fresh, green, floral, fruity odour with rose/geranium-like notes and excellent lift.

VIOTRIL V-06276:
Identity: Compounded specialty
Odour: A strong, floral, tenacious, violet, orris-like odour with a green note reminiscent of violet leaves.

YLANG AB 388A:
Identity: Compounded specialty.
Odour: An unusual power in the top note gives way to a long lasting floral spicy complex which is very typical of the fragrance of the flower.

QUEST INTERNATIONAL: Fragrance Ingredients(Continued):

ALGINOL MD P22:
Identity: A versatile extract originating from the dried thallus of seaweed. (Fucus Vesiculosus L).
Odour and Usage: The product possesses a very powerful sea-shore odour. It can thus bring an elusive ozone/sea connotation to oakmoss, woody, aldehydic or cistus-labdanum notes, when used sparingly. In addition, interesting modifications to traditional colognes can be produced when ALGINOL MD P22 is used in trace amounts.

BEESWAX ABSOLUTE P354:
Identity: This product is obtained from selected qualities of white beeswaxes. (Apis Mellifera).
Odour and Usage: This material has a constant mellow, hay, floral and honey note. It is recommended for use in low concentrations (1% - 5%) to round off synthetic floral bases such as rose and jasmin, and gives a natural effect to mass market fragrances containing such bases.

BENZOIN HYPERSOLUBLE SEMI-LIQUID P85:
Identity: Obtained from Benzoin Siam. (Styrax Tonkinensis Craig or Styrax Benzoin Dryander)
Odour and Usage: This material has an odour which is very pleasant, warm and long lasting. It blends well and gives roundness to the chemicals of the balsamic-vanilla group such as vanillin, coumarin, benzaldehyde, phenylethyl alcohol etc. Particularly suited for use in white toilet soaps and as a non-discolouring vanilla substitute.

BIRCH TAR MD P576:
Identity: This product is produced by molecular distillation of Birch Tar Oil. (Betula Pendula Roth)
Odour and Usage: BIRCH TAR MD P576 possesses an odour reminiscent of wood smoke and leather. This product has a very crisp clean character free of any offensive aniline notes. It has for a long time found favour with perfumers for its Russian leather character which may be used with advantage in fougeres, chypres and men's fragrances such as after-shave lotions. It may be employed to reinforce and give added character to labdanum notes.

BROUTEA P123:
Identity: This product is manufactured from the twigs and leaves of the orange Bigaradier tree by a similar process to the one used for Petitgrain Water Absolute (Absolute Eaux de Brouts). It may replace usefully this very expensive material. (Citrus Aurantium L.)
Odour and Usage: The powerful herbaceous floral note blends well with bergamot, broom, lavender, etc. BROUTEA P123 is also very interesting as it is not a crop product and so does not present the risks of short crop years.

QUEST INTERNATIONAL: Fragrance Ingredients(Continued):

CISTUS ABSOLUTE FRENCH MD P19:

Identity: This absolute is obtained by molecular co-distillation of the concrete obtained after solvent extraction of the leaves and stems of the Cistus herb.

(Cistus Ladaniferus L.)

Odour and Usage: The product possesses a very fine characteristic odour with a sweet ambery, powerful and tenacious character. It is a typical balsamic/ambery note and is markedly smoother than the same product derived from Spanish Concrete, thus blending excellently with musks, mosses, woody or sea notes.

CISTUS LABDANUM ABSOLUTE P659:

Identity: This absolute is processed by selective extraction of dried Cistus twigs cropped in Mediterranean countries.

(Cistus Ladaniferus L.)

Odour and Usage: A powerful, tenacious, warm, herbaceous odour, reminiscent of ambra in its somewhat animal and leather notes. Blends well with modern aldehydic and ambre-type perfumes, oriental, woody and musky bases.

CISTUS OIL FRENCH P20:

Identity: Among all Cistus Labdanum derivatives, CISTUS OIL FRENCH P20 is distinguished as being the heart of the fragrance which permeates the Esterel Massif in Southern France throughout the summer. This essential oil is produced by steam distillation of dried Cistus twigs.

(Cistus Ladaniferus L.)

Odour and Usage: This Cistus oil possesses tremendous power, even at low dosages. Its characteristic warmth and richness blends well with such diverse notes as clary sage, alginols (seaweed extracts), ionones, vetiver, patchouli, cedarwood derivatives, etc.

CLOVE BUD OIL RECTIFIED EXTRA DQ P353:

Identity: Produced from the flower buds of Eugenia Caryophyllata.

(Caryophyllus aromaticus L.)

Odour and Usage: Produced directly from clove buds in the Grasse factory, this very high quality product lacks the harsh notes of the locally distilled oils. Dry down is uniformly sweet and long lasting.

CLOVE BUD OIL RECT. EXTRA DQ P353 is eminently suitable for use in toothpastes, food flavours or for its traditional perfumery uses--to sweeten and intensify floral notes.

QUEST INTERNATIONAL: Fragrance Ingredients(Continued):

COFFEE ROASTED ABSOLUTE MD P352:

Identity: Obtained by molecular distillation of the resinoid resulting from the solvent extraction of a special mixture of various species of coffee beans.

Odour and Usage: The care taken during the roasting stage and the selective extraction process makes this product very powerful, warm and attractive. This material finds its place in musk bases, floral absolutes and tobacco fragrances.

ELEMI RESINOID P8:

Identity: This resinoid is obtained from Elemi gum. (Canarium indicum L.)

Odour and Usage: This material possesses a fresh and lemony topnote which blends very well with citrus and also spice notes.

The good stability compared with many natural products of ELEMI RESINOID P8, together with its very economic price makes it a particularly attractive material.

EUCALYPTUS ABSOLUTE P555:

Identity: This absolute is obtained by alcohol washing of the concrete resulting from solvent extraction of the fresh leaves and twigs of Eucalyptus Globulus.

(Eucalyptus Globulus Labill).

Odour and Usage: The product has a woody and slightly fruity odour.

FIR BALSAM ABSOLUTE MD P61:

Identity: This is a molecular codistillate of fir needle extract, benzyl benzoate and 2-phenoxy ethanol.

(Abies Balsamea L).

Odour and Usage: This product has a very powerful, sweet, fresh, balsamic and fruity note and is useful for high class compositions giving them roundness and originality. Its liquid and colorless presentation allow an easy use give a favourable comparison to similar products on the market.

FOLIA P59:

Identity: This specialty is derived from the distillation of Petitgrain oil Bigarade over bitter orange flowers.

(Citrus Aurantium L.)

Odour and Usage: This procuct is characterized by a powerful floral note with herbal under tones. FOLIA P59 can be used to extend orange flower products. It strengthens the natural character of neroli and also rose, ylang, jasmin and muguet creations. It blends well with linalyl and geranyl esters, also linalool and inexpensive rose type chemicals.

QUEST INTERNATIONAL: Fragrance Ingredients(Continued):

GALBANUM OIL P637:

Identity: This essential oil is obtained by a modern distillation process of the natural oleo-gum-resin.

(Ferula Galbaniflua Boiss et Buhse).

Odour and Usage: Galbanum products have risen, by virtue of their green top notes and resinous background notes, to become one of the essential components of modern perfumes. The odour of GALBANUM OIL P637 possesses a clean character and remarkable power. It performs well everywhere that the typical and powerful green top note is required.

GALBANUM SLC P305:

Identity: This product is derived from the gum by a selective extraction process which concentrates the odorous principles of the gum, while replacing the non-odorous waxes with an inert solvent.

(Ferula Galbaniflua Boiss et Buhse).

Odour and Usage: Galbanum products have become some of the essential components of many modern perfumes, by virtue of their green top note and resinous background notes.

JASMIN ABSOLUTE EGYPTIAN P243:

Identity: This product is produced by alcohol washing of jasmin concrete which is itself produced in Egypt by solvent extraction of the freshly picked jasmin flower.

(Jasminum Officinale, variety Grandiflorum).

Odour and Usage: A very powerful honey-like sweet floral odour with fruity herbaceous undertones and very reminsicent of the flowers themselves. This product, of which Egypt is currently the major world producer, is one of the most important materials used in perfumery and is the most popular floral absolute. Its effectiveness at very low concentrations means that this pure jasmin absolute is rarely too expensive to use.

JASMIN ABSOLUTE FRENCH P403:

Identity: This finest quality jasmin is an exceptional example of the increasingly scarse authentic jasmin absolute French. It is produced in France by alcoholic washing of French jasmin concrete which is itself produced by solvent extraction of fresh French jasmin flowers.

(Jasminum Officinale, variety Grandiflorum).

Usage: JASMIN ABSOLUTE FRENCH P403 is one of the most prestigious materials in perfumery. Employed in the highest class fine fragrances and in low concentrations in other toiletries.

QUEST INTERNATIONAL: Fragrance Ingredients(Continued):

LABDANUM ABSOLUTE H MD P12:

Identity: Labdanum gum derivative obtained by mild extraction followed by molecular distillation to give a very fine ingredient.

(Cistus Ladaniferus L).

Odour and Usage: This product has remarkable power and it has a particularly clean note. The dominant amber character of this product blends well with the expected odour types--oakmoss, sea notes, etc and in addition interesting accords are obtained in combination with such diverse ingredients as vetiver, patchouli and celery.

LEMON OIL TERPENELESS C P51:

Identity: This product is processed from selected Expressed Italian Lemon oils by fractionation and careful vacuum concentration by distillation.

(Citrus Lemon (Linn.) Burm. F).

Perfume Usage: The production process assures the customer of an alcohol soluble product which can be used in high class alcoholic perfumes as a freshener and top note ingredient.

NEROLI OIL TUNISIAN P517:

Identity: This product is produced in the Tunisian factory by steam distillation of the freshly picked flowers of the Orange Bigaradier tree.

(Citrus Aurantium L., sub species amara.)

Odour and Usage: A very powerful, light and refreshing top note which is distinctive and original. It is used extensively in high class colognes where it blends excellently with all citrus oils and most floral absolutes.

OAKMOSS ABSOLUTE GREEN JUGO P40:

Identity: Obtained by alcohol washing of Yugoslavian Oakmoss Concrete itself obtained by solvent extraction of the dried Yugoslavian Oakmoss in the Grasse factory.

(Evernia Prunastri L.)

Odour and Usage: An excellent fixative for many perfumes, this product can be used to bring a natural character to floral notes, as well as fougere, chypre, forest notes, etc.

OAKMOSS ABSOLUTE MD P195:

Identity: The concrete is obtained from Yugoslavian Oakmoss, which is then co-distilled with triethylene glycol under molecular vacuum. The process preserves the whole fragile constituents of the oakmoss odour.

(Evernia Prunastri L).

Odour and Usage: An excellent fixative for many perfumes; it will be appreciated blended with fresh notes, bergamot, lavandin, and with floral notes as well as fougere, chypre or forest type fragrances.

QUEST INTERNATIONAL: Fragrance Ingredients(Continued):

OLIBANUM ALBAROME P26:

Identity: Albaromes are obtained by a special distillation process, which enables Quest International to produce liquid resinoids from gums. They differ from ordinary resinoids by their specially refined note and almost colourless appearance.

(Boswellia species.)

Odour and Usage: The balsamic and powdery note of OLIBANUM ALBAROME P26 is perticularly suitable for use in modern perfumes, oriental creations, etc.

OLIBANUM OIL P573:

Identity: The Olibanum oil is obtained by steam distillation of Somalian Olibanum gums.

(Boswellia species).

Odour and Usage: Based on experience in selecting the correct material, from No. 1 to the upgraded tears, Quest has developed several grades of olibanum oil in order to give customers the quality suitable for their applications. The odour of the resulting oils is tenacious, fresh-terpeney with a certain pepperiness mellowed with a rich balsamic undertone. OLIBANUM OIL P573 uses fine quality tears.

OPOPONAX ABSOLUTE MD P5:

Identity: OPOPONAX ABSOLUTE MD P5 has been created to give a colourless version of OPOPONAX ABSOLUTE. Produced by steam distillation of the exudation from the bark of Commiphora erythraea var. glabrescens Engler.

(Commiphora erythraea var.)

Odour and Usage: Sweet balsamic spicy odour. Useful in many compounds to avoid discoloration of the final product. Moreover, the production process eliminates the allergen contained in Opoponax gum.

ORANGE FLOWER ABSOLUTE P400:

Identity: This product is produced by alcohol washing of the Orange Flower Concrete manufactured by Quest's own Tunisian factory by petroleum ether extraction of the freshly picked flowers of the Orange Bigaradier tree.

(Citrus Aurantium, subspecies amara).

Odour and Usage: A very powerful and intensely floral, heavy, rich odour while also being fresh and delicate.

The product finds a ready use in expensive colognes, particularly citrus ones, as well as floral fine fragrances.

QUEST INTERNATIONAL: Fragrance Ingredients(Continued):

PATCHOULI OIL REDISTILLED P15:

Identity: This product is obtained by a mild process from Indonesian Patchouli oil and Indonesian Patchouli leaves processed into oil in the Tunisian factory.

(Pogostemon Cablin Benth., also known as Pogostemon Patchouli pellet.)

Odour and Usage: The odour value of Patchouli oils is an even more salient diagnostic feature than the differences which appear on analysis. PATCHOULI OIL REDISTILLED P15 is one of the most useful Patchouli products in Quest's range since it combines much of the finest physical and olfactive qualities of their "own distillation" oil with a most competitive price. It has been rendered iron-free to permit the widest possible use in fragrances and cosmetics.

PATCHOULI OIL MD P578:

Identity: This is an iron-free, clear, amber liquid form obtained by rectification of imported Indonesian Patchouli Oil.

(Pogostemon Cablin benth., also known as Pogostemon Patchouli pellet).

Odour and Usage: Molecular distillation is a mild process which does not modify the richness and the tenacity of the selected Patchouli oil. Used widely in fragrances and cosmetics.

PEPPER BLACK OIL P315:

Identity: This essential oil is produced from the whole dried fruit of the pepper vine by steam distillation in the Grasse factory.

(Piper Nigrum L.)

Odour and Usage: The spice note of PEPPER BLACK OIL P315 can give a twist to carnation bases and rose compositions. It also produces interesting effects in floral and woody perfimes.

PEPPERMINT AMERICAN TRIPLE RECTIFIED:

Identity: Peppermint oil is produced from the partially dried herb and then rectified to remove the volatile terpene components.

(Metha Piperita).

Odour and Usage: Quest rectify a selection of the worlds major peppermint oils by removing the volatile terpene components. An example is PEPPERMINT AMERICAN TRIPLE RECTIFIED. This material is a fine grade sweet and fresh piperita product. It finds application in dentals where usage can be up to 40% and in sugar confectionary mainly in boiled sweets and fondants. There is some use in double mint chewing gum.

QUEST INTERNATIONAL: Fragrance Ingredients(Continued):

PERU BALSAM OIL P34:

Identity: PERU BALSAM OIL P34 is obtained from crude Peru balsam by molecular distillation. The resulting product possesses all the olfactory qualities of the originating balsam, without the sensitizing components.

(Myroxylon Pereirae (Royle) Klotzsch).

Odour and Usage: This processing produces a pale yellow mobile liquid which is more convenient to use and has greater solubility in the usual perfumery materials. The good fixative value and considerable tenacity of PERU BALSAM OIL P34 is particularly suitable in ambery or powdery creations and with floral notes.

ROSE DE MAI ABSOLUTE FRENCH P575:

Identity: This product, one of the most extensively used of all floral absolutes, is produced by alcohol washing of the concrete obtained by solvent extraction of the fresh flowers of Rosa Centifolia. Both of these production stages take place in Quest's own factory in Grasse, France.

(Rosa Centifolia).

Odour and Usage: It has a very tenacious warm, deep, floral-rosy odour with less pronounced honey-like notes. Very important for fine fragrances particularly floral bases, chypres, orientals etc. and is also used in trace quantities to round off synthetic compositions.

REWO: REWOTERIC Amphoteric Surfactants:

Overview of REWO Amphoterics:

Product:

Glycine Derivatives:
REWOTERIC AM 2L:
N-lauric acid amidoethyl-N-2-hydroxyethylglycinate

REWOTERIC AM 2C NM:
N-coconut fatty acid amidoethyl-N-2-hydroxy-ethylglycinate

Salt-free amphoteric tensides:
REWOTERIC AM KSF40:
N-coconut fatty acid amidoethyl-N-2-hydroxy-ethylaminopropionate

Anionic mixtures:
REWOTERIC AM CA:
Mixture of N-coconut fatty acid amidoethyl-N-2-hydroxyethyl glycinate and sodium lauryl ether sulphate

REWOTERIC AM G30:
Mixture of N-lauric fatty acid amidoethyl-N-2-hydroxyethylglycinate and sodium lauryl sulphate

Betaine:
REWOTERIC AM DML:
N-lauryl-N,N-dimethyl-N-carboxymethyl ammonium betaine

REWOTERIC AM B13:
N-Coconut fatty acid amidopropyl-N,N-dimethyl-N-carboxymethyl ammonium betaine

REWOTERIC AM R40:
N-ricinoleic fatty acid amido-propyl-N,N-dimethyl-N-carboxymethyl ammonium betaine

REWOTERIC AM TEG:
N-alkyl-N,N-bis (2-hydroxyethyl)-N-carboxymethyl-ammonium betaine

Sulphobetaines:

REWOTERIC AM CAS:
N(N-coconut fatty acid amidopropyl)N,N-dimethyl-ammonium-N-(2-hydroxy-propyl) sulphonate

REWO: REWOTERIC Amphoteric Surfactants(Continued):

<u>Product:</u>

REWOTERIC AM 2L:
Appearance: low-viscosity liquid
concentration %: 50
Minimum active content %: 39
Gardner colour: max. 5
pH value: 8.0-8.5
NaCl content %: 9-10

REWOTERIC AM 2C NM:
Appearance: low viscosity liquid
concentration %: 50
Minimum active content %: 39
Gardner colour: max. 5
pH value: 8.0-8.6
NaCl content %: 9-10

REWOTERIC AM KSF 40:
Appearance: clear liquid
concentration %: 40
Minimum active content %: 40
Gardner colour: max. 5
pH value: 9-10
NaCl Content %: max. 0,05

REWOTERIC AM CA:
Appearance: viscous liquid
concentration %: 30
Minimum active content %: 25
Gardner color: max. 3
pH value: 8.0-8.5
NaCl content %: max. 4

REWOTERIC AM G30:
Appearance: clear liquid
concentration %: min. 47
Minimum active content %: ca. 38
Gardner colour: ca. 4
pH value: 7.5-8.5
NaCl content %: 6.5-7.4

REWOTERIC AM DML:
Appearance: clear liquid
concentration %: 40
Minimum active content %: 31
Gardner color: max. 1
pH value: 7-8
NaCl content %: max. 8

REWO: REWOTERIC Amphoteric Surfactants(Continued):

Product:

REWOTERIC AM B 13:
Appearance: viscous liquid
concentration %: 35
Minimum active content %: 30
Gardner colour: ca. 2
pH value: 5-6
NaCl content %: ca. 5

REWOTERIC AM R40:
Appearance: viscous liquid
concentration %: 40
Minimum active content %: 34
Gardner colour: max. 3
pH value: 5.5-6.0
NaCl Content %: ca. 5

REWOTERIC AM TEG:
Appearance: viscous liquid
concentration %: 40
Minimum active content %: 35
Gardner colour: max. 6
pH value: 4.5-5.5
NaCl content %: ca. 5

REWOTERIC AM CAS:
Appearance: clear liquid
concentration %: 50
Minimum active content %: 43
Gardner colour: max. 4
pH value: 7-8
NaCl content %: ca. 7

RHEOX, INC.: BENTONE Organophilic Clay Rheological Additives:

Outstanding rheological properties can be developed in solvent or oil based systems by using BENTONE organoclay rheological additives. BENTONE additives are the reaction products of an organic quaternary amine with either hectorite or bentonite clay. This treatment makes them capable of swelling and gelling organic systems. Chemically, BENTONE rheological additives are inert and do not interact with most organic systems. BENTONE additives can be used to gel various hydrocarbon and natural oils, solvents and synthetic liquids.

Advantages:

- Produces reproducible thixotropic viscosity
- Suspends pigments, pearl essence and active ingredients
- Controls application, reduces dripping and running
- Imparts good leveling
- Reduces syneresis
- Imparts heat stability, reducing high temperature running and dripping
- Improves emulsion stability

BENTONE 27 Rheological Additive:

CTFA: Stearalkonium Hectorite

Application: Intermediate to high polarity organic systems such as esters and triglycerides. For nail lacquers, antiperspirants and lip products.

Characteristics: Low iron content
Light in color

BENTONE 38 Rheological Additive:

CTFA: Quaternium-18 Hectorite

Application: Low to intermediate polarity organic systems such as silicones and mineral oil. For antiperspirants, creams, lotions, eye products and suntan products.

Characteristics: Low iron content
Light in color

BENTONE 34 Rheological Additive:

CTFA: Quaternium-18 Bentonite

Application: Low to intermediate polarity organic systems such as silicones and mineral oil. For antiperspirants, creams and lotions.

Characteristics: Higher iron
Darker in color

RHEOX, INC.: BENTONE GEL Rheological Additives:

Fully dispersed and activated BENTONE organophilic clays are available in the form of smooth, homogeneous mastergels in common oils and organic solvents.

Advantages:
- the correct BENTONE additive is chosen for the solvent
- the optimum level of the correct polar activator is incorporated
- the highest yield is obtained, due to the full development of the rheological properties

BENTONE GEL MIO rheological additive:
CTFA: Mineral Oil & Quaternium-18 Hectorite & Propylene Carbonate
Applications: Hair grooming preparations, creams, lotions, lip products, suntan products

BENTONE GEL MIO A-40 rheological additive:
CTFA: Mineral Oil & Quaternium-18 Hectorite & SDA 40
Applications: Creams, lotions, hair grooming products

BENTONE GEL IPM rheological additive:
CTFA: Isopropyl Myristate & Stearalkonium Hectorite & Propylene Carbonate
Applications: Antiperspirants, creams, lotions

BENTONE GEL CAO rheological additive:
CTFA: Castor Oil & Stearalkonium Hectorite & Propylene Carbonate
Applications: Lip products

BENTONE GEL LOI rheological additive:
CTFA: Lanolin Oil & Isopropyl Palmitate & Stearalkonium Hectorite & Propylene Carbonate & Propyl Paraben
Applications: Lip products, creams, lotions

BENTONE GEL SS71 rheological additive:
CTFA: Petroleum Distillate & Quaternium-18 Hectorite & Propylene Carbonate
Applications: Eye products

BENTONE GEL VS-5 rheological additive:
CTFA: Cyclomethicone & Quaternium-18 Hectorite & SDA 40
Applications: Antiperspirants, eye products, creams, lotions

Special BENTONE mastergels can be made to meet a formulator's or manufacturer's individual requirements.

RHEOX, INC.: Hydrophilic Clay Rheological Additives:

NL's thickeners for water based products are three purified hectorite clays: BENTONE EW, BENTONE LT and MACALOID. In a wide variety of water based creams, lotions, shampoos, makeup and other cosmetics, the hydrophilic clays efficiently provide thixotropic viscosity and pigment suspension. They contribute to emulsion stability and smooth product application.

Hectorite thickeners are preferred over bentonites for their higher efficiency, whiter color and lower iron content. They impart an excellent texture, without the gummy or stringy consistency often obtained with organic gums.

Advantages:

- High efficiency
- Thixotropic
- Suspend pigments and active ingredients
- Effective over wide pH range
- Improve oil-in-water emulsion stability
- Synergistic viscosity build with organic gums
- Do not require neutralization
- White color
- Low iron content
- No quartz or cristobalite

BENTONE EW rheological additive:

CTFA: Hectorite

Applications: Creams & lotions, antidandruff shampoo, suntan products, shaving preparations, and hair conditioners

Characteristics: A highly beneficiated hectorite clay. Gives high viscosity. Works in a broad pH range: 5-11.

BENTONE LT rheological additive:

CTFA: Hectorite & Hydroxyethylcellulose

Application: Eye products, creams & lotions, powdered hair bleach, makeup

Characteristics: A modified hectorite clay. Gives the highest viscosity. Effective in a pH range: 6-10

MACALOID rheological additive:

CTFA: Hectorite

Application: Antidandruff shampoo, lotions, shaving cream, peroxide cream, antiperspirants

Characteristics: A beneficiated hectorite clay. Gives excellent suspension at moderate viscosities. Efficient in the broadest pH range: 4.5-12

RHEOX, INC.: THIXCIN R Rheological Additive:

When the highest degree of thixotropic flow in aliphatic systems is desired, THIXCIN R is the rheological additive to choose. In addition to the usual rheological control functions, it gives water repellancy, acts as a stiffening agent for lipsticks and ointments and helps stabilize emulsions. If high temperature resistance is needed, BENTONE additives are preferred.

THIXCIN R is a castor oil derivative. It requires heat (95-130F/35-55C) and high shear for activation.

Advantages:

- Very efficient viscosity build
- Greatest thixotropic body
- Superior film build at moderate temperatures
- Excellent leveling
- Imparts water repellency
- Controls liquid penetration
- Adds stiffening to stick products
- Controls suspension of pigments

THIXCIN R rheological additive:

CTFA: Trihydroxystearin

Application: Aliphatic solvent systems such as mineral oils and waxes, i.e., lipsticks, ointments, creams and lotions

Not suitable for aromatic and oxygenated solvents.

Characteristics: Highest thixotropy, superior leveling and film build, waterproofs

RHONE POULENC: ALIPAL CO HF series anionics:

Sodium or ammonium salts of sulfated alkylphenoxypoly (ethyleneoxy) ethanol.
M = NH4+ or Na+
CTFA name: Ammonium Nonoxynol-4 Sulfate

Physical form and description:
Clear, viscous liquids differing primarily in their degree of ethoxylation and activity. (ALIPAL CO-436 contains 12-16% ethanol).
Neutral pH.
Mild aromatic or alcohol odor.
HF-433 is a non-alcohol version of CO-436.

Properties:
High foaming detergents with excellent wetting, dispersing and emulsifying properties.
Readily soluble in cold or hot water, mildly acid or strongly alkaline solutions.
Stable to hard water, alkali and most metal ions.
Foaming properties retained in presence of substantial levels of emollients.
Mild, non-irritating to skin, particularly HF-433. Good lime-soap dispersion.

Uses:
Detergents in:
hand and face cleansing creams
liquid emulsion germicidal skin cleaners
mild shampoos and bubble baths (HF-433)
scrub soaps.

RHONE POULENC: EMULPHOR EL, ON series nonionics:

Polyoxyethylated vegetable oils and fatty alcohols

EMULPHOR EL-620:
CTFA name: PEG-30 Castor Oil
EL-719:
CTFA name: PEG-40 Castor Oil
ON-870:
CTFA name: Oleth-20

Physical form and description:
Vary from viscous liquids to waxes, depending on the n ratio and the oil or alcohol base.
Generally 100% active (EMULPHOR ON-877 is 70% aqueous solution).

Properties:
Predominantly hydrophilic, water soluble emulsifiers, dispersants and solubilizers. Also soluble in a number of organic solvents. Stable to acids, alkalis and hard water.
Low oral toxicity, non-irritant to human skin.

Uses:
Excellent emulsifiers for oils and waxes in the preparation of creams, lotions, hair grooming aids, etc. (ON-870/EL-719).
Solubilizers for essential oils and perfumes (ON-870).
Non-greasy emollients for men's hair grooming aids (ON-870/EL-719).
Pigment dispersants for lipstick, makeup. Emollients for creams/lotions, hair grooming aids.

RHONE POULENC: GAFAC series anionics:

Complex phosphate esters of nonionic surfactants of the ethylene oxide adduct type.
Various CTFA names.

Physical form and description:
Comprehensive range of viscous liquid products, mostly in the free acid form# and of 100% activity.
They vary in the hydrophobic base used (aliphatic or aromatic) and the n ratio.
Colorless and odorless.

#GAFAC MC-470 Surfactant is the partial sodium salt.

Properties:
Excellent cosmetic emulsifiers and solubilizers, with good emulsion spontaneity, stability and redispersibility in hard water.
Emulsifying properties maintained over broad pH range.
Medium to high foaming, with no cloud point.
Broad compatibility with cosmetic ingredients, e.g. amphoterics, cationic resins, etc.
Antistatic properties.
Low skin and eye irritation.
Contributes to improved wet-combability, softness and shine (MC-470).

Uses:
Emulsifiers for cosmetic oil, creams and lotions (RM-510/710, RE-870 + RD-510).
Components of conditioning shampoos (in combination with amphoterics); especially MC-470.

RHONE POULENC: GAFAMIDE CDD 518 superamide:

Fatty acid dialkanolamide
CTFA name: Cocamide DEA

Physical form and description:
Viscous oily liquid

Very high amide content (typically at least 90%)
Low in free amine and free fatty acid (below 1%)
Neutral or faint fatty odor.

Properties:
Highly purified, predominately nonionic detergent, offering excellent wetting, emulsifiying and soil-suspending properties.
Pronounced synergistic foam boosting and stabilization when added to nonionic/anionic surfactant systems.
Thickening effect makes it an excellent viscosity builder.
Super-fatting characteristics, preventing excessive degreasing of the skin.
Fully biodegradable.

Uses:
Foam booster, stabilizer and viscosity modifier in:
Shampoos.
Bubble baths.
Waterless hand cleaners.
Skin protective agent in:
Hand cleaner pastes and gels.
Liquid dishwashing compounds.
Laundry detergents.
Toilet soaps.

RHONE POULENC: IGEPAL CO, CA, DM series nonionics:

Ethoxylated alkylphenols

IGEPAL CO's:
CTFA name: Nonoxynol Series
IGEPAL CA's:
CTFA name: Octoxynol Series
IGEPAL DM's:
CTFA name: Nonyl Nonoxynol Series

Physical form and description:

Vary from slightly viscous oils (n=1.5 to 15) to low melting waxes (n=20 to 100).

All the products are anhydrous (100% active) and cover the range of HLB values from 4.8 to 19.0. Many of the waxes are also available as 70% active aqueous solutions.

Properties:

Oil and/or water soluble detergents, emulsifiers and dispersants.

Nonionic and non-electrolytic, stable to hydrolysis by acids or alkalis. Compatible with anionic or cationic agents, and with positively or negatively charged colloids.

Properties range from low foaming (n=4 to 8) to high foaming (n=10 or 30).

Excellent lime-soap dispersants (n=9 to 100).

Uses:

Color dispersants in hair dyes (IGEPAL CO-430/630).

Emulsifiers for mineral oils, chlorinated hydrocarbons (CO-430 to CO-530).

Water-soluble emulsifiers for vegetable and essential oils and other fatty compounds (CO-710 to CO-880).

Components of soap-syndet toilet bars for hard water (DM-970).

RHONE POULENC: IGEPON A, T series anionics:

Sulfoalkyl amides (I) and esters (II).

IGEPON AC-78:
CTFA names: Sodium Cocoyl Isethionate
IGEPON T's:
CTFA names: Sodium Methyl Oleoyl Taurate
IGEPON TC-42:
CTFA names: Sodium Methyl Cocoyl Taurate

Physical form and description:
White powder (high activity) or paste (low activity).
Very low salt content.
Neutral pH range.
Neutral or mild fatty odor.

Properties:
Outstanding detergency and wetting properties, coupled with dispersing and suspending action.
High, stable foam.
Effective in hard water and concentrated electrolyte solutions
Resistant to hydrolysis.
Excellent lime-soap dispersancy.
Soap-like mildness, softening and lubricating effect on skin and hair.
Readily biodegradable.

Uses:
Ideal base for detergent bars (IGEPON AC-78 Surfactant).
High-foaming shampoos with conditioning properties (AC-78/TC-42).
Salt-free detergent in dentifrices (AC-78).
Mild detergent for bubble baths and liquid skin cleansers.
Dry powder shampoos (AC-78).
Deodorant soaps (AC-78)
Pet products (AC-78)
Fragrance bars (AC-78)
Skin treatment cleaning bars (AC-78)
Medicated bars (AC-78)

ROCHE CHEMICAL DIVISION: Biotin:

Biotin is a member of the water-soluble, B-Complex vitamin group. The term, water soluble is somewhat misleading, for its solubility in water is very low (0.02%).

Physical-Chemical Properties:
- Empirical Formula: $C_{10}H_{16}N_2O_3S$
- Molecular Weight: 244.31
- Melting Range: 229-232C
- Specific Rotation @ 25C: +89 to +93
- Assay: 97.5% (as is)
- Heavy Metals: Max 0.001% (10 ppm)

Solubility:
- One gram dissolves in about 5000 ml. of water and about 1300 ml of ethyl alcohol (at 25C).
- Its solubility is greater in hot water and in dilute alkali.
- Biotin is insoluble in other common organic solvents.

Stability:
- Biotin, when pure, is stable to air and heat.
- Aqueous solutions of Biotin are quite stable in the pH range of 5 to 8 at room temperature.
- Stability decreases outside of the 5-8 pH range, especially if solutions are hot.
- Aqueous solutions are susceptible to mold growth.

Potential Uses in Cosmetics & Toiletries:

Biotin has been incorporated in a number of hair care and skin products. Although Biotin deficiency symptoms are rarely seen, because of the observed effects on the skin, it has been referred to as a "skin factor".

It has also been postulated that Biotin might be involved in blocking androgen receptor sites and thereby prevent the alleged activity of d-hydrotestosterone causing alopecia (baldness). However, no data is available showing that Biotin can reverse alopecia or prevent its occurrence in humans, either topically or systemically.

ROCHE Biotin Product Available for Topical Formulations:

Biotin:
CTFA Designation: Biotin

ROCHE CHEMICAL DIVISION: Citric Acid, USP, FCC:

Citric acid is a white, practically odorless, crystalline product that has a strong acidic taste. It is found widely in nature and occurs in numerous plants and fruits and in milk. Relatively large amounts are found in lemons (3.6% of the juice) and in rosehips (1.4%).

Physical-Chemical Properties:
- Empirical Formula: C8H8O7
- Molecular Weight: 192.13
- Assay: 99.5-100.5%
- Water (Karl Fischer): 0.5% minimum
- Residue on ignition: 0.05% maximum
- Readily Carbonizable Substances: Meets USP/FCC Test
- Oxalate: Meets USP/FCC Test
- Heavy Metals: 10 ppm maximum
- Sulfate: Meets USP Test
- Arsenic: 3.0 ppm maximum

Solubility:
- Very soluble in water (1 gm dissolves in about 0.5 ml water).
- Freely soluble in alcohol (1 gm dissolves in about 2 ml alcohol).
- Sparingly soluble in ether (1 gm dissolves in about 30 ml of ether).

Stability:
Citric acid anhydrous is slightly hydroscopic. Avoid excessive exposure to heat and humidity to prevent caking.

FDA Status:
Citric acid is generally recognized as safe (GRAS) for multipurpose food uses and as a sequestrant.

Cosmetic Use:
- Adjustment of pH in hair, skin care and other cosmetic products.
- Citric acid in combination with its salts acts as a buffer.
- Citric acid and citrates can complex metal ions and thus stabilize cosmetics.

ROCHE Citric Acid Products Available:

Citric Acid, USP, FCC:
Anhydrous, Fine Granular
CTFA Designatuion: Citric Acid

Citric Acid, USP, FCC:
Anhydrous, Granular
CTFA Designation: Citric Acid

Sodium Citrate, USP, FCC:
Dihydrate, Fine Granular
CTFA Designation: Sodium Citrate

ROCHE CHEMICAL DIVISION: **Panthenol (Provitamin B5):**

Panthenol is the alcohol analog of the vitamin pantothenic acid. It is a stable, biologically active form of this B-Complex vitamin. Two forms are available for topical formulations: a liquid, d-Panthenol, and a crystalline powder, dl-Panthenol.

Physical-Chemical Properties:

A. Dexpanthenol (USP Grade)
   Empirical Formula: C9H19NO4
   Molecular Weight: 205.25
   Appearance: Viscous liquid
   Color: Maximum APHA 60
   Solution in Water: Clear and complete at 5%
   Specific Rotation at 25C:+29.0 to +31.5 (dry basis)
   Refractive Index at 20C: 1.4950 to 1.5020
   Water Content: Maximum 1.0%
   Residue on Ignition: Maximum 0.1%
   Heavy Metals: Maximum 10 ppm
   Aminopropanol: Maximum 0.6%
   Assay: Minimum 98.0% (dry basis)

B. dl-Panthenol (Cosmetic Grade)
   Empirical Formula: C9H19NO4
   Molecular Weight: 205.25
   Appearance: White to creamy white, crystalline powder
   Odor: Characteristic
   Solution in water: Clear and complete at 10%; maximum APHA 200
   Melting range: 64.5-69.0C (within 3)
   Residue on ignition: Maximum 0.1%
   Heavy metals: Maximum 20 ppm
   Aminopropanol: Maximum 0.2%
   Assay: Minimum 98.5% (dry basis)

Solubility:

d-panthenol:
- Soluble in water and alcohol.
- Insoluble in fats and oils.

dl-Panthenol:
- Freely soluble in water, alcohol and propylene glycol.
- Soluble in chloroform and ether.
- Slightly soluble in glycerin.
- Insoluble in fats and oils.

Stability:

d-panthenol:
- Slightly hygroscopic.
- Stable to heat, but will racemize on prolonged heating.
- Aqueous topical formulations are stable in the pH range of 4.0-7.0 (optimum pH is 6.0)
- Hydrolysis occurs at an increasing rate as the pH varies from the optimum

dl-panthenol:
- Stable, but should be protected from extended exposure to moisture.
- Its stability characteristics are similar to those of d-panthenol

ROCHE CHEMICAL DIVISION: Panthenol (Provitamin B5)(Continued):

Role in Hair-Care Products:

- Provitamin B5 converts to Pantothenic Acid in hair and skin.
- Deep penetration into the hair shaft (Cortex).
- Long-Lasting moisture control, independent of the atmospheric conditions.
- Prevention of over-drying of hair and scalp caused by blow-dryers.
- Reduces considerably (30-40%) the formation of split ends, due to combing and brushing.
- Reduction of hair tangling by smoothing of the cuticle.
- Clean conditioning without build-up.
- Repairs damage caused by chemical and mechanical treatment (perming, overprocessing, combing, brushing and coloring).
- Slows doen the "aging process" (wear and tear of hair) caused by over-shampooing, brushing, and combing (cuticle damage, cortex weakening).
- Increases the tensile strength of hair (especially beneficial to bleached hair).
- Imparts sheen and luster.
- Cumulative benefits--Panthenol absoprtion increases with repeated use.
- Deposition in hair is dependent on hair types: higher deposition in tinted and bleached hair, lower on virgin hair.

Role in Skin Care Products:

- Quick, deep penetrating moisturizer (from within).
- Stimulates cellurar proliferation, and possibly aids in tissue repair (exact mechanism is unknown).
- Promotes normal Keratinization.
- Recommended levels of panthenol in topical products: 1-5%
- In experimental therapeutics:
  - Claimed to promote wound healing in surgical, traumatic, or thermally induced wounds.
  - Used in treating atopic dermatitis, chronic contact dermatitis, diaper rashes and certain types of itching.
  - Claimed to act as an anti-inflammatory agent.
  - Used in treating acute sunburns, and in promoting pigment formation following tanning exposure.

ROCHE Panthenol Product Available for Topical Formulations:

d-Panthenol, liquid:
Regular Type
CTFA Designation: Panthenol

dl-Panthenol:
Cosmetic Grade, powder
CTFA Designation: Panthenol

dl-Panthenol, liquid(75%):
Cosmetic Grade
CTFA Designation: Panthenol

ROCHE CHEMICAL DIVISION: ROCHE Vitamin A and Vitamin A with Vitamin D3 Products Available for Topical Formulations:

Vitamin A Palmitate Type PIMO/BH:
Contains 1,000,000 I.U./gm with 5 mg each of BHA and BHT per 1,000,000 units as an antixoidant.
CTFA Designation: Retinyl Palmitate (and) corn oil (and) BHA (and) BHT.

Vitamin A Palmitate:
Type PI.8 contains 1,650,000 to 1,800,000 I.U./gm
No antioxidants added.
CTFA Designation: Retinyl Palmitate (and) corn oil

Vitamin A Palmitate:
Type P1.8/BH
1,650,000 to 1,800,000 I.U./gm.
Stabilized with BHA/BHT
CTFA Designation: Retinyl Palmitate (and) corn oil (and) BHA (and) BHT.

AQUAPALM (Vitamin A Palmitate Isomerized):
Contains 1,500,000 I.U. of Vitamin A per gram plus 5 mg each of BHA and BHT per 1,000,000 units as antioxidants.
CTFA Designation: Retinyl Palmitate (and) corn oil (and) BHA (and) BHT.

Vitamin A Palmitate with Vitamin D3 in Vegetable Oil:
Available as 5 to 1 ratio, 1,000,000 I.U. of Vitamin A and 200,000 I.U. of Vitamin D3 per gm.
CTFA Designation: Retinyl Palmitate (and) Cholecalciferol (and) corn oil

Vitamin A Alcohol Blend:

Vitamin A Alcohol blended with Tween 20, 1,500,000 I.U./gm
CTFA Designation: Retinol (and) Polysorbate 20

ROCHE CHEMICAL DIVISION: Vitamin A:

Vitamin A, according to the U.S.P. official monograph, contains a suitable from of retinol (vitamin A alcohol) or its esters formed from edible fatty acids, principally acetic and palmitic acids.

Physical-Chemical Properties of Retinyl Palmitate:
- Empirical formula: $C_{36}H_{60}O_2$
- Molecular Weight: 524.9
- Specific gravity: 0.90-0.95 @ 25C
- Appearance: Clear, light yellow to amber liquid with a slight characteristic odor.

Solubility:
- Insoluble in water and glycerin.
- Very soluble in chloroform and ether.
- Soluble in vegetable oils.
- Clear aqueous dispersions can be prepared with polysorbate 80.

Stability
- Vitamin A is sensitive to oxidation.
- Retinol is more sensitive than fatty acid esters of vitamin A.
- Oxidation is catalyzed by trace metals, notably iron and copper.
- Sensitive to UV light.
- Isomerized by acids.
- Sensitive to heat.

Physiological Function:
- Essential for vision.
- Regulates the growth and activity of epithelial cells.
- Essential for growth and maintenance of bones, glands, teeth, nails, and hair.
- Its effect is dosage dependent:
  - High levels have keratolytic action (detergent-like)
  - Low levels are anti-keratinizing.

Potential Topical Benefits:
- Absorbed by the skin.
- Skin "Normalizer", maintaining skin soft and plump.
- Anti-keratinizing.
- Improves the water barrier properties of skin.
- Possible use in after-sun tan products because of its "Normalizing" properties.
- Recommeneded levels for skin care products are 5,000-10,000 I.U./gm.

ROCHE CHEMICAL DIVISION: Vitamin C (As Ascorbyl Palmitate):

Ascorbyl Palmitate is white or yellowish white crystalline powder having a slight odor. It exhibits the full biological activity of Vitamin C.

Physical-Chemical Properties:
- Empirical Formula: C22H38O7
- Molecular Weight: 414.54
- Identity Test: Positive
- Melting Range: 107C-117C
- Specific Rotation: +21 to +24
- Loss on Drying: Maximum 2.0%
- Residue on Ignition: Maximum 0.1%
- Heavy Metals: maximum 0.001% (10 ppm)
- Arsenic: Maximum 0.0003% (3 ppm)
- Assay (dry basis): 95.0-100.5%

Solubility:
- Very slightly soluble in water (0.00018 gm/100 water at room temperature.
- Soluble in alcohol (1.0 gm/4.5 ml alcohol)
- Solubility in oily substances is low and varies with the nature of the oil or solvent

Stability:

Ascorbyl Palmitate is fairly stable to air if protected from humidity, but somewhat sensitive to light and heat. The bulk product should be stored between 46 and 59F.

Potential Uses in Cosmetics & Toiletries:

Objectionable odors and flavors can develop in products containing vegetable and animal fats and oils due to oxidative rancidity and enzymatic rancidity. Enzymes usually can be controlled by destruction via heat, whereas oxidation rancidity calls for antioxidants, synergists and other means.

Since substrates which are susceptible to such oxidative breakdown are usually oil soluble and not water soluble, oil soluble antioxidants are called for. Ascorbyl Palmitate is especially useful in this regard, owing to its ability to scavenge oxygen, whether the oxygen be dissolved in the product, or in the head space above the product.

Ascorbyl palmitate has long been used as an antioxidant in vegetable oils, where it has been found to be more effective than either BHA or BHT. In animal fats, ascorbyl palmitate acts synergistically with other antioxidants, such as dl-alpha-tocopherol.

Although the air-oxidation of unsaturated compounds such as vegetable oils may be very slow at room temperature, once hydroperoxides are formed, their subsequent breakdown to free-radicals leads to a very rapid decomposition of the substrate.

Concentrations of ascorbyl palmitate in the range of 100-500 ppm usually suffice.

ROCHE CHEMICAL DIVISION: **Vitamin E:**

In nature, vitamin E appears as tocopherols. Of the several tocopherols, the d-alpha form has the highest biological potency.

Physical-Chemical Properties:

A. dl-alpha Tocopherol (USP Grade)

Empirical Formula: $C_{29}H_{50}O_2$

Molecular Weight: 430.7

Biological Activity: 1 mg=1.1 International Units

Vitamin E (dl-alpha Tocopherol) meets all requirements of the USP and FCC when tested according to those compendia.

Identity Tests: Positive

Acidity: Satisfactory

Heavy Metals: Maximum 0.001% (10 ppm)

Assay (USP, GLC Method): 96.0-102.0%

B. dl-alpha Tocopherol Acetate (USP and Cosmetic Grade)

Empirical Formula: $C_{31}H_{52}O_3$

Molecular Weight: 472.76

Biological Activity: 1 mg=1 International Unit

Vitamin E (dl-alpha tocopherol acetate) meets all requirements of the USP and FCC when tested according to those compendia.

Identity Tests: Positive

Acidity: Satisfactory

Heavy Metals: Maximum 0.001% (10 ppm)

Assay (U.S.P. Grade): 96.0-102.0%

(USP, GLC Method)

Cosmetic Grade: 91.0-95.0%

Solubility:

dl-alpha Tocopherol:

- Insoluble in water.
- Freely soluble in alcohol.
- Miscible in ether, chloroform, acetone, vegetable oils, isopropryl myristate, isopropyl palmitate and other cosmetic oils.

dl-alpha Tocopheryl Acetate:

- Insoluble in water.
- Freely soluble in alcohol.
- Miscible with ether, chloroform, acetone, vegetable oils, isopropyl palmitate, isopropyl myristate and other cosmetic oils.

Stability:

dl-alpha Tocopherol:

- Unstable to heat in the presence of oxygen.
- Relatively stable to acids and alkalies in the absence of oxygen
- Oxidizes and darkens on exposure to light and air.
- Ferric or silver salts and rancid fats accelerate degradation.

ROHM AND HAAS: KATHON CG Microbiocide:

The active ingredients of KATHON CG biocide are two isothiazolones identified by the IUPAC system of nomenclature as: 5-chloro-2-methyl-4-isothiazolin-3-one and 2-methyl-4-isothiazolin-3-one.

The CTFA-adopted names for these active substances are:
Methyl Chloro isothiazoline-CAS number 26172-55-4
Methyl isothiazolinone-CAS number 2682-20-4

Typical Physical and Chemical Properties:

Active Ingredients:

5-Chloro-2-methyl-4-isothiazolin-3-one:	1.15%
2-Methyl-4-isothiazolin-3-one	0.35%

Inert Ingredients:

Magnesium salts (chloride and nitrate)	23.0%
Water	75.5%

Appearance: Clear liquid
Color: Light amber
Odor: Mild
Specific Gravity @ 20C: 1.21
pH (as manufactured): 3.5-5.0

Suggested Applications:

KATHON CG biocide is suggested as a preservative for:
- shampoos and hair conditioners
- hair and body gels
- bubble baths
- surfactants and raw materials for cosmetics

Microbiostatic Properties:

KATHON CG biocide exhibits outstanding antimicrobial activity against gram-positive and gram-negative bacteria, fungi and yeasts.

Efficacy in Cosmetics:

The antimicrobial performance of KATHON CG biocide in most cosmetics is excellent. Long-term micro-biological protection is obtained employing use levels up to 15 ppm A.I.

SANDOZ CHEMICALS CORP.: CARTARETIN F-4 and F-23 Water Soluble Cationic Polymers:

CARTARETIN F-4 and F-23 are cationic polymers supplied as aqueous solutions. They are soluble in water and certain blends of water and lower alcohols. They are highly substantive to hair from polymer solutions or properly formulated shampoo systems.

F-4:

CTFA: Adipic Acid/Dimethylaminohydroxy propyl Diethylenetriamine Copolymer
Clear, yellowish liquid
Solids: 30%±1
pH: (as is): 8.5±0.5
Type: Cationic polyamide-amine polymer
Solubility: Water and alcohol

F-23:

CTFA: Adipic Acid/Dimethylaminohydroxy propyl Diethylenetriamine Copolymer
Clear, yellowish liquid
Solids: 23%±1
pH: (as is): 8.0±5
Type: Cationic polyamide-amine polymer
Solubility: Water and alcohol

SANDOZ CHEMICALS CORP.: **SANDOPAN TFL Conc.:**

A mild amphoteric surfactant
CTFA Designation: Amphoteric-7

A specialty surfactant exhibiting properties of a weak ampholyte. Chemically, it is a sulfonated fatty amide-imide.

Physical Properties:

Appearance: Hazy, light amber gel
Odor: Mild, fatty
pH (20% solution): 8-9
Cloud Point: None
Specific Gravity: 1.14 20C
Non-volatiles: 56±0.5%

SANDOZ CHEMICALS CORP: SANDOPAN Carboxylated Surfactants:

Most of these SANDOPAN surfactants are extremely mild to eyes and skin, recommending them for personal care formulations.

<u>Trade Name:</u>

SANDOPAN DTC:
CTFA: Sodium Trideceth-7-Carboxylate
Light yellow gel
% Solids: 75±2%

SANDOPAN DTC-100:
CTFA: Same as above
Clear yellow liquid
% Solids: 70±2%

SANDOPAN DTC Acid:
CTFA: Trideceth-7-Carboxylic Acid
Clear liquid
% Solids: 90±2%

SANDOPAN DTC Linear P:
CTFA: Sodium C12-15 Pareth-6-Carboxylate
White, opaque semi-pourable gel
% Solids: 70±5%

SANDOPAN DTC Linear P Acid:
CTFA: C12-15 Pareth-6-Carboxylic Acid
Clear liquid
% Solids: 90±5%

SANDOPAN LS-24:
CTFA: Sodium Laureth-13-Carboxylate
Clear to slightly hazy gel
% Solids: 69±2%

SANDOPAN JA-36:
CTFA: Trideceth-19-Carboxylic Acid
Clear to slightly hazy liquid
% Solids: 90±2%

SANDOPAN RS-8:
CTFA: Sodium C16-20 Ethoxylate Carboxylate
Off white, firm paste
% Solids: 70±2%

SANDOPAN MA-18:
CTFA: Alkylaryl Ethoxylate Carboxylic Acid
Clear liquid
% Solids: 90±3%

SANDOPAN KST:
CTFA: Sodium Ceteth-13-Carboxylate
Solid
% Solids: 97±2%

SANDOPAN B liquid:
CTFA: Carboxylated, C4 Paraffinic Ethoxylate
Clear light, amber liquid
% Solids: 40±5%

SANDOZ CHEMICALS CORP.: VELSAN:

A new range of emollient esters for adding a "velvety" feeling to cosmetic products

Suggested Uses for VELSAN Products:
- Skin Care
- Sun Care
- Hair Care
- Pigmented Products
- Bath Products

Distinguishing Characteristics:

VELSAN P 8-3: Light yellow oil. Shows water solubility and light velvety feel.

VELSAN P 8-16: Semi-solid light yellow paste. Oil soluble for a heavy emollient feel without oiliness.

VELSAN D8P-3: Light yellow oil. Non comedogenic emulsion that helps to solubilize cosmetic actives.

VELSANS are effective emollients that offer a wide and unique array of benefits to the cosmetic formulator. VELSANS can provide emulsion viscosity, solubilization, wetting and softer feel all in one ingredient.

Properties:

VELSAN P 8-3:
- Light yellow oil
- CTFA Name: Isopropyl C12-15 Pareth-9-Carboxylate
- Specific Gravity: 1.01
- HLB: 18
- Acute Oral Toxicity: >5 g/kg
- Skin Irritation (Rabbit): Not a primary irritant.
- Eye Irritation (Rabbit): Mild Irritant (Reversible)

VELSAN P 8-16:
- Semi-solid yellow paste
- CTFA Name: Cetyl C12-15 Pareth-9-Carboxylate
- Specific Gravity: 0.95
- HLB: 9
- Acute Oral Toxicity: >5 g/kg
- Skin Irritation (Rabbit): Not a primary irritant.
- Eye Irritation (Rabbit): Not an irritant

VELSAN D8P-3:
- Light yellow oil
- CTFA Name: Isopropyl PPG-2-Isodeceth-7 Carboxylate
- Specific Gravity: 1.02
- HLB: 14
- Acute Oral Toxicity: >5g/kg
- Skin Irritation (Rabbit): Not a primary irritant.
- Eye Irritation (Rabbit): Practically Non-irritant

SCHER CHEMICALS, INC.: Amido-Amines:

SCHERCODINE L:
CTFA: Lauramidopropyl Dimethylamine
Fatty Acid Source: Lauric
Light Tan Solid
MW (av): 284
Melting Point, C: 35-40
Free Amine, % (max): 1
Alkali Value: 98
Amide, % (min): 196-206
Cationic surfactant, imtermediate for betaine amphoterics.

SCHERCODINE C:
CTFA: Cocamidopropyl Dimethylamine
Fatty Acid Source: Coconut
Tan Soft Solid
MW (av): 304
Free Amine, % (max): 1
Alkali Value: 98
Amide, % (min): 177-187
Good foaming, cationic surfactant for hair and bath preparations.

SCHERCODINE M:
CTFA: Myristamidopropyl Dimethylamine
Fatty Acid Source: Myristic
Light Tan Wax
MW (av): 312
Melting Point, C: 45-50
Free Amine, % (max): 1
Alkali Value: 98
Amide, % (min): 180-190
Cationic O/W emulsifier. Good conditioner and viscosity builder.

SCHERCODINE P:
CTFA: Palmitamidopropyl Dimethylamine
Fatty Acid Source: Palmitic
Tan Hard Wax
MW (av): 340
Melting Point, C: 55-60
Free Amine, % (max): 1
Alkali Value: 98
Amide, % (min): 160-170
Substantive conditioner and emulsifier in creams, lotions, rinses.

SCHER CHEMICALS, INC.: Amido-Amines(Continued):

SCHERCODINE S:
CTFA: Stearamidopropyl Dimethylamine
Fatty Acid Source: Stearic
Tan Hard Wax
MW (av): 368
Melting Point, C: 65-70
Free Amine, % (max): 1
Alkali Value: 98
Amide, % (min): 145-155
Softener, emulsifier and conditioner in hair and skin preparations.

SCHERCODINE I:
CTFA: Isostearamidopropyl Dimethylamine
Fatty Acid Source: Isostearic
Light Amber Liquid
MW (av): 394
Free Amine, % (max): 98
Amide, % (min): 150-160
Versatile, liquid, O/W cationic emulsifier. Good lubricant, especially for hair rinses and conditioners.

SCHERCODINE O:
CTFA: Oleamidopropyl Dimethylamine
Fatty Acid Source: Oleic
Amber Liquid
MW (av): 366
Free Amine, % (max): 1
Alkali Value: 98
Amide, % (min): 150-160
An emollient conditioner for hair and skin preparations. Offers excellent lubricating and moisturizing properties.

SCHERCODINE T:
CTFA: Tallamidopropyl Dimethylamine
Fatty Acid Source: Tall Oil
Amber Liquid
MW (av): 366
Free Amine, % (max): 1
Alkali Value: 98
Amide, % (min): 150-160
Excellent conditioner for cationic emulsions. Offers good substantivity and thickening properties.

SCHERCODINE B:
CTFA: Behenamidopropyl Dimethylamine
Fatty Acid Source: Behenyl
Tan Hard Wax
MW (av): 394
Melting Point, C: 63-68
Free Amine, % (max): 1
Alkali Value: 98
Amide, % (min): 135-145
Cationic emulsifier offering good conditioning properties for skin and hair preparations.

A wide variety of other fatty acid sources is available.

SCHER CHEMICALS, INC.: Amido-Amine Salts#:

KATEMUL IG-70:

CTFA: Isostearamidopropyl Dimethylamino Glycolate
Amber Viscous Liquid
Dry Solids, % (min): 70
pH: 7

A water-soluble salt of amido-amine neutralized with glycolic acid. A cationic emulsifier, and conditioner for skin and hair preparations.

KATEMUL IGU-70:

CTFA: Isostearamidopropyl Dimethylamino Gluconate
Amber Viscous Liquid
Dry Solids, % (min): 70
pH: 7

A water-soluble salt of amido-amine neutralized with gluconic acid. It offers substantivity and mildness to creams, lotions, and bath preparations.

A wide variety of salts from the entire range of amido-amines neutralized with organic or inorganic acids are available.

SCHER CHEMICALS, INC.: Amine Oxides:

SCHERCAMOX DML:
CTFA: Lauramine Oxide
Fatty Amine Source: Lauryl Dimethyl
Clear Liquid
Free Amine, % (max): 0.5
Free Peroxide, % (max): 0.3
Amine Oxide, % (min): 29
Wetting agent, foamer and foam stabilizer in shampoos, bath preparations, and shave creams.

SCHERCAMOX DMC:
CTFA: Cocamine Oxide
Fatty Amine Source: Cocoyl Dimethyl
Clear Yellow Liquid
Free Amine, % (max): 1
Free Peroxide, % (max): 0.5
Amine Oxide, % (min): 29
Wetting agent. Foam stabilizer and viscosity enhancer.

SCHERCAMOX DMM:
CTFA: Myristamine Oxide
Fatty Amine Source: Myristyl Dimethyl
Clear Liquid
Free Amine, % (max): 0.5
Free Peroxide, % (max): 0.3
Amine Oxide, % (min): 29
Wetting and foaming agent. Foam booster in shampoos, bubble baths and dishwashing compounds.

SCHERCAMOX DMA:
CTFA: Myristamine Oxide
Fatty Amine Source: Alkyl Dimethyl
Viscous Liquid
Free Amine, % (max): 0.5
Free Peroxide, % (max): 0.3
Amine Oxide, % (min): 29
Applications: Same as for DMM

SCHERCAMOX C-AA:
CTFA: Cocamidopropylamine Oxide
Fatty Amine Source: Coco Amido Propyl Dimethyl
Clear to Hazy Liquid
Free Amine, % (max): 0.5
Free Peroxide, % (max): 0.5
Amine Oxide, % (min): 35
Wetting agent, detergent, foam booster, conditioner and viscosity builder in shampoos, hair coloring systems and bath preparations.

SCHERCAMOX DMS:
CTFA: Stearamine Oxide
Fatty Amine Source: Stearyl Dimethyl
White Paste
Free Amine, % (max): 1.5
Free Peroxide, % (max): 0.5
Amine Oxide, % (min): 25
Conditioner and softener in hair rinses and shampoos. Foam stabilizer.

SCHER CHEMICALS, INC.: Betaine Amphoterics:

SCHERCOTAINE CAB-A:
CTFA: Cocamidopropyl Betaine (and) Ammonium Chloride
Fatty Acid Source: Coconut
Clear Light Yellow Liquid
Dry Solids, % (min)#: 45
Salt Content, % (max): 5
pH: 5-7
Mild surfactant with higher foam than the sodium counterpart and decreased defatting properties.

SCHERCOTAINE CAB:
CTFA: Cocamidopropyl Betaine
Fatty Acid Source: Coconut
Clear Light Yellow Liquid
Dry Solids, % (min)#: 45
Salt Content, % (max): 6.5
pH: 5-7
Detergent, wetting agent, excellent wetting and cloud point depressant used in non-irritating shampoos and bubble baths.

SCHERCOTAINE CAB-K:
CTFA: Cocamidopropyl Betaine (and) Potassium Chloride
Fatty Acid Source: Coconut
Appearance: Clear Light Yellow Liquid
Dry Solids, % (min)#: 45
Salt Content, % (max): 3.5
pH: 5-7
Surfactant with increased solubility and lower cloud point; viscosity stabilizer in natural soap systems.

SCHERCOTAINE CAB-Z:
CTFA: Cocamidopropyl Betaine (and) Zinc Chloride
Fatty Acid Source: Coconut
Clear Amber Liquid
Dry Solids, % (min)#: 35
Salt Content, % (max): 3.5 (as Cl)
pH: 5-7
Applications: Same as for UAB-Z.

SCHERCOTAINE SCAB-A:
CTFA: Cocamidopropyl Hydroxy Sultaine (and) Ammonium Chloride
Fatty Acid Source: Coconut
Appearance: Clear Light Yellow Liquid
Dry Solids, % (min)#: 50
Salt Content, % (max): 5.0
pH: 5-7
Low cloud point surfactant with higher foam and decreased defatting properties.

Lower concentrations are available.

SCHER CHEMICALS, INC.: Betaine Amphoterics(Continued):

SCHERCOTAINE SCAB:

CTFA: Cocamidopropyl Hydroxy Sultaine
Fatty Acid Source: Coconut
Clear Light Amber Liquid
Dry Solids, % (min)#: 50
Salt Content, % (max): 6.0
pH: 5-7

Very low cloud point detergent, wetting agent and excellent foamer.

SCHERCOTAINE SCAB-K:

CTFA: Cocamidopropyl Hydroxy Sultaine (and) Potassium Chloride
Fatty Acid Source: Coconut
Clear Light Yellow Liquid
Dry Solids, % (min)#: 50
Salt Content, % (max): 6.0
pH: 5-7

Low cloud point surfactant; viscosity stabilizer in natural soap systems.

SCHERCOTAINE MAB:

CTFA: Myristamidopropyl Betaine
Fatty Acid Source: Myristic
Clear Light Yellow Liquid
Dry Solids, % (min)#: 30
Salt Content, % (max): 5.0
pH: 5-7

Detergent, wetting agent, and thickening agent with anti-static properties for cosmetic and toiletry preparations.

SCHERCOTAINE PAB:

CTFA: Palmitamidopropyl Betaine
Fatty Acid Source: Palmitic
Soft Light Yellow Gel
Dry Solids, % (min)#: 35
Salt Content, % (max): 5.5
pH: 5-7

Thickening agent, good hair and skin conditioner for lotions and cream rinses.

SCHERCOTAINE IAB:

CTFA: Isostearamidopropyl Betaine
Fatty Acid Source: Isostearic
Soft Amber Gel
Dry Solids, % (min)#: 35
Salt Content, % (max): 4.5
pH: 5-7

Detergent with excellent conditioning properties. Recommended for mild shampoos and emollient body treatments.

Lower concentrations are available.

SCHER CHEMICALS, INC.: Betaine Amphoterics(Continued):

SCHERCOTAINE UAB:
CTFA: Undecylenamidopropyl Betaine
Fatty Acid Source: Undecylenic
Clear Amber Liquid
Dry Solids, % (min)#: 35
Salt Content, % (max): 5.5
pH: 5-7
Surfactant with possible germicidal/bactericidal activity; recommended for shampoos.

SCHERCOTAINE UAB-Z:
CTFA: Undecylenamidopropyl Betaine (and) Zinc Chloride
Fatty Acid Source: Undecylenic
Clear Amber Liquid
Dry Solids, % (min)#: 35
Salt Content, % (max): 3.5 (as Cl)
pH: 5-7
Surfactant with possible germicidal/bactericidal activity; recommended for facial scrubs.

SCHERCOTAINE APAB:
CTFA: Apricotamidopropyl Betaine
Fatty Acid Source: Apricot Oil
Clear Amber Liquid
Dry Solids, % (min)#: 35
Salt Content, % (max): 4.0
pH: 5-7
Mild detergent with good conditioning and emolliency properties. Viscosity enhancer.

SCHERCOTAINE WOAB:
CTFA: Wheat Germamidopropyl Betaine
Fatty Acid Source: Wheat Germ Oil
Clear Amber Liquid
Dry Solids, % (min)#: 35
Salt Content, % (max): 4.0
pH: 5-7
Mild surfactant derived from a natural source. Good conditioner; imparts good body to hair.

SCHERCOTAINE MKAB:
CTFA: Minkamidopropyl Betaine
Fatty Acid Source: Mink Oil
Clear Amber Liquid
Dry Solids, % (min)#: 35
Salt Content, % (max): 4.0
pH: 5-7
Mild surfactant. Good emolliency properties. Imparts good body and feel to hair.

Lower concentrations are available.

SCHER CHEMICALS, INC.: 1-1 Fatty Acid Diethanolamides:

SCHERCOMID SL-Extra:
CTFA: Lauramide DEA
Fatty Acid Source: Lauric
White Crystalline Solid
Ionic Nature: N
Acid Value (max): 1
Alkali Value: 20-40
Amide, % (min): 87
Good thickener, foam stabilizer, wetting agent and detergent used in shampoos, lotions, liquid dishwashing compounds.

SCHERCOMID SLM-S:
CTFA: Lauramide DEA
Fatty Acid Source: Lauric-Myristic
White Crystalline Solid
Ionic Nature: N
Acid Value (max): 1
Alkali Value: 20-40
Amide, % (min): 87
Very good thickener, foam stabilizer, wetting agent and detergent used in cosmetic and toiletry products.

SCHERCOMID SLM-LC:
CTFA: Lauramide DEA
Fatty Acid Source: Lauric-Myristic
Clear Amber Liquid
Ionic Nature: N
Acid Value (max): 1
Alkali Value: 30-50
Amide, % (min): 85
A concentrated liquid amide with good wetting properties. Also a good thickener and foam stabilizer.

SCHERCOMID SL-ML:
CTFA: Lauramide DEA
Fatty Acid Source: Modified Lauric-Myristic
Clear Light Amber Liquid
Ionic Nature: N
Acid Value (max): 1
Alkali Value: 20-40
Amide, % (min): 87
Detergent, auxiliary skin and hair conditioner, foam stabilizer. Excellent foam-building characteristics.

SCHER CHEMICALS, INC.: 1-1 Fatty Acid Diethanolamides (Continued):

SCHERCOMID SCE:
CTFA: Cocamide DEA
Fatty Acid Source: Coconut
Clear Light Amber Liquid
Ionic Nature: N
Acid Value (max): 1
Alkali Value: 20-40
Amide, % (min): 87
Detergent, low-irritating thickener and foam stabilizer for cosmetic preparations and hcusehold cleaners.

SCHERCOMID SCO-Extra:
CTFA: Cocamide DEA
Fatty Acid Source: Coconut Oil
Clear Light Amber Liquid
Ionic Nature: N
Acid Value (max): 3
Alkali Value: 20-40
Amide, % (min): 80
Emulsifier, foam stabilizer, wetting agent for household and industrial detergents.

SCHERCOMID SO-A:
CTFA: Oleamide DEA
Fatty Acid Source: Oleic
Clear Amber Liquid
Acid Value (max): 5
Alkali Value: 40-60
Amide, % (min): 85
W/O emulsifier, lubricant, and conditioner.

SCHERCOMID SO-T
CTFA: Tallamide DEA
Fatty Acid Source: Tall Oil
Clear Amber Liquid
Acid Value (max): 15
Alkali Value: 40-50
Amide, % (min): 85
W/O emulsifier.

SCHER CHEMICALS, INC.: 1-1 Fatty Acid Diethanolamides (Continued):

SCHERCOMID SLE:
CTFA: Linoleamide DEA
Fatty Acid Source: Linoleic
Appearance: Clear Amber Liquid
Ionic Nature: N
Acid Value (max): 1
Alkali Value: 20-40
Amide, % (min): 87
Good W/O emulsifier and emulsion stabilizer for O/W emulsions, and outstanding thickener.

SCHERCOMID SLS:
CTFA: Soyamide DEA
Fatty Acid Source: Soya Oil
Clear Amber Liquid
Ionic Nature: N
Acid Value (max): 2
Alkali Value: 20-40
Amide, % (min): 82
Applications: Same as for SLE.

SCHERCOMID SI:
CTFA: Isostearamide DEA
Fatty Acid Source: Isostearic
Clear Light Amber Liquid
Ionic Nature: N
Acid Value (max): 2
Alkali Value: 15-40
Amide, % (min): 87
Detergent, emulsifier, corrosion inhibitor, thickener with high salt tolerance.

SCHER CHEMICALS, INC.: Imidazolinium Amphoterics#:

SCHERCOTERIC CY-2:
CTFA: Caprylamphocarboxyglycinate
Fatty Acid Source: Caprylic
Hydrophilic Group (Carboxylate): Di
Clear Amber Liquid
Dry Solids, % (min): 50
Salt Content, % (max): 11
Non-foaming surfactant with good cleaning properties for household and industrial cleaners.

SCHERCOTERIC CY-SF-2:
CTFA: Caprylamphocarboxypropionate
Fatty Acid Source: Caprylic
Hydrophilic Group (Carboxylate): Di
Soft Amber Paste
Dry Solids, % (min): 75
Salt Content, % (max): 0
Applications: Same as for CY-2

SCHERCOTERIC MS:
CTFA: Cocoamphoglycinate
Fatty Acid Source: Coconut
Hydrophilic Group (Carboxylate): Mono
Clear Amber Viscous Liquid
Dry Solids, % (min): 45
Salt Content, % (max): 9
Mild detergent, excellent foam height and stability. For shampoos and industrial cleaners.

SCHERCOTERIC MS-2:
CTFA: Cocoamphocarboxyglycinate
Fatty Acid Source: Coconut
Hydrophilic Group (Carboxylate): Di
Clear Amber Viscous Liquid
Dry Solids, % (min): 45
Salt Content, % (max): 10
Applications: Same as for MS

SCHERCOTERIC MS-SF:
CTFA: Cocoamphopropionate
Fatty Acid Source: Coconut
Hydrophilic Group (Carboxylate): Mono
Appearance: Clear Amber Liquid
Dry Solids, % (min): 38
Salt Content, % (max): 0
Mild surfactant for shampoos, industrial cleaners, dishwashing compounds.

Amphoterics from a wide variety of other fatty acid homologs are available.

SCHER CHEMICALS, INC.: Imidazolinium Amphoterics# (Continued):

SCHERCOTERIC MS-SF Conc.:
CTFA: Cocoamphopropionate
Fatty Acid Source: Coconut
Hydrophilic Group (Carboxylate): Mono
Amber Paste
Dry Solids, % (min): 75
Salt Content, % (max): 0
Applications: Same as for MS-SF.

SCHERCOTERIC MS-SF-2:
CTFA: Cocoamphocarboxypropionate
Fatty Acid Source: Coconut
Hydrophilic Group (Carboxylate): Di
Clear Amber Liquid
Dry Solids, % (min): 38
Salt Content, % (max): 0
Low-irritation detergent for shampoos and skin cleansers.

SCHERCOTERIC MS-SF-2 Conc.:
CTFA: Cocoamphocarboxypropionate
Fatty Acid Source: Coconut
Hydrophilic Group (Carboxylate): Di
Soft Amber Paste
Dry Solids, % (min): 75
Salt Content, % (max): 0
Applications: Same as for MS-SF-2.

SCHERCOTERIC MS-EP:
CTFA: Cocoamphopropylsulfonate
Fatty Acid Source: Coconut
Hydrophilic Group (Carboxylate): Mono (Sulfonate)
Appearance: Clear Amber Liquid
Dry Solids, % (min): 45
Salt Content, % (max): 6
Excellent surfactant. Remarkable flash foam with low skin irritation and low cloud point.

SCHERCOTERIC IS-SF-2:
CTFA: Isostearoamphocarboxypropionate
Fatty Acid Source: Isostearic
Hydrophilic Group (Carboxylate): Di
Appearance: Clear Amber Liquid
Dry Solids, % (min): 38
Salt Content, % (max): 0
Especially mild surfactant for shampoos, rinses, skin cleansers.

Amphoterics from a wide variety of other fatty acid homologs are available.

SCHER CHEMICALS, INC.: Imidazolinium Amphoterics# (Continued):

SCHERCOTERIC I-AA:
CTFA: Isostearoamphopropionate
Fatty Acid Source: Isostearic
Hydrophilic Group (Carboxylate): Mono
Appearance: Amber Viscous Liquid
Dry Solids, % (min): 34
Salt Content, % (max): 0
Specialty surfactant for cosmetic and industrial cleaners.

SCHERCOTERIC O-AA:
CTFA: Oleoamphopropionate
Fatty Acid Source: Oleic
Hydrophilic Group (Carboxylate): Mono
Appearance: Clear Amber Liquid
Dry Solids, % (min): 80
Salt Content, % (max): 0
Specialty surfactant for dry-cleaning industry, applicable for other industrial cleaners.

Amphoterics from a wide variety of other fatty acid homologs are available.

SCHER CHEMICALS, INC.: Monoalkanolamides:

SCHERCOMID AME:
CTFA: Acetamide MEA
Fatty Acid Source: Acetic
Clear Straw-Colored Liquid
Acid Value (max): 10
Alkali Value (max): 15
Amide, % (min): 95
Humectant, skin and hair conditioner intermediate, coupling agent, pigment dispersant and solubilizer.

SCHERCOMID LME:
CTFA: Lactamide MEA
Fatty Acid Source: Lactic
Clear Yellow Liquid
Acid Value (max): 20
Alkali Value (max): 20
Amide, % (min): 90
Humectant, skin and hair conditioner, emollient.

SCHERCOMID CME:
CTFA: Cocamide MEA
Fatty Acid Source: Coconut
Light Tan Wax
Ionic Nature: N
Acid Value (max): 2
Alkali Value (max): 15
Amide, % (min): 90
Detergent, foam stabilizer for powder detergents.

SCHERCOMID SME:
CTFA: Stearamide MEA
Fatty Acid Source: Stearic
Light Tan Wax
Ionic Nature: N
Acid Value (max): 2
Alkali Value (max): 15
Amide, % (min): 90
Detergent, emulsifier, pearlescing agent, and thickener for stick deodorant.

SCHERCOMID OME:
CTFA: Oleamide MEA
Fatty Acid Source: Oleic
Tan Wax
Acid Value (max): 10
Alkali Value (max): 20
Amide, % (min): 85
W/O emulsifier, conditioner and thickener.

SCHER CHEMICALS, INC.: Monoalkanolamides(Continued):

SCHERCOMID OMI:
CTFA: Oleamide MIPA
Fatty Acid Source: Oleic
Clear Amber Liquid to Soft Solid
Acid Value (max): 5-15
Alkali Value (max): 7-17
Amide, % (min): 85
Detergent, emulsifier, conditioner, and W/O emulsifer for mineral oil, isopropyl palmitate, isopropyl myristate, butyl stearate.

SCHERCOMID HT-60:
CTFA: PEG-50 Tallow Amide
Fatty Acid Source: Tallow
Hard Tan Wax
Ionic Nature: N
Acid Value (max): 2
Alkali Value (max): 10
Detergent, emulsifier, dispersant, and thickener with excellent foam-building characteristics.

SCHER CHEMICALS, INC.: Monohydric Alcohol Di/Tri Esters:

SCHERCEMOL DIA:
CTFA: Diisopropyl Adipate
Clear Colorless Liquid
Acid Value (max): 2
Saponification Value: 480-500
Iodine Value (max): Nil
Freezing/Melting Pt., C: -1
Penetrating emollient and solvent for creams and lotions.

SCHERCEMOL DIS:
CTFA: Diisopropyl Sebacate
Clear Colorless Liquid
Acid Value (max): 1
Saponification Value: 380-400
Iodine Value (max): Nil
Freezing/Melting Pt., C: 0
Emollient, solubilizer, and coupling agent in creams, lotions and bath oils.

SCHERCEMOL DID:
CTFA: Diisopropyl Dimerate
Yellow Liquid
Acid Value (max): 3
Saponification Value: 165-185
Iodine Value (max): 15
Freezing/Melting Pt., C: -9
Light, penetrating emollient for skin formulations. Good dispersant in pigmented products. Excellent ingredient for lipstick and lip gloss preparations.

SCHERCEMOL DISM:
CTFA: Diisostearyl Malate
Colorless Liquid
Acid Value (max): 2
Saponification Value: 160-170
Iodine Value (max): 3
Freezing/Melting Pt., C: -5
A good binder for pigmented products. Excellent additive for imparting gloss and sheen in makeup and hair preparations.

SCHERCEMOL DISF:
CTFA: Diisostearyl Fumarate
Colorless Liquid
Acid Value (max): 2
Saponification Value: 160-175
Iodine Value (max): Nil
Freezing/Melting Pt., C: -5
Excellent lubricant. Good conditioning properties.

SCHER CHEMICALS, INC.: Monohydric Alcohol Di/Tri Esters (Continued):

SCHERCEMOL DISD:
CTFA: Diisostearyl Dimerate
Yellow Liquid
Acid Value (max): 5
Saponification Value: 90-110
Iodine Value (max): 20
Freezing/Melting Pt., C: -3
Heavy moisturizing emollient. Excellent binder for pigmented products. Recommended for rich night creams, lipsticks and makeup formulations.

SCHERCEMOL TISC:
CTFA: Triisostearyl Citrate
Colorless Liquid
Acid Value (max): 3
Saponification Value: 150-165
Iodine Value (max): 3
Freezing/Melting Pt., C: -5
High viscosity ester, imparts gloss. Recommended in lipstick and lip gloss preparations.

SCHERCEMOL TIST:
CTFA: Triisostearyl Trimerate
Dark Amber Liquid
Acid Value (max): 5
Saponification Value: 90-110
Iodine Value (max): 30
Freezing/Melting Pt., C: -10
A highly viscous ester with super gloss and moisturizing characteristics. It has some of the properties of lanolin oil; offering emolliency, shine, viscosity and good binding properties.

SCHER CHEMICALS, INC.: Monohydric Alcohol Esters:

SCHERCEMOL 318:
CTFA: Isopropyl Isostearate
Clear Yellow Liquid
Acid Value (max): 1
Saponification Value: 160-180
Iodine Value (max): 3.0
Freezing/Melting Pt., C: -28
Low cloud point emollient for creams, lotions and bath oils.

SCHERCEMOL 85:
CTFA: Octyl Neopentanoate
Clear Liquid
Acid Value (max): 2
Saponification Value: 250-265
Iodine Value (max): Nil
Freezing/Melting Pt., C: -5
Low viscosity, light penetrating emollient with high solubil- in hydro-alcoholic systems.

SCHERCEMOL 105:
CTFA: Isodecyl Neopentanoate
Clear Liquid
Acid Value (max): 2
Saponification Value: 220-250
Iodine Value (max): Nil
Freezing/Melting Pt., C: -5
Light liquid emollient, reduces tackiness in skin and hair preparations.

SCHERCEMOL 185:
CTFA: Isostearyl Neopentanoate
Clear Straw-Colored Liquid
Acid Value (max): 2
Saponification Value: 135-155
Iodine Value (max): 12
Freezing/Melting Pt., C: -20
A low cloud point emollient for bath oils, creams and lotions that imparts freeze-thaw stability. Binder for pigment systems in makeup preparations.

SCHERCEMOL OPG:
CTFA: Octyl Pelargonate
Clear Straw-Colored Liquid
Acid Value (max): 2
Saponification Value: 200-215
Iodine Value (max): Nil
Freezing/Melting Pt., C: -10
A light penetrating emollient for use in non-oily type skin and makeup preparations. Good binder for powder systems.

SCHER CHEMICALS, INC.: Monohydric Alcohol Esters(Continued):

SCHERCEMOL OP:
CTFA: Octyl Palmitate
Clear Straw-Colored Liquid
Acid Value (max): 2
Saponification Value: 145-160
Iodine Value (max): Nil
Freezing/Melting Pt., C: 0
Liquid emollient. Provides slip and lubricity to skin preparations. A good anti-tack agent in anti-perspirants, creams and lotions.

SCHERCEMOL DO:
CTFA: Decyl Oleate
Clear Yellow Liquid
Acid Value (max): 3
Saponification Value: 130-140
Iodine Value (max): 65
Freezing/Melting Pt., C: -10
Emollient, lubricant and penetrant with unusual pigment-dispersing properties. Recommended for makeup, makeup removers and lipsticks without the drawback of unsaturation.

SCHERCEMOL IDO:
CTFA: Isodecyl Oleate
Clear Liquid
Acid Value (max): 5
Saponification Value: 130-140
Iodine Value (max): 65
Freezing/Melting Pt., C: 10
Emollient, lubricant and penetrant with unusual pigment dispensing properties. Recommended for makeup and makeup removers.

SCHERCEMOL MP:
CTFA: Myristyl Propionate
Clear Straw-Colored Liquid
Acid Value (max): 2
Saponification Value: 190-210
Iodine Value (max): Nil
Liquid emollient for anti-perspirants, body oils, creams and lotions.

SCHERCEMOL MM:
CTFA: Myristyl Myristate
White Waxy Solid
Acid Value (max): 2
Saponification Value: 120-135
Iodine Value (max): Nil
Freezing/Melting Pt., C: 40
Solid emollient for creams and lotions. Viscosity builder.

SCHER CHEMICALS, INC.: Monohydric Alcohol Esters(Continued):

SCHERCEMOL MEL-3:
CTFA: Myreth-3 Laurate
Clear Straw-Colored Liquid
Acid Value (max): 3
Saponification Value: 100-120
Iodine Value (max): Nil
Freezing/Melting Pt., C: 15
Emollient, solubilizer, and coupling agent in creams, lotions and bath oils.

SCHERCEMOL MEM-3:
CTFA: Myreth-3 Myristate
Clear Straw-Colored Liquid
Acid Value (max): 3
Saponification Value: 95-115
Iodine Value (max): Nil
Freezing/Melting Pt. C: 23
Applications: Same as for MEL-3.

SCHERCEMOL MEP-3:
CTFA: Myreth-3 Palmitate
Soft Cream-Colored Wax
Acid Value (max): 3
Saponification Value: 85-100
Iodine Value (max): Nil
Freezing/Melting Pt., C: 29
Applications: Same as for MEL-3

SCHERCEMOL MS:
CTFA: Myristyl Stearate
Cream Colored Liquid
Acid Value (max): 2
Saponification Value: 110-125
Iodine Value (max): Nil
Freezing/Melting Pt., C: 45
Waxy emollient for creams and lotions

SCHERCEMOL CO:
CTFA: Cetyl Octanoate
Clear Liquid
Acid Value (max): 3
Saponification Value: 140-155
Iodine Value (max): Nil
Freezing/Melting Pt., C: 10
Good solvency properties, applicable in make-up removers.

SCHER CHEMICALS, INC.: Monohydric Alcohol Esters(Continued):

SCHERCEMOL CM:
CTFA: Cetyl Myristate
Cream-Colored Waxy Solid
Acid Value (max): 2
Saponification Value: 110-125
Iodine Value (max): Nil
Freezing/Melting Pt., C: 51
Solid emollient, lubricant, and body builder. Imparts a non-oily slip to creams, lotions and makeup preparations.

SCHERCEMOL CP:
CTFA: Cetyl Palmitate
Cream-Colored Waxy Solid
Acid Value (max): 2
Saponification Value: 110-125
Iodine Value (max): Nil
Freezing/Melting Pt., C: 54
Synthetic spermaceti wax.

SCHERCEMOL CS:
CTFA: Cetyl Stearate
Cream-Colored Waxy Solid
Acid Value (max): 2
Saponification Value: 105-120
Iodine Value: Nil
Freezing/Melting Pt., C: 52
Waxy emollient for creams and lotions. Thickener and body builder.

SCHERCEMOL ICS:
CTFA: Isocetyl Stearate
Clear Straw-Colored Liquid
Acid Value (max): 2
Saponification Value: 105-120
Iodine Value (max): Nil
Freezing/Melting Pt., C: -5
Light liquid emollient; imparts elegant feel to makeup, lotions and bath preparations.

SCHERCEMOL 1688:
CTFA: Cetearyl Octanoate
Clear Colorless Liquid
Acid Value (max): 1
Saponification Value: 135-150
Iodine Value (max): Nil
Freezing/Melting Pt, C: 3
Light, moisturizing emollient for use in bath and skin preparations where a silky, water-resistant barrier is desired.

SCHER CHEMICALS, INC.: Monohydric Alcohol Esters(Continued):

SCHERCEMOL 1818:
CTFA: Isostearyl Isostearate
Clear Yellow Liquid
Acid Value (max): 2
Saponification Value: 95-110
Iodine Value (max): 13
Freezing/Melting Pt., C: -5
Emollient in creams and lotions. Co-solvent and solubilizer in perfumes.

SCHERCEMOL OLO:
CTFA: Oleyl Oleate
Clear Amber Liquid
Acid Value (max): 2
Saponification Value: 95-110
Iodine Value (max): 95
Freezing/Melting Pt, C: -5
Applications: Same as for 1818.

SCHERCEMOL SE:
CTFA: Stearyl Erucate
Cream-Colored Soft Solid
Acid Value (max): 2
Saponification Value: 85-110
Iodine Value (max): 60
Freezing/Melting Pt., C: 32
Emollient wax with melting point at skin temperature. It has the look and feel of real cocoa butter.

SCHERCEMOL ISE:
CTFA: Isostearyl Erucate
Clear Yellow Liquid
Acid Value (max): 2
Saponification Value: 90-105
Iodine Value (max): 70
Freezing/Melting Pt., C: 16
Rich, non-greasy, lubricating emollient for skin and bath preparations.

SCHER CHEMICALS, INC.: Monohydric Alcohol Esters(Continued):

SCHERCEMOL BE:
CTFA: Behenyl Erucate
Cream-Colored Soft Solid
Acid Value (max): 2
Saponification Value: 80-95
Iodine Value (max): 55
Freezing/Melting Pt., C: 45
Emollient wax with melting point at mid-40C range. Recommended for use in wax systems where its solubility characteristics help to prevent bleeding and syneresis.

SCHERCEMOL EE:
CTFA: Erucyl Erucate
Clear Yellow Liquid
Acid Value (max): 2
Saponification Value: 80-95
Iodine Value (max): 100
Freezing/Melting Pt., C: 22
A super-rich emollient ester for use in skin, hair and suntanning preparations.

SCHEROBA OIL:
CTFA: Isostearyl-Erucyl Erucate
Clear Yellow Liquid
Acid Value (max): 2
Saponification Value: 85-100
Iodine Value (max): 85
Freezing/Melting Pt., C: 15
As an alternative to jojoba oil, this ester has similar physical and chemical properties with the added advantages of low price, product consistency, and availability.

SCHER CHEMICALS, INC.: Polyhydric Alcohol Esters:

SCHERCEMOL EGMS:
CTFA: Glycol Stearate
White to Cream-Colored Solid
Acid Value (max): 5
Saponification Value: 170-190
Iodine Value (max): 1
Freezing/Melting Pt., C: 58
Emulsifier, opacifier and pearling agent in hair and skin preparations.

SCHERCEMOL PGDP:
CTFA: Propylene Glycol Dipelargonate
Clear Straw-Colored Liquid
Acid Value (max): 5
Saponification Value: 300-320
Iodine Value (max): Nil
Freezing/Melting Pt., C: -25
Low freezing point emollient and co-solvent for perfumes, bath oils, creams and lotions.

SCHERCEMOL PGML:
CTFA: Propylene Glycol Laurate
Clear Yellow Liquid
Acid Value (max): 5
Saponification Value: 225-240
Iodine Value (max): 1
Freezing/Melting Pt., C: 10
Emollient and solvent in lotions and lipsticks.

SCHERCEMOL PGMS:
CTFA: Propylene Glycol Stearate
White to Cream-Colored Solid
Acid Value (max): 4
Saponification Value: 175-190
Iodine Value (max): 1
Freezing/Melting Pt., C: 35
Primary emulsifier for lotions and low viscosity creams.

SCHERCEMOL GMS:
CTFA: Glycerol Stearate
White to Cream-Colored Solid
Acid Value (max): 3
Saponification Value: 160-180
Iodine Value (max): 1
Freezing/Melting Pt., C: 58
Primary emulsifier for creams and lotions.

SCHER CHEMICALS, INC.: Polyhydric Alcohol Esters(Continued):

SCHERCEMOL GMIS:
CTFA: Glycerol Isostearate
Clear Straw-Colored Liquid to Soft Solid
Acid Value (max): 5
Saponification Value: 160-180
Iodine Value (max): 10
Freezing/Melting Pt., C: 5
Emulsifier and emollient for creams and lotions.

SCHERCEMOL NGDC:
CTFA: Neopentyl Glycol Dicaprate
Clear Liquid
Acid Value (max): 3
Saponification Value: 255-270
Iodine Value (max): Nil
Freezing/Melting Pt., C: 2
Good solvency properties; applicable in make-up removers.

SCHER CHEMICALS, INC.: Sunscreens:

DIPSAL:
CTFA: Dipropylene Glycol Salicylate
PPG-2 Salicylate
Clear Yellow Liquid
Acid Value (max): 3
Saponification Value: 225-240
Iodine Value (max): Nil
UV Absorber, Sunscreen

Ethylene Glycol Salicylate:
CTFA: Glycol Salicylate
Light Amber Crystalline Solid
Acid Value (max): 3
Saponification Value: 300-330
Iodine Value (max): Nil
UV Absorber, Sunscreen

SCHER CHEMICALS, INC.: Specialty Quats (Mono):

SCHERCOQUAT CAS:
CTFA: Cocamidopropyl Ethyldimonium Ethosulfate
Fatty Acid Source: Coconut
Hydrophilic Group: Ethyl Sulfate
MW (av): 445
Amber Viscous Liquid
Dry Solids, % (min): 98
Concentrated liquid cationic surfactant with excellent water solubility and foam height. Static control properties.

SCHERCOQUAT SAS:
CTFA: Stearamidopropyl Ethyldimonium Ethosulfate
Fatty Acid Source: Stearic
Hydrophilic Group: Ethyl Sulfate
MW (av): 508
Yellow Liquid
Dry Solids, % (min): 80
An excellent conditioner for hair rinses. Provides body and bounce to hair; improves shine.

SCHERCOQUAT SAB:
CTFA: Stearamidopropalkonium Chloride
Fatty Acid Source: Stearic
Hydrophilic Group: Benzyl Chloride
MW (av): 480
Yellow Liquid
Dry Solids, % (min): 80
Applications: Similar to SAS.

SCHERCOQUAT IAS:
Isostearamidopropyl Ethyldimonium Ethosulfate
Fatty Acid Source: Isostearic
Hydrophilic Group: Ethyl Sulfate
MW (av): 550
Amber Viscous Liquid
Dry Solids, % (min): 90
Highly concentrated liquid quaternary. Good compatibility with most anionic surfactants and good water solubility. Recommended for clear conditioning shampoos where it contributes body, combability, and antistatic properties.

SCHER CHEMICALS, INC.: Specialty Quats (Mono)(Continued):

SCHERCOQUAT IEP:
CTFA: (Quaternium-62) or Isostearamidopropyl Epoxypropyldimonium Chloride
Fatty Acid Source: Isostearic
Hydrophilic Group: Epoxypropyl Chloride
MW (av): 486
Amber Viscous Liquid
Dry Solids, % (min): 80

A unique type of specialty quaternary. Good water solubility due to the epoxide ring. Good compatibility with many anionic surfactants.

SCHERCOQUAT IIS:
CTFA: Isostearyl Ethyl Imidonium Ethosulfate
Fatty Acid Source: Isostearic
Hydrophilic Group: Ethyl Sulfate
MW (av): 532
Dark Amber Viscous Liquid
Dry Solids, % (min): 98

Very highly concentrated quaternary. Extremely effective conditioner. Recommended for all skin and hair products where it enhances softness to skin and fullness and body to hair.

SCHERCOQUAT DAS:
CTFA: Quaternium-61
Fatty Acid Source: Dimer
Hydrophilic Group: Ethyl Sulfate
MW (av): 1050
Amber Viscous Liquid
Dry Solids, % (min): 90

Highly concentrated liquid quaternary of the unique dimer acid, which enhances its performance as a hair and skin conditioner, especially in low irritation cosmetic preparations. Possesses the ability to exhaust from dilute solutions to negatively charged substrates due to its double positive charge.

SCHERCOQUAT DAB:
CTFA: Quaternium-63
Fatty Acid Source: Dimer
Hydrophilic Group: Benzyl Chloride
MW (av): 998
Appearance: Amber Viscous Liquid
Dry Soids, % (min): 90

Strongly cationic quaternary due to the presence of the double positive charge. This intensifies substantivity to substrate. Recommended for creams, lotions, shampoos, and hair preparations.

SCHER CHEMICALS, INC.: Specialty Quats (Mono)(Continued):

SCHERCOQUAT SOAS:
CTFA: Soyamidopropyl Ethyldimonium Ethosulfate
Fatty Acid Source: Soya Oil
Hydrophilic Group: Ethyl Sulfate
MW (av): 516
Appearance: Amber Viscous Liquid
Dry Solids, % (min): 90

Low-cost, highly concentrated liquid quaternary. Effective in hair conditioners, offering good slip, shine and combability.

SCHERCOQUAT FOAS#:
CTFA: Safflоweramidopropyl Ethyldimonium Ethosulfate
Fatty Acid Source: Safflower Oil
Hydrophilic Group: Ethyl Sulfate
MW (av): 520
Amber Viscous Liquid
Dry Solids, % (min): 90
Applications: Similar to SOAS

SCHERCOQUAT ROAS#:
CTFA: Rapeseedamidopropyl Ethyldimonium Ethosulfate
Fatty Acid Source: Rapeseed Oil
Hydrophilic Group: Ethyl Sulfate
MW (av): 560
Amber Viscous Liquid
Dry Solids, % (min): 90

An excellent, low-cost conditioning agent, especially for dry and over-processed hair. In anionic systems it often acts as a viscosity builder.

SCHERCOQUAT ROEP#:
CTFA: Rapeseedamidopropyl Epoxypropyl Dimonium Chloride
Fatty Acid Source: Rapeseed Oil
Hydrophilic Group: Epoxypropyl Chloride
MW (av): 533
Dark Amber Viscous Liquid
Dry Solids, % (min): 80

A low-cost, water-soluble liquid conditioner with good compatibility with anionic and nonionic surfactants. Recommended for use in conditioning shampoos and hair sprays.

These quaternaries are also available as Benzyl Chloride Quats. Scher can supply many others based on a wide variety of vegetable oils.

SCHER CHEMICALS, INC.: Specialty Quats (Mono)#(Continued):

SCHERCOQUAT BAS:
CTFA: Behenamidopropyl Ethyldimonium Ethosulfate
Fatty Acid Source: Behenic
Hydrophilic Group: Ethyl Sulfate
MW (av): 548
Appearance: Amber Liquid
Dry Solids, % (min): 50
Strong conditioning properties due to long chain length. Excellent for dry and over-processed hair.

SCHERCOQUAT APAS:
CTFA: Apricotamidopropyl Ethyldimonium Ethosulfate
Fatty Acid Source: Apricot Kernel Oil
Hydrophilic Group: Ethyl Sulfate
MW (av): 515
Amber Viscous Liquid
Dry Solids, % (min): 90
Natural, mild conditioner, imparts good slip and shine.

SCHERCOQUAT WOAS:
CTFA: Wheat Germamidopropyl Ethyldimonium Ethosulfate
Fatty Acid Source: Wheat Germ Oil
Hydrophilic Group: Ethyl Sulfate
MW (av): 528
Appearance: Amber Viscous Liquid
Dry Solids, % (min): 90
Mild conditioner from a natural source which has Vitamin E.

SCHERCOQUAT MKAS:
CTFA: Minkamidopropyl Ethyldimonium Ethosulfate
Fatty Acid Source: Mink Oil
Hydrophilic Group: Ethyl Sulfate
MW (av): 505
Appearance: Amber Viscous Liquid
Dry Solids, % (min): 90
Mild skin and hair conditioner. Imparts velvety feel and lustrous sheen.

These quaternaries are also available as Benzyl Chloride Quats. Scher Chemicals is able to supply many others based on a wide variety of vegetable oils.

SCHER CHEMICALS, INC.: Specialty Quats (Bis):

SCHERCOQUAT 21AE:
CTFA: Bis Isostearamidopropyl Ethoxyethyl Dimonium Chloride
Fatty Acid Source: Isostearic
Hydrophilic Group: Ethyl Ether
MW (av): 932
Amber Viscous Liquid
Dry Solids, % (min): 85

Good cationic emulsifier. Heavier conditioning properties than monoquats, rinses clean with no build-up. Good water solubility and compatibility with cationic surfactants.

SCHERCOQUAT 21AP:
CTFA: Bis Isostearamidopropyl Hydroxypropyl Dimonium Chloride
Fatty Acid Source: Isostearic
Hydrophilic Group: Hydroxy Propyl
MW (av): 918
Appearance: Amber Viscous Liquid
Dry Solids, % (min): 85

Similar to 21AE, but has slightly heavier conditioning properties and slightly less water solubility.

SCHER CHEMICALS, INC.: Sulfosuccinates:

SCHERCOPOL LPS:
CTFA: Disodium Laureth-sulfosuccinate
Fatty Acid Source: Lauric
Clear Yellow Liquid
Dry Solids, % (min): 39
Sodium Bisulfite, % (max): 0.3
pH: 5-7
Mild, high-foaming surfactant; viscosity enhancer when used with alkylsulfates and sulfonates.

SCHERCOPOL CMS-Na:
CTFA: Disodium Cocamido MEA Sulfosuccinate
Fatty Acid Source: Coconut
Clear Yellow Liquid
Dry Solids, % (min): 29
Sodium Bisulfite, % (max): 0.3
pH: 5-7
Mild detergent for shampoos, bubbble baths, liquid dishwashing compounds, rug and upholstery shampoos.

SCHERCOPOL OMS-Na:
CTFA: Disodium Oleamido MEA Sulfosuccinate
Fatty Acid Source: Oleic
Light Amber Liquid
Dry Solids, % (min): 34
Sodium Bisulfite, % (max): 0.3
pH: 5-7
Non-irritating, high-foaming surfactant for facial scrubs, bubble baths, and mild shampoos.

SCHERCOPOL OMES-NA:
CTFA: Disodium Oleamido PEG-2 Sulfosuccinate
Fatty Acid Source: Oleic
Light Amber Liquid
Dry Solids, % (min): 34
Sodium Bisulfite, % (max): 0.3
pH: 5-7
Applications: Same as for OMS-Na.

SCHERCOPOL OMES-A:
CTFA: Diammonium Oleamido PEG-2 Sulfosuccinate
Fatty Acid Source: Oleic
Light Amber Liquid
Dry Solids, % (min): 39
Sodium Bisulfite, % (max): 0.3
pH: 6-8
Applications: Same as for OMS-Na.

SCHER CHEMICALS, INC.: Sulfosuccinates(Continued):

SCHERCOPOL DOS-70:
CTFA: Dioctyl Sodium Sulfosuccinate
Fatty Acid Source: Succinic
Viscous Liquid
Dry Solids, % (min): 70
Sodium Bisulfite, % (max): 0.3
pH: 6-8
Good wetting agent and surface tension depressant.

SCHERCOPOL DOS-PG-85:
CTFA: Dioctyl Sodium Sulfosuccinate
Fatty Acid Source: Succinic
Viscous Liquid
Dry Solids, % (min): 85
Sodium Bisulfite, % (max): 0.3
pH: 6-8
Applications: Same as for DOS-70.

SCHERCOPOL RMS-Na:
CTFA: Disodium Ricinoleamido MEA Sulfosuccinate
Fatty Acid Source: Ricinoleic
Clear Yellow Liquid
Dry Solids, % (min): 40
Sodium Bisulfite, % (max): 0.3
pH: 5-7
In conjunction with activated ingredients, good in antidandruff shampoos.

SCHERCOPOL UMS-Na:
CTFA: Disodium Undecylenamido MEA Sulfosuccinate
Fatty Acid Source: Undecylenic
Pearlescent Yellow Liquid
Dry Solids, % (min): 40
Sodium Bisulfite, % (max): 0.3
pH: 5-7
Applications: Same as for RMS-Na.

SCHWEIZERHALL, INC.: Cosmetic Raw Materials:

Product Name:

Benzocaine
(Ethyl-p-aminobenzoate)

Benzoic Acid

Benzyl Alcohol, FCC, NF & Kosher

D-(+)-Camphor

DL-(+)-Camphor, USP

L-Cysteine Hydrochloride

L-Cysteine Free Base

Citric Acid

Collagen (Soluble)

1,3-Dihydroxyacetone

Ethyl Nicotinate

D-(-)-Fructose, USP

Fumaric Acid

2-Hydroxy-4-methoxybenzophenone
(Benzophenone 3)

Inositol, FCC

D-Mannitol, FCC

Phenol, USP

Potassium sorbate

Sorbic Acid, NF, FCC

D-(+)-Xylose, USP

SHELL CHEMICAL CO.: NEODOL Alcohols:

91:

Carbon chains present: C9/C10/C11
Molecular weight: 160
Active content, %w: 100
Melting range, F: 3-25
Pour point, F: 10
Color, APHA (Pt-Co): 0-5
Sp. gravity, 77/77F: 0.829
Viscosity, cSt @ 100F: 9
Acid value, eq/100g: <0.001
Carbonyl value, ppm as C=O: 25
Hydroxyl value, eq/100g: 0.62
Hydroxyl number, mgKOH/g: 350
Flash point, PMCC (ASTM-D93), F: 228
Water, %w: 0.02

1:

Carbon chains present: C11
Molecular weight: 173
Active content, %w: 100
Melting range, F: 42-57
Pour point, F: 52
Color, APHA (Pt-Co): 0-5
Sp. gravity, 77/77F: 0.831
Viscosity, cSt @ 100F: 11
Acid value, eq/100g: <0.001
Carbonyl value, ppm as C=O: 25
Hydroxyl value, eq/100g: 0.58
Hydroxyl number, mgKOH/g: 324
Flash point, PMCC (ASTM-D93), F: 250
Water, %w: 0.02

23:

Carbon chains present: C12/C13
Molecular weight: 194
Active content, %w: 100
Melting Range: 45-72
Pour point, F: 63
Color, APHA (Pt-Co): 0-5
Sp. gravity, 77/77F: 0.833
Viscosity, cSt @ 100F: 14
Acid value, eq/100g: <0.001
Carbonyl value, ppm as C=O: 37
Hydroxyl value, eq/100g: 0.52
Hydroxyl number, mgKOH/g: 289
Flash point, PMCC (ASTM-D93), F: 279
Water, %w: 0.02

SHELL CHEMICAL CO.: NEODOL Alcohols(Continued):

25:

Carbon chains present: C12/C13/C14/C15
Molecular weight: 203
Active content, %w: 100
Melting range, F: 54-77
Pour point, F: 66
Color, APHA (Pt-Co): 0-5
Sp. gravity, 77/77F: 0.834
Viscosity, cst @ 100F: 15
Acid value, eq/100g: <0.001
Carbonyl value, ppm as C=O: 41
Hydroxyl value, eq/100g: 0.49
Hydroxyl number, mgKOH/g: 276
Flash point, PMCC (ASTM-D93), F: 286
Water, %w: 0.02

45:

Carbon chains present: C14/C15
Molecular weight: 218
Active content, %w: 100
Melting range, F: 59-97
Pour point, F: 84
Color, APHA (Pt-Co): 0-5
Sp. gravity, 77/77F: 0.820
Viscosity, cSt @ 100F: 18
Acid value, eq/100g: <0.001
Carbonyl value, ppm as C=O: 50
Hydroxyl value, eq/100G: 0.46
Hydroxyl number, mgKOH/g: 257
Flash point, PMCC (ASTM-D93), F: 315
Water, %w: 0.02

SHELL CHEMICAL CO.: NEODOL Ethoxylates:

91-2.5:
EO groups/alcohol, mole/mole, avg.: 2.7
Molecular weight: 281
Active content, %w: 100
EO content, %w: 42.3
Melting range, F: -31 to -2
Color, APHA (Pt-Co): 5-10
Sp. gravity, 77/77F: 0.925
Viscosity, cSt @ 100F: 12
Acid value, eq/100g: <0.001
Hydroxyl value, eq/100g: 0.36
Hydroxyl number, mgKOH/g: 200
HLB No. (hydrophile/lipophile balance): 8.5
Flash point, PMCC (ASTM D-93), F: 255
Cloud point, 1% aq. soln., F: 35.8
Pour point, F: 9
pH, 1% aq. soln.: 6.0
Water, %w: 0.02

91-6:
EO groups/alcohol, mole/mole, avg.: 6.1
Molecular weight: 428
Active content, %w: 100
EO content, %w: 62.7
Melting range, F: 21-52
Color, APHA (Pt-Co): 5-10
Sp. gravity, 77/77F: 0.984
Viscosity, cSt @ 100F: 23
Acid value, eq/100g: <0.001
Hydroxyl value, eq/100g: 0.23
Hydroxyl number, mgKOH/g: 131
HLB No. (hydrophile/lipophile balance): 12.5
Flash point, PMCC (ASTM D-93), F: 289
Cloud point, 1% aq. soln., F: 12.5
Pour point, F: 43
pH, 1% aq. soln.: 6.0
Water, %w: 0.02

SHELL CHEMICAL CO.: NEODOL Ethoxylates(Continued):

91-8:

EO groups/alcohol, mole/mole, avg.: 8.2
Molecular weight: 519
Active content, %w: 100
EO content, %w: 69.5
Melting range, F: 45-68
Color, APHA (Pt-Co): 5-10
Sp. gravity, 77/77F: 1.008
Viscosity, cSt @ 100F: 39
Acid value, eq/100g: <0.001
Hydroxyl value, eq/100g: 0.19
Hydroxyl number, mgKOH/g: 108
HLB No. (hydrophile/lipophile balance): 13.9
Flash point, PMCC (ASTM D-93), F: 318
Cloud point, 1% aq. soln., F: 176
Pour point, F: 59
pH, 1% aq. soln.: 6.0
Water, %w: 0.02

23-1:

EO groups/alcohol, mole/mole, avg.: 1.0
Molecular weight: 238
Active content, %w: 100
EO content, %w: 18.5
Melting range, F: 27-48
Color, APHA (Pt-Co): 5-10
Sp. gravity, 77/77F: 0.873
Viscosity, cSt @ 100F: 13
Acid value, eq/100 g: <0.001
Hydroxyl value, eq/100g: 0.42
Hydroxyl number, mgKOH/g: 236
HLB No. (hydrophile/lipophile balance): 3.7
Flash point, PMCC (ASTM D-93), F: 289
Cloud point, 1% aq. soln., F: 13.6
Pour point, F: 41
pH, 1% aq. soln.: 10.1
Water, %w: 0.02

SHELL CHEMICAL CO.: NEODOL Ethoxylates(Continued):

23-3:

EO groups/alcohol, mole/mole, avg.: 2.9
Molecular weight: 322
Active content, %w: 100
EO content, %w: 39.6
Melting range, F: 19-37
Color, APHA (Pt-Co): 5-10
Sp. gravity, 77/77F: 0.922
Viscosity, cSt @ 100F: 14
Acid value, eq/100g: <0.001
Hydroxyl value, eq/100g: 0.31
Hydroxyl number, mgKOH/g: 174
HLB No. (hydrophile/lipophile balance): 7.9
Flash point, PMCC (ASTM D-93), F: 306
Cloud point, 1% aq. soln., F: 33.1
Pour point, F: 34
pH, 1% aq. soln.: 6.0
Water, %w: 0.02

23-5:

EO groups/alcohol, mole/mole, avg.: 5.0
Molecular weight: 413
Active content, %w: 100
EO content, %w: 53.3
Melting range, F: 27-61
Color, APHA (Pt-Co): 5-10
Sp. gravity, 77/77F: 0.965
Viscosity, cSt @ 100F: 23
Acid value, eq/100g: <0.001
Hydroxyl value, eq/100g: 0.24
Hydroxyl number, mgKOH/g: 136
HLB No. (hydrophile/lipophile balance): 10.7
Flash point, PMCC (ASTM D-93), F: 315
Pour point, F: 45
pH, 1% aq. soln.: 6.0
Water, %w: 0.02

SHELL CHEMICAL CO.: NEODOL Ethoxylates(Continued):

23-6.5:
EO groups/alcohol, mole/mole, avg.: 6.7
Molecular weight: 488
Active content, %w: 100
EO content, %w: 60.4
Melting range, F: 39-70
Color, APHA (Pt-Co): 5-10
Sp. gravity, 77/77F: 0.984
Viscosity, cSt @ 100F: 29
Acid value, eq/100g: <0.001
Hydroxyl value, eq/100g: 0.20
Hydroxyl number, mgKOH/g: 115
HLB No. (hydrophile/lipophile balance): 12.1
Flash point, PMCC (ASTM D-93, F): 334
Cloud point, 1% aq. soln., F: 113
Pour point, F: 59
pH, 1% aq. soln.: 6.0
Water, %w: 0.02

23-6.5T:
EO groups/alcohol, mole/mole, avg.: 7.6
Molecular weight: 529
Active content, %w: 100
EO content, %w: 63.2
Melting range, F: 36-66
Color, APHA (Pt-Co): 10-15
Sp. gravity, 77/77F: 0.993
Viscosity, cSt @ 100F: 33
Acid value, eq/100g: <0.001
Hydroxyl value, eq/100g: 0.19
Hydroxyl number, mgKOH/g: 106
HLB No. (hydrophile/lipophile balance): 12.6
Flash point, PMCC (ASTM D-93), F: 448
Cloud point, 1% aq. soln., F: 147
Pour point, F: 61
pH, 1% aq. soln.: 6.5
Water, %w: 0.02

SHELL CHEMICAL CO.: NEODOL Ethoxylates(Continued):

23-12:
EO groups/alcohol, mole/mole, avg.: 11.9
Molecular weight: 719
Active content, %w: 100
EO content, %w: 72.8
Melting range, F: 63-90
Color, APHA (Pt-Co): 10-20
Sp. gravity, 77/77F: 1.006
Viscosity, cSt @ 100F: 53
Acid value, eq/100g: <0.001
Hydroxyl value, eq/100g: 0.14
Hydroxyl number: 78
HLB No. (hydrophile/lipophile balance): 14.6
Flash point, PMCC (ASTM D-93), F: 399
Cloud Point, 1% aq. soln., F: 177
Pour point, F: 79
pH, 1% aq. soln.: 10.1
Water, %w: 0.02

25-3:
EO groups/alcohol, mole/mole, avg.: 3.0
Molecular weight: 338
Active content, %w: 100
EO content, %w: 39.0
Melting range, F: 27-45
Color, APHA (Pt-Co): 5-10
Sp. gravity, 77/77F: 0.921
Viscosity, cSt @ 100F: 19
Acid value, eq/100g: <0.001
Hydroxyl value, eq/100g: 0.30
Hydroxyl number: 166
HLB No. (hydrophile/lipophile balance): 7.8
Flash point, PMCC (ASTM D-93), F: 315
Cloud point, 1% aq. soln., F: 30
Pour point, F: 37
pH, 1% aq. soln.: 7.1
Water, %w: 0.02

SHELL CHEMICAL CO.: NEODOL Ethoxylates(Continued):

25-7:

EO groups/alcohol, mole/mole, avg.: 7.3
Molecular weight: 524
Active content, %w: 100
EO content, %w: 61.3
Melting range, F: 36-70
Color, APHA (Pt-Co): 5-10
Sp. gravity, 77/77F: 0.965
Viscosity, cSt @ 100F: 34
Acid value, eq/100 g: <0.001
Hydroxyl value, eq/100g: 0.19
Hydroxyl number, mgKOH/g: 107
HLB No. (hydrophile/lipophile balance): 12.3
Flash point, PMCC (ASTM D-93), F: 367
Cloud point, 1% aq. soln., F: 121
Pour point, F: 66
pH, 1% aq. soln.: 6.0
Water, %w: 0.02

25-9:

EO groups/alcohol, mole/mole, avg.: 8.9
Molecular weight: 597
Active content, %w: 100
EO content, %: 65.6
Melting range, F: 57-77
Color, APHA (Pt-Co): 5-10
Sp. gravity, 77/77F: 0.982
Viscosity, cSt @ 100F: 41
Acid value, eq./100g: <0.001
Hydroxyl value, eq/100g: 0.17
Hydroxyl number, mgKOH/g: 94
HLB No. (hydropohile/lipophile balance): 13.1
Flash point, PMCC (ASTM D-93), F: 370
Cloud point, 1% aq. soln., F: 163
Pour point, F: 70
pH, 1% aq. soln.: 6.0
Water, %w: 0.02

SHELL CHEMICAL CO.: NEODOL Ethoxylates(Continued):

25-12:
EO groups/alcohol, mole/mole, avg.: 11.9
Molecular weight: 729
Active content, %w: 100
EO content, %w: 71.8
Melting range, F: 68-86
Color, APHA (Pt-Co): 5-10
Sp. gravity, 77/77F: 0.999
Viscosity, cSt @ 100F: 53
Acid value, eq/100g: <0.001
Hydroxyl value, eq/100g: 0.14
Hydroxyl number, mgKOH/g: 77
HLB No. (hydrophile/lipophile balance): 14.4
Flash point, PMCC (ASTM D-93), F: 433
Cloud point, 1% aq. soln., F: 173
Pour point, F: 81
pH, 1% aq. soln.: 6.0
Water, %w: 0.02

45-2.25:
EO groups/alcohol, mole/mole, avg.: 2.29
Molecular weight: 319
Active content, %w: 100
EO content, %w: 31.6
Melting range, F: 48-68
Color, APHA (Pt-Co): 5-10
Sp. gravity, 77/77F: 0.903
Viscosity, cSt @ 100F: 19
Acid value, eq/100g: <0.001
Hydroxyl value, eq/100g: 0.31
Hydroxyl number, mgKOH/g: 176
HLB No. (hydrophile/lipophile balance): 6.3
Flash point, PMCC (ASTM D-93), F: 336
Cloud point, 1% aq. soln., F: 21
Pour point, F: 59
pH, 1% aq. soln.: 6.5
Water, %w: 0.02

SHELL CHEMICAL CO.: NEODOL Ethoxylates(Continued):

45-7:
EO groups/alcohol, mole/mole, avg.: 7.1
Molecular weight: 529
Active content, %w: 100
EO content, %w: 59.0
Melting range, F: 48-75
Color, APHA (Pt-Co): 5-10
Sp. gravity, 77/77F: 0.959
Viscosity, cSt @ 100F: 35
Acid value, eq/100g: <0.001
Hydroxyl value, eq/100g: 0.19
Hydroxyl number, mgKOH/g: 106
HLB No. (hydrophile/lipophile balance): 11.8
Flash point, PMCC (ASTM D-93), F: 365
Cloud point, 1% aq. soln., F: 112
Pour point, F: 66
pH, 1% aq. soln.: 6.0
Water, %w: 0.02

45-7T:
EO groups/alcohol, mole/mole, avg.: 7.9
Molecular weight: 567
Active content, %w: 100
EO content, %w: 61.3
Melting range, F: 46-73
Color, APHA (Pt-Co): 10-15
Sp. gravity, 77/77F: 0.966
Viscosity, cSt @ 100F: 39
Acid value, eq/100g: <0.001
Hydroxyl value, eq/100g: 0.18
Hydroxyl number, mgKOH/g: 99
HLB No. (hydrophile/lipophile balance): 12.3
Flash point, PMCC (ASTM D-93), F: 441
Cloud point, 1% aq. soln., F: 131
Pour point, F: 66
pH, 1% aq. soln.: 6.8
Water, %w: 0.02

SHELL CHEMICAL CO.: NEODOL Ethoxylates(Continued):

45-13:

EO groups/alcohol, mole/mole, avg.: 13.0
Molecular weight: 790
Active content, %w: 100
EO content, %w: 72.4
Melting range, F: 77-93
Color, APHA (Pt-Co): 5-10
Sp. gravity, 77/77F: 1.003
Viscosity, cSt @ 100F: 59
Acid value, eq/100g: <0.001
Hydroxyl value, eq/100g: 0.13
Hydroxyl number, mgKOH/g: 71
HLB No. (hydrophile/lipophile balance): 14.5
Flash point, PMCC (ASTM D-93), F: 480
Cloud point, 1% aq. soln., F: 178
Pour point, F: 86
pH, 1% aq. soln.: 6.4
Water, %w: 0.02

SHEREX CHEMICAL CO.: Personal Care Surfactants and Specialty Products:

Nonionics:
Ethoxylated Monodiglycerides:

Product Name:

VARONIC LI-63:
CTFA Adopted Name: PEG-30 Glyceryl Monococoate
Approx. HLB: 15.9
Form at 20C: Paste MP=27
% Solids: 100
Recommended for low-irritation shampoos with alkyl sulfates, alkylether sulfates, and olefin sulfonates.

VARONIC LI-67:
CTFA Adopted Name: PEG-78 Glyceryl Monococoate
Approx. HLB: 18.0
Form at 20C: Solid MP=42
% Solids: 100
Recommended for low-irritation shampoos with sulfates, alkylether sulfates and olefin sulfonates. Viscosity builder.

VARONIC LI-67(75%):
CTFA Adopted Name: PEG-78 Glyceryl Monococoate
Approx. HLB: 18.0
Form at 20C: Liquid
% Solids: 75
Same as LI-67. Solvent is water.

VARONIC LI-48:
CTFA Adopted Name: PEG-82 Glyceryl Monotallowate
Approx. HLB: 18.0
Form at 20C: Solid MP=41
% Solids: 100

VARONIC LI-420 (80%):
CTFA Adopted Name: PEG-200 Glyceryl Monotallowate
Approx. HLB: 19.0
Form at 20C: Solid MP=53
% Solids: 80
Recommended for low-irritation shampoos with sulfates, alkylether sulfates and olefin sulfonates. Viscosity builder.

VARONIC LI-42:
CTFA Adopted Name: PEG-20 Glyceryl Monotallowate
Approx. HLB: 13.0
Form at 20C: Paste MP=27
% Solids: 100
Low-irritation emulsifier for creams and lotions.

SHEREX CHEMICAL CO.: Personal Care Surfactants and Specialty Products(Continued):

Nonionics(Continued):
Ethoxylated Alcohols:

Product Name:

AROSURF 66-E2:
CTFA Adopted Name: Isosteareth-2
Approx. HLB: 4.6
Chemical Description: PEG-2 Isostearyl Ether
Form at 20C: Liquid
% Solids: 100
Emulsion stabilizer, cosmetic emollient.

AROSURF 66-E10:
CTFA Adopted Name: Isosteareth-10
Approx. HLB: 12.0
Chemical Description: PEG-10 Isostearyl Ether
Form at 20C: Liquid
% Solids:100

AROSURF 66-E20:
CTFA Adopted Name: Isosteareth-20
Approx HLB: 18.0
Chemical Description: PEG-20 Isostearyl Ether
Form at 20C: Paste
% Solids: 100
Emulsifier for creams and lotions.

AROSURF 66-PE12:
CTFA Adopted Name: PPG-3 Isosteareth-9
Approx. HLB: 12.2
Chemical Description: PPG-3-PEG-9 Isostearyl Ether
Form at 20C: Liquid
% Solids: 100
Emulsifier, dispersant, bath-oil spreading agent.

SHEREX CHEMICAL CO.: Pesonal-Care Surfactants and Specialty Products(Continued):

Nonionics(Continued):
Ethoxylated Amines:

Product Name:

VARONIC K202:
CTFA Designation: PEG-2 Cocamine
Approx. HLB: 6.2
Chemical Description: Coco Amine Ethoxylate
Moles EO: 2
% Solids: 99

VARONIC K205:
CTFA Designation: PEG-5 Cocamine
Approx. HLB: 11.0
Chemical Description: Coco Amine Ethoxylate
Moles EO: 5
% Solids: 99

VARONIC K205LC:
CTFA Designation: PEG-5 Cocamine
Approx. HLB: 11.0
Chemical Description: Coco Amine Ethoxylate
Moles EO: 5
% Solids: 99

VARONIC K210:
Approx. HLB: 13.8
Chemical Description: Coco Amine Ethoxylate
Moles EO: 10
% Solids: 99

VARONIC K210LC:
Approx. HLB: 13.8
Chemical Description: Coco Amine Ethoxylate
Moles EO: 10
% Solids: 99

VARONIC K215:
CTFA Designation: PEG-15 Cocamine
Approx. HLB: 15.4
Chemical Description: Coco Amine Ethoxylate
Moles EO: 15
% Solids: 99

VARONIC K215LC:
CTFA Designation: PEG-15 Cocamine
Approx. HLB: 15.4
Chemical Description: Coco Amine Ethoxylate
Moles EO: 15
% Solids: 99

Used and recommended for use as dye levelers. VARONIC ethoxylated amines improve wetability and reduce dye affinity at the surface, permitting the dye to migrate more evenly into substrates such as hair, resulting in brighter colors. Coemulsifier.

SHEREX CHEMICAL CO.: Personal-Care Surfactants and Specialty Products(Continued):

Nonionics(Continued):
Ethoxylated Amines(Continued):

Product Name:

VARONIC L202:
CTFA Designation: PEG-2 Soyamine
Approx. HLB: 4.7
Chemical Description: Soya Amine Ethoxylate
Moles EO: 2
Form at 25C: Liquid
% Solids: 99

VARONIC L205:
CTFA Designation: PEG-5 Soyamine
Approx. HLB: 8.4
Chemical Description: Soya Amine Ethoxylate
Moles EO: 5
Form at 25C: Liquid
% Solids: 99

VARONIC L205LC:
CTFA Designation: PEG-5 Soyamine
Approx. HLB: 8.4
Chemical Description: Soya Amine Ethoxylate
Moles EO: 5
Form at 25C: Liquid
% Solids: 99

VARONIC L230LC:
Approx. HLB: 16.4
Chemical Description: Soya Amine Ethoxylate
Moles EO: 30
Form at 25C: Paste
% Solids: 99

VARONIC L230LC(80%):
Approx. HLB: 16.4
Chemical Description: Soya Amine Ethoxylate
Moles EO: 30
Form at 25C: Liquid
% Solids: 99

Used and recommended for use as dye levelers. VARONIC ethoxylated amines improve wetability and reduce dye affinity at the surface, permitting the dye to migrate more evenly into substrates such as hair, resulting in brighter colors. Co-emulsifier.

SHEREX CHEMICAL CO.: Personal-Care Surfactants and Specialty Products(Continued):

Nonionics:(Continued):
Ethoxylated Amines(Continued):

VARONIC T202:
CTFA Designation: PEG-2 Tallow Amine
Approx. HLB: 5.1
Chemical Description: Tallow Amine Ethoxylate
Moles EO: 2
Form at 25C: Paste
% Solids: 99

VARONIC T205:
CTFA Designation: PEG-5 Tallow Amine
Approx. HLB: 9.2
Chemical Description: Tallow Amine Ethoxylate
Moles EO: 5
Form at 25C: Liquid
% Solids: 99

VARONIC T205LC:
CTFA Designation: PEG-5 Tallow Amine
Approx. HLB: 9.2
Chemical Description: Tallow Amine Ethoxylate
Moles EO: 5
Form at 25C: Liquid
% Solids: 99

VARONIC T210:
Approx. HLB: 12.6
Chemical Description: Tallow Amine Ethoxylate
Moles EO: 10
Form at 25C: Liquid
% Solids: 99

VARONIC T215:
CTFA Designation: PEG-15 Tallow Amine
Approx. HLB: 14.4
Chemical Description: Tallow Amine Ethoxylate
Moles EO: 15
Form at 25C: Liquid
% Solids: 99

VARONIC T215LC:
CTFA Designation: PEG-15 Tallow Amine
Approx. HLB: 14.4
Chemical Description: Tallow Amine Ethoxylate
Moles EO: 15
Form at 25C: Liquid
% Solids: 99

Used and recommended for use as dye levelers. VARONIC ethoxylated amines improve wetability and reduce dye affinity at the surface, permitting the dye to migrate more evenly into substrates such as hair, resulting in brighter colors. Co-emulsifier.

SHEREX CHEMICAL CO.: Personal-Care Surfactants and Specialty Products(Continued):

Nonionics(Continued):
Ethoxylated Amines(Continued):

Product Name:

VARONIC U202:
Approx. HLB: 5.0
Chemical Description: Cetyl/Stearyl Amine Ethoxylate
Moles EO: 2
Form at 25C: Solid
% Solids: 99

VARONIC U205:
Approx. HLB: 9.2
Chemical Description: Cetyl/Stearyl Amine Ethoxylate
Moles EO: 5
Form at 25C: Paste
% Solids: 99

VARONIC U205LC:
Approx. HLB: 9.2
Chemical Description: Cetyl/Stearyl Amine Ethoxylate
Moles EO: 5
Form at 25C: Paste
% Solids: 99

VARONIC U215:
Approx. HLB: 14.3
Chemical Description: Cetyl/Stearyl Amine Ethoxylate
Moles EO: 15
Form at 25C: Solid
% Solids: 99

VARONIC U250:
Approx. HLB: 17.9
Chemical Description: Cetyl/Stearyl Amine Ethoxylate
Moles EO: 50
Form at 25C: Solid
% Solids: 99

Used and recommended for use as dye levelers. VARONIC ethoxylated amines improve wetability and reduce dye affinity at the surface, permitting the dye to migrate more evenly into substrates such as hair, resulting in brighter colors. Co-emulsifier.

SHEREX CHEMICAL CO.: Personal-Care Surfactants and Specialty Products(Continued):

Alkanolamides:
1:1 Diethanolamides:

Product Name:

VARAMIDE ML-1:
CTFA Adopted Name: Lauramide DEA
Chemical Description: Lauric Acid Diethanolamide
Form at 20C: White Wax
% Solids: 100
Water dispersible, a thickener, and a foam stabilizer for shampoo, bubble bath and hand laundry detergent, ML-1 gives the highest viscosity and stability of the superamides.

VARAMIDE MA-1:
CTFA Adopted Name: Cocamide DEA
Chemical Description: Refined Coconut Oil Diethanolamide
Form at 20C: Clear Liquid
% Solids: 100
Water dispersible, a foam stabilizer, and a thickener, MA-1 is the basic liquid superamide for shampoos, bubble bath and dishwash. A low-cost equivalent to lauric superamide, MA-1 does not require melting and is readily dispersible in water. MA-1 gives higher viscosity and foam stability.

VARAMIDE ML-4:
CTFA Adopted Name: Lauramide DEA
Chemical Description: Lauric/Myristic (70/30) Diethanolamide
Form at 20C: White Wax
% Solids: 100
Thickener and foam stabilizer for shampoo formulations requiring high-formulation viscosity. Conditioning agent.

2:1 Diethanolamides:

VARAMIDE L-1:
CTFA Adopted Name: Lauramide DEA
Chemical Description: Lauric Acid Diethanolamide
Form at 20C: Paste
% Solids: 100
Detergent and foam stabilizer, thickener for shampoos.

1:1 Monoethanolamides:

VARAMIDE C-212:
CTFA Adopted Name: Cocamide MEA
Chemical Description: Coco Acid Monoethanolamide
% Solids: 100

VARAMIDE L-203:
CTFA Adopted Name: Lauramide MEA
Chemical Description: Lauric Acid Monoethanolamide
% Solids: 100
Foam stabilizer, foam builder in anionic systems. Peak viscosity is obtained with minimum chloride levels. Recommended for stick formulations.

SHEREX CHEMICAL CO.: Personal-Care Surfactants and Specialty Products(Continued):

Amine Oxides:

Product Name:

VAROX 365:
CTFA Adopted Name: Lauramine Oxide
Chemical Description: Lauryl Dimethylamine Oxide
Form at 20C: Liquid
% Solids: 30

VAROX 1770:
CTFA Adopted Name: Cocamidopropyl Amine Oxide
Chemical Description: Cocamidopropyl Dimethylamine Oxide
Form at 20C: Liquid
% Solids: 35
Foam booster and stabilizer, conditioner for hair-care products and detergents.

VAROX 185-E:
CTFA Adopted Name: Dihydroxyethyl C12-C15 Alkoxypropyl Amine Oxide
Chemical Description: C12-C15 Alkyl Propoxy BIS-2-Hydroxyethylamine Oxide
Form at 20C: Liquid
% Solids: 40
Foam booster and stabilizer with fatty alcohol sulfates and alcohol ether sulfates.

SHEREX CHEMICAL CO.: Personal-Care Surfactants and Specialty Products(Continued):

Concentrates:

Product Name:

VARIFOAM A:
Chemical Description: Blend of Alkyl Sulfates, Amphoterics and Ethoxylated Glycerides
Form at 20C: Liquid
% Solids: 55
High-foaming, low-irritation shampoo base.

VARIFOAM SXC:
Chemical Description: Blend of Amphoterics, Alkyl Sulfates and Alkanolamides
Form at 20C: Liquid
% Solids: 38
High-foaming, cost-effective shampoo base. Can be diluted with water to as low as 13% solids and still retain good viscosity response to sodium chloride.

VARIFOAM CRC:
Chemical Description: Blend of Fatty Alcohols, Quaternaries and Tertiary Amines
Form at 20C: Waxy Solid
% Solids: 100
Cost-effective concentrate for formulating creme rinses and conditioners.

Quaternaries:
Monoalkyl Trimethyl Quaternaries:

Product Name:

VARISOFT TSC:
CTFA Adopted Name: Steartrimonium Chloride
Chemical Description: Stearyl Trimethyl Ammonium Chloride
Form at 20C: Liquid
% Solids: 25
Base for hair conditioners. Solvent is water.

VARISOFT E-228:
CTFA Adopted Name: Cetrimonium Chloride
Chemical Description: Cetyl Trimethyl Ammonium Chloride
Form at 20C: Liquid
% Solids: 24-27
Base for formulation of hair conditioners and creme rinses. Imparts softness and manageability to hair without greasy feeling. Can Be used alone or with VARISOFT DHT.

VARISOFT E-290:
CTFA Adopted Name: Cetrimonium Chloride
Chemical Description: Cetyl Trimethyl Ammonium Chloride
Form at 20C: Liquid
% Solids: 28-30
Base for formulation of hair conditioners and creme rinses. Same as E-228, but higher activity.

SHEREX CHEMICAL CO.: Personal-Care Surfactants and Specialty Products(Continued):

Quaternaries(Continued):
Alkyl Dimethyl Benzyl Quaternaries:

Product Name:

VARISOFT SDC:
CTFA Adopted Name: Stearalkonium Chloride
Chemical Description: Stearyl Dimethyl Benzyl Ammonium Chloride
Form at 20C: Paste
% Solids: 24-26
Base for hair conditoners and antistats. Imparts softness, manageability and anti-static properties to hair. Contains isopropanol.

VARISOFT SDC-W:
CTFA Adopted Name: Stearalkonium Chloride
Chemical Description: Stearyl Dimethyl Benzyl Ammonium Chloride
Form at 20C: Paste
% Solids: 24-26
Base for hair conditioners and antistats. Same as above, but contains no isopropanol.

VARISOFT SDC-85S:
CTFA Adopted Name: Stearalkonium Chloride and Stearyl Alcohol
Chemical Description: Stearyl Dimethyl Benzyl Ammonium Chloride
Form at 20C: Flakes
% Solids: 100
Base for hair-conditioner emulsifier for creams and lotions. Supplied as flakes diluted with stearyl alcohol.

Dialkyl Dimethyl Quaternaries:

AROSURF TA-100:
CTFA Adopted Name: Distearyldimonium Chloride
Chemical Description: Distearyl Dimethyl Ammonium Chloride
Form at 20C: Powder
% Solids: 100
Base for hair conditioners. Imparts soft feel to hair and antistatic properties.

ADOGEN 432-CG:
CTFA Adopted Name: Quaternium 31 Dicetyldimonium Chloride
Chemical Description: Di (C12-C18) Alkyl Dimethyl Ammonium Chloride
Form at 20C: Liquid
% Solids: 67-69
Base for hair conditioners and creme rinses. Supplied in isopropanol. Can be easily diluted with water, even cold water.

SHEREX CHEMICAL CO.: Personal-Care Surfactants and Specialty Products(Continued):

Quaternaries(Continued):
Dialkyl Dimethyl Quaternaries(Continued):

<u>Product Name:</u>

ADOGEN 432-ET:
CTFA Adopted Name: Quaternium 31 Dicetyldimonium Chloride
Chemical Description: Di (C12-C18) Alkyl Dimethyl Ammonium Chloride
Form at 20C: Liquid
% Solids: 67-69
Base for hair conditioners. Same as 432-CG, but supplied in ethanol.

ADOGEN 442-100P:
CTFA Adopted Name: Quaternium 18
Chemical Description: Dimethyl Di (Hydrogenated Tallow) Ammonium Chloride
Form at 20C: Powder
% Solids: 100
Base for hair conditioners. Powder version of DHT for ease of handling and formulating.

VARISOFT DHT:
CTFA Adopted Name: Quaternium 18
Chemical Description: Dimethyl Di (Hydrogenated Tallow) Ammonium Chloride
Form at 20C: Paste
% Solids: 75
Base for creme rinses and hair conditioners. Supplied in aqueous isopropanol. Has antistatic properties in hair conditioners.

Methyl Dialkoxy Alkyl Quaternary:

VARIQUAT 638:
CTFA Adopted Name: PEG-2 Cocomonium Chloride
Chemical Description: Methyl Bis (2-Hydroxyethyl) Cocammonium Chloride
Form at 20C: Liquid
% Solids: 74-75
Base for hair conditioners and antistats, making clear creme rinse.

SHEREX CHEMICAL CO.: Personal-Care Surfactants and Specialty Products(Continued):

Sulfosuccinates:

Product Name:

VARSULF SBFA-30:
CTFA Adopted Name: Disodium Laureth Sulfosuccinate
Chemical Description: Disodium Lauryl Polyglycol Ether Sulfosuccinate
Form at 20C: Liquid
% Solids: 40
Low-irritation foaming agent for shampoos, bubble bath and body cleansers. Has some conditioning and moisturizing effect.

VARSULF SBL-203:
CTFA Adopted Name: Disodium Lauramido MEA Sulfosuccinate
Chemical Description: Disodium Lauryl Monoethanolamido Sulfosuccinate
Form at 20C: Liquid/Paste
% Solids: 40
Low-irritation foaming and cleansing agents. Highest foaming sulfosuccinate.

VARSULF S-1333:
CTFA Adopted Name: Disodium Ricinolamido MEA Sulfosuccinate
Chemical Description: Disodium Ricinoleic Monoethanolamido Sulfosuccinate
Form at 20C: Liquid
% Solids: 40
Mild surfactant for cleansing agents and lotions, reduces irritancy of surfactants. Skin friendly, improved skin feel.

VARSULF SBU-185:
CTFA Adopted Name: Disodium Undecylenamido MEA Sulfosuccinate
Chemical Description: Disodium Undecilenic Monoethanolamido Sulfosuccinate
Form at 20C: Liquid
% Solids: 40
Surfactant for shampoos and ingredient for anti-dandruff shampoos.

SHEREX CHEMICAL CO.: Personal-Care Surfactants and Specialty Products(Continued):

Amphoterics:
Glycinates:

Product Name:

VARION 2C:
CTFA Adopted Name: Cocoamphocarboxy Glycinate
Chemical Description: N-cocoamido-ethyl-N-2-hydroxy-ethyl-N-carboxy-methyl-glycine, Sodium Salt.
Form at 20C: Liquid
% Solids: 50
High-foaming surfactant for low-irritation shampoos, skin cleansers, medicated cosmetics, conditioning.

VARION 2L:
CTFA Adopted Name: Lauroampho-carboxy Glycinate
Chemical Description: N-lauricamido-ethyl-N-2-hydroxy-ethyl-N N-carboxy-ethyl-glycine, Sodium Salt
Form at 20C: Liquid
% Solids: 50
High-foaming surfactant for low-irritation shanpoos and skin cleansers.

VARION TEG 40%:
CTFA Adopted Name: Dihydroxyethyl Tallow Glycinate
Chemical Description: Tallow Dihydroxyethyl Betaine
Form at 20C: Liquid
% Solids: 40
High-foaming mild surfactant for shampoos and bubble bath. Conditioner.

Betaines:

VARION CADG-HS:
CTFA Adopted Name: Cocamidopropyl Betaine
Chemical Description: Cocoamidopropyl Dimethyl Betaine
Form at 20C: Liquid
% Solids: 35
Foam booster for shampoos and viscosity builder, low-irritation skin cleanser, lime soap dispersant. Aids in the deposition of protein and cationic polymers on the hair.

VARION CADG-LS:
CTFA Adopted Name: Cocamidopropyl Betaine
Chemical Description: Cocoamidopropyl Dimethyl Betaine
Form at 20C: Liquid
% Solids: 35
Low-salt version of CADG-HS

SHEREX CHEMICAL CO.: Personal-Care Surfactants and Specialty Products(Continued):

Amphoterics(Continued):
Betaines(Continued):

VARION CADG-W:
CTFA Adopted Name: Cocamidopropyl Betaine
Chemical Description: Cocoamidopropyl Dimethyl Betaine
Form at 20C: Liquid
% Solids: 35
Economic version of VARION CADG made from whole coconut oil.

VARION CDG:
CTFA Adopted Name: Lauryl Betaine
Chemical Description: Lauryl Dimethyl Betaine
Form at 20C: Liquid
% Solids: 35
High-foaming mild detergent for shampoos and bubble bath.

Sulfobetaines:

VARION CAS:
CTFA Adopted Name: Cocamidopropyl Hydroxy Sultaine
Chemical Description: 3-(3-Cocoamido-propyl Dimethylamino) 2-hydroxy-propane Sulphonate
Form at 20C: Liquid
% Solids: 50
High-foaming mild detergent for shampoos and bubble bath. Better response to salt than regular betaines.

VARION CAS-W:
CTFA Adopted Name: Cocamidopropyl Hydroxy Sultaine
Chemical Description: 3-(3-Cocoamido-propyl Dimethylamino) 2-hydroxy-propane Sulphonate
Form at 20C: Liquid
% Solids: 50
Economic version of VARION CAS made from whole coconut oil.

SHEREX CHEMICAL CO.: Personal-Care Surfactants and Specialty Products(Continued):

Fatty Acid Esters:

Product Name:

STARFOL OO:
CTFA Adopted Name: Oleyl Oleate
Chemical Description: Ester of Oleyl Alcohol and Oleic Acid
Form at 20C: Light Amber Oil
% Solids: 100

Emollient for creams and skin preparations. STARFOL OO is an excellent solvent for many lipid-soluble products. It is non-irritating to the skin and exhibits good solubilization properties.

STARFOL OO is recommended in the formulation of baby oils and other skin-care preparations. Because of its solvency of fats and oils of vegetable and animal origin, it acts as a vehicle for lipid-soluble substances.

STARFOL BB:
CTFA Adopted Name: Behenyl Behenate
Chemical Description: Ester of Behenyl Alcohol and Behenic Acid
Form at 20C: Light Amber Waxy Solid MP=60
% Solids: 100

Emollient wax. STARFOL BB is recommended for use in wax systems where its solubility characteristics help to prevent bleeding and syneresis. It is useful in the preparation of sun-care and stick products. STARFOL BB improves the consistency of stick formulations.

STARFOL IS:
CTFA Adopted Name: Isostearyl Isostearate
Chemical Description: Ester of Isostearyl Alcohol and Isostearic Acid
Form at 20C: Light Amber Liquid
% Solids: 100

Emollient in creams and lotions, solubilzer and solvent for fragrances.

STARFOL IS is an excellent solvent for many lipid-soluble products. The combination of a branched fatty acid with a branched fatty alcohol provides a more luxurious feel to the skin than the commonly utilized isopropyl esters. It imparts non-greasy emolliency to skin-care formulations. STARFOL IS is recommended for body oils, creams, and lotions where a quality emollient oil is desired.

SHEREX CHEMICAL CO.: Personal-Care Surfactants and Specialty Products(Continued):

Fatty Acid Esters(Continued):

STARFOL OS:

CTFA Adopted Name: Octyldodecyl Stearate
Chemical Description: Ester of Octyldodecyl Alcohol and Stearic Acid
Form at 20C: Light Amber Liquid
% Solids: 100
Emollient for creams and lotions.

STARFOL OS is the ester of a branched 20 carbon alcohol and stearic acid. It is easily incorporated into the oil phase of oil/water emulsions. STARFOL OS imparts a luxurious feel to the skin without greasiness.

STARFOL WAX CG:

CTFA Adopted Name: Cetyl Esters
Chemical Description: Predominately Ester of Cetyl Alcohol and Palmitic Acid
Form at 20C: White Flakes MP=48
% Solids: 100

STARFOL WAX CG is recommended for creams and lotions. Its chemical and physical properties are strikingly similar to those of natural Spermaceti Wax. It is generally useful as an ingredient in skin-care formulations and lipsticks.

STARFOL CP:

CTFA Adopted Name: Cetyl Palmitate
Chemical Description: Ester of Cetyl Alcohol and Palmitic Acid
Form at 20C: Cream-Colored Flakes
% Solids: 100

Emollient. STARFOL CP is recommended for creams and lotions and for lipstick formulations. It is generally useful as an ingredient in skin-care products.

SHEREX CHEMICAL CO.: Personal Care Surfactants and Specialty Products(Continued):

Fatty Acids:

Product Name:

HYDROFOL Acid 1655 CG-NF:
CTFA Adopted Name: Stearic Acid
Chemical Description: 48% Stearic
52% Palmitic Acid (Triple-Pressed Stearic Acid)
Form at 20C: Flake
MP=54.5-59C

HYDROFOL Acid 1690:
CTFA Adopted Name: Palmitic Acid
Chemical Description: 90% Palmitic Acid
Form at 20C: Flake
MP=58-62C

HYDROFOL Acid 1855:
CTFA Adopted Name: Stearic Acid
Chemical Description: 55% Stearic Acid
Form at 20C: Flake
MP=54.4-56.5C

HYDROFOL Acid 1895:
CTFA Adopted Name: Stearic Acid
Chemical Description: 95% Stearic Acid
Form at 20C: Flake
MP=67-69C

Used in creams and lotions as emulsifiers in the form of its triethanol amine salts. The free fatty acid forms part of the emollient system.

Glycerine:

High-Gravity Glycerine:
CTFA Adopted Name: Glycerin
Chemical Description: Glycerine 99.5% min.
Form at 20C: Liquid
Used as a humectant in skin and hair-care formulations.

SHEREX CHEMICAL CO.: Personal Care Surfactants and Specialty Products(Continued):

Fatty Alcohols:

<u>Product Name:</u>

ADOL 52-NF:
CTFA Adopted Name: Cetyl Alcohol
Chemical Description: Cetyl Alcohol Natural
Form at 20C: Flake MP=45-50C

ADOL 520-NF:
CTFA Adopted Name: Cetyl Alcohol
Chemical Description: Cetyl Alcohol Synthetic
Form at 20C: Flake MP=45-50C

ADOL 62-NF:
CTFA Adopted Name: Stearyl Alcohol
Chemical Description: Stearyl Alcohol Natural
Form at 20C: Flake MP=56-60C

ADOL 620-NF:
CTFA Adopted Name: Stearyl Alcohol
Chemical Description: Stearyl Alcohol Synthetic
Form at 20C: Flake MP=55-60C

Emollients and emulsion stabilizers for creams and lotions and other skin-care products. Skin lubricants, impart velvety feel.

ADOL 66:
CTFA Adopted Name: Isostearyl Alcohol
Chemical Description: Isostearyl Alcohol Natural
Form at 20C: Liquid Cloud Point 8C

Used in skin-care products where emolliency, emulsion stability and thickening is desired. Reduces greasiness and after tack. Used where chemical stability is a concern and liquidity desired.

ADOL 90-NF:
CTFA Adopted Name: Oleyl Alcohol
Chemical Description: Oleyl Alcohol Natural
Form at 20C: Liquid Cloud Point 6C
Used in creams and lotions. Imparts a feeling of smoothness, freshness and suppleness to the skin.

SONNEBORN DIVISION: MULTIWAX Microcrystalline Waxes:

MULTIWAX designation:

180-M:
Melting point F, C: 180/190 (82/88)
Needle Penetration @ 77F (25C): 15/20
Viscosity SUS 210F (CST @ 99C): 75/90 (14.3/18.0)
Color: Visual: Light Yellow
ASTM D-1500: 1.0/2.0
Flash Point COC F, C: 530 min. (277 min.)

ML-445:
Melting point F, C: 170/180 (77/82)
Needle Penetration @ 77F (25C): 25/35
Viscosity SUS 210F (CST @ 99C): 75/90 (14.3/18.0)
Color: Visual: Light Yellow
ASTM D-1500: 1.0/2.0
Flash Point COC F, C: 525 min. (274 min.)

W-445:
Melting Point F, C: 170/180 (77/82)
Needle Penetration @ 77F (25C): 25/35
Viscosity SUS 210F (CST @ 99C): 75/90 (14.3/18.0)
Color: Visual: White
ASTM D-1500: +16 min.
Flash Point COC F, C: 525 min. (274 min.)

HS:
Melting Point F, C: 160/170 (71/77)
Needle Penetration @ 77F (25C): 25/30
Viscosity SUS 210F (CST @ 99C): 75/90 (14.3/18.0)
Color: Visual: Light Yellow
ASTM D-1500: 1.0
Flash Point COC F, C: 525 min. (274 min.)

W-835:
Melting Point F, C: 165/175 (74/79)
Needle Penetration @ 77F (25C): 60/80
Viscosity SUS 210F (CST @ 99C): 75/90 (14.3/18.0)
Color: Visual: White
ASTM D-1500: +16 min.
Flash Point COC F, C: 475 min. (246 min.)

X-145A:
Melting Point F, C: 150/160 (66/71)
Needle Penetration @ 77F (25C): 35/45
Viscosity SUS 210F (CST @ 99C): 75/90 (14.3/18.0)
Color: Visual: Light Yellow
ASTM D-1500: 1.0/2.0
Flash point COC F, C: 500 min. (260 min.)

SONNEBORN DIVISION: MULTIWAX Microcrystalline Waxes(Continued):

WITCO Designation:

180-M:
Hot melt adhesives; hot melt coatings for corrugated board; electrical insulation and impregnation; chewing gum base; protective coatings.

ML-445:
Laminating agent for paper, film and foil; hot melt adhesives; chewing gum base; electrical insulation and impregnation; anti-checking agents; sealants and caulking compound; rust proofing compounds; waterproofing and protective coatings; crayons, dental waxes; candles; artificial flowers, paraffin wax modifier; sculpture wax.

W-445:
Hair dressings, medicated creams and unguents, chewing gum base, artificial flowers, dental waxes, lubricants, candles and hot melt adhesives.

HS:
Strike-through foil/tissue laminations, heat seal laminations, glassine laminations, heat-sealable laminations.

W-835:
Silk screen printing, cold creams, cleansing creams, hair pomades, pharmaceuticals, art restoration, crayons, layout and paste-up adhesive.

X-145A:
Cheese coatings, lamination of cellophane and plastic film, waterproofing and protective linings.

MULTIWAX complies with NF Edition XVI specifications and is suitable for use in cosmetics and pharmaceuticals.

SONNEBORN DIVISION: SONNEBORN Petrolatums:

Properties of standard SONNEBORN petrolatums, USP grade

PROTOPET Petrolatums, USP:
Medium consistency and melting point, amber to white, useful in cosmetics, pharmaceutical ointments and various industrial applications.

Product:

ALBA:
Melting Point USP: C (F): 54/60 (130/140)
Consistency USP: 180/210
Viscosity: Saybolt @ 210F: 60/80
Centistoke @ 100C: 10/16
Color: Lovibond: 1.0Y
ASTM D-1500: L0.5

White 1S:
Melting Point USP: C (F): 54/60 (130/140)
Consistency USP: 180/210
Viscosity: Saybolt @ 210F: 60/80
Centistoke @ 100C: 10/16
Color: Lovibond: 1.5Y
ASTM D-1500: L0.5

White 2L:
Melting Point USP: C (F): 54/60 (130/140)
Consistency USP: 180/210
Viscosity: Saybolt @ 210F: 60/80
Centistoke @ 100C: 10/16
Color: Lovibond: 8Y0.6R
ASTM D-1500: 1.0

White 3C:
Melting Point USP: C (F): 54/60 (130/140)
Consistency USP: 180/210
Viscosity: Saybolt @ 210F: 60/80
Centistoke @ 100C: 10/16
Color: Lovibond: 25Y1.0R
ASTM D-1500: L1.5

Yellow 2A:
Melting Point USP C (F): 54/60 (130/140)
Consistency USP: 180/210
Viscosity: Saybolt @ 210F: 60/80
Centistoke @ 100C: 10/16
Color: Lovibond: 30Y2.5R
ASTM D-1500: L2.0

SONNEBORN DIVISION: SONNEBORN Petrolatums(Continued):

Properties of standard SONNEBORN petrolatums, USP grade (Continued):

PERFECTA Petrolatum:

Cosmetic and pharmaceutical grade, lightest color, medium consistency and high melting point.

Product:

PERFECTA USP:
Melting Point USP: C (F): 57/60 (135/140)
Consistency USP: 180/210
Viscosity: Saybolt @ 210F: 60/80
Centistoke @ 100C: 9/14
Color: Lovibond: 0.3Y
ASTM D-1500: L0.5

FONOLINE Petrolatums, USP:

Soft, low melting point, white and yellow, for consumer use as petroleum jelly, ointments incorporating large quantities of solids, and industrial applications.

White:
Melting Point USP C (F): 53/58 (127/137)
Consistency USP: 210/245
Viscosity: Saybolt @ 210F: 55/75
Centistoke @ 100C: 9/14
Color: Lovibond: 1.7Y
ASTM D-1500: 0.5

Yellow:
Melting Point USP C (F): 53/58 (127/137)
Consistency USP: 210/245
Viscosity: Saybolt @ 210F: 55/75
Centistoke @ 100C: 9/14
Color: Lovibond: 30Y2.5R
ASTM D-1500: L2.0

All above grades have no odor and meet FDA 21 CFR 172.880 for UV absorbance. These products are available conforming to the requirements of BP, DAB, VII French Codex, Italian, Japanese and other pharmacopoeias. Special grades can be formulated to meet particular customer specifications.

SONNEBORN DIVISION: SONNEBORN Petrolatums(Continued):

Properties of SONNEBORN mineral jellies:

SONNEBORN mineral jellies are compounded from FDA quality white mineral oils, petrolatums and paraffin waxes.

Mineral Jellies:

SONO JELL No. 4:
Melting Point USP C (F): 38/52 (100/125)
Consistency USP: 100/125
Viscosity: Saybolt @ 210F: 35/50
Centistoke @ 100C: 2.6/5.7
Color: Lovibond: 0.5Y
ASTM D-1500: L.05

SONO JELL No. 9:
Melting Point USP C (F): 42/49 (107/120)
Consistency USP: 150/170
Viscosity: Saybolt @ 210F: 35/50
Centistoke @ 100C: 2.6/5.7
Color: Lovibond: 0.5Y
ASTM D-1500: L0.5

Mineral Jelly No. 10:
Melting Point USP C (F): 38/43 (100/110)
Consistency USP: 280/340
Viscosity: Saybolt @ 210F: 35/50
Centistoke @ 100C: 2.6/5.7
Color: Lovibond: 1.0Y
ASTM D-1500: L0.5

Mineral Jelly No. 14:
Melting Point USP C (F): 38/52 (100/125)
Consistency USP: 200/225
Viscosity: Saybolt @ 210F: 35/50
Centistoke @ 100C: 2.6/5.7
Color: Lovibond: 1.0Y
ASTM D-1500: L0.5

Mineral Jelly No. 17:
Melting Point USP C (F): 36/49 (95/120)
Consistency USP: 340/400
Viscosity: Saybolt @ 210F: 35/50
Centistoke @ 100C: 2.6/5.7
Color: Lovibond: 1.0Y
ASTM D-1500: L0.5

The above grades meet FDA 21 CFR 178.880 for UV absorbance. They also meet USP purity requirements. However, they do not qualify as USP, because they do not meet some of the USP physical requirements; i.e., specific gravity, consistency and/or melting point.

SONNNEBORN DIVISION: SONNEBORN Petrolatums(Continued):

Properties of standard SONNEBORN petrolatums, technical grade:

Product:

TECH PET M:
Melting Point C (F): 51/65 (125/150)
Consistency: 125/175
Color: dark brown

TECH PET F:
Melting Point C (F): 57/66 (135/150)
Consistency: 160/190
Color: dark brown

DARK GREEN No. 2:
Melting Point C (F): 60/68 (140/155)
Consistency: 160/180
Color: dark brown

PETROLATUM RPB:
Melting Point C (F): 71/77 (160/170)
Consistency: 35/45
Color: dark brown

Petrolatums: Applications and functions:

Cosmetics, toiletry, personal care:
- Carrier
- Lubricant
- Moisture Barrier
- Protective agent
- Softener

Typical Applications:
Skin lotions, creams, sun-protection products, petroleum jelly, hair-care products, hand cleaners

SONNEBORN petrolatums are homogeneous semi-solid mixtures of oily and waxy hydrocarbons. They are characterized by viscosity and shear strength, chemical and biological inertness and non-polar hydrocarbon properties. These characteristics make them particularly suitable for industrial applications where lubricity and moisture resistance are important, as well as for food and cosmetic use.

In fact, because of their high quality, SONNEBORN "pets" are used extensively in the food-processing, pharmaceutical, cosmetic and toiletry industries, as well as in a wide range of products requiring inertness and safety.

STEPAN CO.: Cationics:

Methosulfate:

STEPANQUAT 6585:
STEPANQUAT 6585-ET:
CTFA: Dipalmethyl Hydroxyethylmonium Methosulfate
Active %: 83-87% Solids
White soft paste
Mild cationic surfactants designed for cream rinses and conditioners

Stearyl Dimethyl Benzyl Ammonium Chloride:

AMMONYX 4:
CTFA: Stearalkonium Chloride
Active %: 17-19
Paste
AMMONYX 4B:
CTFA: Stearalkonium Chloride
Active %: 16-18
Paste
AMMONYX 4-IPA:
CTFA: Stearalkonium Chloride
Active %: 17-19
Paste
AMMONYX 485:
CTFA: Stearalkonium Chloride
Active %: 85 Min.
Powder
AMMONYX 4002:
CTFA: Stearalkonium Chloride
Active %: 94 Min.
Powder
AMMONYX CA-Special:
CTFA: Stearalkonium Chloride
Active %: 20.0-22.5
Paste

These products possess pronounced conditioning, softening, and emolliency characteristics. Suggested applications include hair rinses, skin creams and lotions. Used as a cationic emulsifier.

STEPAN CO.: Cationics/Anionics(Continued):

Cationics(Continued):

Cetyl Trimethyl Ammonium Chloride:

AMMONYX CETAC:
CTFA: Cetrimonium Chloride
Active %: 24-26
Liquid
AMMONYX CETAC-30:
CTFA: Cetrimonium Chloride
Active %: 29 min.
Liquid
These products possess pronounced conditioning, softening, and emolliency characteristics. Suggested applications include hair rinses, skin creams and lotions. Used as a cationic emulsifier.

Oleyl Dimethyl Benzyl Ammonium Chloride:

AMMONYX KP:
CTFA: Olealkonium Chloride
Active %: 50 Min.
Physical Form @ 25C: Liquid
Used in clear hair rinses exhibiting excellent conditioning and anti-static properties.

STEPAN CO.: Anionics:

Alkyl Sulfates:
STEPANOL WA-Extra:
CTFA: Sodium Lauryl Sulfate
Active %: 28-30
Clear Liquid
STEPANOL WAC:
CTFA: Sodium Lauryl Sulfate
Active %: 28-30
Clear Liquid
STEPANOL WA-Special:
CTFA: Sodium Lauryl Sulfate
Active %: 28-30
Clear Liquid
STEPANOL WAQ:
CTFA: Sodium Lauryl Sulfate
Active %: 28-30
Viscous Liquid
STEPANOL WA-Paste:
CTFA: Sodium Lauryl Sulfate
Active %: 28-30
Paste
Clear liquid and gel shampoos and bath products. WAQ and WA Paste are also applicable to paste and cream shampoos. Various grades offer choice in developing optimum formulations.

STEPAN CO.: Anionics(Continued):

Alkyl Sulfates(Continued):

STEPANOL WAT:
CTFA: TEA Lauryl Sulfate
Active %: 39.5-41.0
Clear Liquid

STEPANOL DEA:
CTFA: DEA Lauryl Sulfate
Active %: 33-35
Clear Liquid
Shampoo and bath products. Easily formulated into high active premium products to provide excellent performance.

STEPANOL AM:
CTFA: Ammonium Lauryl Sulfate
Active %: 28-30
Viscous Liquid

STEPANOL AM-V:
CTFA: Ammonium Lauryl Sulfate
Active %: 27-29
Viscous Liquid to Gel
Low pH products with excellent foam. AM-V offers higher viscosities; AM provides improved handling properties.

STEPANOL ME-Dry:
CTFA: Sodium Lauryl Sulfate
Active %: 93 min.
White Powder

STEPANOL WA-100:
CTFA: Sodium Lauryl Sulfate
Active %: 97 min.
White Powder
Powdered bubble baths, cleansing creams. WA-100 with low unsulfated matter is also used in toothpastes.

STEPAN CO.: Anionics(Continued):

Alcohol Ether Sulfates:

STEOL CA-130:
CTFA: Ammonium Laureth Sulfate
Active %: 26-28
Clear Liquid
STEOL CS-130:
CTFA: Sodium Laureth Sulfate
Active %: 24.5-26.5
Clear Liquid
STEOL CA-230:
CTFA; Ammonium Laureth Sulfate
Active %: 24-26
Clear Liquid
STEOL CS-230:
CTFA: Sodium Laureth Sulfate
Active %: 24.5-26.5
Clear Liquid
STEOL CA-330:
CTFA: Ammonium Laureth Sulfate
Active %: 27-29
Clear Liquid
STEOL CS-330:
CTFA: Sodium Laureth Sulfate
Active %: 27.5-29.5
Clear Liquid
STEOL 4N:
CTFA: Sodium Laureth Sulfate
Active %: 28-30
Clear Liquid
STEOL 7N:
CTFA: Sodium Laureth Sulfate
Active %: 28-30
Clear Liquid
STEOL CA-460:
CTFA: Ammonium Laureth Sulfate
Active %: 58-60
Clear Liquid
STEOL CS-460:
CTFA: Sodium Laureth Sulfate
Active %: 58.5-60.5
Clear Liquid

Excellent surfactants for shampoos and bath products. Lower ethylene oxide content products have viscosity characteristics similar to SLS, while higher ethylene oxide materials offer improved mildness. The ammonium ether sulfates are used in low pH systems.

STEPAN CO.: Anionics(Continued):

Alpha Olefin Sulfonate:

BIO-TERGE AS-40:
CTFA: Sodium C14-16 Olefin Sulfonate
Active %: 34-40
Yellow Liquid
Shampoos, hand soaps and bath products. More stable than alcohol sulfates over a broad pH range. Provides excellent flash foam.

Linear Alkylbenzene Sulfonate:

BIO-SOFT N-300:
CTFA: TEA-Dodecylbenzenesulfonate
Active %: 59-61
Clear Liquid
Oily hair shampoo. Provides effective removal of soil without stripping hair.

Sarcosinate:

MAPROSYL 30:
CTFA: Sodium Lauroyl Sarcosinate
Active %: 29.5-30.5
Liquid
A soap-like detergent which provides excellent wetting and foaming. Its low cloud point and extreme mildness make it an excellent candidate for personal care and household products.

Sulfoacetates:

LATHANOL LAL:
Sodium Lauryl Sulfoacetate
Active %: 70
White Powder
Power bubble baths and shampoos. Cleansing creams, cream and paste shampoos, and syndet bars. Viscosity modifier in AOS formulas.

STEPAN CO.: Anionics(Continued):

Sulfosuccinates:

ANIONYX 12S:
CTFA: Disodium Monooleamido Sulfosuccinate
Active %: 19-21
Liquid

STEPAN-MILD SL3:
CTFA: Disodium Laureth Sulfosuccinate
Active %: 30-34
Liquid
Low irritation shampoos, bubble baths and dishwashing detergents

Blends:

BIO-TERGE 804:
CTFA: Sodium C14-16 Olefin Sulfonate/Sodium Laureth Sulfate Lauramide DEA
Active %: 50
Amber Liquid

STEPANOL AEG:
CTFA: Ammonium Lauryl Sulfate/Ammonium Laureth Sulfate/ Cocamidopropyl Betaine/Cocamide DEA
Active %: 40-44
Liquid

STEPANOL AEM:
CTFA: Ammonium Laureth Sulfate/Cocamide MEA
Active %: 48
Liquid

STEPANOL LX:
CTFA: DEA Lauryl Sulfate/DEA Lauraminoproprionate/Sodium Lauraminopropionate
Active %: 33-37
Liquid

STEPANOL 317:
CTFA: Proprietary Blend
Active %: 100
Amber Liquid

STEPANOL 360:
CTFA: Sodium Lauryl Sulfate/Lauramide DEA
Active %: 24-25
Liquid
Surfactant concentrates for shampoos and bath products. These products offer ease of handling and simple mixing and dilution attributes.

STEPAN CO.: Nonionics/Amphoterics:

Alkanolamides:

NINOL 30-LL:
CTFA: Lauramide DEA
Active %: 100
Liquid
NINOL 40-CO:
CTFA: Cocamide DEA
Active %: 100
Liquid
NINOL 49-CE:
CTFA: Cocamide DEA
Active %: 100
Liquid
NINOL 50-LL:
CTFA: Lauramide DEA
Active %: 100
Liquid to Gel
NINOL 51-LL:
CTFA: Lauramide DEA
Active %: 100
Liquid
NINOL 55-LL:
CTFA: Lauramide DEA
Active %: 100
Liquid
NINOL 70-SL:
CTFA: Lauramide DEA
Active %: 100
Solid
NINOL 96-SL:
CTFA: Lauramide DEA
Active %: 100
Solid
NINOL L-9:
CTFA: Lauramide DEA
Active %: 100
Paste to Solid
NINOL GR:
CTFA: Cocamide DEA
Active %: 100
Liquid
Foam boosters, stabilizers and viscosity builders/modifiers for shampoos, hand soaps and bath products as well as for a variety of household and industrial formulations.

STEPAN CO.: Nonionics/Amphoterics(Continued):

Ethoxylated Alkanolamides:

AMIDOX C-2:
CTFA: PEG-2 Cocamide
Active %: 100
Liquid

AMIDOX C-5:
CTFA: PEG-5 Cocamide
Active %: 100
Liquid

AMIDOX L-2:
CTFA: PEG-2 Lauramide
Active %: 100
Solid

AMIDOX L-5:
CTFA: PEG-5 Lauramide
Active %: 100
Solid

Mild, effective emulsifiers for fragrances and essential oils. Also, impart viscosity and foam enhancement.

Amine Oxides:

AMMONYX CO:
CTFA: Cetamine Oxide
Active %: 29-31
Liquid

AMMONYX LO:
CTFA: Lauramine Oxide
Active %: 29-31
Liquid

AMMONYX DMCD-40:
CTFA: Lauramine Oxide
Active %: 40-42
Liquid

AMMONYX MO:
CTFA: Myristamine Oxide
Active %: 29-31
Liquid

AMMONYX MCO:
CTFA: Myristamine Oxide
Active %: 29-31
Liquid

AMMONYX SO:
CTFA: Stearamine Oxide
Active %: 24.5-31.5
Paste

AMMONYX CDO:
CTFA: Cocamidopropylamine Oxide
Active %: 29.5-31.5
Liquid

Conditioner, emulsifier, viscosity modifier with wetting, foaming, and foam stabilization properties.

STEPAN CO.: Nonionics/Amphoterics(Continued):

Betaines:

AMPHOSOL CA:
CTFA: Cocamidopropyl Betaine
Active %: 30
Liquid

AMPHOSOL CG:
CTFA: Cocamidopropyl Betaine
Active %: 30
Liquid
Shampoos, bubble baths and liquid hand soaps. Provide good foaming and stabilization with excellent wetting properties. Compatible with anionic, cationic and nonionic surfactants.

STEPAN CO.: Polymeric Resins:

Hair Care Fixatives:

STEPANHOLD R-1:
CTFA: PVP/Ethyl Methacrylate/Methacrylic Acid Copolymer
Active %: 50 (solids)
Viscous Fluid
Hair sprays, mousses, setting gels. Low hydroscopicity. Water solubility controlled by degree of neutralization.

STEPANHOLD EXTRA:
CTFA: PVP/Ethyl Methacrylate/Methacrylic Acid Copolymer
Active %: 40 (solids)
Viscous Fluid
Designed especially for super-hold formulations. Provides firmer hold than STEPANHOLD R-1.

STEPAN CO.: Esters:

Emollients:

Isopropyl Myristate(IPM):
CTFA: Isopropyl Myristate
Acid Value (Max.): 1.0
Color (APHA) (Max.): 20
Physical Form at 25C: Liquid
Melting Point C: -3

Isopropyl Palmitate(IPP):
CTFA: Isopropyl Palmitate
Acid Value (Max.): 1.0
Color (APHA) (Max.): 20
Physical Form at 25C: Liquid
Melting Point C: 13

Isopropyl Myristate and Isopropyl Palmitate are outstanding emollients, blending agents, solubilizers and good liquid vehicles for highly pigmented products. Extensively used in pre-shave and after-shave products, liquid and cream make-ups, lipsticks, bath oils, creams, lotions and hair preparations.

Isobutyl Stearate (IBS):
Isobutyl Stearate
Acid Value (Max.): 1.0
Color (APHA) (Max.): 35
Physical Form at 25C: Liquid
Melting Point C: 15

Isobutyl Stearate imparts good slip properties to formulations containing mineral oil and can function as a wetting agent for pigments. It is used in lipsticks, bath oils, nail polishes and removers, skin cleansers, creams, and lotions.

Octyl Isononanoate:
CTFA: Octyl Isononanoate
Acid Value (Max.): 1.0
Color (APHA) (Max.): 20
Physical Form at 25C: Liquid
Melting Point C: -34

Octyl Isononanoate has the driest, non-oily-skin feel properties of the emollient esters. Its dry-feel emollient and skin breathing properties find application in creams, lotions, makeup, lipsticks, and anti-perspirants.

STEPAN CO.: Esters(Continued):

Emollients(Continued):

Octyl Palmitate:

CTFA: Octyl Palmitate
Acid Value (Max.): 3.0
Color (APHA) (Max.): 20
Physical Form at 25C: Liquid
Melting Point C: 0

Octyl Palmitate enhances the gloss in stick makeup and hair grooming products. It is used in suntan and bath oil formulas, as a binder for pressed powders, and as an emollient in all types of creams and lotions.

Cetyl Palmitate:

CTFA: Cetyl Palmitate
Acid Value (Max.): 2.0
Color (APHA): White
Flakes
Melting Point C: 50-55

Cetyl Palmitate is a waxy ester that imparts good skin feel properties to cosmetics. It is used as a base material in stick cosmetic products and as an emollient thickener in creams and lotions.

Isocetyl Stearate(ICS):

CTFA: Isocetyl Stearate
Acid Value (Max.): 2.0
Color (APHA) (Max.): 100
Physical Form at 25C: Liquid
Melting Point C: 0

Isocetyl Stearate is a very rich emollient recommended for use in makeup formulations.

NEOBEE M-20:

CTFA: Propylene Glycol Dicaprylate/Dicaprate
Acid Value (Max.): 0.10
Color (APHA) (Max.): 450
Physical Form at 25C: Liquid
Melting Point C: -20

NEOBEE M-5:

CTFA: Caprylic/Capric Triglyceride
Acid Value (Max.): 0.10
Color (APHA) (Max.): 450
Liquid
Melting Point C: -5

NEOBEE 0:

CTFA: Caprylic/Capric Triglyceride
Acid Value (Max.): 0.10
Color (APHA) (Max.): 450
Liquid
Melting Point C: 0

The NEOBEE Oils are a series of high quality vegetable derived emollient oils. Their safety, excellent stability, and controlled range of lubricities have made them widely used in the cosmetic and pharmaceutical fields in applications such as bath oils, creams, lotions, lipsticks and glosses.

STEPAN CO.: Esters(Continued):

Base:

WECOBEE FW:
CTFA: Hydrogenated Vegetable Oil
Acid Value (Max.): 0.20
Color (APHA) (Max.): 400
Solid
Melting Point C: 35

WECOBEE S:
CTFA: Hydrogenated Vegetable Oil
Acid Value (Max.): 0.20
Color (APHA) (Max.): 400
Flake
Melting Point C: 42

WECOBEE SS:
CTFA: Hydrogenated Vegetable Oil
Acid Value (Max.): 0.20
Color (APHA) (Max.): 400
Flake
Melting Point C: 42

WECOBEE M:
CTFA: Hydrogenated Vegetable Oil
Acid Value (Max.): 0.20
Color (APHA) (Max.): 400
Solid
Melting Point C: 33.5

The WECOBEES are a series of triglycerides derived from edible vegetable oils. They exhibit most of the desirable features of cocoa butter, but few of its shortcomings. The WECOBEES are extremely stable and uniform in composition and exhibit excellent mold release characteristics. They have been widely used as cocoa butter replacements and in lipsticks and glosses, pomades, solid fragrances, anti-perspirant sticks, suppositories and in emollient creams and lotions.

Emulsifiers, Opacifiers:

Glycerol Monostearate Pure (GMS):
CTFA: Glyceryl Stearate
HLB Value: 3.8
Acid Value (Max.): 3.0
Color (APHA) (Max.): White
Physical Form at 25C: Flakes
Melting Point C: 56.5-58.5

Emulsifier-opacifier and bodying agent. Used in creams, lotions, anti-perspirants, hair care products and sun screens.

STEPAN CO.: Esters(Continued):

Emulsifiers, Opacifiers(Continued):

Glycerol Distearate 386F:
CTFA: Glyceryl Distearate
HLB Value: 2.4
Acid Value (Max.): 5.0
Color (APHA) (Max.): White
Waxy Flakes
Melting Point C: 55-60
Alternative to Glycerol Monostearate Pure offering lower HLB value.

Glycerol Monostearate S.E. (GMS-S.E.):
CTFA: Glyceryl Stearate SE
Acid Value (Max.): 20
Color (APHA) (Max.): White to Cream
Flakes
Melting Point C: 56.5-59.5
Anionic modified. Recommended for use in oil-in-water emulsions that are in the pH range of 5 to 9.

Glycerol Monostearate S.E. acid stable (GMS-S.E.A.S.):
CTFA: Glyceryl Stearate (and) PEG-100 Stearate
HLB Value: 11.2
Acid Value (Max.): 3.0
Color (APHA) (Max.): White to Cream
Physical Form at 25C: Flakes
Melting Point C: 54-58
Nonionic. Recommended for low pH (3 to 5). Used as emulsifier, self-emulsifying cream base, hair and skin conditioner. Good electrolyte stability.

Glycerol Monostearate (GMO):
CTFA: Glyceryl Oleate
HLB Value: 3.8
Acid Value (Max.): 3.0
Color (APHA) (Max.): Yellow
Physical Form at 25C: Liquid
Melting Point C: 20
Effective water-in-oil emulsifier. Used in bath oil as emollient and spreading agent, in makeup as pigment dispersant and in vanishing and moisturizing cream to impart slip.

STEPAN CO.: Esters(Continued):

Emulsifiers, Opacifiers(Continued):

Glycerol Monolaurate (GML):
CTFA: Glyceryl Laurate
HLB Value: 4.9
Acid Value (Max.): 5.0
Color (APHA) (Max.): White
Solid
Melting Point C: 54

Glycerol Monolaurate functions as a primary emulsifier for water-in-oil emulsions. In addition to its emulsifier function, Glycerol Monolaurate can impart a lasting emollient feel to formulations.

Glycerol Dilaurate (GDL):
CTFA: Glyceryl Dilaurate
HLB Value: 4.0
Acid Value (Max.): 5.0
Color (APHA) (Max.): White
Physical Form at 25C: Solid
Melting Point C: 30

Glycerol Dilaurate is a semi-solid recommended for use in free-flowing lotions where the glycerol laurate emolliency is desired.

Pearlescent Agents, Auxiliary Emulsifiers:

Ethylene Glycol Monostearate Pure (EGMS):
CTFA: Glycol Stearate
HLB Value: 2.9
Acid Value (Max.): 2.0
Color (APHA) (Max.): White to Cream
Flakes
Melting Point C: 56-60

Pearlescent agent in shampoos and liquid hand soaps. It also functions as a bodying agent and emulsion stabilizer in those systems.

Ethylene Glycol Distearate (EGDS):
CTFA: Glycol Distearate
HLB Value: 1.5
Acid Value (Max.): 15.0
Color (APHA) (Max): White to Cream
Flakes
Melting Point C: 60-63

Pearlizer, emollient and emulsifier. Suggested for use when no additional viscosity response is desired such as high-solids formulations.

STEPAN CO.: Esters(Continued):

Pearlescent Agents, Auxiliary Emulsifiers(Continued):

Ethylene Glycol Amido Stearate(EGAS):
CTFA: Glycol Stearate (and) Stearamide AMP
Acid Value (Max.): 5.0
Color (APHA) (Max.): White to Cream
Flakes
Melting Point C: 56.5-58.5
Pearlescent and bodying agent in shampoos and liquid hand soaps. Also imparts a soft, smooth skin feel to formulations.

Diethylene Glycol Monostearate (DGMS):
CTFA: PEG-2 Stearate
HLB Value: 4.3
Acid Value (Max.): White to Cream
Flakes
Melting Point C: 44.5-47.5
Opacifier in shampoos and lotions. Imparts a luxurious emolliency and adds body to those formulations.

Propylene Glycol Monostearate Pure (PGMS):
CTFA: Propylene Glycol Stearate
HLB Value: 3.4
Acid Value (Max.): 3.0
Color (APHA) (Max): White to Cream
Flakes
Melting Point C: 33.5-38.5
Melting point near that of body temperature and so is used in suppositories, lipsticks and sunscreens. Also functions as auxiliary emulsifier and opacifier.

Propylene Glycol Monolaurate (PGML):
CTFA: Propylene Glycol Laurate
HLB Value: 3.2
Acid Value (Max.): 3.0
Color (APHA) (Max.): Clear
Liquid
Melting Point C: 33.5-38.5
Light color and low odor liquid emollient and auxiliary emulsifier, imparts a soft, velvety skin feel to cosmetic products.

STEPAN CO.: Esters(Continued):

Emulsifiers, Viscosity Builders:

PEG 200-6000 Mono and Dilaurates:
CTFA: PEG-4 to PEG-150 Laurate and Dilaurate
HLB Value: 5.9-19.3
Acid Value (Max.): 5-10
Color (APHA) (Max): Light Yellow to Cream
Liquids/Soft Solids/Waxes
Melting Point C: 5-61

PEG 200-6000 Mono and Dioleates:
CTFA: PEG-4 and PEG-150 Oleate and Dioleate
HLB Value: 5.0-19.1
Acid Value (Max.): 5-10
Color (APHA) (Max.): Light Amber to Cream
Liquids/Soft Solids/Waxes
Melting Point C: -15-59

PEG 200-6000 Mono and Distearates:
CTFA: PEG-4 to PEG-150 Stearate and Distearate
HLB Value: 4.8-19.1
Acid Value (Max.): 5-10
Color (APHA) (Max.): White to Cream
Solids/Waxes
Melting Point C: 28-61

Nonionic emulsifiers covering wide HLB range. Non-toxic and non-irritants. Viscosity modifiers, emollients, opacifiers, spreading agents, wetting and dispersing agents. Can be used in lotions, creams, make-up, bath oils, ointments, shampoos, conditioners, suppositories, sunscreen products.

STEPAN CHEMICAL CO.: DREWMULSE Mono and Diglycerides:

The DREWMULSE series are the most universally used of all lipophilic emulsifiers. A wide range is available in liquid, semi-plastic and solid form to provide the formulator with a variety of O/W emulsifiers. Most of these mono-diglycerides are sanctioned by 21 CFR 182.4505 and are available in a Kosher grade.

Product:

10K:
Glyceryl Mono-Shortening

55K:
Glyceryl Mono-Shortening

85K:
Glyceryl Monooleate
Vitamins & Minerals

200K:
Glyceryl Monostearate
Suppositories

900K:
Glyceryl Monostearate
Suppositories

8731-S:
Triglyceryl Mono-Shortening

D-4661:
Mixed Vegetable Oils

GMC-8:
Glyceryl Caprylate/Caprate
Topical Ointments & Creams
Vitamins & Minerals

HM 100:
Glyceryl Stearate & PEG 40 Stearate
Topical Ointments & Creams
Suppositories

STEPAN CHEMICAL CO.: KESSCO Alcohol Esters:

KESSCO alcohol esters are high purity emollients which are used in all types of cosmetic, toiletry and pharmaceutical formulations. In addition to their excellent emollient properties, these esters also function as lubricants, solvents for dyes and other cosmetic materials, and as plasticizers in aerosol hair sprays. Alcohol esters are used in cosmetic products to reduce the oily, greasy feel of mineral oil and petrolatum. Where applicable, KESSCO esters are available with purity that meets the National Formulary requirements.

Isopropyl Myristate, Isopropyl Palmitate and KESSCO 639 are soluble in alcohols, animal and vegetable oils and mineral oils. They have good solvent action on dyes, waxes, lanolin and many other cosmetic materials.

Isopropyl Myristate leaves a dry, soft, non-oily feel due to its ready absorption into the skin. Isopropyl Palmitate has similar properties though to a lesser extent. KESSCO 639, a blend of Isopropyl Myristate and Isopropyl Palmitate, has intermediate properties.

The Isopropyl esters are outstanding emollients, blending agents, solubilizers and good liquid vehicles for highly pigmented products. They are extensively used in pre-shave and after-shave products, liquid and cream makeups, lipstick, bath oils, creams, lotions and hair preparations.

Butyl Stearate Cosmetic and Isobutyl Stearate are relatively inexpensive emollient esters having high purity, low iodine value and low odor. They are often used in formulations containing mineral oil to impart good "slip" properties and as wetting agents for pigments.

Butyl Stearate Cosmetic and Isobutyl Stearate are used in lipsticks, bath oils, nail lacquers and removers, skin cleansers and in creams and lotions.

Isocetyl Stearate is a liquid with a low freezing point (0C.), a high degree of saturation and a relatively high molecular weight with a moderate viscosity (32 cps @ 25C.). It is a very rich emollient with a dry, velvety feel and is recommended for use in high quality applications and in makeup formulations.

Octyl Palmitate is guaranteed to be of 90% or greater purity. It imparts a dry, light silky skin feel and is an excellent substitute for Isopropyl Myristate when this effect is desired.

STEPAN CHEMICAL CO.: KESSCO Alcohol Esters(Continued):

Octyl Palmitate enhances the gloss in stick makeup and hair grooming products. It is used in suntan and bath oil formulas, as a binder for pressed powders, and as an emollient in all types of creams and lotions.

Octyl Isononanoate has the lowest freeze point of the KESSCO alcohol esters (below -30C.) and the driest, non-oily skin feel properties. It is rapidly becoming the preferred emollient in skin care and makeup products where normal "breathing" of the skin through the cosmetic is desired.

Octyl Isononanoate is recommended for use in hair sprays as a resin plasticizer, also aiding in the prevention of nozzle clog. Its dry emollient and skin breathing properties find application in lipsticks, creams, lotions and makeup formulations. Octyl Isononanoate can be used as a partial replacement for the expensive volatile silicone oils, particularly in anti-perspirants where it aids in reducing tackiness caused by the active ingredients.

Octyl Oxystearate exhibits the phenomenon of preventing the defatting effects on the skin of surfactants and/or detergents. Not only is the defatting controlled, but in some instances actual refatting of the skin occurs. Octyl Oxystearate is recommended for use in all cosmetic and cleansing formulations where high levels of surfactants and/or detergents are found.

Isopropyl Myristate:
Acid Value: Max.: 1.0
Iodine Value: Max.: 1.0
Color: APHA: Max.: 20
Specific Gravity @ 25/20C.: 0.849-0.855
Liquid
Flash Point F. COC: 305
Freezing Point C.: -3
Viscosity @ 25C.: cps: 4.8
Refractive Index @ 25C: 1.433
Mid-Boiling Point @ 4 mm: 160C.

Isopropyl Palmitate:
Acid Value: Max.: 1.0
Iodine Value: Max.: 1.0
Color APHA: Max.: 20
Specific Gravity @ 25/20C.: 0.849-0.855
Liquid
Flash Point F. COC: 325
Freezing Point C.: 13
Viscosity @ 25C. cps: 6.7
Refractive Index @ 25C.: 1.437
Mid-Boiling Point @ 4 mm: 170C.

STEPAN CHEMICAL CO.: KESSCO Alcohol Esters(Continued):

KESSCO 639:
Acid Value: Max.: 1.0
Iodine Value: Max.: 1.0
Color APHA: Max.: 30
Specific Gravity @ 25/20C.: 0.849-0.855
Liquid
Flash Point F. COC: 305
Freezing Point C.: 7
Viscosity @ 25C. cps: 5.9
Refractive Index @ 25C: 1.436
Mid-Boiling Point @ 4 mm: 170C.

Butyl Stearate Cosmetic:
Acid Value: Max.: 1.0
Iodine Value: Max.: 0.5
Color APHA: Max.: 40
Specific Gravity @ 25/20C: 0.850-0.860
Liquid
Flash Point F. COC: 370
Freezing Point C.: 19
Viscosity @ 25C. cps: 7.0
Refractive Index @ 25C: 1.442
Mid-Boiling Point @ 4 mm: 200C.

Isobutyl Stearate:
Acid Value: Max.: 1.0
Iodine Value: Max.: 1.0
Color APHA: Max.: 35
Specific Gravity @ 25/20C.: 0.849-0.855
Form @ 25C.: Liquid
Flash Point F. COC: 360
Freezing Point C.: 15
Viscosity @ 25C. cps: 8.5
Refractive Index @ 25C: 1.441
Mid-Boiling Point @ 4 mm: 200C.

Isocetyl Stearate:
Acid Value: Max.: 3.0
Iodine Value: Max.: 5.0
Color APHA: Max.: 200
Specific Gravity @ 25/20C: 0.853-0.859
Form @ 25C.: Liquid
Flash Point F. COC: 450
Freezing Point C.: 0
Viscosity @ 25C. cps: 32.0
Refractive Index @ 25C.: 1.452

STEPAN CHEMICAL CO.: KESSCO Alcohol Esters(Continued):

Octyl Palmitate:
Acid Value: Max.: 3.0
Iodine Value: Max.: 1.0
Color APHA: Max.: 35
Specific Gravity @ 25/20C.: 0.854-0.858
Form @ 25C.: Liquid
Flash Point F. COC: 395
Freezing Point C.: 0
Viscosity @ 25C. cps: 11.3
Refractive Index @ 25C.: 1.4453

Octyl Isononanoate:
Acid Value: Max.: 1.0
Iodine Value: Max.: 1.0
Color APHA: Max.: 20
Specific Gravity @ 25/20C: 0.853-0.859
Form @ 25C.: Liquid
Flash Point F. COC: 260
Freezing Point C: -34
Viscosity @ 25C. cps: 4.3
Refractive Index @ 25C.: 1.434

Octyl Oxystearate:
Acid Value: Max.: 1.0
Iodine Value: Max.: 5.0
Color APHA: Max.: Yellow
Specific Gravity @ 25/20C.: 0.889-0.895
Form @ 25C.: Liquid
Flash Point F. COC: 425
Freezing Point C.: 12
Viscosity @ 25C. cps: 84.2
Refractive Index @ 25C.: 1.4565

STEPAN CHEMICAL CO.: KESSCO Glycerol Esters:

KESSCO Glycerol Esters are widely used in a variety of cosmetic formulations. They are excellent primary and auxiliary emulsifiers and stabilizers, have a pronounced bodying effect in cosmetic systems, and are excellent opacifiers and emollients.

Many of the glycerol esters are used as bases for creams and lotions. They enhance both the appearance and feel of formulations.

Glycerol Monostearate Pure is made from high purity triple pressed stearic acid and contains no soaps. Because it simultaneously acts as an emulsifier, opacifier and bodying agent, this grade of glycerol monostearate has traditionally been the work horse of the industry.

Major applications for Glycerol Monostearate Pure include creams, lotions, antiperspirants, hair care products and sunscreens. Pharmaceutical uses include topical creams, ointments and lotions.

Glycerol Monostearate 860 is a food grade product with a slightly higher melting point but with similar fuunctionality to that of Glycerol Monostearate Pure. It is used in chewing gums as an emulsifier and to reduce tackiness.

Glycerol Distearate is offered as an alternative to Glycerol Monostearate Pure, having an extremely low HLB value.

Stepan offers three self-emulsifying Glycerol Monostearates: Glycerol Monostearate SE, Glycerol Monostearate 24 SE and Glycerol Monostearate SE, Acid Stable. The SE grades allow the formulator to utilize glycerol esters as primary emulsifiers for oil-in-water systems.

Glycerol Monostearate SE and 24 SE are anionic modified and are recommended for use in oil-in-water emulsions that are in the pH range of 5 to 9.

Glycerol Monostearate SE, Acid Stable is nonionic and recommended for low pH (3 to 5) emulsions where an anionic emulsifier is unstable. Glycerol Monostearate SE, Acid Stable finds wide application as an emulsifier, self emulsifying cream base, hair and skin conditioner and in roll-on antiperspirants. In systems containing electrolytes, such as cream hair rinses and antiperspirants, it is relatively insensitive to salt content.

STEPAN CHEMICAL CO.: KESSCO Glycerol Esters(Continued):

Glycerol Monoloeate and Glycerol Dioleate are effective water-in-oil emulsifiers. They are used in bath oils as lubrivants and spreading agents, in makeup as pigment dispersants and in vanishing and moisturizing creams to impart slip. It is recommended that an antioxidant be used in conjunction with these products in cosmetic formulations.

Glycerol Monolaurate (KESSCO 675) is made from commercially pure lauric acid and has a melting point above body temperature, making it suitable for use in suppositories, vaginal creams, etc. Self emulsifying properties of KESSCO 675 make it convenient for use a primary emulsifier for water-in-oil emulsions. In some formulations KESSCO 675 imparts an emolliency that lasts through several washings.

Glycerol Dilaurate is a semi-solid recommended for use in free flowing lotions where the glycerol laurate emolliency is desired.

Specifications:

Glycerol Monostearate Pure:
Acid Value: Max.: 3.0
Iodine Value: Max.: 0.5
White
HLB Value: 3.8
Flakes
Flash Point F. COC: 410
Melting Point, C: 56.5-58.5

Glycerol Monostearate 860 Food Grade:
Acid Value: Max.: 3.0
Iodine Value: Max.: 2.0
White
HLB Value: 3.8
Flakes
Flash Point F. COC: 450
Melting Point, C: 58.5-61.5

Glycerol Distearate:
Acid Value: Max.: 5.0
Iodine Value: Max.: 1.0
Color: White
HLB Value: 2.4
Waxy Flake
Flash Point F. COC: 470
Melting Point, C: 55-60

STEPAN CHEMICAL CO.: KESSCO Glycerol Esters(Continued):

Glycerol Monostearate S.E.:
Acid Value: Max.: 20.0
Iodine Value: Max.: 0.5
White to Cream
Flakes
Flash Point F. COC: 400
Melting Point, C: 56.5-59.5

Glycerol Monostearate 24 S.E.:
Acid Value: Max.: 20.0
Iodine Value: Max.: 3.0
White to Cream
Flakes
Flash Point F. COC: 372
Melting Point, C: 56-60

Glycerol Monostearate S.E., Acid Stable:
Acid Value: Max.: 3.0
Iodine Value: Max.: 1.0
White to Cream
HLB Value: 11.2
Flakes
Flash Point F. COC: 460
Melting Point, C: 54-58

Glycerol Monooleate:
Acid Value: Max.: 5.0
Iodine Value: Max.: 77.0
Specific Gravity: 25/20C.: 0.945-0.953
Yellow
HLB Value: 3.8
Liquid
Flash Point F. COC: 435
Melting Point, C: <20
Viscosity @ 25C., cps: 204

Glycerol Dioleate:
Acid Value: Max.: 5.0
Iodine Value: Max.: 82.0
Specific Gravity: 25/20C.: 0.923-0.929
Yellow-Amber
HLB Value: 2.9
Liquid
Flash Point F. COC: 520
Melting Point, C: 0
Viscosity @ 25C., cps: 90

STEPAN CHEMICAL CO.: KESSCO Glycerol Esters(Continued):

Glycerol **Monolaurate** (KESSCO 675):
Acid Value: Max.: 5.0
Iodine Value: Max.: 1.0
Color: White
HLB Value: 4.9
Solid
Flash Point F. COC: 425
Melting Point, C: 53.9

Glycerol Dilaurate:
Acid Value: Max.: 5.0
Iodine Value: Max.: 2.0
White
HLB Value: 4.0
Solid
Flash Point F. COC: 480
Melting Point, C: 30.0

STEPAN CHEMICAL CO.: KESSCO Glycol Esters:

KESSCO Glycol Esters are used in cosmetic and pharmaceutical formulations as opacifiers, emollients and emulsifiers. Some glycol esters produce brilliant pearlescence in Liquid Hand Soaps and Shampoos and when used in creams and lotions they impart a soft velvety skin feel.

Ethylene Glycol Monostearate Pure (CTFA name: Glycol Stearate) is the most commonly used pearlescent agent in shampoos amd liquid hand soaps. It also functions as a bodying agent and emulsion stabilizer in these systems.

Ethylene Glycol Monostearate 70 is recommended for use in low solids content formulations because it tends to increase viscosity. Conversely, Ethylene Glycol Distearate is recommended when no additional viscosity is desired such as in formulations containing relatively high solids concentrations (18% and higher). KESSCO EGAS (CTFA name: Glycol Amido Stearate) functions as a pearling and bodying agent and imparts a soft, smooth skin feel to formulations due to the presence of a small amount of amide.

Diethylene Glycol Monostearate and Diethylene Glycol Distearate are used as opacifiers in shampoos and lotions. They impart a luxurious emolliency and add body to these types of free flowing systems.

Diglycol Stearate SE and Diglycol Stearate Neutral are excellent emulsifiers and opacifiers. Diglycol Stearate SE functions as a primary emulsifier while Diglycol Stearate Neutral is used as an auxiliary emulsifier. Both produce rich creams and lotions.

Propylene Glycol Monostearate Pure and Propylene Glycol Distearate have melting points near that of body temperatures and so are used in suppositories, lipsticks and sunscreens. They also function as auxiliary emulsifiers and opacifiers.

Propylene Glycol Monostearate 8615 is anionic modified making it self-emulsifying. It has a higher melting point than the other Propylene Glycol Stearates and acts as a primary emulsifier. Propylene Glycol Monostearate 534 is a food grade product with a high monoester content (approximately 70%). It tends to aid in building viscosity in suppositories, creams and lotions.

Propylene Glycol Monolaurate E is a low color and odor liquid emollient and auxiliary emulsifier. It imparts a soft velvety skin feel to cosmetic products.

STEPAN CHEMICAL CO.: KESSCO Glycol Esters(Continued):

Ethylene Glycol Monostearate Pure:
Acid Value Max.: 2.0
Iodine No. Max.: 0.5
Flakes
Flash Pt. F. COC: 390
Melting Point C: 56-60
HLB Value: 2.9

Ethylene Glycol Monostearate 70:
Acid Value Max.: 2.0
Iodine No. Max.: 0.5
Flakes
Flash Pt. F. COC: 370
Melting Point C: 52-56
HLB Value: 2.9

Ethylene Glycol Distearate:
Acid Value Max.: 15.0
Iodine No. Max.: 0.5
Flakes
Flash Pt. F. COC: 390
Melting Point C: 60-63
HLB Value: 1.5

Ethylene Glycol Amido Stearate:
Acid Value Max.: 5.0
Iodine No. Max.: 0.5
Flakes
Flash Pt. F. COC: 360
Melting Point C: 56.5-58.5

Diethylene Glycol Monostearate:
Acid Value Max.: 5.0
Iodine No. Max.: 0.5
Flakes
Flash Pt. F. COC: 395
Melting Point C.: 44.5-47.5
HLB Value: 4.3

Diethylene Glycol Distearate:
Acid Value Max.: 10.0
Iodine No. Max.: 0.5
Flakes
Flash Pt. F. COC: 360
Melting Point C.: 42-48
HLB Value: 2.8

STEPAN CHEMICAL CO.: KESSCO Glycol Esters(Continued):

Diglycol Stearate S.E.:
Acid Value Max.: 103.0
Iodine No. Max.: 7.0
Flakes
Flash Pt. F. COC: 345
Melting Point C.: 48-53

Diglycol Stearate Neutral:
Acid Value Max.: 103.0
Iodine No. Max.: 7.0
Flakes
Flash Pt. F. COC: 365
Melting Point C.: 42-48
HLB Value: 2.9

Propylene Glycol Monostearate Pure:
Acid Value Max.: 3.0
Iodine No. Max.: 5.0
Flakes
Flash Pt. F. COC: 390
Melting Point C.: 33.5-38.5
HLB Value: 3.4

Propylene Glycol Distearate:
Acid Value Max.: 10.0
Iodine No. Max.: 1.0
Flakes
Flash Pt. F. COC: 430
Melting Point C.: 36-38
HLB Value: 2.2

Propylene Glycol Monostearate 8615:
Acid Value Max.: 20.0
Iodine No. Max.: 3.0
Flakes
Flash Pt. F. COC: 379
Melting Point C.: 57-62

Propylene Glycol Monolaurate E:
Acid Value Max.: 3.5
Iodine No. Max.: 1.0
Oily Liquid
Flash Pt. F. COC: 370
Melting Point C.: 10
HLB Value: 3.2

Propylene Glycol Monostearate 534:
Acid Value Max.: 3.0
Iodine No. 2.0
Flakes
Flash Pt. F. COC: 390
Melting Point C.: 34.5-39.5
HLB Value: 2.9

STEPAN CHEMICAL CO.: KESSCO Polyethylene Glycol Esters:

The Polyethylene Glycol (PEG) series of esters offers graduated hydrophilic and lipophilic surface active properties making them useful as primary and auxiliary nonionic surfactants with stability over a wide range of formulating conditions. The PEG esters in the range of 200 to 1540 in particular are the most versatile in regard to emulsification properties. The PEG series includes mono and diesters of lauric, oleic and stearic acids.

In addition to being effective emulsifiers the PEG esters are highly emollient and are excellent solubilizers in bath oils and fragrance compositions. They provide washability in anhydrous formulations such as hair preparations and ointments. The higher molecular weight PEG distearates are outstanding thickening agents in aqueous systems.

The lower molecular weight PEG esters are examples of oil soluble surface active agents which, aside from their value as emulsifiers, are also of use where surface phenomena in non-aqueous systems must be considered. The use of surface active agents in non-aqueous systems has attained importance in those applications involving fundamental molecular forces and where those phenomena exert influence on stability, viscosity, wetting, absorption, foaming and other physical properties. The mono and diesters of PEG 200 and 300, and the diesters of PEG 400 are the most important for these properties.

An extremely important characteristic of PEG esters is that they are non-toxic and are neither eye nor skin irritants. The use of emulsifiers with such low irritation indices serves to lower the overall irritation level of a formulation.

The PEG esters range in HLB value from 5.0 (PEG 200 Distearate) to 19.2 (PEG 6000 Monolaurate). This wide range permits their use as primary emulsifiers for creams and lotions in both water-in-oil and oil-in-water systems and as solubilizers in clear microemulsions.

STEPAN CHEMICAL CO.: KESSCO Polyethylene Glycol Esters (Continued):

The PEG esters are extremely versatile and convenient to use because of the variety of functions they perform in a given formulation. The following is a listing of typical applications for some of the Peg esters:

End Use: Makeup
Function: Pigment wetting and dispersing
PEG Ester: PEG 200 to 400 Mono and Dioleates

End Use: Bath oils, Bath lotions & After Bath products
Function: Emulsifiers and lubricants
PEG Ester: PEG 400 and 600 Mono and Dioleates

End Use: Ointments
Function: Viscosity builders
PEG Ester: PEG 1540 and 4000 Monostearate

End Use: Lotions, Shampoos & Cream rinses
Function: Viscosity builders
PEG Ester: PEG 1540-6000 Distearate

End Use: Creams and Lotions
Function: Primary and auxiliary emulsifiers
PEG Ester: PEG 200-1540 esters
Function: Emollients
PEG Ester: PEG 200-600 Mono and Dioleates and Laurates

End Use: Hair Care Products
Function: Opacifiers
PEG Ester: PEG 200 and 400 Mono and Distearate
Function: Conditioners
PEG Ester: PEG Stearates

End Use: Clear Bath Oils
Function: Spreading Agents
PEG Ester: PEG 200 and 400 Dilaurate

End Use: Suppositories
Function: Melting Point control
PEG Ester: PEG 400 and 600 Dilaurate

STEPAN CHEMICAL CO.: KESSCO Polyethylene Glycol Esters (Continued):

Product Name:

PEG 200 Monolaurate:
HLB: 9.3
Liquid
MP/FP C.: <5
Acid Value: 5
Saponification Value: 132-142
Lt. Yellow
Specific Gravity: 0.9833

PEG 200 Dilaurate:
HLB: 5.9
Liquid
MP/FP C.: <9
Acid Value: 10
Saponification Value: 176-186
Lt. Yellow
Specific Gravity: 0.9520

PEG 300 Monolaurate:
HLB: 11.4
Liquid
MP/FP C.: <8
Acid Value: 5
Saponification Value: 104-114
Lt. Yellow
Specific Gravity: 1.0100

PEG 300 Dilaurate:
HLB: 7.9
Liquid
MP/FP C.: <13
Acid Value: 10
Saponification Value: 148-158
Lt. Yellow
Specific Gravity: 0.9703

PEG 400 Monolaurate:
HLB: 13.0
Liquid
MP/FP C.: 12
Acid Value: 5
Sapionification Value: 86-96
Lt. Yellow
Specific Gravity: 1.0242

STEPAN CHEMICAL CO.: KESSCO Polyethylene Glycol Esters (Continued):

Product Name:

PEG 400 Dilaurate:
HLB: 9.7
Liquid
MP/FP C.: 18
Acid Value: 10
Saponification Value: 127-137
Lt. Yellow
Specific Gravity: 0.9884

PEG 600 Monolaurate:
HLB: 14.6
Liquid
MP/FP C.: 23
Acid Value: 5
Saponification Value: 64-74
Lt. Yellow
Specific Gravity: 1.0505

PEG 600 Dilaurate:
HLB: 11.7
Soft Solid
MP/FP C.: 24
Acid Value: 10
Saponification Value: 102-112
Cream
Specific Gravity: 0.9820

PEG 1000 Monolaurate:
HLB: 16.6
Soft Solid
MP/FP C.: 40
Acid Value: 5
Saponification Value: 41-51
Cream
Specific Gravity: 1.035

PEG 1000 Dilaurate:
HLB: 14.2
Soft Solid
MP/FP C.: 38
Acid Value: 10
Saponification Value: 68-78
Cream
Specific Gravity: 1.015

STEPAN CHEMICAL CO.: KESSCO Polyethylene Glycol Esters (Continued):

Product Name:

PEG 1540 Monolaurate:
HLB: 17.5
Wax
MP/FP C.: 46
Acid Value: 5
Saponification Value: 26-36
Cream
Specific Gravity: 1.060

PEG 1540 Dilaurate:
HLB: 15.8
Wax
MP/FP C.: 42
Acid Value: 10
Saponification Value: 48-56
Cream
Specific Gravity: 1.040

PEG 4000 Monolaurate:
HLB: 19.0
Wax
MP/FP C.: 55
Acid Value: 5
Saponification Value: 9-18
Cream
Specific Gravity: 1.075

PEG 4000 Dilaurate:
HLB: 18.1
Wax
MP/FP C.: 52
Acid Value: 5
Saponification Value: 20-30
Cream
Specific Gravity: 1.065

PEG 6000 Monolaurate:
HLB: 19.3
Wax
MP/FP C.: 61
Acid Value: 5
Saponification Value: 7-13
Color: Cream

STEPAN CHEMICAL CO.: KESSCO Polyethylene Glycol Esters (Continued):

Product Name:

PEG 6000 Dilaurate:
HLB: 18.7
Wax
MP/FP C.: 57
Acid Value: 9
Saponification Value: 12-20
Cream
Specific Gravity: 1.077

PEG 200 Monostearate:
HLB: 8.1
Solid
MP/FP C.: 31
Acid Value: 5.0
Saponification Value: 120-129
Wt. to Cream
Specific Gravity: 0.9360

PEG 200 Distearate:
HLB: 4.8
Solid
MP/FP C.: 34
Acid Value: 10.0
Saponification Value: 153-162
Wt. to Cream
Specific Gravity: 0.9060

PEG 300 Monostearate:
HLB: 10.3
Solid
MP/FP C.: 28
Acid Value: 5.0
Saponification Value: 97-105
Wt. to Cream
Specific Gravity: 0.9660

PEG 300 Distearate:
HLB: 6.9
Solid
MP/FP C.: 32
Acid Value: 10.0
Saponification Value: 130-139
Wt. to Cream

STEPAN CHEMICAL CO.: KESSCO Polyethylene Glycol Esters (Continued):

<u>Product Name:</u>

PEG 400 Monostearate:
HLB: 11.7
Solid
MP/FP C.: 32
Acid Value: 5.0
Saponification Value: 83-92
Wt. to Cream
Specific Gravity: 0.9780

PEG 400 Distearate:
HLB: 8.5
Solid
MP/FP C.: 36
Acid Value: 10
Saponification Value: 115-124
Wt. to Cream
Specific Gravity: 0.9390

PEG 600 Monostearate:
HLB: 13.5
Solid
MP/FP C.: 37
Acid Value: 5.0
Saponification Value: 61-70
Wt. to Cream
Specific Gravity: 1.0000

PEG 600 Distearate:
HLB: 10.7
Solid
MP/FP C.: 39
Acid Value: 10.0
Saponification Value: 93-102
Wt. to Cream
Specific Gravity: 0.9670

PEG 1000 Monostearate:
HLB: 15.7
Wax
MP/FP C.: 41
Acid Value: 5.0
Saponification Value: 40-48
Cream
Specific Gravity: 1.030

STEPAN CHEMICAL CO.: KESSCO Polyethylene Glycol Esters (Continued):

Product Name:

PEG 1000 Distearate:
HLB: 13.3
Wax
MP/FP C.: 40
Acid Value: 10.0
Saponification Value: 65-74
Cream
Specific Gravity: 1.005

PEG 1540 Monostearate:
HLB: 16.9
Wax
MP/FP C.: 47
Acid Value: 5.0
Saponification Value: 27-36
Cream
Specific Gravity: 1.050

PEG 1540 Distearate:
HLB: 14.6
Wax
MP/FP C.: 45
Acid Value: 10.0
Saponification Value: 49-58
Cream
Specific Gravity: 1.015

PEG 4000 Monostearate:
HLB: 18.7
Wax
MP/FP C.: 56
Acid Value: 5.0
Saponification Value: 10-18
Cream
Specific Gravity: 1.075

PEG 4000 Distearate:
HLB: 17.6
Wax
MP/FP C.: 51
Acid Value: 5.0
Saponification Value: 19-27
Cream
Specific Gravity: 1.060

STEPAN CHEMICAL CO.: KESSCO Polyethylene Glycol Esters (Continued):

Product Name:

PEG 6000 Monostearate:
HLB: 19.1
Wax
MP/FP C.: 61
Acid Value: 5.0
Saponification Value: 7-13
Cream
Specific Gravity: 1.080

PEG 6000 Distearate:
HLB: 18.4
Wax
MP/FP C.: 55
Acid Value: 9.0
Saponification Value: 14-20
Cream
Specific Gravity: 1.075

PEG 200 Monooleate:
HLB: 8.2
Liquid
MP/FP C.: <-15
Acid Value: 5
Saponification Value: 115-124
Lt. Amber
Specific Gravity: 0.9742

PEG 200 Dioleate:
HLB: 5.0
Liquid
MP/FP C.: <-15
Acid Value: 10
Saponification Value: 148-153
Lt. Amber
Specific Gravity: 0.9405

PEG 300 Monooleeate:
HLB: 10.2
Liquid
MP/FP C.: <-5
Acid Value: 5
Saponification Value: 94-102
Lt. Amber
Specific Gravity: 0.998

STEPAN CHEMICAL CO.: KESSCO Polyethylene Glycol Esters (Continued):

Product:

PEG 300 Dioleate:
HLB: 6.9
Liquid
MP/FP C.: <-5
Acid Value: 10
Saponification Value: 128-137
Lt. Amber
Specific Gravity: 0.9609

PEG 400 Monooleate:
HLB: 11.6
Liquid
MP/FP C.: <10
Acid Value: 5
Saponification Value: 80-89
Lt. Amber
Specific Gravity: 1.0135

PEG 400 Dioleate:
HLB: 8.3
Liquid
MP/FP C.: <7
Acid Value: 10
Saponification Value: 113-122
Lt. Amber
Specific Gravity: 0.977

PEG 600 Monooleate:
HLB: 13.6
Liquid
MP/FP C.: 23
Acid Value: 5
Saponification Value: 60-69
Lt. Amber
Specific Gravity: 1.0381

PEG 600 Dioleate:
HLB: 10.6
Liquid
MP/FP C.: 19
Acid Value: 10
Saponification Value: 92-102
Lt. Amber
Specific Gravity: 1.0038

STEPAN CHEMICAL CO.: KESSCO Polyethylene Glycol Esters (Continued):

Product Name:

PEG 1000 Monooleate:
HLB: 15.9
Soft Solid
MP/FP C.: 39
Acid Value: 5
Saponification Value: 40-49
Cream
Specific Gravity: 1.035

PEG 1000 Dioleate:
HLB: 13.2
Soft Solid
MP/FP C.: 37
Acid Value: 10
Saponification Value: 64-74
Cream
Specific Gravity: 1.005

PEG 1540 Monooleeate:
HLB: 17.0
Wax
MP/FP C.: 45
Acid Value: 5
Saponification Value: 28-37
Cream
Specific Gravity: 1.050

PEG 1540 Dioleate:
HLB: 14.9
Wax
MP/FP C.: 44
Acid Value: 10
Saponification Value: 45-55
Cream
Specific Gravity: 1.025

PEG 4000 Monooleate:
HLB: 18.7
Wax
MP/FP C.: 55
Acid Value: 5
Saponification Value: 10-18
Cream
Specific Gravity: 1.075

STEPAN CHEMICAL CO.: KESSCO Polyethylene Glycol Esters (Continued):

Product Name:

PEG 4000 Dioleate:
HLB: 17.7
Wax
MP/FP C.: 49
Acid Value: 5
Saponification Value: 19-27
Cream
Specific Gravity: 1.060

PEG 6000 Monooleate:
HLB: 19.1
Wax
MP/FP C.: 59
Acid Value: 5
Saponification Value: 7-13
Cream
Specific Gravity: 1.085

PEG 6000 Dioleate:
HLB: 18.4
Wax
MP/FP C.: 56
Acid Value: 9
Saponification Value: 13-21
Cream
Specific Gravity: 1.070

STEPAN CHEMICAL CO.: KESSCO Specialty Products:

Synthetic Waxes:

Synthetic Spermaceti N.F., KESSCO 653 and KESSCO 654 are waxy materials that impart good skin feel properties to cosmetic products. They are used as a base material or stiffening agent in stick cosmetic products, and as emollient thickeners in creams and lotions.

Synthetic Spermaceti N.F. meets the National Formulary requirements for synthetic spermaceti, and has the same melting point as the natural product. KESSCO 653 contains 90% cetyl palmitate as does natural spermaceti, has a higher melting point and is the most widely used of the cetyl ester waxes.

KESSCO 654 (cetyl myristate) has the lowest melting point and the softest, most pleasing feel on the skin.

KESSCOWAX B:

KESSCOWAX B was developed specifically as a replacement for natural beeswax used in beeswax-borax cosmetic cold cream formulations. When formulated into a borax cold cream system, KESSCOWAX B yields a pure white product with a consistency and feel almost indistinguishable from the same product formulated with natural beeswax. KESSCOWAX B reacts with borax like natural beeswax in cold cream formulations. KESSCOWAX B is a light colored, low odor product, which is supplied in flake form.

KESSCOLIN Water Absorption Base:

KESSCOLIN Water Absorption Base offers the advantages of lanolin and many of its derivatives while avoiding many of the disadvantages of the naturally derived product. It exceeds the capacity of lanolin to absorb water and forms more stable water-in-oil emulsions than lanolin or lanolin-containing water absorption bases. Its emollient and moisturizing properties make it an ideal base for conditioning and moisturizing creams and lotions as well as for nonallergenic ointment bases. KESSCOLIN may be substituted for lanolin and formulations; it may also be substituted for USP hydrophilic petrolatum. KESSCOLIN is odor free and has better color stability than lanolin, allowing formulation of uniform products batch after batch.

Myristyl Myristate:

Myristyl Myristate is a waxy solid emollient, melting at body temperature. When used at low percentages in o/w emulsions, it improves viscosity and gives "richness" to the emulsion. It is used in creams, lotions and makeup items.

STEPAN CHEMICAL CO.: KESSCO Specialty Products(Continued):

KESSCO Glycerol Acetate Esters:
Acetin, Diacetin and Triacetin are outstanding polar solvents for essential oils and are carriers for flavors and fragrances. They are colorless, odorless and have very low freeze points. Triacetin is an excellent plasticizer and is used in tablet coating applications.

KESSCO Cetyl Alcohol NF:
KESSCO Cetyl Alcohol NF conforms to the specifications of the National Formulary. It is used in cosmetics to provide body to creams, lotions and sticks. Cetyl Alcohol imparts a very soft skin feel to cosmetic products.

Synthetic Spermaceti, N.F.:
Melting Range, C.: 43-47
Acid Value Max.: 2.0
Saponification No.: 109-117
White
Flakes

KESSCO 653:
Melting Range C.: 51-55
Acid Value Max.: 2.0
Saponification No.: 109-117
White
Flakes

KESSCO 654:
Melting Range C.: 47-53
Acid Value Max.: 2.0
Saponification No.: 116-124
White to Cream
Flakes

KESSCOWAX B:
Melting Range C.: 60-65
Acid Value Max.: 12-30
Saponification No.: 190-200
White to Cream
Flakes

STEPAN CHEMICAL CO.: KESSCO Specialty Products(Continued):

KESSCOLIN Water Absorption Base:
Melting Range C.: 44-50
Acid Value Max.: 1 Max.
Saponification No.: 11-16
Yellow
Solid
Color APHA: 125 Hazy
Moisture %: 0.23
Melting Point: 47C.
Flash Point: 425F.
Viscosity: 50C.: 55.3 cps
100C.: 9.05 cps

Myristyl Myristate:
Melting Range C.: 36-39
Acid Value Max.: 3 Max.
Saponification No.: 120-130
White
Flakes

Acetin:
Melting Range C.: -18
Acid Value Max.: 0.5% as Acetic
Saponification No.: 475-525
Color: 75 APHA Max.
Form: Liquid

Diacetin:
Melting Range C.: -35
Acid Value Max.: 0.5% as Acetic
Saponification No. 580-630
Color: 50 APHA Max.
Liquid

Triacetin:
Melting Range C.: -50
Acid Value Max.: 0.005% as Acetic
Saponification No.: 765-805
Color: 50 APHA Max.
Liquid

Cetyl Alcohol:
Melting Range C.: 48-53
Acid Value Max.: 0.5 Max.
Saponification No.: 1.0 Max.
Color: 40 APHA Max.
Flakes

STEPAN CHEMICAL CO.: NEOBEE Triglycerides and Propylene Glycol Derivatives:

The NEOBEES are a series of vegetable-derived products that are stable to oxidation and exhibit bland taste and odor. All of the NEOBEE line is considered GRAS under FDA 21 CFR 170.30 and is available in a Kosher grade.

Product:

M-5:
Caprylic/Capric Triglycerides

M-20:
Propylene Glycol Dicaprylate/Dicaprate

O:
Coconut Oil Derived Triglycerides

18:
High Oleic Safflower Oil

62:
Soya Stearine

1053:
Coconut Oil Derived Triglycerides

1054:
Coconut Oil Derived Diesters of Propylene Glycol

1062:
Coconut Oil Derived Triglycerides

STEPAN CHEMICAL CO.: Coconut Oils:

These products are highly refined and deodorized natural and hydrogenated coconut oils with superior stabilities. They are highly resistant to oxidative rancidity and are bland in taste.

76:
Refined, Bleached, Deodorized
Emollient for Creams and Lotions

92:
Refined, Bleached, Deodorized
Emollient for Creams and Lotions

110:
Refined, Bleached, Deodorized
Emollient for Creams and Lotions

STEPAN CHEMICAL CO.: WECOBEE:

The WECOBEE series are processed to achieve specific melting points with accompanying bland odor and taste qualities. They are often used as a cocoa butter substitute and are stable to oxidation. The WECOBEES are all GRAS under FDA 21 CFR 170.30 and are available in a Kosher grade.

Product:

FS:
Suppositories
Ointment Bases

FW:
Suppositories
Ointment Bases
Cream Bases

M:
Suppositories
Ointment Bases
Cream Bases

S:
Suppositories
Ointment Bases

SS:
Suppositories
Ointment Bases

W:
Suppositories
Cream Bases

The WECOBEES are triglycerides derived from lauric based fats such as coconut and palm kernel oils

SUTTON LABORATORIES, INC.: GERMABEN II Preservative System:

GERMABEN II is a convenient, economical, liquid preservative system for use in cosmetics. It is readily soluble at levels of 1% in both aqueous solution and oil-water emulsions, and provides adequate preservation against microorganisms without the use of any additional preservative.

Composition:

GERMALL II	30%
Methylparaben	11%
Propylparaben	3%
Propylene Glycol	56%

Use Levels:

Incorporation of GERMABEN II at a level of 1% of the finished formulation results in a product preservation system of 0.3% GERMALL II + 0.11% methylparaben + 0.03% propylparaben. This preservative combination has been found to be exceptionally effective in a wide range of cosmetic products. Levels of 0.5-1.0% have proved to be effective in emulsions and levels of 0.25%-0.5% have been effective in most shampoos tested.

Formulation Method:

The GERMABEN II preservation system is conveniently incorporated into hot or cold mix shampoos and into creams or lotions at any stage of their formulation. The preferred method of incorporation is slow addition to the finished formulation just prior to addition of the perfume.

Ingredient Labeling:

Propylene glycol (and)
Diazolidinyl Urea (and)
Methylparaben (and)
Propylparaben

Specifications for GERMABEN II:

Appearance: Clear viscous liquid
Color: Pale to light yellow (APHA 100 Maximum)
Odor: Characteristically mild
% Nitrogen (Kjeldahl): 5.8-6.4
Specific Gravity: 1.1731-1.1839
Assay for Total Solids (by Specific Gravity): 42.5-45.5%
Residue on Ignition (Sulfated): 0.50% Maximum

GERMABEN II:

Combines GERMALL II with methylparaben, propylparaben and propylene glycol in effective water-soluble proportions. Its liquid form permits cold formulation of shampoos, liquid soaps and other cosmetic products. GERMABEN II has been used successfully to preserve troublesome formulations containing proteins, soluble collagens, aloe plant and herb extracts.

SUTTON LABORATORIES, INC.: GERMABEN II-E Preservative:

GERMALL II is the latest and most active member of the GERMALL family of preservatives. The combination of GERMALL II with methylparaben, propylparaben, and propylene glycol to give GERMABEN II, a complete liquid preservative system, has been enthusiastically received by cosmetic chemists. Because some chemists have requested a GERMABEN II composition containing a greater proportion of parabens, while still retaining the GERMABEN advantages of cost-effectiveness, ease of handling, and versatility, Sutton is now offering GERMABAN II-E, a new liquid preservative system, for the cosmetic industry. GERMABEN II-E is especially useful for creams and lotions that present special preservation problems. It will find application also in many other types of cosmetic formulations.

Composition:

GERMALL II	20%
Methylparaben	10%
Propylparaben	10%
Propylene Glycol	60%

Use Levels:

Incorporation of GERMABEN II-E at a level of 1% of the finished formulation results in a product preservation system of 0.20% GERMALL II + 0.10% methylparaben + 0.10% propylparaben. In certain formulations, less than 1% GERMABEN II-E can sometimes be used, while maintaining the same 2:1:1 ratio. For example, some emulsions are well-preserved using only 0.5% GERMABEN II-E, which provides a preservative system of 0.10% GERMALL II + 0.05% methylparaben + 0.05% propylparaben.

Formulation Method:

Although GERMABEN II-E may be incorporated conveniently at almost any stage of emulsion formulation, the preferred method of incorporation is slow addition to the formulation after emulsification and just prior to the addition of the fragrance. In fact, many perfumes are effectively incorporated into cosmetics by pre-dissolving them in the GERMABEN II-E, and then adding the mixture slowly to the finished formulation.

Specifications:

Appearance: Clear liquid
Color: Pale to light yellow (APHA 100 Maximum)
Odor: Characteristically mild
% Nitrogen (Kjeldahl): 3.8-4.4
Specific Gravity: 1.1353-1.1438
Assay for Total Solids (by Specific Gravity): 38.5-41.5%
Residue on Ignition (Sulfated): 0.50% Maximum

SUTTON LABORATORIES, INC.: GERMALL II Preservative:

GERMALL II is the latest and most active member of the family of Imidazolidinyl Urea preservatives. It is a superior Imidazolidinyl Urea because it has a wider spectrum of activity, including activity against a troublesome "house" microorganisms. Not only is GERMALL II more active against gram-negative bacteria such as Pseudomonas, but also it has increased activity against yeast and mold. It is therefore an excellent preservative for shampoos, either alone or in combination with parabens. Creams and lotions preserved with a GERMALL II-paraben combination system will retain activity against yeast and mold even when paraben activity has been diminished by interaction with nonionics or proteins, or has migrated into the oil phase. The GERMALL II-paraben combination is the preservative system of choice for creams and lotions.

Molecular formula:
$C_8H_{14}N_4O_7$
Chemical Name:
N-(Hydroxymethyl)-N-(1,3-dihydroxymethyl-2,5-dioxo-4-imidazolidinyl)-N'-(hydroxymethyl) Urea
CTFA adopted name:
Diazolidinyl Urea
Formulating Method:
GERMALL II should be used in cosmetics and toiletries at the same concentrations as those recommended for GERMALL 115: 0.1-0.5% alone or in combination with the parabens. GERMALL II may be added to the final product either as a free-flowing powder or as a water concentrate at temperatures below 60C.

Effective Preservation:
GERMALL II is the newest and most active member of the imidazolidinyl urea family of preservatives.

Safe and Easy to Use For Shampoos, Creams, and Lotions:
GERMALL II is water-soluble, stable, and colorless. It is effective over a wide pH range and does not require special handling.

Eliminates Preservation Problems and is Cost Effective:
GERMALL II is more cost effective than most preservatives being offered for shampoos or other cosmetic products. It solves many problems of cosmetic preservation, such as: differences between test microorganisms and mutated types; inactivation effects of emulsifiers, proteins, and other cosmetic ingredients on antimicrobials; unexpected appearance in the production plant or water supply of mutant gram-negative bacteria; inadvertant use of a contaminated raw material; occasional lapse in GMP by production personnel; or gradual partition of a preservative from the water phase.

Broad Spectrum of Activity:
GERMALL II has a broad spectrum of activity against standard gram-positive and gram-negative bacteria.

SUTTON LABORATORIES, INC.: GERMALL 115 Preservative System:

The combination of GERMALL 115, methylparaben, and propylparaben takes advantage of a synergistic action which exists between GERMALL 115 and parabens to provide a generally applicable wide range, versatile, preservative system. The basic system which has been successful in numerous cosmetic products is:

0.30% GERMALL 115
0.20% Methylparaben
0.10% Propylparaben

Every cosmetic formulation needs a preservative system which has been hand-tailored to meet its specific requirements. In order to hand-tailor the basic GERMALL 115-paraben preservative system to a specific cosmetic formulation, factors such as the following must be considered:

- The quantity of parabens may be limited by the nature of the formulation. For wholly aqueous systems, for example, the propylparaben content should be reduced because of its low solubility in water.

- GERMALL 115 should always be incorporated into the water phase. Methods for incorporating the parabens may differ, depending on the nature of the formulation and on personal preference.

- Where large quantities of non-ionic emulsifiers, proteins, or other paraben-deactivating components are present, the GERMALL 115 content may have to be increased (eg. to 0.5%).

- If challenge tests include challenging with an unusually resistant strain, the GERMALL 115 content may have to be increased (eg. to 0.5%).

- Certain surfactants (eg. sodium lauryl sulfate, DUPONOL XL) enhance the preservative potency of GERMALL 115, and may permit a reduction in the amount of GERMALL 115 needed.

Specifications for GERMALL 115:
CTFA Adopted Name: Imidazolidinyl Urea
NF Designation: Imidurea, NF
(USP XXI - NF XVI, p. 1569)
Color: White
Odor: None
Appearance: Fine, free-flowing powder
% Nitrogen: 26.0-28.0
pH (1% Aqueous Solution): 6.0 to 7.5
Solubility: Clear, colorless, absence of insolubles
APHA Color: Not darker than APHA 20
Turbidity: Not more than 6.0 NTU
Loss on Drying: 3.0 Maximum
Residue on Ignition: 3.0 Maximum
Heavy Metals: 10 ppm maximum
Infrared Spectrogram: To match standard

SUTTON LABORATORIES, INC.: SUTTOCIDE A Sodium Hydroxymethylglycinate:

Specifications for SUTTOCIDE A, 50% Solution:

SUTTOCIDE A (C3H6NO3Na, MW 127.10)
CTFA Adopted Name: Sodium Hydroxymethylglycinate
Appearance: Clear liquid
Color: Nearly colorless to pale yellow (APHA 100 Maximum)
Odor: Mild characteristic odor
Specific gravity: 1.28 - 1.30
Assay (by Specific gravity): 49.0 - 52.0%
pH: 10.0 - 12.0
% Nitrogen (Kjeldahl): 5.5 -6.1%

SUTTON LABORATORIES, INC.: Allantoin:

Current Status:

The U.S. Food and Drug Administration has classified Allantoin in Category I (Safe and Effective) as an active skin protectant. Based on the wide use and clinical acceptance of Allantoin, as well as on published reports in the literature, the FDA has approved the following statements for drug products containing Allantoin:

1. For the temporary protection of minor cuts, scrapes, burns, and sunburn.
2. Helps prevent and temporarily protects chafed, chapped, cracked, or windburned skin and lips.

Action of Allantoin:

Allantoin produces its desirable effects by promoting and speeding up the healthy natural processes of the body. It helps the skin to help itself.

Cosmetic Uses of Allantoin:

Alantoin has long been known to enhance the efficacy and desirability of cosmetic creams and lotions by acting as a skin healing and softening agent. It has also been incorporated into shampoos, lipsticks, shaving creams, treatment lines, suntan products, bath foams, hair gels and rinses, baby powders, and various aerosol preparations. Although the OTC Panel approved Allantoin as safe and effective as a skin protectant for OTC use, they decided that the data available was insufficient to permit final classification of its effectiveness as a wound-healing agent.

TIC GUMS, INC.: Classical Gum Arabic (Acacia) TIC PRETESTED 1 FCC Powder:

Classical Gum Arabic (Acacia) is the dried exudate only from stems and branches of Acacia senegal harvested exclusively in the Sudan. Classical Gum Arabic 1 Powder is a slightly acidic, highly branched polysaccharide with a molecular weight ranging from 250,000 to 1,000,000.

To differentiate from other Acacia species, all TIC Classical Arabic Assays show the optical rotation figure conforming to Acacia senegal. This assures you of receiving only the highest quality material, conforming to the most rigorous FCC standards.

Classical Gum Arabic 1 is the most soluble of the exudate hydro-colloids. Classical Gum Arabic suspensions exhibit neutonian flow up to 14% concentrations; above that level suspensions become pseudo-plastic. Classical Gum Arabic solutions reduce the surface tension of water and also the interfacial tension of oil-water systems. Classical Gum Arabic is used to prepare stable emulsions.

Main Characteristics:
- Cold Water Soluble
- Low Viscosity
- Emulsifier
- Film Former
- Binder in Tablets
- pH Stable (3 to 9)
- Kosher Approved

Physical & Chemical:
- A. Moisture: 15% max.
- B. Viscosity: 300 cps. max. (30%)
- C. pH: 4-5
- D. Mesh: 90% min. thru USS #140
- E. Starch: Negative
- F. Solution Color: Lemon
- G. Optical Rotation: -25 to -33

Application:
- Coatings: Sugar
- Coatings: Ground Nuts, Snacks
- Confection Glaze
- Encapsulation: Flavors
- Clouding Agent
- Emulsion: Beverage
- Binder: Tablet
- Lithographic Solutions

TIC GUMS, INC.: TIC PRETESTED Colloid 710H Powder:

TIC PRETESTED Colloid 710H Powder is a highly purified kappa carrageenan extracted from Euchema cottonii.

COLLOID 710H hydrates in cold water to produce viscous solutions. Heated to 180F, COLLOID 710H forms clear, strong gels. It is an excellent stabilizer for both milk and water based products.

Produced without alcohol treatment, COLLOID 710H exhibits far superior water gel characteristics to when used with setting salts. It is synergistic with Locust Bean Gum, with increased gel strength and lower syneresis characteristics. Solutions made with COLLOID 710H exhibit excellent stability under pH's as low as 3.5.

For greatest stability under low pH conditions, add COLLOID 710H to cold water and heat to 180 degrees F without any acids or salts. Cool to 160F before adding acidifying agents and/or salts.

Main Characteristics:
Heat to 180F. Strong, Clear
Gels That Are F/T Stable.
Easy to Disperse, Reactive
With Other Gums and Salts
Kosher Approved

Physical & Chemical:
A. Moisture: 10% max.
B. Viscosity Forms gel upon heating
C. pH 7-8
D. Mesh: 45-55% thru USS #200
E. Gel Strength 400 g. min. (1.5%)
F. Sulfite: Less than 10 ppm

Application:
Ice Cream Mixes
Puddings
Dessert Flan
Custards
Cosmetic Gels
Processed Meats
Frozen Doughs
Frozen Sauce

TIC GUMS, INC.: TIC PRETESTED Tragacanth "C" FCC Powder:

Gum Tragacanth TIC PRETESTED "C" USP Powder is the exudate from the roots and stems of shrubs from the Astragalus species.

Gum Tragacanth is a mixture of the salt of a complex acid poly- saccharide, a neutral polysaccharide, and a small amount of glycoside. It consists of a water soluble polymer (traganthin) and an insoluble fraction (bassorin). The insoluble fraction forms bonds with oil groups, thus making Gum Tragacanth ideal for use in emulsions. The molecular weight of Gum Tragacanth is approximately 850,000.

Gum Tragacanth exhibits excellent acid, heat, and salt tolerance. For maximum benefit, allow Gum Tragacanth to hydrate for 10 to 20 minutes alone in water before adding other ingredients. Gum Tragacanth is slow to hydrate, a feature which for many processes is of benefit. Since Gum Tragacanth is hydroscopic, always store under cool and dry conditions.

Main Characteristics:
- Cold Water Soluble
- Emulsifier
- Medium Viscosity
- Excellent Binder
- Suspension
- Kosher Approved

Physical & Chemical:
- A. Moisture: 15% max.
- B. Viscosity (1%) 250 cps. min.
- C. pH 4-7
- D. Mesh: 90% min. thru USS #140
- E. KARAYA Negative
- F. Sulfite: Less than 10 ppm

Application:
- Candy Lozenges
- Shampoo/Conditioner
- Dressings: Pourable
- Ceramic Finish
- Car Polish/Wax Emulsions
- Pharmaceutical Suspensions
- Emulsion: Bakery
- Toothpaste

TRI-K INDUSTRIES, INC.: Product List by C.T.F.A Nomenclature:

Acetamide MEA

Allantoin and Complexes:
- Allantoin
- Aluminum Chlorhydroxy Allantoinate (ALCLOXA)
- Aluminum Dihydroxy Allantoinate (ALDIOXA)
- Allantoin Acetyl Methionine
- Allantoin Glycyrrhetinic Acid
- Allantoin Encapsulated

Aloe Vera Gel, Oil & Powder:

Aloe Flowers Extract:

ARIANOR Semi-Permanent Hair Dyes:
- Basic Blue 99: (Steel Blue)
- Basic Brown 16: (Mahogany)
- Basic Brown 17: (Sienna Brown)
- Basic Red 76: (Madder Red)
- Basic Yellow 57: (Straw Yellow)

Biologicals:
- Bone Marrow Extract
- E.F.A. Complex: (TRI-K E.F.A.)
- Fibronectin: (FIBRONEX)
  - (and) Collagen: (COLLNECTIN)
  - (and) Plasma: (PLASMANECTIN)
  - (and) Procollagen (PRONECTIN)
- Gamma Linolenic Acid
- Hydrolyzed Mucopolysaccharides: TRI-K HMP
- Liver Extract: C.P.R. 110
- Neural Extract: NEURAL LIPID EXTRACT
- Pancreas Extract (Pwdr. & Sol.): KALLIKREIN
- Placenta Extract
- Striated Muscle (Powder & Extract)
- Sodium Hyaluronate (Powder; 0.5% & 1% Sol.)
- Spleen Extract: TRI-DERM S.E.
- Thymus Extract: TRI-DERM T

Cetyl Betaine: (DETAINE PB)
Chelating Agents: (KELATE Series)
Copper Disodium EDTA: (KELATE CU)
Cocoa Butter, USP: (Deodorized & Non-Deodorized)
Gamma Oryzanol
Herb and Plant Extracts (Alban Muller International):
Lactamide MEA
Lactic Acid 88%, USP

TRI-K INDUSTRIES, INC.: Product List by C.T.F.A. Nomenclature (Continued):

Natural Oils:

- Almond
- Apricot
- Avocado
- Black Currant
- Borage
- Camellia
- Canola
- Coconut
- Corn
- Evening Primrose
- Grapeseed
- Hazelnut
- Jojoba
- Kukui Nut
- Macadamia Nut
- Mink
- Olive
- Passion Fruit
- Peanut
- Rice Bran
- Rose Hip Seed
- Safflower
- Sesame
- Soybean
- Sunflower
- Tea Tree
- Walnut
- Wheat Germ

Organic Silicon Compounds & Silanol Derivatives (Exsymol Co.):

Panthenol (d & dl)

Pantethine

Preservatives:

- 2-Bromo-2-Nitropropane-1,3-Diol: (TRISTAT BNP)
- Imidazolidinyl Urea, NF: (TRISTAT IU)
- Butylparaben, NF: (TRISEPT B)
- Ethylparaben, NF: (TRISEPT E)
- Methylparaben, NF, FCC: (TRISEPT M)
- Propylparaben, NF, FCC: (TRISEPT P)
- Phenoxyethanol
- Potassium Sorbate, NF, FCC (Granular & Powder): (TRISTAT K)
- Sorbic Acid, NF, FCC: (TRISTAT)
- Sodium Benzoate, NF, FCC
- Sodium Dehydroacetate, FCC: (TRISEPT SDHA)

Proteins & Protein Derivatives:

- Bovine Serum Albumin: (TRI-K B.S.A. 30)
- Collagen Amino Acids: (TRI-K CAA)
- Collagen Powder: (COLLAGEN SD)
- Hydrolyzed Animal Protein: (40%, 55%, 100%) (TRI-TEIN Series)
- Hydrolyzed Animal Elastin (10%, 30%, 100%): (TRI-LASTIN Series)
- Hydrolyzed Hair Keratin: (TRI-K HKP)
- Hydrolyzed Milk Protein: (TRI-TEIN Milk Polypeptide)
- Hydrolyzed Soy Protein: (TRI-K SOYPRO)
- Procollagen: (TRI-COL PC-1)
- Silk Amino Acids: (TRI-TEIN SILK AA)
- Soluble Animal Collagen: (TRI-COL SP-1)
- Steartrimonium Hydrolyzed Animal Collagen: (TRI-QUAT S)

TRI-K INDUSTRIES, INC.: Product List by C.T.F.A. Nomenclature (Continued):

Rice Bran Wax, FCC:
Saccharin, Insoluble: (Calcium Saccharin)
Saccharin, Soluble: (Sodium Saccharin)
Sodium Bromate
Sodium P.C.A.: (Sodium DL-2-Pyrrolidone-5-Carboxylate)
Soluble Sulfur: (NEWSULFUR W & O)
Squalane, Natural: (TRILANE)
Tocopheryl Acetate, USP: (Vitamin E Acetate)
Tocopheryl Linoleate: (Vitamin E Linoleate)
Sun Screening Agents:
- Benzophenones
- Drometrizole: (JF 77)
- Micro Titanium Dioxide, Powder
- Micro Titanium Dioxide, Solution (SUNVEIL)

Specialty Products:
- Dihydroxyacetone: (DHA)
- TRI-K CMF: (Cell Moisturizing Factor)
- TRI-K HMF: COMPLEX: (Hair Moisturizing Factor)
- T-BASE: (Mineral Oil (&) PEG-30 Lanolin (&) Cetyl Alcohol)
- T-WAX: (Emulsifying Wax NF)
- Liposomes
- Magnesium Ascorbyl Phosphate: (ASCORBYL PM)
- Morus Root Bark Extract: SOHAKUI Extract
- Quaternium-51: (TAKANAL)

Tanning Accelerators:
- Glucose Tyrosinate
- Tanogen
- Tyrosilane

UNION CARBIDE CHEMICALS AND PLASTICS CO., INC.: CARBOWAX Polyethylene Glycols:

CARBOWAX Polyethylene Glycols and CARBOWAX Methoxy Polyethyene Glycols are a family of the linear polymers formed by the addition reaction of ethylene oxide.

Each CARBOWAX Polyethylene Glycol or PEG is designated by a number that represents its average molecular weight. For example, CARBOWAX PEG consists of a distribution of polymers of varying molecular weights with an average of 600, which corresponds to an average number of repeating oxyethylene groups ("n") of 13.

CARBOWAX Polyethylene Glycols are available in average molecular weights ranging from 200 to 8000 and CARBOWAX Methoxy Polyethylene Glycols are available in average molecular weights ranging from 350 to 5000.

Typical Applications:

Cosmetics and Toiletries:
- Creams and Lotions
- Dentifrices
- Deoodorant, Antiperspirant, and Cologne Sticks
- Emulsifier Intermediates
- Glycerine Replacement
- Hair Dressings and Pomades
- Lip Sticks
- Makeup and Bath Powders

Typical Physical Properties:
CARBOWAX Polyethylene Glycols:
200:
- Range of Average Molecular Weight: 190-210
- Density, g/cm3: 20C: 1.1239
- Solubility in Water at 20C: Complete
- Viscosity at 210F., Centistokes: 4.3
- Average Number of Repeating Oxyethylene Units: 4.2
- Surface Tension at 25C dynes/cm: 44.5
- Refractive Index n20D: 1.459
- Average Liquid Specific Heat, cal/g/C: 0.52
- Heat of Combustion at 25C, BTU/lb: 10.180
- CTFA: PEG-4

300:
- Average Molecular Weight: 285-315
- Density, g/cm3: 20C: 1.1250
- Melting or Freezing Range, C: -15 to -8
- Solubility in Water at 20C: Complete
- Viscosity at 210F, centistokes: 5.8
- Average Number of Repeating Oxyethylene Units: 6.4
- Surface Tension at 25C dynes/cm: 44.5
- Refractive Index n20/D: 1.463
- Average Liquid Specific Heat, cal/g/C: 44.5
- Heat of Fusion cal/g: 37
- Heat of Combustion at 25C, BTU/lb: 10.180
- CTFA Nomenclature: PEG-6

UNION CARBIDE CHEMICALS AND PLASTICS CO., INC.: CARBOWAX Polyethylene Glycols(Continued):

CARBOWAX Polyethylene Glycols(Continued):

400:
Range of Average Molecular Weight: 380 to 420
Density, g/cm3: 20C: 1.1254
Melting or Freezing Range, C: 4 to 8
Solubility in Water at 20C, % by weight: Complete
Viscosity at 210F, centistokes: 7.3
Average Number of Repeating Oxyethylene Units: 8.7
Surface Tension at 25C dynes/cm: 44.5
Refractive Index n20/D: 1.465
Average Liquid Specific Heat, cal/g/C: 0.49
Heat of Fusion, cal/g: 36
Heat of Combustion at 25C, BTU/lb: 11,060
CTFA: PEG-8

540 Blend:
Range of Average Molecular Weight: 500 to 600
Density, g/cm3: 60C: 1.0930
Melting or Freezing Range, C: 38 to 41
Solubility in Water at 20C, % by weight: 73
Viscosity at 210F, centistokes: 15.1
Average Liquid Specific Heat, cal/g/C: 0.54
Heat of Fusion cal/g: 37
Heat of Combustion at 25C, BTU/lb: 11,090
CTFA: PEG-6 (and) PEG-32

600:
Range of Average Molecular Weight: 570 to 630
Density, g/cm3: 20C: 1.1257
Melting or Freezing Range, C: 20 to 25
Solubility in Water at 20C, % by weight: Complete
Viscosity at 210F, centistokes: 10.8
Average Number of Repeating Oxyethylene Units: 13.2
Surface Tension at 25C dynes/cm: 44.5
Refractive Index n20/D: 1.467
Average Liquid Specific Heat, cal/g/C: 0.59
Heat of Fusion cal/g: 35
Heat of Combustion at 25C, BTU/lb: 11,130
CTFA: PEG-12

900:
Range of Average Molecular Weight: 855 to 900
Density, g/cm3: 60C: 1.0927
Melting or Freezing Range, C: 32 to 36
Solubility in Water at 20C, % by weight: 86
Viscosity at 210F, centistokes: 15.3
Average Number of Repeating Oxyethylene Units: 20.0
Average Liquid Specific Heat, cal/g/C: 0.54
Heat of Fusion cal/g: 36
Heat of Combustion at 25C, BTU/lb: 11,200
CTFA: PEG-20

UNION CARBIDE CHEMICALS AND PLASTICS CO., INC.: CARBOWAX Polyethylene Glycols(Continued):

1000:
Range of Average Molecular Weight: 950 to 1050
Density, g/cm3: 1.0926
Melting or Freezing Range, C: 37 to 40
Solubility in Water at 20C, % by weight: 80
Viscosity at 210F, centistokes: 17.2
Average Number of Repeating Oxyethylene Units: 22.3
Average Liquid Specific Heat, cal/g/C: 0.54
Heat of Fusion cal/g: 37
Heat of Combustion at 25C, BTU/lb: 11,250
CTFA: PEG-20

1450:
Range of Average Molecular Weight: 1300 to 1600
Density, g/cm3: 1.0919
Melting or Freezing Range, C: 43 to 46
Solubility in Water at 20C, % by weight: 72
Viscosity at 210F, centistokes: 26.5
Average Number of Repeating Oxyethylene Units: 32.5
Average Liquid Specific Heat, cal/g/C: 0.54
Heat of Fusion cal/g: 37
Heat of Combustion at 25C, BTU/lb: 11,320
CTFA: PEG-32

3350:
Range of Average Molecular Weight: 3000 to 3700
Density, g/cm3: 60C: 1.0926
Melting or Freezing Range, C: 54 to 58
Solubility in Water at 20C, % by weight: 67
Viscosity at 210F, centistokes: 90.8
Average Number of Repeating Oxyethylene Units: 75.7
Average Liquid Specific Heat, cal/g/C: 0.50
Heat of Fusion cal/g: 43
Heat of Combustion at 25C, BTU/lb: 11,350
CTFA: PEG-75

4600:
Range of Average Molecular Weight: 4400 to 4800
Density, g/cm3: 60C: 1.0926
Melting or Freezing Range, C: 57 to 61
Solubility in Water at 20C, % by weight: 65
Viscosity at 210F, centistokes: 184
Average Number of Repeating Oxyethylene Units: 104.1
Average Liquid Specific Heat, cal/g/C: 0.55
Heat of Fusion cal/g: 45
Heat of Combustion at 25C, BTU/lb: 11,375
CTFA: PEG-100

UNION CARBIDE CHEMICALS AND PLASTICS CO., INC.: CARBOWAX Polyethylene Glycols(Continued):

8000:
Range of Average Molecular Weight: 7000 to 9000
Density, g/cm3: 60C: 1.0845
Melting or Freezing Range, C: 60 to 63
Solubility in Water at 20C, % by weight: 63
Viscosity at 210F, centistokes: 822
Average Number of Repeating Oxyethylene Units: 181.4
Average Liquid Specific Heat, cal/g/C: 0.55
Heat of Fusion cal/g: 46
Heat of Combustion at 25C, BTU/lb: 11,390
CTFA: PEG-150

Compound 20M:
Range of Average Molecular Weight: 15,000 to 20,000
Density, g/cm3: 60C: 1.0540
Melting or Freezing Range, C: 61 to 64
Solubility in Water at 20C, % by weight: 65
Viscosity at 210F, centistokes: 18,650
Average Number of Repeating Oxyethylene Units: 2 moles joined with an epoxide
Surface Tension at 25C dynes/cm: 52.0
Average Liquid Specific Heat, cal/g/C: 0.59
Heat of Fusion cal/g: 41
Heat of Combustion at 25C, BTU/lb: 11,460
CTFA: PEG-350

CARBOWAX Methoxy Polyethylene Glycols:
350:
Range of Average Molecular Weight: 335 to 365
Density, g/cm3: 20C: 1.0891
Melting or Freezing Range, C: -5 to 10
Solubility in Water at 20C, % by weight: Complete
Viscosity at 210F, centistokes: 3.9
Average Number of Repeating Oxyethylene Units: 7.2
Surface Tension at 25C dynes/cm: 40.0
Refractive Index n20/D: 1.455
Average Liquid Specific Heat cal/g/C: 0.52
Heat of Combustion at 25C, BTU/lb: 11,309
CTFA: PEG-6 Methyl Ether

550:
Range of Average Molecular Weight: 525 to 575
Density, g/cm3: 20C: 1.1039
Melting or Freezing Range, C: 15 to 25
Solubility in Water at 20C, % by weight: Complete
Viscosity at 210F, centistokes: 6.6
Average Number of Repeating Oxyethylene Units: 11.8
Surface Tension at 25C dynes/cm: 37.5
Refractive Index n20/D: 1.455
Average Liquid Specific Heat cal/g/C: 0.55
Heat of Combustion at 25, BTU/lb: 10,596
CTFA: PEG-10 Methyl Ether

UNION CARBIDE CHEMICALS AND PLASTICS CO., INC.: CARBOWAX Polyethylene Glycols(Continued):

750:
Range of Average Molecular Weight: 715 to 785
Density, g/cm3: 60C: 1.0760
Melting or Freezing Range, C: 27 to 32
Solubility in Water at 20C, % by weight: Complete
Viscosity at 210F, centistokes: 10.3
Average Number of Repeating Oxyethylene Units: 16.3
Surface Tension at 25C dynes/cm: 40.7
Refractive Index n20/D: 1.459
Average Liquid Specific Heat, cal/g/C: 0.47
Heat of Fusion cal/g: 44
Heat of Combustion at 25C, BTU/lb: 11,425
CTFA: PEG-16 Methyl Ether

2000:
Range of Average Molecular Weight: 1900 to 2100
Density, g/cm3: 60C: 1.0871
Melting or Freezing Range, C: 49 to 54
Solubility in Water at 20C, % by weight: 68
Viscosity at 210F, centistokes: 45.5
Average Number of Repeating Oxyethylene Units: 44.7
Average Liquid Specific Heat, cal/g/C: 0.53
Heat of Fusion cal/g: 52
Heat of Combustion at 25C, BTU/lb: 11,379
CTFA: PEG-40 Methyl Ether

5000:
Range of Average Molecular Weight: 4750 to 5250
Density, g/cm3: 60C: 1.0907
Melting or Freezing Range, C: 57 to 63
Solubility in Water at 20C, % by weight: 64
Viscosity at 210F, centistokes: 320
Average Number of Repeating Oxyethylene Units: 112.9
Average Liquid Specific Heat, cal/g/C: 0.49
Heat of Fusion cal/g: 52
Heat of Combustion at 25C, BTU/lb: 11,344
CTFA: PEG-100 Methyl Ether

R. T. VANDERBILT CO., INC.: Bactericides/Fungicides:

VANCIDE is the group name for a line of bactericides and fungicides used in agricultural, veterinary, cosmetic and pharmaceutical products. They also have application in textiles, detergents, drywall plaster and industrial deodorants. VANCIDE products are highly effective with relatively low toxicity.

VANCIDE 51:
Sodium dimethyldithiocarbamate and sodium 2-mercaptobenzothiazole supplied as a 30% solution. Industrial bactericide/ fungicide for textiles and for agriculture.

VANCIDE 51Z:
Zinc dimethyldithiocarbamate and zinc 2-mercaptobenzothiazole. Available as a water-insoluble powder. Used as a mildew-proofing agent and preservative in drywall plaster.

VANCIDE 51Z Dispersion:
50% liquid dispersion of VANCIDE 51Z. Used to prevent mildew on textiles and in cellulose sponges.

VANCIDE 89:
N-trichloromethylthio-4-cyclohexene-2,2-dicarboximide. Technical grade of captan used as an industrial preservative, a wallpaper adhesive preservative and in veterinary products for skin diseases.

VANCIDE 89RE:
Purified captan. Antimicrobial and preservative for cosmetics and topical pharmaceuticals. Available as a white powder.

VANCIDE MZ-96:
Ziram wettable powder, used to preserve starch and synthetic latex adhesive formulations against bacterial degradation. Also used in joint compounds for drywall construction.

VANCIDE TH:
Hexahydro-1,3,5-triethyl-s-triazine. Industrial preservative, prevents bacterial action in cutting oils, synthetic rubber latex, starch-based adhesives, latex paint and aqueous slurries. It is soluble in acetone, ethyl alcohol, ether and water, moderately soluble in hydrocarbon solvents.

R.T. VANDERBILT CO., INC.: Dispersing Agents:

DARVAN products include complex polymerized organic salts of sulfonic acids of the alkyl-aryl type and salts of acrylic polymers. They are highly effective dispersing agents used to disperse finely divided solids or liquids and keep them dispersed in aqueous media. All DARVAN products are water-soluble.

DARVAN No. 1:
Sodium salts of polymerized alkyl naphthalene sulfonic acid in granular form
DARVAN No. 1SD:
Powder form of DARVAN No. 1
DARVAN No. 2:
Sodium lignosulfate powder
DARVAN No. 6:
Sodium salts of polymerized naphthalene sulfonic acids in granular form
DARVAN No. 6SD:
Powder form of DARVAN No. 6
DARVAN No. 7:
Aqueous solution of sodium polymethacrylate
DARVAN No. 9:
Sodium salts of polymerized naphthalene sulfonate in powder form
DARVAN No. 9L:
Aqueous solution of DARVAN No. 9
DARVAN No. 404:
Calcium Lignosulfonate

Some specific uses of DARVAN products are for dispersing pigments in finger paints and crayons, and for dispersing insecticides in agricultural sprays, dusts and flowables.

R.T. VANDERBILT CO., INC.: Anionic Surfactants:

VANSEAL surfactants are biodegradable, anionic surfactants derived from natural fatty acids and the amino acid sarcosine. These products exhibit outstanding mildness, lather building and conditioning properties, and unusual compatibility with cationic materials.

VANSEAL LS: Lauroyl sarcosine
VANSEAL CS: Cocoyl sarcosine
VANSEAL OS: Oleoyl sarcosine
VANSEAL NALS-30: Sodium lauroyl sarcosinate, 30%
VANSEAL NALS-95: Sodium lauroyl sarcosinate, 95%
VANSEAL NACS-30: Sodium cocoyl sarcosinate, 30%
VANSEAL 35: Industrial grade of sodium cocoyl sarcosinate, 24%

Some specific uses of VANSEAL products are soaps, bath gels, shampoos, shaving creams, dentifrices, rug shampoos, oven cleaners, dishwashing products, textile and leather processing.

R. T. VANDERBILT CO., INC.: Emulsion Stabilizer/Suspending Agent:

VEEGUM:

Magnesium aluminum silicate, is described in USP-NF. It is used as a suspending agent, emulsion stabilizer and viscosity modifier. It is supplied as an insoluble flake that forms a colloidal dispersion in water.

VEEGUM is the group name for all grades and also designates the regular grade of VEEGUM. Vanderbilt considers the regular grade of VEEGUM the most suitable and economical for the greatest number of applications. The key property of each grade follows:

VEEGUM - regular grade
VEEGUM HV - high viscosity
VEEGUM K - acid stability
VEEGUM F - microfine powder
VEEGUM D - fast dispersing, fluoride compatible
VEEGUM T - technical grade
VEEGUM HS - high electrolyte stability
VEEGUM PRO - superior soap and surfactant compatibility

VEEGUM is used in cosmetics, pharmaceuticals, veterinary products, chemical specialties and household products. Some of the largest uses of VEEGUM are in toothpastes, antacids, pharmaceutical tablets, antiperspirants and oven cleaners. Other important applications are in thixotropic, water-based paints, paper coatings, ceramic glazes and agricultural flowables.

R.T. VANDERBILT CO., INC.: Xanthan Gum Thickener/Suspending Agent:

Xanthan gum is a versatile polysaccharide thickener and stabilizer produced by the fermentation and extraction of the naturally occuring plant bacteria xanthomonas campestris.

RHODIGEL - Cosmetic/pharmaceutical grade
RHODIGEL 200 - Fine grind pharmaceutical grade
RHODOPOL 23 - Industrial grade
RHODOPOL 50 MD - Rapid dispersing industrial grade
RHODOPOL 50 MC - Cationic compatible industrial grade

R.T. VANDERBILT CO., INC.: Lipoamino Acids and Their Salts:

LIPACIDE lipoamino acids are cosmetic/pharmaceutical raw materials derived from natural animal collagen or keratin. They provide antimicrobial activity, anti-inflammatory and anti-pruritic properties and can limit the rate of sebaceous excretion in humans. As such, they are effective ingredients in antimicrobial and antiacne skin creams and lotions, antidandruff shampoos and anti-inflammatory skin treatments.

LIPOPROTEOL products are the salts of LIPACIDE lipoamino acids. They are mild surfactants with excellent lathering and wetting properties and substantivity to the hair and skin. They are useful in frequent-use shampoos, bath gels, soaps and shaving creams.

LIPACIDE CCO: Capryloyl collagenic acid

LIPACIDE UCO: Undecylenoyl collagen acid

LIPACIDE PCO: Palmitoyl collagenic acid

LIPACIDE PK: Palmitoyl keratinic acid

LIPACIDE DPHP: Dipalmitoyl hydroxyproline

LIPOPROTEOL LCO: Lauroyl collagenic acid, mixed sodium and triethanolamine salts

LIPOPROTEOL LK: Lauroyl keratinic acid, triethanolamine salt

R.T. VANDERBILT CO., INC.: RHODIGEL Xanthan Gum:

RHODIGEL is produced by a fermantation process using the natural plant organism Xanthomonas campestris. The resulting product is a water-soluble, anionic polysaccharide. This structure is one of over 1,000 repeating units which form the polysaccharide chain with a Molecular Weight of approximately two million.

RHODIGEL Xanthan Gum is produced in two grades. RHODIGEL is the standard grade product and RHODIGEL 200 is a fine mesh, ground version of RHODIGEL. Typical physical properties of these two grades are shown below:

RHODIGEL:
- Appearance: Cream-White Powder
- Moisture Content, %: 12 max.
- pH (1.0 wt% Aqueous Solution): 6 to 8
- Viscosity, cps: 1200-6000
- Mesh Size:
  - % thru 60 mesh: 100
  - % thru 80 mesh: 95 min.

RHODIGEL 200:
- Appearance: Cream-White Powder
- Moisture Content, %: 12 max.
- pH (1.0% wt% Aqueous Solution): 6 to 8
- Viscosity, cps: 1200-1600
- Mesh Size:
  - % thru 80 mesh: 100
  - % thru 200 mesh: 92 min.

Performance Benefits:
- High Efficiency Thickener
- Dissolves Easily in Cold or Warm Water
- Produces Highly Pseudoplastic Solutions
- Aqueous Solutions Have High Yield Values
- Excellent Stabilizing Agent for Emulsions, Suspensions and Foams
- Stable Viscosity Over a Broad pH Range
- Little Viscosity Change With Temperature
- Synergistic Rheological Performance with other Organic Gums and Inorganic Suspending Agents
- Compatible with Acids, Bases, Salts, and most Pharmaceutical and Cosmetic Ingredients

R.T. VANDERBILT CO., INC.: VEEGUM:

VEEGUM is a complex colloidal magnesium aluminun silicate derived from natural smectite clays and refined by a special process. It is supplied as soft white flakes with a surprising ability to swell in water forming an opaque colloidal dispersion. It is thixotropic, tending to gel when at rest, but flowing freely when shaken. VEEGUM is ordinarily used in aqueous dispersions to capitalize on this most unusual property.

Chemical Analysis:

VEEGUM is an inorganic, complex, colloidal, magnesium aluminum silicate. The typical chemical analysis of VEEGUM, conventionally expressed as oxides, is as follows:

Silicon dioxide: 63.0%
Magnesium oxide: 10.5%
Aluminum oxide: 10.5%
Ferric oxide: 0.9%
Calcium oxide: 2.3%
Sodium oxide: 2.4%
Potassium oxide: 1.2%
Ignition Loss: 7.5%

This analysis is for the regular grade of VEEGUM only.

Typical Properties:

Appearance: Small flakes
Odor: None
Taste: None
Color: White to Tan
Texture: Soft, considerable slip
Moisture content: Less than 8% at time of shipment
Solubility: Insoluble in water or alcohol; swells to many times original volume in water to form colloidal dispersions
Viscosity: 5% VEEGUM-water dispersion: 250 cps±25%
pH: Slightly alkaline: the pH of a 5% VEEGUM-water dispersion is about 9.5
Acid demand: 6 to 8 mls. 0.1N HCl required to reduce the pH of 1 gram of VEEGUM to pH 4
Bacterial count: All grades of VEEGUM are controlled for a maximum allowable number of bacteria and no E. coli, S. aureus, Salmonella sp. or P. aeruginosa at time of shipment.
Density: 2.6 Mg/m3

VAN DYK: CERAPHYL and CERASYNT:

CERAPHYL 28:
CTFA Name: Cetyl Lactate
CAS Number: 35274-05-6
Physical Form: White solid
Odor: Faint, characteristic
Specifications:
Acid Value (@ 25C): 2.0 maximum
Saponification Number: 174-189
Specific Gravity (27C): 0.893-0.905
Applications:
CERAPHYL 28 is one of a series of lactate esters designed for use in cosmetic and pharmaceutical formulas. It is a lubricant and an emollient, imparting sheen and silkiness to skin and hair. CERAPHYL 28 is non-greasy, non-oily, non-drying, and has a Required HLB of 13-15. A solid at room temperature, it liquifies readily on the skin. CERAPHYL 28 is also useful as tack-reducer in deodorant sticks and as a solvent for dyes in lipstick formulas.

CERAPHYL 31:
CTFA Name: Lauryl Lactate
CAS Number: 6283-92-7
Physical Form: Light yellow liquid
Odor: Faint, Characteristic
Specifications:
Acid Value (@ 25C): 2.0 maximum
Saponification Number: 210-225
Specific Gravity (@ 25C): 0.910-0.922
Refractive Index (@ 25C): 1.4417-1.4456
Applications:
CERAPHYL 31 is a highly effective emollient and skin lubricant. It is especially recommended as an anti-tack agent in antiperspirant creams and lotions, and other preparations requiring improved slip. CERAPHYL 31 has a Required HLB of 10, and is not recommended for use in formulas with high pH.

CERAPHYL 41:
Chemical Name: Linear Alcohol Lactate
CTFA Name: C12-C15 Alcohols Lactate
Physical Form: White to straw-colored liquid
Specifications:
Acid Value: 2.0 Maximum
Saponiication Number: 195-210
Specific Gravity (25C): 0.900-0.920
Refractive Index: 1.4430-1.4450
Applications:
CERAPHYL 41 is an excellent choice for alcoholic and hydroalcoholic skin preparations when a non-greasy emolliency afterfeel is desired.
CERAPHYL 41 reduces the tacky, greasy feel of formulations which contain high levels of petrolatum or mineral oil. It is also useful as a de-tackifier in antipersiprants.
CERAPHYL 41 imparts high sheen in hair preparations, and is an excellent plasticizer for hair spray films.

VAN DYK: CERAPHYL and CERASYNT(Continued):

CERAPHYL 45:

Chemical Name: bis (2-ethylhexyl) malate
CTFA Name: Dioctyl Malate
CAS Number: 56235-92-8
Physical Form: Clear Liquid, Colorless to Pale Yellow
Odor: Characteristically Mild
Molecular Formula: C20H38O5
Molecular Weight: 358.52

Specifications:

Acid Value: 5.0 Max
Saponification Number: 310 min.
Iodine Number: 1.0 Max.
Specific Gravity (25C): 0.960-0.970
Refractive Index: 1.4480-1.4500

Applications:

CERAPHYL 45 is an emollient that exhibits an unusual silky feel. It is non-sensitizing and non-comedogenic, so it is ideal for hypoallergenic products of all types.

Because CERAPHYL 45 is a hydroxydiester, it shows broad spectrum solubility. It is an excellent fragrance coupler, and it solubilizes Benzophenone-3 as well as other difficult-to-solubilize materials.

Its unusual chemical structure promotes clear hydro-alcoholic gels and lotions which can accomodate more water, thus reducing the cost of the formula.

CERAPHYL 45 is easy to emulsify, and has a Required HLB of 12. It will reduce the greasiness of mineral oil and petrolatum and the tackiness of carbomer formulas. Used in antiperspirants, it reduces the stickiness of aluminum salts. In hair products, CERAPHYL 45 imparts conditioning and wet-comb benefits.

CERAPHYL 50:

CTFA Name: Myristyl Lactate
CAS Number: 1323-03-1
Physical Form: Water-white to pale-yellow liquid or soft solid (depending on temperature)

Specifications:

Acid Value (25C): 2.0 maximum
Saponification Number: 166-181
Specific Gravity (25C): 0.892-0.904

Applications:

CERAPHYL 50 is particularly recommended for use in all types of alcoholic preparations, such as shaving lotions and body rubs, because of the soft, silky, water-resistant film it leaves on the skin. It imparts lubricity, sheen, and an enhanced creamy texture to lipsticks. In hair products, it adds manageability and feel to the hair. Required HLB = 12. Not stable at high pH.

VAN DYK: CERAPHYL and CERASYNT(Continued):

CERAPHYL 50S:
CTFA Name: Myristyl Lactate
CAS Number: 1323-03-1
Physical Form: Water-white to pale-yellow liquid or soft solid (depending on temperature)
Specifications:
Acid Value (25C): 2.0 maximum
Saponification Number: 166-181
Specific Gravity (25C): 0.890-0.910
Applications:
CERAPHYL 50S exhibits excellent skin feel when used in stick makeup formulations. An added benefit is the noticeable improvement it imparts to the consistency and creaminess of the stick products. When used in alcoholic preparations, it ameliorates the harsh drying and defatting effects of the alcohol while leaving a lubricious feel on the skin.
Required HLB = 12. Not stable at high pH.

CERAPHYL 55:
CTFA Name: Tridecyl Neopentanoate
CAS Number: 106436-39-9
Physical Form: Clear, light yellow liquid
Odor: Characteristically mild
Molecular Formula: C18H36O2
Molecular Weight: 284
Specifications:
Acid Value: 2.0 Max.
Saponification Number: 190 Min.
Specific Gravity (25C): 0.850-0.860
Refractive Index: 1.4345-1.4365
Applications:
CERAPHYL 55 is a non-comedogenic branched ester with a Required HLB of 8. It enhances the application properties of creams & lotions by imparting non-oily, non-occlusive lubricity and elegant skin feel. In pigmented products, CERAPHYL 55 improves gloss and spreading characteristics.

VAN DYK: CERAPHYL and CERASYNT(Continued):

CERAPHYL 60:
U.S. Patent No. 3,766,267
Chemical Name: Gluconamidopropyl Dimethyl-2-hydroxyethyl Ammonium Chloride
CTFA Name: Quaternium-22
Physical Form: Clear Liquid
Specifications:
Color: Yellow to light amber
% Ionic Chloride: 5.8-6.2
% Solids: 58-62
% Water: 38-42
pH as is: 4.0-5.0
Specific Gravity @ 25C: 1.170-1.210
Applications:

This cationic, water-soluble Emollient Moisturizer shows pronounced substantivity to skin and hair. It leaves a rich emollient skin-feel when applied in the form of hydro-alcoholic solutions such as Fresheners, Colognes, or After Shaves, and has pronounced humectant properties which can be demonstrated to continue even on absorbed CERAPHYL 60 films which remain after rinse-off of products containing it.

Formulation of various hair preparations such as shampoos, wave sets and cold waves with CERAPHYL 60 have shown considerable improvement of performance. When added to typical anionic shampoo or bubble bath formulations, CERAPHYL 60 is compatible and does not adversely affect foaming. In some cases, foam was actually improved 10-15% (as measured by the Ross-Miles test). Addition to cold waves improves hair texture.

CERAPHYL 65:
U.S. Patent #4,012,098
Chemical Name: Mink-Amidopropyl Dimethyl 2-Hydroxyethyl Ammonium Chloride
CTFA Name: Quaternium-26
CAS Number: 68953-64-0
Specifications:
Acid Value: 20.0 maximum
Alkali Number: 25.0 maximum
Solids: 53-65%
Ionic Chloride: 4.0-5.0%
Applications:

Hair products: Provides anti-tangle, anti-static, and conditionng properties. Use in shampoos at 2-3%, conditioners at 2-3%, setting lotions at 0.50-0.75%, and hair dressings at 0.5-1.0%.

Emulsions: CERAPHYL 65 is a substantive, water-soluble emollient. When neutralized (ie. with lactic acid) it is also an efficient cationic emulsifier.

VAN DYK: CERAPHYL and CERASYNT(Continued):

CERAPHYL 70:
U.S. Patent #4,038,294
Chemical Name: Stearic Amidopropyl Dimethyl-Myristyl Acetate Ammonium Chloride
CTFA Name: Quaternium-70 (and) Propylene Glycol
Physical Form: Soft Gel
Specifications:
Saponification Number: 45-60
Alkali Number: 5.0 Maximum
% Ionic Chloride: 2.6-3.4
Melting Point C (U.S.P. #2): 27-32
Total Solids: 48-58
Applications:
As an after shampoo conditioner at 2.5-4.0% (put on-rinse off type) and as a skin emollient at 0.5-2.0%. This compound has anti-tangle, anti-static, emulsifying and emollient properties. When used in a creme rinse, it leaves the hair with slight "body" and with a clean feel after rinsing.

CERAPHYL 85:
U.S. Patent 4,342,706
CTFA Name: Stearamidopropyl Dimethyl Cetearyl Ammonium Tosylate and Propylene Glycol
CAS Number: 87616-36-2 and 83554-37-4
Physical Form: Cream colored waxy solid
Specifications:
Acid Value: 12 Maximum
Saponification Number: 20 Maximum
Alkali Number: 5 Maximum
Melting Point: 44-48C
Applications:
CERAPHYL 85 is a unique quaternary conditioning agent which has application in hair care as well as skin care products. It is quite lipophilic while being noticeably substantive to skin and hair protein. Unlike many quats, it has very low eye irritation potential.

Product	CERAPHYL 85 Benefit
Hair Conditioner	Improved wet comb and manageability
Liquid Soap	Substantivity; Emollient skin feel
Cationic Emulsions	Emulsifier; Substantive emollient
Make-up Products	Emolliency, smooth rub-in, afterfeel

VAN DYK: CERAPHYL and CERASYNT(Continued):

CERAPHYL 140:
Chemical Name: Decyl Oleate
CTFA Name: Decyl Oleate
Physical Form: White to straw liquid
Specifications:
Acid Number: 5.0 Maximum
Saponification Number: 132-142
Specific Gravity @ 25C.: 0.855-0.865
Refractive Index @ 25C.: 1.4530-1.4555
Iodine Number: 57-60
Applications:
Emollient and co-solvent especially recommended for creams, lotions, and other preparations requiring the improvement of "slip". Useful in moisturizing preparations, bath oils, as a lubricant; in liquid make-up for smoothness and application; in pressed powder to hold moisture and impart creamy texture.

CERAPHYL 140A:
CTFA Name: Isodecyl Oleate
CAS Number: 59231-34-4
Description: White-to-straw colored liquid; characteristic-ally mild odor
Specifications:
Acid Number: 5.0 maximum
Saponification Number: 130-145
Specific Gravity (25C): 0.858-0.864
Refractive Index (25C): 1.4540-1.4560
Iodine Number: 50-65
Applications:
CERAPHYL 140A is an emollient and co-solvent for various cosmetic systems. It is particularly useful in liquid makeup for imparting viscosity control and "return" characteristics at low temperature.

CERAPHYL 230:
Chemical Name: Diisopropyl Adipate
CTFA Name: Diisopropyl Adipate
Physical Form: Water White Liquid
Specifications:
Acid Number: 2.0 Maximum
Saponification Number (2 Hrs.): 465-500
Specific Gravity @ 25C: 0.950-0.962
Refractive Index @ 25C: 1.4216-1.4245
Applications:
Useful in aqueous alcohol systems as an emollient and coupling agent with high solubility, such as in shave lotions and hair tonics.

Useful in high oil content products to reduce the greasy feel of mineral oil. Imparts spreadability to bath oils and hair pom-ades. Useful in reducing the stringiness of petrolatum-based products.

Useful in hair sprays as a plasticizer. In "Carbopol" clear gel items it minimized "roll-up" in the presence of electro-lytes.

VAN DYK: CERAPHYL and CERASYNT(Continued):

CERAPHYL 368:
Chemical Name: 2-Ethyl Hexyl Palmitate
CTFA Name: Octyl Palmitate
Physical Form: Water white liquid
Specifications:
Acid Number: 3.0 Maximum
Saponification Number (1 Hr.): 146-156
Refractive Index @ 25C: 1.4445-1.4465
Specific Gravity @ 25C: 0.850-0.856
Applications:
The non-oily feel of this ester makes it most useful in aerosol antiperspirants, bath oils, and liquid make-up products. It also provides gloss in stick make-up products. It is also an excellent binder for pressed powders.

CERAPHYL 375:
CTFA Name: Isostearyl Neopentanoate
CAS Number: 58958-60-4
Physical Form: Clear, pale yellow liquid
Specifications:
Acid Value: 2.0 Maximum
Saponification Number: 144-161 (6 hours)
Specific Gravity (25C): 0.858-0.870
Refractive Index (25C): 1.4467-1.4497
Color (Gardner): 2 Maximum
Applications:
Binder for pressed powder makeups. Pigment dispersing agent, especially in eye makeup products. Improves spreading and play-time in highly pigmented products. Also for general emolliency where a completely bland, completely synthetic material is desired.

CERAPHYL 424:
CTFA Name: Myristyl Myristate
CAS Number: 3234-85-3
Physical Form: White to slightly yellow waxy solid, with a bland characteristic odor
Specifications:
Acid Value: 3.0 maximum
Saponification Number: 120-130
Melting Point: 36-39C
Applications:
CERAPHYL 424 is particularly useful in creams and lotions, and also many makeup items, because it melts at body temperature. At low percentages in O/W lotions, it improves formula viscosity and "richness".
CERAPHYL 424 has a Required HLB of 8-10.

VAN DYK: CERAPHYL and CERASYNT(Continued):

CERAPHYL 494:

CTFA Name: Isocetyl Stearate
CAS Number: 25339-09-7
Physical Form: White to light yellow liquid
Odor: Bland, characteristic

Specifications:

Acid Value: 5.0 maximum
Saponification Number: 95-110
Specific Gravity (25C): 0.845-0.865
Refractive Index (25C): 1.446-1.456

Applications:

CERAPHYL 494 is a liquid branched-chain fatty ester with a Required HLB of 8. It can be used as an all-purpose lubricant in creams and lotions, where it will impart a dry, emollient feel.

CERAPHYL 847:

CTFA Name: Octyldodecyl Stearoyl Stearate
CAS Number: 90052-75-8
Description: Light-to-straw colored liquid; characteristically mild odor. (A small amount of white solid may form occasionally, in which case the product should be heated slightly before use and stirred to ensure homogeneity.)

Specifications:

Acid Number: 10.0 maximum
Saponification Number: 115-135
Specific Gravity (25C): 0.860-0.880
Refractive Index (25C): 1.447-1.467

Applications:

CERAPHYL 847 is an excellent pigment dispersant and binder. It imparts a smooth creamy application and no taste to lipsticks, and its low odor of toxicity makes it suitable for eye products and other makeup formulas. In emulsions, CERAPHYL 847 imparts long-lasting lubricity and a rich, cushioned feel to the skin.

CERAPHYL 847 has a Required HLB of 6.

CERAPHYL GA:

Chemical Name: Carboxylated Vegetable Tri-Glyceride
CTFA Name: Maleated Soybean Oil
CAS Registry Number: 68648-66-8
Physical Form: Amber yellow oily, slightly viscous liquid with a mild characteristic odor.

Tentative Specifications:

Acid Number/T.H.F./@ 25C: 62-72
Acid Number/MEOH/@ 25C: 60-70
Saponification Index @ 25C: 1.4750-1.4850
Moisture, %: <0.1%
Iodine Value: 87-107

VAN DYK: CERAPHYL and CERASYNT(Continued):

CERAPHYL GA(Continued):

Applications:

CERAPHYL GA is an unusual skin softener/moisturizer derived from refined soybean oil. It provides a rich, full-bodied feel during rub-out and imparts a non-greasy, non-tacky silky emollience to the skin. The perceptibly moist, smooth afterfeel persists long after application.

Several literature references cite CERAPHYL GA's skin-plasticizing properties and indicate that it may offer the formulator a novel alternative to traditional moisturization methods, such as occlusion and hygroscopicity.

When neutralized with an appropriate material (such as Triethanolamine) CERAPHYL GA also acts as a secondary o/w emulsifier while retaining its skin-softening properties. With careful selection of ingredients and processing conditions, therefore, multi-functional CERAPHYL GA can form the backbone of products manufactured using low-energy emulsification techniques.

CERAPHYL GA has a required HLB of 10-12, and is suitable for use in all types of creams and lotions. It is also functional in products such as bath oils and nail treatments, which are designed to alleviate dryness. In hair products CERAPHYL GA improves wet-combability and gloss, and reduces static. A level of 3-5% CERAPHYL GA is recommended for most skin care applications, and 0.5-2% is recommended for hair care products.

CERAPHYL ICA:

CTFA Name: Isocetyl Alcohol
CAS Number: 36311-34-9
Physical Form: Clear, colorless, low-odor liquid
Molecular Weight: 242

Specifications:

Acid Number: 5.0 Max.
Saponification Number: 10.0 Max.
Hydroxyl Number: 195-230
Iodine Value: 10.0 Max.
Specific Gravity (25C): 0.830-0.840

Applications:

CERAPHYL ICA is a non-comedogenic branched fatty alcohol which imparts a non-greasy, non-occlusive emolliency to emulsions. It is ideally suited for "oil-free" cosmetic formulations, and high-pH applications. Required HLB 12-14. CERAPHYL ICA is an effective pigment dispersant and binder. It couples mineral oil and castor oil, resulting in more uniform and stable stick products. It has low cloud and freezing points, and is an excellent carrier and extender for flavor and fragrance oils. It promotes hydroalcoholic solubility, making it a candidate for clear gel formulas.

VAN DYK: CERAPHYL and CERASYNT(Continued):

CERAPHYL IPL:
Chemical Name: Isopropyl Linoleate
CTFA Name: Isopropyl Linoleate
Physical Form: Straw to yellow liquid
Specifications:
Acid Number: 3.5 Maximum
Saponification Number: 170-180
Specific Gravity @ 25C: 0.860-0.870
Refractive Index: 1.4520-1.4540
Iodine Number: 125 Minimum
Applications:
Isopropyl Linoleate is an excellent skin conditioner which is rapidly absorbed imparting lubrication and smoothness. As a conditioner for the hair, it results in lustre and softness and is an effective superfatting agent in detergent systems.

CERASYNT D:
Chemical Name: Stearamide MEA Stearate
CTFA Name: Stearamide MEA Stearate
Physical Form: Cream Flakes
Specifications:
Acid Number: 10 - 20
Saponification Number: 97 - 107
Iodine Number: 0.5 Maximum
Melting Point C: 76 - 82
Applications:
Thickener and opacifier for liquid cream shampoos, particularly concentrates. Auxiliary emulsifier for hydrocarbon propellent gas in aqueous aerosol systems such as shave creams.

CERASYNT GMS:
Chemical Name: Glyceryl Monostearate
CTFA Name: Glyceryl Stearate
Physical Form: White to Cream Flakes
Specifications:
Acid Value: 3.0 Maximum
Saponification Number: 162-175
Iodine Number: 2.0 Maximum
Melting Point C: 56-59
Applications:
A Nonionic secondary O/W emulsifier for creams and lotions. It also effectively increases the viscosity of emulsions.

VAN DYK: CERAPHYL and CERASYNT(Continued):

CERASYNT IP:

Chemical Name: Glycol Stearate (And) Other Ingredients
CTFA Name: Glycol Stearate (And) Other Ingredients
Physical Form: White to cream flakes

Specifications:

Acid Number: 5.0 Maximum
Saponification Number: 174-184
Iodine Number: 0.5 Maximum
Melting Point C: 56.5-58.5

Applications:

An opacifier and pearling agent for liquid cream and cream paste shampoos. Imparts minimal decrease to the foaming and lathering qualities of the detergent system.

CERASYNT LP:

CTFA Name: Glycol Stearate (and) Sodium Laureth Sulfate (and) Hexylene Glycol
Physical Form: White Opaque Pourable Liquid

Tentative Specification:

pH (as is): 6.5±0.5
Viscosity: <5,000 cps
Percent Water: 60.0% - 70.0%

Application:

CERASYNT LP is used to opacify and pearlize shampoo and other surfactant-based products. The use of CERASYNT LP allows the formulator to produce EGMS- and EGDS-type pearls without heating, which can help make the manufacture of pearlized surfactant based products more cost efficient. A suggested use level of 3-10% is recommended, but the formulator can vary these levels to fit their specific goals. This convenient product is pourable, readily dispersible in water at room temperature, and exhibits excellent stability as a raw material and in finished products.

CERASYNT M:

Chemical Name: Ethylene Glycol Monostearate Pure
CTFA Name: Glycol Stearate
Physical Form: White to cream wax (flakes)

Specifications:

Acid Number: 5 Maximum
Saponification Number: 185-195
Melting Range: 56-60

Applications:

Emulsifier for O/W type solid creams and lotions with high opacity. Thickening agent for liquid and solid cream shampoos.

VAN DYK: CERAPHYL and CERASYNT(Continued):

CERASYNT MN:
Chemical Name: Ethylene Glycol Monostearate SE
CTFA Name: Glycol Stearate SE
Physical Form: White to cream flakes
Specifications:
Acid Number: 5.0 Maximum
Saponification Number: 181-191
Iodine Number: 0.5 Maximum
Melting Point C: 57-60
Applications:
Emulsifier for O/W type creams and lotions. Viscosity builder for emulsions containing high percentage of water in aqueous phase.

CERASYNT PA:
CTFA Name: Propylene Glycol Stearate
Chemical Name: Propylene Glycol Monostearate
CAS Number: 1323-39-3
Physical Form: Flakes, white to cream-colored
Odor: Bland, typical
Specifications:
Acid Value: 5.0 maximum
Saponification Number: 181-191
Iodine Number: 0.5 maximum
Melting Point: 35-38C
Applications:
Emulsifier for lotions and soft creams.
Particularly useful in liquid and cream makeup products.
HLB = 3.

CERASYNT Q:
Chemical Name: Glyceryl Monostearate Self-Emulsifying (Anionic)
CTFA Name: Glyceryl Stearate SE
Physical Form: White to Cream Flakes
Specifications:
Acid Number: 10 Maximum
Saponification Number: 150-160
Melting Point IIC: 57-59
Iodine Number: 1.0 Maximum
Applications:
Auxiliary emulsifier for soap O/W emulsions.

VAN DYK: CERAPHYL and CERASYNT(Continued):

CERASYNT SD:

Chemical Name: Glyceryl Monostearate, Pure
CTFA Name: Glyceryl Stearate
Physical Form: White to cream flakes

Specifications:

Acid Number: 2.0 Maximum
Saponification Number: 165-177
Iodine Number: 0.5 Maximum
Melting Point C: 55.0 - 57.5

Applications:

Deodorized non-ionic O/W emulsifier for creams and lotions. Useful in lotions where thixotropic effect is advantageous.

CERASYNT WM:

Chemical Name: Glyceryl Monostearate, SE (Anionic)
CTFA Name: Glyceryl Stearate (and) Stearyl Alcohol (and) Sodium Lauryl Sulfate
Physical Form: White to cream flakes

Specifications:

Acid Number: 5.0 Maximum
Saponification Number: 140 - 150
Iodine Number: 0.5 Maximum
Melting Point C: 55 - 57
% Ash: 1.6 - 1.9

Applications:

Emulsifier for O/W type creams, lotions and ointments. Used in preparations containing high concentrations of electrolytes such as antiperspirants.

CERASYNT 303:

Chemical Name: Dialkyl Amino Ethyl Stearate
CTFA Name: Diethylaminoethyl stearate
Physical Form: Straw to amber-colored liquid to semi solid
Odor: Amine

Specifications:

Acid Number: 30-40
Saponification Number: 150-160
Specific Gravity (@ 25C): 0.860-0.880
pH (3% solution @ 25C): 9.5-10.5
Alkaline Number: 127-137

Applications:

When neutralized with certain acids (Phosphoric for one) to pH 3-4, CERASYNT 303 becomes an oil in water emulsifier. This enables the preparation of acid pH emulsions without the use of nonionic emulsifiers, thus simplifying preservation requirements. When used with neutral glyceryl monostearate, the water dispersible CERASYNT 303-Phosphoric Acid salt forms a self-emulsifying system.

VAN DYK: CERASYNT and EMULSYNT:

CERASYNT 840:
Chemical Name: Polyethylene Glycol 1000 Monostearate
CTFA Name: PEG-20 Stearate
Physical Form:
Acid Number: 5.0 Maximum
Saponification Number: 40-50
Iodine Number: 0.25 Maximum
Melting Point C: 39.5-42.5
Applications:
Emulsifier for O/W type creams and lotions. As a superfatting agent in detergent shampoos. Increases viscosity and stability in cream lotions. Stable with medicaments in ointment type preparations. Useful vehicle for products in stick form required to melt at body temperature.

CERASYNT 945:
Chemical Name: Glyceryl Monostearate, SE (Non-Ionic)
CTFA Name: Glyceryl Stearate (and) Laureth-23
Physical Form: White to cream flakes
Specifications:
Acid Number: 5.0 Maximum
Saponification Number: 142-152
Iodine Number: 0.5 Maximum
Melting Point C: 53-55
Applications:
Nonionic acid stabilized emulsifier for cosmetic and pharmaceutical creams and lotions. Particularly applicable for anti-perspirant lotions, depilitory creams, hair straighteners, and emulsions containing a wide variety of medicaments. Forms opaque gels with mineral oil. Excellent emulsifier base for cleansing creams.

EMULSYNT GDL:
Chemical Name: Glyceryl Dilaurate
CTFA Name: Glyceryl Dilaurate
Physical Form: White to off-white soft solid
Specifications:
Acid Number: 5.0 Maximum
Saponification Number: 219-229
Applications:
EMULSYNT GDL is a water dispersible lipid which melts slightly below body temperature and has a distinct emulsifying characteristic as well as emollient feel which makes it a unique ingredient for creams and lotions.

VAN DYK: EMULSYNT and ESCALOL:

EMULSYNT 1055:
Chemical Name: Poloxyalkylene Oleate-Laurate
CTFA Name: Polyglyceryl-4 Oleate (And) PEG-8 Propylene Glycol Cocoate
Physical Form: Light amber liquid
Specifications:
Acid Number: 5.0 Maximum
Saponification Number: 142-152
Iodine Number: 58-68
Specific Gravity @ 25C: 0.960-0.980
Applications:
A very effective W/O emulsifier for cosmetic and pharmaceutical lotions and creams. Excellent stabilizer as auxiliary emulsifier for O/W systems.

ESCALOL 507:
Chemical Name: 2-Ethylhexyl p-Dimethylaminobenzoate
CTFA Name: Octyl Dimethyl PABA
CAS Number: 21245-02-3
Physical Form: Pale yellow mobile liquid
Odor: Very mild, characteristic
Molecular Formula: $C_{17}H_{25}O_2N$
Molecular Weight: 277
Specifications:
Acid Value: 1.0 maximum
Saponification Number: 195-215
Specific Gravity (25C): 0.990-1.000
Refractive Index (25C): 1.5390-1.5430
Applications:
Non-staining Categeory I UV-B sunscreen.
Approved use level: 1.8-8.0% in the U.S.
Required HLB: 10-12

ESCALOL 557:
Chemical Name: 2-Ethylhexyl p-Methoxycinnamate
CTFA Name: Octyl Methoxycinnamate
Molecular Formula: $C_{18}H_{26}O_3$
Molecular Weight: 290.4
CAS Number: 5466-77-3
Physical Form: Pale yellow liquid with slight, characteristically mild, odor.
Specifications:
Acid Number: 1.0 Max.
Saponification Number: 189 Min.
Refractive Index (25C): 1.542-1.548
Specific Gravity (25C): 1.005-1.013
Purity (GLC): 98% Min.
Applications:
Non-staining Category I UV-B sunscreen with worldwide acceptance and broad compatibility with most cosmetic ingredients.
Required HLB = 6-8

VAN DYK: ESCALOL and FOAMOLE:

ESCALOL 567:
Chemical Name: 2-Hydroxy-4-Methoxybenzophenone
CTFA Name: Benzophenone-3
CAS #: 131-57-7
Molecular Weight: 228.25
Physical Form: Slightly yellowish, fine crystalline powder
Molecular Formula: C14H12O3
Specifications:
Melting Point: 62C. Min.
Assay: 97.0-103.0% (USPXXI)
Loss on Drying: 2.0% Maximum
Applications:
This category I sunscreen exhibits broad UV absorbance in the UVB range and into the UVA range. Its worldwide acceptance makes it a natural choice, particularly when high SPF's are desired.

ESCALOL 587:
Chemical Name: 2-Ethylhexyl Salicylate
CTFA Name: Octyl Salicylate
CAS Number: 118-60-5
Physical Form: Colorless to pale yellow liquid
Odor: Typical, bland
Molecular Formula: C15H22O3
Molecular Weight: 250.34
Tentative Specifications:
Acid Value: 2 max.
Saponification Number: 200-230
Specific Gravity (25C): 1.013-1.022
Refractive Index: 1.495-1.505
Applications:
Category I UV B sunscreen
Approved use level: 3-5% in the U.S.

FOAMOLE A:
Chemical Name: Linoleic Alkanolamide (1:1)
CTFA Name: Linoleamide DEA
Physical Form: Amber Liquid
Specifications:
Acid Number: 5.0 Maximum
% Amine as DEA: 7.0 Maximum
Alkali Number: 21-35
Specific Gravity @ 25C: 0.972-0.982
Applications:
Excellent conditioning agent for hair dressings and shampoos. Also acts as a thickening agent and foam stabilizer.

VAN DYK: FOAMOLE:

FOAMOLE B:

Chemical Name: Minkamidopropyl Dimethylamine
CTFA Name: Minkamidopropyl Dimethylamine
Physical Form: Amber liquid to flowing gel

Typical Assay:

Alkali Number: 125-140
Acid Value: 5.0 Maximum
pH (1% Aq. Soln.): 9.5-10.5

Applications:

As a super-fatting agent in shampoos and other hair care products when used at 2-4% level. This fatty tertiary amino amide is water dispersible and surfactant and propylene glycol soluble. FOAMOLE B reacts with acids (i.e. lactic, phosphoric) to form a water soluble cationic surfactant which has emulsifying and emollient properties. In combination with neutral Glyceryl Monostearate, the FOAMOLE B-lactate or phosphate derivative forms stable, acid pH creams and lotions without the need for ethoxylated nonionics, thus enabling the use of parabens for preservation. The preferable pH for the lactate or phosphate salt to remain stable (no pH drift) is 3.5-4.2.

FOAMOLE M:

Chemical Name: Coconut Monoethanolamide
CTFA Name: Cocamide MEA
Physical Form: Cream colored flakes

Specifications:

Acid Number: 2.0 Maximum
Alkali Number: 12.0 Maximum
Melting Point C: 70-74

Applications:

Foam builder, stabilizer and thickener for shampoos, bubble baths, and other detergents. Emulsifier for creams and lotions, particularly useful in cream hair dyes.

VAN DYK: BI-LITE Pearlescent Pigments:

The BI-LITE pearls were developed as a less expensive alternative to pure bismuth oxychloride. They consist of a chemical deposition of bismuth oxychloride on platelets of muscovite mica. This "sandwich" or laminate structure provides excellent crystalline strength to withstand rigorous processing and handling while still maintaining a high degree of pearlescent luster typical of bismuth oxychloride type pearls. The mica also regulates the particle size and reduces the density of the pearl in comparison to 100% bismuth oxychloride. The optical reflectance is accentuated by the high refractive indices of mica and bismuth oxychloride, 1.59 and 2.15 respectively.

BI-LITE 20:

Bismuth Oxychloride Deposited on Mica
A translucent pearlescent pigment providing high luster, moderate density, and good skin adhesion.

Typical Analysis:
Assay BiOCl: 40%
Density: 3.8 gm/ml
Appearance: White pearlescent, free flowing powder
Bulk Density (Scott): 2-5 gm/cu inch
Particle Size: Average 15-25 microns
90% less than 44 microns
Lead: 20 ppm Maximum
Arsenic: 3 ppm Maximum
Mercury: 1 ppm Maximum
Microbiological: Total Count: 100 colonies/gram Maximum
Pathogens: Negative

Applications:
BI-LITE 20 can be used in pressed powders, creams and lipsticks or wherever a satiny, translucent pearlescent effect is desired. Alone or in combination with other pearlescent pigments, BI-LITE 20 offers excellent economy and performance. The deposition provides a non-separating, moderate density pearlescent pigment. Crystal strength is heightened by the laminated structure to permit rigorous processing without the loss of luster.

Formulation Considerations:
BI-LITE 20 is recommended in pressed powders at concentrations up to 25%. For higher concentrations, BI-LITE ULTRAPRESS, or PEARL-GLO pearls are recommended for ease of compressibility.

Regulatory Information: U.S. Patent No. 3,597,250
FDA CRMCS No. R0010979

VAN DYK: BI-LITE ULTRALITE:

Bismuth Oxychloride Deposited on Mica

A highly transparent pearlescent pigment providing low density, satiny luster, low coverage, and skin adhesion.

Typical Analysis:
- Assay BiOCl: 10%
- Density: 2.9 gm/ml
- Appearance: White pearlescent, free flowing powder
- Bulk Density (Scott): 2-4 gm/cu inch
- Particle Size: Average 15-30 microns
  75% less than 44 microns
- Lead: 20 ppm Maximum
- Arsenic: 3 ppm Maximum
- Mercury: 1 ppm Maximum
- Microbiological: Total Count: 100 colonies/gram, Maximum
  Pathogens: Negative

Applications:

BI-LITE ULTRALITE pigment is used in low and moderate concentrations as a cosmetic filler with the added qualities of skin adhesion, slip, and brightness of bismuth oxychloride. The deposition provides a non-separating, low density, highly transparent pearlescent pigment. Crystal strength is heightened by the laminated structure of the mica substrate.

Formulation Considerations:

BI-LITE ULTRALITE may be used in formulations which require moderate levels of pearlescence and cannot use high density pearlescent pigments. BI-LITE ULTRALITE may also be readily incorporated in wax, oil and water based cosmetic products.

Regulatory Information:
- U.S. Patent No. 3,597,250
- FDA CRMCS No. R0013103

VAN DYK: BI-LITE ULTRAPRESS:

Bismuth Oxychloride Deposited on Mica with a Surface Modification

A translucent, highly compressible pearlescent pigment providing a satiny luster, moderate opacity, and good skin adhesion.

Typical Analysis:
- Assay BiOCl: 40%
- Density: 3.7 gm/ml
- Appearance: White, pearlescent, free flowing powder
- Bulk Density (Scott): 4-6 gm/cu inch
- Particle Size: Average 15-25 microns
  85% less than 44 microns
- Lead: 20 ppm Maximum
- Arsenic: 3 ppm Maximum
- Mercury: 1 ppm Maximum
- Microbiological: Total Count: 100 colonies/gram Maximum
  Pathogens: Negative

Applications:

The outstanding compressibility of BI-LITE ULTRAPRESS pigment makes it especially useful in pressed powders, where a satiny luster and relatively low cost are desired.

Formulation Considerations:

BI-LITE ULTRAPRESS was developed primarily for pressed powder formulations, however, it can be used effectively in all types of cosmetic systems. It has exceptional flow and blending properties and mixes readily with waxes, oils and emollients.

Regulatory Information:
- U.S. Patent No. 3,597,250 and 3,656,982
- FDA CRMCS No. R0010979

VAN DYK: BI-LITE ULTRAWHITE:

Titanium Dioxide Bonded to Bismuth Oxychloride, Deposited on Mica

A pearlescent pigment providing satiny luster with high covering power (opacity), and good skin adhesion at low cost.

Typical Analysis:
- Assay BiOCl: 36%
- Assay TiO2: 5%
- Density: 3.5 gm/ml
- Appearance: White, pearlescent, free flowing powder
- Bulk Density (Scott): 2-4 gm/cu inch
- Particle Size: Average 15-25 microns
  85% less than 44 microns
- Lead: 20 ppm Maximum
- Arsenic: 3 ppm Maximum
- Mercury: 1 ppm Maximum
- Microbiology: Total Count: 100 colonies/gram, Maximum
  Pathogens: Negative

Applications:

BI-LITE ULTRAWHITE pearlescent pigment is used in pressed powders and creams where satiny luster and especially high high opacity are needed at low cost.

Formulation Considerations:

Added covering power or shade lightening may be accomplished using BI-LITE ULTRAWHITE alone, or in combination with other pearlescent pigments. A satiny luster may be imparted in powder or wax systems.

Regulatory Information:
- U.S. Patent No. 3,597,250 and 3,647,492
- FDA CRMCS No. R0010978

VAN DYK: CHROMA-LITE Pearlescent Pigments:

CHROMA-LITE pearlescent pigments are uniquely bonded combinatoons of colored pigments and BI-LITE 20 (Mica (and) Bismuth Oxychloride). The colored pigments are completely extended and bonded to the BI-LITE 20 using calcium stearate. In the same manner as BI-LITE ULTRAWHITE, the CHROMALITE pearls provide intense color and a subdued satiny luster.

CHROMA-LITE Black:

Colored pigment bonded to Bismuth Oxychloride Deposited on Mica.

The CHROMA-LITE pearls are colored pigments chemically bonded to the BI-LITE 20 substrate. The color and pearlescence in each crystalline particle produce a uniquely clear, intense, lustrous color which is readily blended, dispersed or compressed in cosmetic formulations.

Typical Analysis:
- Assay BiOCl: 30%
- Assay Fe2O3: 25%
- Appearance: Intensely colored pearlescent powder
- Bulk Density (Scott): 3-6 gm/cu inch
- Particle Size: 90% less than 44 microns
- Density: 3.5 gm/ml
- Lead: 20 ppm Maximum
- Arsenic: 3 ppm Maximum
- Mercury: 1 ppm Maximum
- Microbiological: Total Count: 100 colonies/gran, Maximum
Pathogens: Negative

Applications:

In pressed powders and cream eye shadows, as well as all other colored make-up products, the CHROMA-LITE pigments offer the advantages of blending, rather than milling color into the product, excellent skin adhesion, good pearlescence, and a unique optical appearance. The bonding of the colored pigment minimizes pigment migration which results in "creasing".

Formulation Considerations:

Rigorous milling should be minimized to avoid loss of pearlescence and pigment separation from the platelet.

Regulatory Information:
- U.S. Patent Nos. 3,597,250 and 3,647,492
- FDA CRMCS No. R0010980

VAN DYK: CHROMA-LITE Bronze:

Colored pigment bonded to Bismuth Oxychloride Deposited on Mica.

The CHROMA-LITE pearls are colored pigments chemically bonded to the BI-LITE 20 substrate. The color and pearlescence in each crystalline particle produce a uniquely clear, intense, lustrous color which is readily blended, dispersed or compressed in cosmetic formulations.

Typical Analysis:

Assay BiOCl: 30-38%
Assay Fe2O3: 25%
Appearance: Intensely colored pearlescent powder
Bulk Density (Scott): 3-6 gm/cu inch
Particle Size: 90% less than 44 microns
Density: 3.5 gm/ml
Lead: 20 ppm Maximum
Arsenic: 3 ppm Maximum
Mercury: 1 ppm Maximum
Microbiological: Total Count: 100 colonies/gram, Maximum
Pathogens: Negative

Applications:

In pressed powders and cream eye shadows, as well as all other colored makeup products, the CHROMA-LITE pigments offer the advantages of blending, rather than milling color into the product, excellent skin adhesion, good pearlescence, and a unique optical appearance. The bonding of the colored pigment minimizes pigment migration which results in "creasing".

Regulatory Information:

U.S. Patent Nos. 3,597,250 and 3,647,492
FDA CRMCS No. R0010980

CHROMA-LITE Light Blue:

Colored pigment bonded to Bismuth Oxychloride Deposited on Mica.

Typical Analysis:

Assay BiOCl: 38%
Assay FeNH4Fe(CN)6: 5%
Appearance: Intensely colored pearlescent powder
Bulk Density (Scott): 3-6 gm/cu inch
Particle Size: 90% less than 44 microns
Density: 3.5 gm/ml
Lead: 20 ppm Maximum
Arsenic: 3 ppm Maximum
Mercury: 1 ppm Maximum
Microbiological: Total Count: 100 colonies/gram, Maximum
Pathogens: Negative

Regulatory Information: U.S. Patent Nos. 3,597,250 and 3,647,492
FDA CRMCS No. R0010980

VAN DYK: CHROMA-LITE Dark Blue:

Colored pigment bonded to Bismuth Oxychloride Deposited on Mica

Typical Analysis:
Assay BiOCl: 35%
Assay FeNH4Fe(CN)6: 13%
Appearance: Intensely colored pearlescent powder
Bulk Density (Scott): 3-6 gm/cu inch
Particle Size: 90% less than 44 microns
Density: 3.5 gm/ml
Lead: 20 ppm Maximum
Arsenic: 3 ppm Maximum
Mercury: 1 ppm Maximum
Microbiological: Total Count: 100 colonies/gram, Maximum
Pathogens: Negative

Regulatory Information: U.S. Patent Nos. 3,597,250 and 3,647,492
FDA CRMCS No. R0010980

CHROMA-LITE Yellow:

Colored pigment bonded to Bismuth Oxychloride Deposited on Mica.

Typical Analysis:
Assay BiOCl: 30-38%
Appearance: Intensely colored pearlescent powder
Bulk Density (Scott): 3-6 gm/cu inch
Particle Size: 90% less than 44 microns
Density: 3.5 gm/ml
Lead: 20 ppm Maximum
Arsenic: 3 ppm Maximum
Mercury: 1 ppm Maximum
Microbiological: Total Count: 100 colonies/gram, Maximum
Pathogens: Negative

Regulatory Information: U.S. Patent Nos. 3,597,250 and 3,647,492
FDA CRMCS No. R0010980

VAN DYK: CHROMA-LITE Green:

Colored pigment bonded to Bismuth Oxychloride Deposited on Mica.

Typical Analysis:
- Assay BiOCl: 30%
- Assay Chromium Oxide Green: 25%
- Appearance: Intensely colored pearlescent powder
- Bulk Density (Scott): 3-6 gm/cu inch
- Particle Size: 90% less than 44 microns
- Density: 3.5 gm/ml
- Lead: 20 ppm Maximum
- Arsenic: 3 ppm Maximum
- Mercury: 1 ppm Maximum
- Microbiological: Total Count: 100 colonies/gram, Maximum
  Pathogens: Negative

Regulatory Information: U.S. Patent Nos. 3,597,250 and 3,647,492
FDA CRMCS No. R0010880

CHROMA-LITE Gold:

Colored pigment bonded to Bismuth Oxychloride Deposited on Mica.

Typical Analysis:
- Assay BiOCl: 30%
- Assay Fe2O3: 25%
- Appearance: Intensely colored pearlescent powder
- Bulk Density (Scott): 3-6 gm/cu inch
- Particle Size: 90% less than 44 microns
- Density: 3.5 gm/ml
- Lead: 20 ppm Maximum
- Arsenic: 3 ppm Maximum
- Mercury: 1 ppm Maximum
- Microbiogical: Total Count: 100 colonies/gram, Maximum
  Pathogens: Negative

Regulatory Information: U.S. Patent Nos. 3,597,250 and 3,647,492
FDA CRMCS No. R0010980

VAN DYK: CHROMA-LITE Magenta:

Colored pigment bonded to Bismuth Oxychloride Deposited on Mica.

Typical Analysis:
Assay BiOCl: 38%
Assay Carmine: 5%
Appearance: Intensely colored pearlescent powder
Bulk Density (Scott): 3-6 gm/cu inch
Particle Size: 90% less than 44 microns
Density: 3.5 gm/ml
Lead: 20 ppm Maximum
Arsenic: 3 ppm Maximum
Mercury: 1 ppm Maximum
Microbiological: Total Count: 100 colonies/gram, Maximum
Pathogens: Negative

Regulatory Information: U.S. Patent Nos. 3,597,250 and 3,647,492
FDA CRMCS No. R0010980

CHROMA-LITE Red:

Colored pigment bonded to Bismuth Oxychloride Deposited on Mica.

Typical Analysis:
Assay BiOCl: 30%
Assay Fe2O3: 25%
Appearance: Intensely colored pearlescent powder
Bulk Density (Scott): 3-6 gm/cu inch
Particle Size: 90% less than 44 microns
Density: 3.5 gm/ml
Lead: 20 ppm Maximum
Arsenic: 3 ppm Maximum
Mercury: 1 ppm Maximum
Microbiological: Total Count: 100 colonies/gram, Maximum
Pathogens: Negative

Regulatory Information: U.S. Patent Nos. 3,597,250 and 3,647,492
FDA CRMCS No. R0010980

VAN DYK: CHROMA-LITE Violet:

Colored pigment bonded to Bismuth Oxychloride Deposited on Mica.

Typical Analysis:
- Assay BiOCl: 30%
- Assay Manganese Violet: 25%
- Appearance: Intensely colored pearlescent powder
- Bulk Density (Scott): 3-6 gm/cu inch
- Particle Size: 90% less than 44 microns
- Density: 3.5 gm/ml
- Lead: 20 ppm Maximum
- Arsenic: 3 ppm Maximum
- Mercury: 1 ppm Maximum
- Microbiological: Total Count: 100 colonies/gram, Maximum
Pathogens: Negative

Regulatory Information: U.S. Patent Nos. 3,597,250 and 3,647,492
FDA CRMCS No. R0010980

CHROMA-LITE Aqua:

Colored pigment bonded to Bismuth Oxychloride Deposited on Mica.

Typical Analysis:
- Assay BiOCl: 30%
- Assay Chromium Hydroxide Green: 25%
- Appearance: Intensely colored pearlescent powder
- Bulk Density (Scott): 3-6 gm/cu inch
- Particle Size: 90% less than 44 microns
- Density: 3.5 gm/ml
- Lead: 20 ppm Maximum
- Arsenic: 3 ppm Maximum
- Mercury: 1 ppm Maximum
- Microbiological: Total Count: 100 colonies/gram, Maximum
Pathogens: Negative

Regulatory Information: U.S. Patent Nos. 3,597,250 and 3,647,492
FDA CRMCS No. R0010980

VAN DYK: CHROMA-LITE Brown:

Colored pigment bonded to Bismuth Oxychloride Deposited on Mica.

Typical Analysis:
- Assay BiOCl: 30-38%
- Assay Fe2O3: 25%
- Appearance: Intensely colored pearlescent powder
- Bulk Density (Scott): 3-6 gm/cu inch
- Particle Size: 90% less than 44 microns
- Density: 3.5 gm/ml
- Lead: 20 ppm Maximum
- Arsenic: 3 ppm Maximum
- Mercury: 1 ppm Maximum
- Microbiological: Total Count: 100 colonies/gram, Maximum
  Pathogens: Negative

Regulatory Information: U.S. Patent Nos. 3,597,250 and 3,647,492
FDA CRMCS No. R0010980

CHROMA-LITE Purple:

Colored pigment bonded to Bismuth Oxychloride Deposited on Mica.

Typical Analysis:
- Assay BiOCl: 30%
- Assay Ultramarine Blue: 25%
- Appearance: Intensely colored pearlescent powder
- Bulk Density (Scott): 3-6 gm/cu inch
- Particle Size: 90% less than 44 microns
- Density: 3.5 gm/ml
- Lead: 20 ppm Maximum
- Arsenic: 3 ppm Maximum
- Mercury: 1 ppm Maximum
- Microbiological: Total Count: 100 colonies/gram, Maximum
  Pathogens: Negative

Regulatory Information: U.S. Patent Nos. 3,597,250 and 3,647,492
FDA CRMCS No. R0010980

VAN DYK: CHROMA-LITE Mauve:

Colored pigment bonded to Bismuth Oxychloride Deposited on Mica.

Typical Analysis:
- Assay BiOCl: 30-38%
- Assay Fe2O3: 25%
- Appearance: Intensely colored pearlescent powder
- Bulk Density (Scott): 3-6 gm/cu inch
- Particle Size: 90% less than 44 microns
- Density: 3.5 gm/ml
- Lead: 20 ppm Maximum
- Arsenic: 3 ppm Maximum
- Mercury: 1 ppm Maximum
- Microbiological: Total Count: 100 colonies/gram, Maximum
  Pathogens: Negative

Regulatory Information: U.S. Patent Nos. 3,597,250 and 3,647,492
FDA CRMCS No. R0010980

VAN DYK: LUSTRA-PEARL Pigments:

VAN DYK: LUSTRA-PEARL Amethyst:

Titanium Dioxide Deposited on Mica

A highly lustrous pearlescent pigment providing exceptional ultraviolet light stability and low density.

Typical Analysis:
- Assay TiO_2: 20%
- Assay Mica: 60%
- Assay Fe_2O_3: 20%
- Density: 2.0-4.0 gm/ml
- Appearance: Mauve colored, free flowing pearlescent powder
- Bulk Density (Scott): 4-6 gm/cu in.
- Particle Size: 90% less than 74 microns
- Lead: 20 ppm Maximum
- Arsenic: 3 ppm Maximum
- Mercury: 1 ppm Maximum
- Microbiological: Total Count: 100 colonies/gram, Maximum
  Pathogens: Negative

Applications:
LUSTRA-PEARL Amethyst pigment is used in pressed powders, creams, lipsticks, nail lacquers and lotions to obtain a shiny sparkle. Alone or in combination with other pigments, LUSTRA-PEARL Amethyst pigment offers excellent economy in cost and performance. The deposition provides a non-separating, moderate density pearlescent pigment. Crystal strength is heightened by the laminated structure to permit rigorous mixing and milling without the loss of luster.

Regulatory Information: CTFA Name: Mica (and) Titanium Dioxide

VAN DYK: LUSTRA-PEARL Glimmer:

Titanium Dioxide Deposited on Mica

A highly lustrous pearlescent pigment providing exceptional ultraviolet light stability and low density.

Typical Analysis:
- Assay TiO_2: 22%
- Density: 2.7 gm/ml
- Appearance: White pearlescent, free flowing powder
- Bulk Density (Scott): 4.0-4.5 gm/cu in.
- Particle Size: Average 8-15 microns
  85% less than 44 microns
- Lead: 20 ppm Maximum
- Arsenic: 3 ppm Maximum
- Mercury: 1 ppm Maximum
- Microbiological: Total Count: 100 colonies/gram, Maximum
  Pathogens: Negative

Applications:

LUSTRA-PEARL Glimmer pigment is used in pressed powders, creams, lipsticks, nail lacquers and lotions to obtain a shiny sparkle. Alone or in combination with other pigments, LUSTRA-PEARL Glimmer pigment offers excellent economy in cost and performance. The deposition provides a non-separating, moderate density pearlescent pigment. Crystal strength is heightened by the laminated structure to permit rigorous mixing and milling without the loss of luster.

Formulation Considerations:

LUSTRA-PEARL Glimmer pigment is recommended in pressed powders at concentrations up to 35%. For higher concentrations, an addition of 5-10% BI-LITE Ultrapress or PEARL-GLO pigments are recommended to aid compressibility.

Regulatory Consideration:

CTFA Name: Mica (and) Titanium Dioxide

VAN DYK: LUSTRA-PEARL Gloss:

Titanium Dioxide Deposited on Mica
A highly lustrous pearlescent pigment providing exceptional ultraviolet light stability and low density.

Typical Analysis:
Assay TiO2: 24%
Density: 2.8 gm/ml
Appearance: White pearlescent, free flowing powder
Bulk Density (Scott): 3.0-4.0 gm/cu in
Particle Size: Average 5-14 microns
85% less than 44 microns
Lead: 20 ppm Maximum
Arsenic: 3 ppm Maximum
Mercury: 1 ppm Maximum
Microbiological: Total Count: 100 colonies/gram, Maximum
Pathogens: Negative

Applications:
LUSTRA-PEARL Gloss pigment is used in pressed powders, creams, lipsticks, nail lacquers and lotions to obtain a glossy sparkle. Alone or in combination with other pigments, LUSTRA-PEARL Gloss pigment offers excellent economy in cost and performance. The deposition provides a non-separating, moderate density pearlescent pigment. Crystal strength is heightened by the laminated structure to permit rigorous mixing and milling without the loss of luster.

Formulation Considerations:
LUSTRA-PEARL Gloss pigment is recommended in pressed powders at concentrations up to 35%. For higher concentrations, an addition of 5-10% BI-LITE Ultrapress or PEARL-GLO pigments are recommended to aid compressibility.

Regulatory Information:
CTFA Name: Mica (and) Titanium Dioxide

VAN DYK: LUSTRA-PEARL Gold:

Titanium Dioxide Deposited on Mica

A highly lustrous pearlescent pigment providing exceptional ultraviolet light stability and low density.

Typical Analysis:
- Assay TiO2: 35%
- Density: 3.4 gm/ml
- Appearance: Yellow-gold pearlescent, free flowing powder
- Bulk Density (Scott): 3.0-4.0 gm/cu in.
- Particle Size: Average 5-15 microns
  85% less than 44 microns
- Lead: 20 ppm Maximum
- Arsenic: 3 ppm Maximum
- Mercury: 1 ppm Maximum
- Microbiological: Total Count: 100 colonies/gram, Maximum
  Pathogens: Negative

Applications:
LUSTRA-PEARL Gold is used in pressed powders, creams, lipsticks, nail lacquers and lotions to obtain a glossy sparkle. Alone or in combination with other pigments, LUSTRA-PEARL Gold pigment offers excellent economy in cost and performance. The deposition provides a non-separating, moderate density pearlescent pigment. Crystal strength is heightened by the laminated structure to permit rigorous mixing and milling without the loss of luster.

Formulation Considerations:
LUSTRA-PEARL Gold pigment is recommended in pressed powders at concentrations up to 45%. For higher concentrations, BI-LITE Ultrapress or PEARL-GLO pigments are recommended to aid compressibility.

Regulatory Information:
CTFA Name: Mica (and) Titanium Dioxide (and) Iron Oxide

VAN DYK: LUSTRA-PEARL Satin:

Titanium Dioxide Deposited on Mica

A highly lustrous pearlescent pigment providing exceptional ultraviolet light stability and low density.

Typical Analysis:

Assay TiO_2: 27%
Density: 3.0 gm/ml
Appearance: White pearlescent, free flowing powder
Bulk Density (Scott): 2.5-3.5 gm/cu in.
Particle Size: Average 4-11 microns
90% less than 14 microns
Lead: 20 ppm Maximum
Arsenic: 3 ppm Maximum
Mercury: 1 ppm Maximum
Microbiological: Total Count: 100 colonies/gram, Maximum
Pathogens: Negative

Applications:

LUSTRA-PEARL Satin pigment is used in pressed powders, creams, lipsticks, nail lacquers and lotions to obtain a glossy satin luster. Alone or in combination with other pigments, LUSTRA-PEARL Satin pigment offers excellent economy in cost and performance. The deposition provides a non-separating, moderate density pearlescent pigment. Crystal strength is heightened by the laminated structure to permit rigorous mixing and milling without the loss of luster.

Formulation Considerations:

LUSTRA-PEARL Satin pigment is recommended in pressed powders at concentrations up to 40%. For higher concentrations, BI-LITE Ultrapress or PEARL-GLO pigments are recommended to aid compressibility.

Regulatory Information:

CTFA Name: Mica (and) Titanium Dioxide

VAN DYK: LUSTRA-PEARL Silk:

Titanium Dioxide Deposited on Mica

A highly lustrous pearlescent pigment providing exceptional ultraviolet light stability and low density.

Typical Analysis:
Assay TiO2: 30%
Density: 3.0 gm/ml
Appearance: White pearlescent, free flowing powder
Bulk Density (Scott): 2.4-3.4 gm/cu in.
Particle Size: Average 3-9 microns
95% less than 44 microns
Lead: 20 ppm Maximum
Arsenic: 3 ppm Maximum
Mercury: 1 ppm Maximum
Microbiological: Total Count: 100 colonies/gram, Maximum
Pathogens: Negative

Applications:
LUSTRA-PEARL Silk pigment is used in pressed powders, creams, lipsticks, nail lacquers and lotions to obtain a silky sheen. Alone or in combination with other pigments, LUSTRA-PEARL Silk pigment offers excellent economy in cost and performance. The deposition provides a non-separating, moderate density pearlescent pigment. Crystal strength is heightened by the laminated structure to permit rigorous mixing and milling without the loss of luster.

Formulation Considerations:
LUSTRA-PEARL Silk pigment is recommended in pressed powders at concentrations up to 45%. For higher concentrations, an addition of 5-10% BI-LITE Ultrapress or PEARL-GLO pigments are recommended to aid compressibility.

Regulatory Information:
CTFA Name: Mica (and) Titanium Dioxide

VAN DYK: PEARL-GLO Pearlescent Pigments:

PEARL-GLO pearlescent pigments are pure bismuth oxychloride pigments designed for cosmetics which require a high degree of luster and opacity. PEARL-GLO pigments provide a deep, satiny luster with high covering power and optimum compressibility. Their superior skin adhesion makes them excellent pigments for cosmetics which are formulated as pressed or loose powders to be worn for long periods.

PEARL-GLO:

Bismuth Oxychloride
A highly lustrous pigment with excellent covering power, skin adhesion and compressibility.

Typical Analysis:
Assay BiOCl: 98% Minimum
Density: 7.7 gm/ml
Appearance: White, pearlescent, free flowing powder
Bulk (Scott): 9-13 gm/cu inch
Particle Size: Average 6-15 microns
95% less than 44 microns
100% less than 74 microns
Lead: 20 ppm Maximum
Arsenic: 3 ppm Maximum
Mercury: 1 ppm Maximum
Microbiological: Total Count: 100 colonies/gram, Maximum
Pathogens: Negative

Applications:
PEARL-GLO is a highly pearlescent pigment which provides an optimum balance of luster, opacity and compressibility in pressed powders, lipsticks and other make-up products. In powder formulations, PEARL-GLO imparts a more satiny pearlescence and greater adhesion to the skin than non-bismuth based pearlescent pigments.

Regulatory Information: FDA CRMCS No. R0010976

VAN DYK: PEARL-GLO M:

Bismuth Oxychloride

A lustrous pearl with excellent covering power, skin adhesion, and compressibility in a UV light resistant form, PEARL-GLO is similar to PEARL-GLO UVR but with higher covering power and smaller particle size.

Typical Analysis:

Assay BiOCl: 98% Minimum
Density: 7.7 gm/ml
Appearance: White, pearlescent, free flowing powder
Bulk Density (Scott): 6-10 gm/cu inch
Particle Size: Average 4-10 microns
98% less than 44 microns
100% less than 74 microns
Lead: 20 ppm Maximum
Arsenic: 3 ppm Maximum
Mercury: 1 ppm Maximum
Microbiological: Total Count: 100 colonies/gram, maximum
Pathogens: Negative

Applications:

For make-up products in transparent packaging, PEARL-GLO M offers the superior satiny pearlescence, opacity, and compressibility of bismuth oxychloride in a light resistant form. Since PEARL-GLO M has a slightly smaller particle size than PEARL-GLO UVR it delivers a somewhat higher degree of opacity. In all types of packaging PEARL-GLO M provides the formulator with extra insurance against light-induced darkening of bismuth oxychloride pearls in pressed powders, anhydrous creams and gels, and stick products.

Regulatory Information: U.S. Patent No. 3,917,671
FDA CRMCS No. R0011891

VAN DYK: PEARL-GLO UVR:

Bismuth Oxychloride

A highly lustrous pigment with excellent covering power, skin adhesion and compressibility. PEARL-GLO UVR is unique among bismuth oxychloride type pearls in being extremely resistant to the graying effect of ultraviolet light, without the addition of UV light absorbers. PEARL-GLO M and PEARL-GLO SF-UVR offer the same advantage, but have different particle sizes.

Typical Analysis:

Assay BiOCl: 98% Minimum
Density: 7.7 gm/ml
Appearance: White, pearlescent, free flowing powder
Bulk Density (Scott): 7-10 gm/cu. inch
Particle Size: Average 6-15 microns
90% less than 44 microns
97% less than 74 microns
100% less than 105 microns
Lead: 20 ppm Maximum
Arsenic: 3 ppm Maximum
Mercury: 1 ppm Maximum
Microbiological: Total Count: 100 colonies/gram, Maximum
Pathogens: Negative

Applications:

For make-up products in transparent packaging, PEARL-GLO UVR offers the superior satiny pearlescence, opacity, and compressibility of bismuth oxychloride in light-resistant form. In all types of packaging, PEARL-GLO UVR provides the formulator with extra insurance against light-induced darkening of bismuth oxychloride pearls in pressed powders, anhydrous cremes, and stick products.

Regulatory Information: U.S. Patent No. 3,917,671
FDA CRMCS No. R0011891

VAN DYK: PEARL GLO F:

Bismuth Oxychloride

A lustrous pigment with excellent covering power, skin adhesion, compressibility and a fine particle size.

Typical Analysis:

Assay BiOCl: 98% Minimum
Density: 7.7 gm/ml
Appearance: White, pearlescent, free flowing powder
Bulk Density (Scott): 7-13 gm/cu inch
Particle Size: Average 1-8 microns
98% less than 44 microns
100% less than 74 microns
Lead: 20 ppm Maximum
Arsenic: 2 ppm Maximum
Mercury: 1 ppm Maximum
Microbiological: Total Count: 100 colonies/gram, Maximum
Pathogens: Negative

Applications:

PEARL-GLO is a highly pearlescent pigment which provides a balance of luster, opacity and compressibility in pressed powders, pencils, sticks and other make-up products. In formulations, PEARL-GLO F imparts a greater adhesion to the skin than other bismuth based pearls. The fine particle size offers an elegant appearance and texture to the finished product.

Regulatory Information: FDA CRMCS No. R0010976

PEARL-GLO SF:

Bismuth Oxychloride

A pearlescent pigment with excellent flowability and dispersability providing exceptional covering power, skin adhesion, compressibility with an extremely fine particle size.

Typical Analysis:

Assay BiOCl: 98% Minimum
Density: 7.7 gm/ml
Appearance: White pearlescent, free flowing powder
Bulk Density (Scott): 10 gm/cu inch
Particle Size: Average 3-8 microns
99.5% less than 44 microns
100% less than 74 microns
Lead: 20 ppm Maximum
Arsenic: 3 ppm Maximum
Mercury: 1 ppm Maximum
Microbiological: Total Count: 100 colonies/gram, Maximum
Pathogens: Negative

VAN DYK: PEARL-GLO SF(Continued):

Applications:

PEARL-GLO SF has applications in various types of eye shadows, blushers, lip and general face make-up formulations. The unique flow, bulk and dispersion characteristics aid pressed powder, extrusion molded and cream type cosmetic products. As with all PEARL-GLO pearls, the satiny luster, skin adhesion and elegant skin feel complement any product in which they are used.

Regulatory Information:

FDA CRMCS No. R0010976

PEARL-GLO SF UVR:

Bismuth Oxychloride

A lustrous pigment with excellent covering power, skin adhesion and dispersibility with an extremely fine particle size. PEARL-GLO SF UVR is similar to the other ultraviolet stable bismuth oxychloride by resisting graying without addition of UV light absorbers.

Typical Analysis:

Assay BiOCl: 98% Minimum
Density: 7.7 gm/ml
Appearance: White, pearlescent, free flowing powder
Bulk Density (Scott): 7-10 gm/cu inch
Particle Size: Average 3-15 microns
90% less than 44 microns
100% less than 105 microns
Lead: 20 ppm Maximum
Arsenic: 3 ppm Maximum
Mercury: 1 ppm Maximum
Microbiological: Total Count: 100 colonies/gram, Maximum
Pathogens: Negative

Applications:

For make-up products in transparent packaging, PEARL-GLO SF UVR offers the superior satiny pearlescence, opacity, and compressibility of bismuth oxychloride with in a light-resistant form and an extremely fine particle size. In all types of packaging, PEARL-GLO SF UVR provides the formulator with extra insurance against light-induced darkening of bismuth oxychloride pearls in pressed powders, anhydrous cremes, and stick products.

Regulatory Information:

U.S. Patent No. 3,917,671
FDA CRMCS No. R0011891

VAN DYK: SPECTRA-PEARL Pigments:

SPECTRA-PEARL BLW:

Colored inorganic pigments bonded to titanium dioxide deposited on Mica.

The SPECTRA-PEARL pigments are chemically bonded colors to a pearlescent pigment substrate. The combination of the pearlescence and color in each crystalline particle produces a unique non-separating, intensely colored, pearlescent pigment with a sparkle appearance. The SPECTRA-PEARL pigments are readily blended or dispersed either wet or dry in cosmetic formulations.

Typical Analysis:

Assay Mica: 66%
Titanium Dioxide: 18%
Assay FeNH4Fe(CN)6: 15%
Density: 2.0-4.0 gm/ml
Appearance: Intensely colored, blue pearlescent powder
Bulk Density (Scott): 4-6 gm/cu in.
Particle Size: 90% less than 74 microns
Lead: 20 ppm Maximum
Arsenic: 3 ppm Maximum
Mercury: 1 ppm Maximum
Microbiological: Total Count: 100 colonies/gram, Maximum
Pathogens: Negative

Applications:

In wax based systems, SPECTRA-PEARL pigments provide a high sparkle appearance which may be used with bismuth oxychloride pigments or alone. In pressed powders, low SPECTRA-PEARL pigment concentrations create a unique surface effect and optical appearance. The bonding of the colored pigment reduces pigment migration which may result in "creasing".

Regulatory Information: U.S. Patent No. 3,647,492
FDA CRMCS No. R0012218

SPECTRA-PEARL GNW:

Colored inorganic pigments bonded to titanium dioxide deposited on Mica.

Typical Analysis:

Assay Mica: 50%
Titanium Dioxide: 14%
Chromium Oxide: 35%
Density: 2.0-4.0 gm/ml
Appearance: Intensely colored, green pearlescent powder
Bulk Density (Scott): 4-6 gm/cu in.
Particle Size: 90% less than 74 microns
Lead: 20 ppm Maximum
Arsenic: 3 ppm Maximum
Mercury: 1 ppm Maximum
Microbiological: Total Count: 100 colonies/gram, Maximum
Pathogens: Negative

VAN DYK: SPECTRA-PEARL BKW:

Colored inorganic pigments bonded to titanium dioxide deposited on Mica.

Typical Analysis:

Assay Mica: 50%
Titanium Dioxide: 14%
Iron Oxide: 35%
Density: 2.0-4.0 gm/ml
Appearance: Intensely colored, pearlescent powder
Bulk Density (Scott): 4-6 gm/cu in.
Particle Size: 90% less than 74 microns
Lead: 20 ppm Maximum
Arsenic: 3 ppm Maximum
Mercury: 1 ppm Maximum
Microbiological: Total Count: 100 colonies/gram, Maximum
Pathogens: Negative

SPECTRA-PEARL PLW:

Colored inorganic pigments bonded to titanium dioxide deposited on Mica.

Typical Analysis:

Assay Mica: 50%
Titanium Dioxide: 14%
Ultramarine Blue: 35%
Density: 2.0-4.0 gm/ml
Appearance: Intensely colored, purple pearlescent powder
Bulk Density (Scott): 4-6 gm/cu in.
Particle Size: 90% less than 74 microns
Lead: 20 ppm Maximum
Arsenic: 3 ppm Maximum
Mercury: 1 ppm Maximum
Microbiological: Total Count: 100 colonies/gram, Maximum
Pathogens: Negative

SPECTRA-PEARL BNW:

Colored inorganic pigments bonded to titanium dioxide deposited on Mica.

Typical Analysis:

Assay Mica: 50%
Titanium Dioxide: 14%
Iron Oxide: 35%
Density: 2.0-4.0 gm/ml
Appearance: Intensely colored, brown pearlescent powder
Bulk Density (Scott): 4-6 gm/cu in.
Particle Size: 90% less than 74 microns
Lead: 20 ppm Maximum
Arsenic: 3 ppm Maximum
Mercury: 1 ppm Maximum
Microbiological: Total Count: 100 colonies/gram, Maximum
Pathogens: Negative

VAN DYK: SPECTRA-PEARL YWG:

Colored inorganic pigments bonded to titanium dioxide deposited on Mica.

Typical Analysis:

Assay Mica: 45%
Titanium Dioxide: 19%
Iron Oxide: 35%
Density: 2.0-4.0 gm/ml
Appearance: Intensely colored, gold pearlescent powder
Bulk Density (Scott): 4-6 gm/cu in.
Particle Size: 90% less than 74 microns
Lead: 20 ppm Maximum
Arsenic: 3 ppm Maximum
Mercury: 1 ppm Maximum
Mirobiological: Total Count: 100 colonies/gram, Maximum
Pathogens: Negative

SPECTRA-PEARL BLG:

Colored inorganic pigments bonded to titanium dioxide deposited on Mica.

Typical Analysis:

Assay Mica: 59%
Titanium Dioxide: 25%
FeNH4Fe(CN)6: 25%
Density: 2.0-4.0 gm/ml
Appearance: Intensely colored, blue pearlescent powder with a gold background
Bulk Density (Scott): 4-6 gm/cu in.
Particle Size: 90% less than 74 microns
Lead: 20 ppm Maximum
Arsenic: 3 ppm Maximum
Mercury: 1 ppm Maximum
Microbiological: Total Count: 100 colonies/gram, Maximum
Pathogens: Negative

SPECTRA-PEARL GNG:

Colored inorganic pigments bonded to titanium dioxide deposited on Mica.

Typical Analysis:

Assay Mica: 45%
Titanium Dioxide: 19%
Chrome Oxide: 35%
Density: 2.0-4.0 gm/ml
Appearance: Intensely colored, green pearlescent powder with a gold background
Bulk Density (Scott): 4-6 gm/cu in.
Particle Size: 90% less than 74 microns
Lead: 20 ppm Maximum
Arsenic: 3 ppm Maximum
Mercury: 1 ppm Maximum
Microbiological: Total Count: 100 colonies/gram, Maximum
Pathogens: Negative

VAN DYK: SPECTRA-PEARL BKG:

Colored inorganic pigments bonded to titanium dioxide deposited on Mica.

Typical Analysis:

- Assay Mica: 45%
- Titanium Dioxide: 19%
- Iron Oxide: 35%
- Density: 2.0-4.0 gm/ml
- Appearance: Intensely colored, black pearlescent powder with a gold background
- Bulk Density (Scott): 4-6 gm/cu in.
- Particle Size: 90% less than 74 microns
- Lead: 20 ppm Maximum
- Arsenic: 3 ppm Maximum
- Mercury: 1 ppm Maximum
- Microbiological: Total Count: 100 colonies/gram, Maximum
  Pathogens: Negative

SPECTRA-PEARL BNG:

Colored inorganic pigments bonded to titanium dioxide deposited on Mica.

Typical Analysis:

- Assay Mica: 45%
- Titanium Dioxide: 19%
- Iron Oxide: 35%
- Density: 2.0-4.0 gm/ml
- Appearance: Intensely colored, brown pearlescent powder with gold background
- Bulk Density (Scott): 4-6 gm/cu in.
- Particle Size: 90% less than 74 microns
- Lead: 20 ppm Maximum
- Arsenic: 3 ppm Maximum
- Mercury: 1 ppm Maximum
- Microbiological: Total Count: 100 colonies/gram, Maximum
  Pathogens: Negative

SPECTRA-PEARL MTW:

Colored inorganic pigments bonded to titanium dioxide deposited on Mica.

Typical Analysis:

- Assay Mica: 65%
- Titanium Dioxide: 25%
- Carmine: 10%
- Density: 2.0-4.0 gm/ml
- Appearance: Intensely colored, bright red pearlescent powder
- Bulk Density (Scott): 4-6 gm/cu in.
- Particle Size: 90% less than 74 microns
- Lead: 20 ppm Maximum
- Arsenic: 3 ppm Maximum
- Mercury: 1 ppm Maximum
- Microbiological: Total Count: 100 colonies/gram, Maximum
  Pathogens: Negative

VAN DYK: SPECTRA-PEARL MTG:

Colored inorganic pigments bonded to titanium dioxide deposited on Mica.

Typical Analysis:
- Assay Mica: 65%
- Titanium Dioxide: 25%
- Carmine: 10%
- Density: 2.0-4.0 gm/ml
- Appearance: Intensely colored, bright red pearlescent powder with a gold background
- Bulk Density (Scott): 4-6 gm/cu in.
- Particle Size: 90% less than 74 microns
- Lead: 20 ppm Maximum
- Arsenic: 3 ppm Maximum
- Mercury: 1 ppm Maximum
- Microbiological: Total Count: 100 colonies/gram, Maximum
  Pathogens: Negative

SPECTRA-PEARL RDW:

Colored inorganic pigments bonded to titanium dioxide deposited on Mica.

Typical Analysis:
- Assay Mica: 45%
- Titanium Dioxide: 20%
- Iron Oxide: 35%
- Density: 2.0-4.0 gm/ml
- Appearance: Intensely colored, red pearlescent powder
- Bulk Density (Scott): 4-6 gm/cu in.
- Particle Size: 90% less than 74 microns
- Lead: 20 ppm Maximum
- Arsenic: 3 ppm Maximum
- Mercury: 1 ppm Maximum
- Microbiological: Total Count: 100 colonies/gram, Maximum
  Pathogens: Negative

SPECTRA-PEARL RDG:

Colored inorganic pigments bonded to titanium dioxide deposited on Mica.

Typical Analysis:
- Assay Mica: 45%
- Titanium Dioxide: 20%
- Iron Oxide: 35%
- Density: 2.0-4.0 gm/ml
- Appearance: Intensely colored, red pearlescent powder with a gold background
- Bulk Density (Scott): 4-6 gm/cu in.
- Particle Size: 90% less than 74 microns
- Lead: 20 ppm Maximum
- Arsenic: 3 ppm Maximum
- Mercury: 1 ppm Maximum
- Microbiological: Total Count: 100 colonies/gram, Maximum
  Pathogens: Negative

WAITAKI INTERNATIONAL BIOSCIENCES: The Cosmetic Industry:

Bone Calcium Supplement:
Potential Use: Natural source of calcium and phosphorus
Source: Bovine Bone
Isolated from bones of government inspected cattle
Description: Off-white powder
Loss on Drying: NMT 4.0%
Residue on Ignition: NMT 68.0%
Fat: NMT 6.0% (Soxhlet/Hexane)
Total Nitrogen: NLT 3.0%
Calcium: NLT 21.0%

Bone Marrow Extract:
Potential Use: Mixture of key components
Source: Bovine Marrow
Isolated frm marrow of government inspected animals
Description: Grey powder
Loss on Drying: NMT 5%
Residue on Ignition: NMT 14%
Fat: NMT 10% (Soxhlet/Hexane)
Protein: NLT 70%
Microbial Total Aerobic Plate count: NMT 1000/g

Bovine Serum Albumin (BSA)--pH 5.2:
Potential Use: Enhance penetration of fatty acids for healthier cells and smoother skin.
Source: Bovine Plasma
Isolated from plasma of government inspected cattle, by modification of the Cohn fractionation process
Description: Off-white to pale green/yellow lyophilized powder
Solubility: Readily soluble at 7% in water
pH (2% in water): 5.2±0.5
Loss on Drying: NMT 5.0%
Residue on Ignition: NMT 2.0%
Heavy Metals: NMT 50 ppm
Protein Content: NLT 97% db
Purity: NLT 98% (cellulose acetate electrophoresis)

Bovine Serum Albumin--pH 7.0:
Potential Use: Enhance penetration of fatty acids for healthier cells and smoother skin
Source: Bovine Plasma
Isolated from plasma of government inspected cattle, by modification of the Cohn fractionation process
Description: Off-white to pale green/yellow lyophilized powder
Solubility: Readily soluble at 7% in water
pH (2% in water): 7.0±0.5
Loss on Drying: NMT 5.0%
Residue on Ignition: NMT 2.0%
Heavy Metals: NMT 50 ppm
Protein: NLT 97% db
Purity: NLT 98% (cellulose acetate electrophoresis)

WAITIKI INTERNATIONAL BIOSCIENCES: The Cosmetic Industry (Continued):

Bovine Serum Albumin--pH 7.0:
Potential Use: Enhance penetration of fatty acids for healthier cells and smoother skin.
Source: Bovine Plasma
Isolated from the plasma of government inspected cattle, by ion exchange chromatography
Description: Off-white to pale green/yellow lyophilised powder (Caprylate free)
Solubility: Readily soluble at 7% in water
pH (2% in water): 7.0±0.5
Loss on drying: NMT 5.0%
Residue on ignition: NMT 3.0%
Protein: NLT 96.0%
Purity: NLT 98.0% (Cellulose acetate electrophoresis)

Bovine Serum Albumin, Fatty Acid Free:
Potential Use: Enhance penetration of fatty acids for healthier cells and smoother skin.
Source: Bovine Plasma
Isolated from plasma of government inspected cattle, by ion exchange chromatography
Description: Off-white to pale green/yellow lyophilised powder (Caprylate free)
Solubility: Readily soluble at 7% in water
pH (2% in water): 7.0±0.5
Loss on drying: NMT 5.0%
Residue on ignition: NMT 3.0%
Protein: NLT 96.0%
Purity: NLT 98.0% (Cellulose acetate electrophoresis)
Free Fatty Acid: NMT 0.01%

Bovine Serum Albumin, pH 7.0, Fatty Acid Reduced:
Potential Use: Enhance penetration of fatty acids for healthier cells and smoother skin.
Source: Bovine Plasma
Isolated from plasma of government inspected cattle, by modification of the Cohn fractionation process
Description: Off-white to pale green/yellow lyophilized powder
Solubility: Readily soluble at 7% in water
pH (2% in water): 7.0±0.5
Loss on Drying: NMT 5.0%
Residue on Ignition: NMT 2.0%
Heavy Metals: NMT 50 ppm
Protein: NLT 97% db
Purity: NLT 98% (Cellulose acetate electrophoresis)
Free Fatty Acids: NMT 0.35%

WAITAKI INTERNATIONAL BIOSCIENCES: The Cosmetic Industry (Continued):

Bovine Serum Albumin (BSA)--30% Solution:

Potential Use: Enhance penetration of fatty acids for healthier cells and smoother skin.
Source: Bovine Plasma
Isolated from plasma of government inspected cattle
Description: An amber or green solution (containing 0.12 M sodium chloride; may contain 0.1% Sodium Azide as preservative)
pH: 7.2±0.1
Protein: 30%±1% (w/v)
Purity: NLT 98% (Cellulose acetate electrophoresis)

Collagen Protein Hydrolysate:

Potential Use: Positively charged collagen hydrolysate for better absorption. No need for quaternization. Superior to hide or bone hydrolysate.
Source: Porcine Collagen
Prepared from skin of gevernment inspected hogs
Description: A light amber powder
Typical Analysis:
Loss on Drying: 4%
Residue on Ignition: 2.5%
pH: 5.2
Solubility: Soluble in water
Protein: 92%
Microbial:
Total Aerobic Plate Count: NMT 500/g
Tested for the Presence of: Coliform--Not detected
Salmonella--Not detected

CPR 110:

Potential Use: Potential to increase the respiration of cells and stimulate healthy regenerative functions, especially under conditions of stress to the skin.
Source: Bovine Liver
Isolated from liver of government inspected cattle
Description: A low molecular weight, light yellow to beige, hygroscopic powder
Solubility: Soluble in water
Typical Analysis:
Loss on Drying: NMT 5.0%
Activity: Oxygen consumption NLT 100% increase compared to liver homogenate control
Microbial:
Total Aerobic Plate Count: NMT 100/g
Tested for the Presence of: E. coli--Not detected
P. aeruginosa--Not detected
Salmonella--Not detected

WAITAKI INTERNATIONAL BIOSCIENCES: The Cosmetic Industry (Continued):

Fetuin--Pedersen:

Potential Use: Concentrated mixture of growth factors
Source: Fetal Serum
Isolated from fetal bovine serum of government inspected animals
Description: Off-white lyophilized powder
Loss on Drying: NMT 5%
Residue on Ignition: NMT 3%
Protein: NLT 80%

Fibronectin--Bovine:

Potential Use: Normal component of young, healthy skin. Needed to anchor basal cells for proper function and growth.
Source: Bovine Plasma
Isolated from plasma of government inspected cattle
Description: A clear, essentially colourless solution containing buffer salts, urea and Fibronectin.
Fibronectin Concentration: NLT 2.0 mg of Fibronectin/ml in phosphate buffered saline (pH 7.2) containing 2M urea. Concentrations based on an E 1%/280 = 12.8
% of Total Protein as Fibronectin: NLT 90% based on FPLC (Gel filtration)
Purity: A single doublet band of approx. 220,000 Dalton molecular mass is evident using SDS-PAGE

Hyaluronic Acid, Sodium Salt:

Potential Use: Wrinkle smoothing agent, natural component of skin, moisturizer
Source: Avian Comb
Isolated from combs of government inspected chickens
Description: White to slightly off-white powder
Clarity (0.1% solution): Essentially clear in aqueous solution
Purity: NLT 80% Hyaluronic Acid, Sodium Salt (Carbazole reaction)
Protein: NMT 1.0% (Lowry)
Molecular Weight: NLT 1,000,000 Daltons (Viscometric)
Reference:
Cleland and Wang, Biopolymers Vol. 9, 799 (1970)

WAITAKI INTERNATIONAL BIOSCIENCES: The Cosmetic Industry (Continued):

Kallikrein:

Potential Use: Maintains blood flow to skin cells, and thus a steady flow of nutrients for healthy skin.
Source: Porcine Pancreas
Isolated from pancreas of government inspected hogs
Description: Beige to off-white crystalline powder
pH: 6.5±1.0
Loss On Drying: NMT 3.0%
Residue on Ignition: NMT 5.0%
Heavy Metals: NMT 20 ppm
Arsenic: NMT 2 ppm
Fat: NMT 0.5%
Assay: 100 I.U./mg db

Linoleic Acid, 96%:

Potential Use: Essential fatty acid
Source: Safflower Oil
Description: A yellow oily liquid.
(May contain 0.01% BHT as antioxidant)
Linoleic Acid: NLT 96%
Oleic Acid: NMT 3%
Saturated Fatty Acids: NMT 1%

Neural Lipid Extract:

Potential Use: Natural moisturizer
Source: Bovine Neural Tissue
Isolated from neural tissue of government inspected cattle
Description: Off-white to cream coloured powder
Loss on Drying: NMT 5%
Protein: NMT 2.0% (Lowry)
Neutral Glycosphingolipids: 20±5%
Microbial:
Total Aerobic Plate Count: NMT 1000/g
Yeast & Mold: NMT 100/g
Tested for Presence of: Coliforms--Not detected
Staphylococci--Not detected
Higher concentrations available upon request

WAITAKI INTERNATIONAL BIOSCIENCES: The Cosmetic Industry (Continued):

Trachea Hydrolysate:

Potential Use: Contains two important components of skin (collagen & glycosaminoglycans).
Source: Bovine Trachea
Isolated from the trachea of government inspected cattle
Description: Pale beige, granular powder
pH: 4.5±0.5
Loss on Drying: NMT 10%
Residue on Ignition: NMT 3.0%
Heavy Metals: NMT 20 ppm
Fat: NMT 0.5% (Soxhlet)
Nitrogen: 11-14%
Microbial:
Total Aerobic Plate Count: NMT 100/g
Tested for the Presence of: E. coli--Not detected
P. aeruginosa--Not detected
S. aureus--Not detected
Enterococci--Not detected
Salmonella--Not detected

Transferrin, Bovine (Apo):

Essentially Iron Free
Potential Use: Necessary universal nutrient for skin cells.
Source: Bovine Plasma
Isolated from plasma of government inspected cattle
Description: Off-white lyophilized powder
Solubility: Readily soluble at 1% in water
pH (1% solution): 4.5±0.5
Loss on Drying: NMT 5%
Residue on Ignition: NMT 2.0%
Iron: NMT 50 ppm
Protein: NLT 95% db
Purity: NLT 98%

Transferrin, Bovine (Holo):

Essentially Iron Saturated
Potential Use: Necessary universal nutrient for skin cells
Source: Bovine Plasma
Isolated from plasma of government inspected cattle
Description: Reddish, lyophilized powder
Solubility: Readily soluble at 1% in water
pH (1% solution): 7.0±0.5
Loss on Drying: NMT 5%
Residue on Ignition: NMT 2.0%
Iron: NLT 1,000 ppm
Protein: NLT 95% db
Purity: NLT 98%

WAITAKI INTERNATIONAL BIOSCIENCES: Neutral Lipids: Typical Chemical Profile:

Concentrate NLC:
- Neutral glycosphingolipids: 40±5%
- Ethanolamine phospholipids: 20±5%
- Serine phospholipids: 12.5±2.5%
- Choline phospholipids: 10±5%
- Sphingomyelin: 7.5±2.5%
- Cholesterol: 2±1%
- Sulphatides: 2±1%

Extract NLE:
- Neutral glycosphingolipids: 20±5%
- Ethanolamine phospholipids: 15±5%
- Serine phospholipids: 12.5±2.5%
- Choline phospholipids: 10±5%
- Sphingomyelin: 7.5±2.5%
- Cholesterol: 5±2%
- Sulphatides: 5±2%

Phosphatide NLP:
- Neutral glycosphingolipids: 20±5%
- Ethanolamine phospholipids: 20±5%
- Serine phospholipids: 7±3%
- Choline phospholipids: 20±5%
- Sphingomyelin: 5±2%
- Cholesterol: 2±1%
- Sulphatides: 2±1%
- Inositolphospholipids: 10±3%

WITCO: Surfactants for Cosmetics and Toiletries:

Surfactant:

EMCOL 4161L:
CTFA Adopted Name: Disodium Oleamido-MIPA-Sulfosuccinate

EMCOL 4100M:
CTFA Adopted Name: Disodium Myristamido MEA-Sulfosuccinate

EMCOL 4400-1:
CTFA Adopted Name: Disodium Lauryl Sulfosuccinate

EMCOL 4300:
CTFA Adopted Name: Disodium Pareth-3 Sulfosuccinate
Typical Applications:
Foaming agents for mild skin cleansers, creme shampoos and bubble baths

EMCOL 4072:
CTFA Adopted Name: Disodium Hydrogenated Cottonseed Glyceride Sulfosuccinate
Typical Applications:
Anionic coemulsifier for mild, acid-pH creams and lotions.

EMCOL CC-42:
CTFA Adopted Name: PPG-40 Diethylmonium Chloride

EMCOL CC-9:
CTFA Adopted Name: PPG-9 Diethylmonium Chloride
Typical Applications:
Antistats, dispersants and conditioners.

EMCOL CC37-18:
CTFA Adopted Name: Coco-Betaine
Foaming agent; foam stabilizer; viscosity modifier for cationic, anionic or nonionic formulations.

EMCOL E-607L:
CTFA Adopted Name: Lapyrium Chloride
Mild cationic emulsifier; detergent.

EMCOL E-607S:
CTFA Adopted Name: Steapyrium Chloride
Mild cationic emulsifier; hair conditioner; skin conditioner for after-shaves and alcoholic lotions.

EMPHOS PS-810:
CTFA Adopted Name: Oleth-3 Phosphate
Antistat; conditioner; emulsifier; lubricant; wetting agent; dispersant.

WITCO: Surfactants for Cosmetics and Toiletries(Continued):

Surfactant:

EMPHOS F27-85:
CTFA Adopted Name: Sodium Hydrogenated Vegetable Glycerides Phosphate
EMPHOS D70-30C:
CTFA Adopted Name: Sodium Glyceryl Oleate Phosphate
Water-in-oil emulsifiers; moisture barriers; dispersants; antistats and conditioners; gelling agents.

EMPHOS CS-1361:
CTFA Adopted Name: Sodium Nonoxynol-6 Phosphate
Antistat and emulsifier for microemulsions.

WITCAMIDE 61:
CTFA Adopted Name: Oleamide MIPA
Viscosity modifier; hair conditioner.

WITCAMIDE 70:
CTFA Adopted Name: Stearamide MEA
WITCAMIDE MAS:
CTFA Adopted Name: Stearamide MEA Stearate
Gellants; opacifiers; viscosity modifiers in cream shampoos, stick-type antiperspirants and specialty makeup products.

WITCAMIDE 511C:
CTFA Adopted Name: Oleamide DEA
Water-in-oil emulsifier for creams and lotions; coupling agent for oil-based systems.

WITCAMIDE 82:
CTFA Adopted Name: Cocamide DEA
WITCAMIDE 5130:
CTFA Adopted Name: Cocamide DEA
WITCAMIDE 5133:
CTFA Adopted Name: Cocamide DEA
WITCAMIDE 5195:
CTFA Adopted Name: Lauramide DEA
WITCAMIDE MEAC:
CTFA Adopted Name: Cocamide MEA
Foam stabilizers; viscosity modifiers; conditioners in shampoos and bath products.

WITCOLATE AE-3:
CTFA Adopted Name: Ammonium Pareth-25-3 Sulfate
WITCOLATE SE-5:
CTFA Adopted Name: Sodium Laureth Sulfate
Primary foaming agents and detergents for shampoos, bath gels and shower gels.

WITCO: Surfactants for Cosmetics and Toiletries(Continued):

Surfactant:

WITCOLATE 60T:
CTFA Adopted Name: TEA-Dodecylbenzene Sulfonate
Foaming agent and detergent for shampoos and bath products.

WITCONATE AOS:
CTFA Adopted Name: Sodium C14-16 Olefin Sulfonate
Base for liquid detergent shampoo and bubble base formulations

WITCONOL APEB:
CTFA Adopted Name: PPG-26-Buteth-26
Low-foaming, acid-stable emulsifier; lubricant; spreading and penetrating agent.

WITCONOL APM:
CTFA Adopted Name: PPG-3 Myristyl Ether
WITCONOL APS:
CTFA Adopted Name: PPG-11 Stearyl Ether
Emollients for creams, lotions and cosmetic gels; acid-stable coupling agents.

WITCONOL 14:
CTFA Adopted Name: Polyglyceryl-4 Oleate
WITCONOL 18L:
CTFA Adopted Name: Polyglyceryl-4 Isostearate
Water-in-oil emulsifiers; lubricants; antifoaming agents; antistats.

WITCONOL 18F:
CTFA Adopted Name: Polyglyceryl-4 Stearate
Oil-in-water emulsifier; thickener; foaming agent.

WITCONOL CD-17:
CTFA Adopted Name: PPG-34
WITCONOL CD-18:
CTFA Adopted Name: PPG-27 Glyceryl Ether
WITCONOL PPG-400:
CTFA Adopted Name: PPG-9
Emollients for bath oils, hair preparations, toners, cleansers.

WITCONOL F26-46:
CTFA Adopted Name: PPG-36 Oleate
Emollient; spreading and coupling agent.

WITCO: Surfactants for Cosmetics and Toiletries(Continued):

<u>Surfactant:</u>

WITCONOL H-31A:
CTFA Adopted Name: PEG-8 Oleate
WITCONOL H-35A:
CTFA Adopted Name: PEG-8 Stearate
Oil-in-water emulsifiers and spreading agents for creams, lotions and bath oils.

WITCONOL L32-45:
CTFA Adopted Name: PEG-150 Distearate
Thickener; emulsion stabilizer.

WITCONOL MST:
CTFA Adopted Name: Glyceryl Stearate
Emulsifier; lubricant; stabilizer

WITCONOL RHT:
CTFA Adopted Name: Glyceryl Stearate SE
Emulsifier; stabilizer.

ZSCHIMMER & SCHWARZ: Basic Material for Cosmetic Industry:

Fatty Alcohol Sulfates:

SULFETAL Cjot 38:

Basic material for the formulation of hair shampoos, bath preparations, and liquid cleaning agents

Chemical composition: Fatty alcohol sulfate (C12-C14) neutralized with mono-isopropanolamine
CTFA adopted name: MIPA-Lauryl Sulfate
CAS No.: 21142-28-9
Aspect: clear, straw coloured, viscous liquid
Ionic character: anionic

Typical Data:
Active ingredient: 38%
Water content: 62%
pH (10%): 6.5-7.0
Viscosity (20C): 5000-10000 mPa-s
Density (20C): 1.02 g/cm3
Cloud point: approx. 8C
free of inorganic salts
Application:
SULFETAL Cjot 38 is an anionic, surface-active basic material for the formulation of liquid hair shampoos, foam baths, as well as cleaning agents.
SULFETAL Cjot 38 can be combined with anionic, nonionic and amphoteric surfactants.
SULFETAL CJOT 38 produces a creamy and stable lather.

SULFETAL KT 400:

Basic material for the formulation of hair shampoos

Chemical composition: Triethanolamine salt of a fatty alcohol sulfate
CTFA adopted name: TEA-Lauryl Sulfate
CAS-No.: 139-96-8
Aspect: clear, colourless to straw-coloured liquid
Ionic character: anionic

Typical data:
Active ingredient: 40%
Unsulfated matter: max. 2%
Colour value: max. 5
pH (10%): 7.0-7.5
Density (20C): 1.05 g/cm3
Cloud point: approx. +5C

Application:
SULFETAL KT 400 can be combined with anionic and non-ionic surfactants and is primarily used in hair shampoos.
It is resistant to water hardness and has good foaming properties even at low temperatures.

ZSCHIMMER & SCHWARZ: Basic Material for Cosmetic Industry (Continued):

Fatty Alcohol Ether Sulfates:

ZETESOL AP:

Basic material for use in hair shampoos and bath preparations as well as in dishwashing agents.

Chemical composition: Ammonium alkyl ether sulfate with an addition of approx. 14% 1,2-propylene glycol
CTFA adopted name: Ammonium Pareth-25 Sulfate (and) Propylene Glycol
Aspect: free-flowing, straw coloured product
Ionic character: anionic

Typical data:
Active ingredient: 60%
Ammonium chloride: max. 2%
Ammonium sulfate: max. 2%
pH (10%): 6.0-6.8
Colour value (Gardner): max. 4
Density (25C): 1.05 g/cm3

Application:
ZETESOL AP is an anionic, surface-active basic material for the formulation of liquid foam baths, hair shampoos, and cleaning agents such as dishwashing agents, household cleaners, etc.

ZETESOL NL:

Basic material for the formulation of bath preparations, hair shampoos and cleaning agents for specific applications

Chemical Composition: Sodium lauryl ether sulfate
CTFA adopted name: Sodium Laureth Sulfate
CAS-No.: 1335-72-4
Aspect: clear, almost colourless liquid
Ionic character: anionic

Typical data:
Active ingredient: 28%
Water content: 70%
pH (10%): 6.0-7.0
Viscosity (20C): 100 mPa-s
Sodium chloride: max. 0.1%
Sodium sulfate: 0.5%
Density (20C): 1.04 g/cm3
Cloud point: 0 C

Application:
ZETESOL NL is an anionic, surface-active basic material for the formulation of liquid foam baths, hair shampoos, as well as cleaning agents. ZETESOL NL can be combined with anionic, nonionic and amphoteric surfactants.

ZSCHIMMER & SCHWARZ: Basic Material for Cosmetic Industry (Continued):

Fatty alcohol ether sulfates(Continued):

ZETESOL 100:

Basic material for the formulation of oil foam baths with high oil content

Chemical Composition: Combination of fatty alcohol ether sulfate with non ionic materials

CTFA adopted name: MIPA-Laureth Sulfate (and) Laureth-4 (and) Cocamide DEA

Aspect: clear, yellow-brown liquid at 25C

Ionic character: anionic

Typical data:

Active ingredient: min. 98%

Water content: max. 2%

pH (10%): 6.5-7.5

Density (20C): 1.00 g/cm3

Application:

ZETESOL 100 is used in the formulation of bath preparations with high oil content. Contents of vegetable and mineral oils of up to 35% give clear products which remain homogeneous even under changing temperatures. Diluted in water the oil gives a coarse dispersion.

ZETESOL 856 T:

Basic material for the formulation of bath preparations, hair shampoos and body cleaning agents

Chemical composition: Lauryl ether sulfate as alkylolamine salt with fatty acid amidoalkyl betaine

CTFA adopted name: MIPA-Laureth Sulfate (and) Cocamidopropyl Betaine

Aspect: clear, viscous, yellow liquid

Ionic character: anionic/amphoteric

Typical data:

Active ingredient: 56%

Water content: 44%

pH (10%): 6.5-7.0

Viscosity (20C): 2700-3300 mPa-s

Density (20C): 1.06 g/cm3

Chloride, sulfate: 3.8%

Cloud point: approx. +15C

Properties:

ZETESOL 856 T can be diluted directly with water without getting gel formation.

ZSCHIMMER & SCHWARZ: Basic Material for the Cosmetic Industry (Continued):

Fatty Alcohol Ether Sulfates(Continued):

ZETESOL 2056:

Basic material for the formulation of foam bathes, hair shampoos, all kinds of cleaning agents

Chemical composition: Lauryl ether sulfate as alkylolamine salt

CTFA adopted name: MIPA-Laureth Sulfate

Aspect: clear, yellowish liquid

Ionic character: anionic

Typical data:

Active ingredient: 56%

Water content: 44%

pH (10%): 6.5-7.0

Chloride, sulfate: 3.5%

Viscosity (20C): 3000 mPa-s

Density (20C): 1.06 g/cm3

Cloud point: approx. 10C

Application:

ZETESOL 2056 is an anionic, surface-active basic material for the formulation of liquid foam bathes, hair shampoos, and cleaning agents.

ZETESOL 2056 can be combined with anionic, non-ionic and amphoteric surfactants.

ZETESOL 2056 has a good lathering power and is resistant to water hardness.

Sulfosuccinic acid semi-esters:

SETACIN F spezial Paste:

Basic material for the formulation of mild body cleaning agents

Chemical composition: Disodium fatty alcohol (C12-C14) sulphosuccinic acid semi-ester

CTFA adopted name: Disodium Lauryl Sulfosuccinate

CAS No.: 13192-12-6

Aspect: white paste

Ionic character: anionic

Typical data:

Active ingredient: 40%

Water content: 60%

pH (10%): 6.0-7.0

Solubility: soluble in warm water

Application:

SETACIN F spezial Paste can be combined with anionic, non-ionic and amphoteric surfactants. Because of its mildness and lathering power it is used in the formulation of body cleaners, e.g. face cleansing cream and cream shampoos.

ZSCHIMMER & SCHWARZ: Basic Material for Cosmetic Industry (Continued):

Sulfosuccinic acid semi-esters (Continued):

SETACIN M:

Basic material for the formulation of bath preparations, hair shampoos, and cleaning agents for specific applications

Chemical composition: Sulfosuccinic half-ester of ethoxylated fatty alcohols and fat/nitrogen condensates, neutralized with organic base

Properties:

Aspect: yellow liquid

Ionic character: anionic

Typical data:

Active ingredient: 42%

Water content: 58%

pH (original product): 6.0-7.0

Solubility: freely soluble in alcohol and water in any ratio

Application:

This product can be combined with anionic, nonionic and amphoteric surfactants.

SETACIN M produces a creamy and stable lather, is very compatible with the skin and has good refatting properties.

SETACIN M is used for the formulation of bath preparations, hair shampoos, rinsing and handcleaning agents to improve the skin compatibility.

When combining SETACIN M with other surfactants an addition quantity of at least 10% of total active matter is recommended.

SETACIN 103 spezial:

Basic material for the formulation of bath preparations, hair shampoos and different types of cleaning agents

Chemical composition: Semi-sulfosuccinate of ethoxylated lauryl alcohol

CTFA adopted name: Disodium Laurethsulfosuccinate

CAS-No.: 39354-45-5

Aspect: nearly colourless liquid

Ionic character: anionic

Typical data:

Active ingredient: 40%

Water content: 60%

pH (10%): 6.0-7.5

Density (20C): 1.11 g/cm3

Solubility: clear soluble in water

Application:

SETACIN 103 spezial can be combined with anionic, non-ionic and amphoteric surfactants, resulting in a stable foam with bubbles of small size. Because of its excellent dermatological properties SETACIN 103 spezial is particularly applied in bath preparations, hair shampoos and liquid soaps.

Also its use in formulations for domestic cleaners, dishwashing agents, light-duty detergents, etc. Improves the mildness to the skin.

ZSCHIMMER & SCHWARZ: Basic Material for Cosmetic Industry (Continued):

Fatty acid alkylolamides and superfatting agents:

OXYPON 306:
Solubilizer and refatting agent for cosmetics
Chemical composition: Ethoxylated mink oil
CTFA adopted name: PEG-13 Mink Oil
Aspect: yellow-brown, liquid to pasty
Ionic character: non-ionic
Typical data:
Active ingredient: 100%
Water content: max. 1%
pH (10%): 5-7
Acid value: max. 3
Density (20C): 1.03 g/cm3
Solubility: soluble in water
Properties:
Because of its hydrophilic properties OXYPON 306 is suitable for the application as refatting agent in body cleansing agents.
OXYPON 306 can also be used as solubilizer for volatile oils and as co-emulsifier.

OXYPON 328:
Refatting agent for cosmetics
Chemical composition: Ethoxylated jojoba oil
CTFA adopted name: PEG-53 Jojoba Oil
Aspect: yellowish wax
Ionic character: non-ionic
Typical data:
Active ingredient: 100%
Water content: max. 1%
pH (10%): 5-7
Acid value: max. 3
Solubility: soluble in water
Properties:
Because of its hydrophilic properties OXYPON 328 is suitable for the application as refatting agent in body cleansing agents.

OXYPON 329:
Refatting agent and solubilizer for cosmetics
Chemical composition: Ethoxylated almond oil
CTFA adopted name: PEG-12 Almond Oil
Aspect: yellowish paste
Ionic character: non-ionic
Typical data:
Active ingredient: 100%
Water content: max. 1%
pH (10%): 5-7
Acid value: max. 3
Solubility: soluble in water
Properties:
Because of its hydrophilic properties OXYPON 329 is suitable for the application as refatting agent in body cleansing agents.

ZSCHIMMER & SCHWARZ: Basic Material for Cosmetic Industry:

Fatty acid alkylolamides and superfatting agents(Continued):

OXYPON 2145:
Refatting agent for cosmetics
Chemical composition: Ethoxylated glyceryl isostearate
CTFA adopted name: PEG-15 Glyceryl Isostearate
Aspect: clear, yellow brown liquid of low viscosity
Ionic character: non-ionic
Typical data:
Active ingredient: 100%
pH (10%): 7-8
Density (25C): 1.05 g/cm3
Acid value: max. 2
Iodine value: max. 5
Viscosity (25C): 330 mPa-s
Solubility: soluble in water, ethanol, isopropanol
Application:
OXYPON 2145 is used for skin refatting mainly in cosmetic products such as bath preparations, shampoos, synthetic toilet soaps (liquids or bars), bath oils (water soluble) etc. Lather quality is not affected. OXYPON 2145 is also applied in dish washing agents and hand cleaners.
The quantity of application should be 5-10% of the surfactant portion.

PURTON CFD:
Additive for hair shampoos, bath preparations, dishwashing and cleaning agents
Chemical composition: Coconut fatty acid diethanolamide
CTFA adopted name: Cocamide DEA
CAS-No.: 61791-31-9
Aspect: straw-coloured, oily liquid
Ionic character: non-ionic
Typical data:
Water content: max. 0.5%
Density (20C): 1.00 g/cm3
Free fatty acid: max. 1%
Free amine: max. 6%
Viscosity (20C): 1400 mPa-s
Solubility: dispersible in water, gives clear solutions in combination with surfactants
Application:
PURTON CFD ameliorates the foaming characteristics of preparations containing surfactants. Besides the lathering power, foam stability and foam structure are improved.
PURTON CFD increases the viscosity of e.g. fatty alcohol ether sulphates (ZETESOLs), fatty alcohol sulphates (SULFETALs), sulfosuccinates (SETACINs) and amphoterics (AMPHOTENSIDs).
PURTON CFD has been tested for dermatological properties. It has good skin and eye tolerance.
According to experience the quantity of application should be 1-3%, relative to the final product.

ZSCHIMMER & SCHWARZ: Basic Material for Cosmetic Industry (Continued):

PURTON SFD:

Additive for bath preparations, shampoos and body cleaning agents

Chemical composition: Linoleic acid diethanolamide

CTFA adopted name: Linoleamide DEA

CAS No.: 56863-02-6

Aspect: light-brown, oily liquid

Ionic character: non-ionic

Typical data:

Content of amide: 90%

Water content: max. 0.5%

Free fatty acid: max. 0.5%

Viscosity (20C): 1300 mPa-s

Density (20C): 0.98 g/cm3

Solubility: dispersible in water, gives clear solutions in combination with surfactants

Application:

Foaming:

PURTON SFD ameliorates the foaming characteristics of preparations containing surfactants. Besides the lathering power, foam stability and foam structure are improved.

Viscosity:

PURTON SFD increases the viscosity of e.g. fatty alcohol ether sulfates (ZETESOLs) and fatty alcohol sulfates (SULFETALs). This property is especially suitable for the formulation of liquid body cleaning agents and bath preparations.

The thickening power of PURTON SFD is higher in comparison to coco fatty acid diethanolamides, due to the fact that it is almost free of C12-fatty acid

Skin compatibility:

PURTON SFD ameliorates the skin feeling during and after the application of body cleaning agents, such as shower baths, foam baths etc.

The scientific literature attributes excellent skin-caring properties to the essential fatty acids incorporated in PURTON SFD. The absence of lauric acid, compared to analogous coconut fatty acid derivatives, may at least partially be responsible for the particularly low irritant behaviour of PURTON SFD.

PURTON SFD has been tested dermatologically. The excellent eyetolerance has especially to be mentioned.

Investigations using a powdered collagen substrate as a model substance showed that PURTON SFD in surfactant containing preparations is adsorbed on proteins. The substantivity is very high. As could be proved by means of thin layer chromatography, also several rinsing procedures with water did not reduce remarkably the adsorbed amount of PURTON SFD.

ZSCHIMMER & SCHWARZ: Basic Material for Cosmetic Industry (Continued):

Fatty alcohol polyglycol ether:
OXETAL types:
Basic materials for the manufacture of washing and cleaning agents as well as for cosmetics and technical products
Chemical composition: Fatty alcohol polyglycol ether
Ionic character: non-ionic
Typical data:
Active ingredient: 100%
Water content: max. 1%
pH (10%): 5-7
Compatibility: compatible with anionic, non-ionic or cationic surfactants
Solubility: fatty alcohols of low ethoxylation are soluble in oil, they become water-soluble with approx. 7 moles ethylene oxides; OXETAL types are soluble in most of organic solvents.
Nomenclature:
The nomenclature of the OXETAL types depends on the fatty alcohol basis and the degree of ethoxylation. The fatty alcohol is identified by the suffix letter, the degree of ethoxylation by the number higher than 100.
OXETAL D: decyl alcohol
OXETAL C..spez.: lauryl alcohol
OXETAL C: coconut fatty alcohol
OXETAL TG: tallow fatty alcohol
OXETAL O: oleyl alcohols of different iodine values
OXETAL ID: isodecyl alcohol
OXETAL T: tridecyl alcohol

OXETAL D 104:
Aspect: colourless liquid
Cloud point: 58-61C
OXETAL T 103:
Aspect: colourless liquid
Cloud point: 74-77C
OXETAL T 106:
Aspect: colourless liquid
Cloud point: 40-44C
OXETAL C 110:
Aspect: white paste
Cloud point: 75-78C
OXETAL TG 111:
Aspect: white wax
Cloud point: 70-75C
OXETAL O 108:
Aspect: white paste
Cloud point: 45-50C
OXETAL O 112:
Aspect: white paste
Cloud point: 83-87C
OXETAL ID 104:
Aspect: colourless liquid
Cloud point: 50-53C

ZSCHMIMMER & SCHWARZ: Basic Material for Cosmetic Industry (Continued):

Emulsifiers:

MULSIFAN CB:

All purpose oil/water emulsifier for cosmetics

Chemical composition: Mixture of special saturated fatty alcohol polyglycol ethers

CTFA adopted name: Beheneth-10

Aspect: white, wax-like

Ionic character: non-ionic

Typical data:

Active ingredient: 100%

Water content: max. 1%

pH (1%): 5-7

HLB-value: 10

Density (50C): 0.94 g/cm3

Setting point (on rotary thermometer): 48C

Solubility: soluble in paraffin oils, ethanol, isopropanol; soluble in peanut oil under heating; dispersible in water

Application:

MULSIFAN CB enables formulation of creams with low content of fatty materials. The emulsifier is especially suitable for vanishing creams, fatty creams, lotions, etc.

MULSIFAN CPA:

Oil/water emulsifier for cosmetics

Chemical composition: Fatty alcohol polyglycol ether

CTFA adopted name: Laureth-4

CAS-No.: 5274-68-0

Aspect: colourless liquid

Ionic character: nonionic

Typical data:

Active ingredient: 100%

Water content: max. 1%

pH (1%): 5-7

Density (20C): 0.94 g/cm3

HLB Value: 9.3

Cloud point: 57-62C

Solubility: soluble in paraffin oils, soya bean oil, ethanol; dispersible in water

Application: Formulation of oil/water emulsions for cosmetics

ZSCHIMMER & SCHWARZ: Basic Material for Cosmetic Industry (Continued):

Emulsifiers(Continued):

MULSIFAN RT 11:
Emulsifier for fatty acids, fatty alcohols, triglycerides of vegetable and animal origin, waxes
Chemical Composition: Fatty alcohol polyglycol ether
Physical form: white to yellowish wax
Ionic character: nonionic
Typical properties:
Active ingredient: 100%
Water content: below 1%
HLB value: 16
Density (50C): 1.03 g/cm3
pH (10%): 5-7
Solubility: soluble in water, ethanol, isopropanol, acetone, benzene, toluene, xylene, chlorinated hydrocarbons
Application:
Emulsifier e.g. for lard, cocofatty acid, stearic acid, Carnauba and Montan waxes and their blends with paraffin wax.
Application quantity:
For triglycerides, waxes, fatty acids, fatty alcohols:
8-15% MULSIFAN RT 11
92-85% wax, fatty acid, etc.

MULSIFAN RT 23:
Emulsifier for white oils and paraffin oils
Chemical composition: Fatty alcohol polyglycol ether
CTFA adopted name: Laureth-5
CAS-No.: 3055-95-6
Properties:
Aspect: colourless liquid
Ionic character: non ionic
Typical data:
Active ingredient: 100%
Water content: max. 1%
HLB-value: 11
pH (10%): 5-7
Density (20C): 0.96 g/cm3
Solubility: soluble in white oils, mineral oils, petrol, benzene, toluene, carbon tetrachloride, kerosene, perchloroethylene, turpentine; dispersible in water
Application:
Formulation of lubricating agents, spin finishes, coning oils and emulsions for cosmetics.
Application quantity: 10-20% MULSIFAN RT 23
90-80% oil

ZSCHIMMER & SCHWARZ: Basic Material for Cosmetic Industry (Continued):

Emulsifiers(Continued):

MULSIFAN RT 27:
Emulsifier for fatty acids, fatty alcohols, triglycerides; for the formulation of polishing, latex and wax emulsions

Chemical composition: Fatty alcohol polyglycol ether
CTFA adopted name: Oleth-25
CAS-No.: 9004-98-2
Properties:
Aspect: white to yellowish wax
Ionic character: non-ionic
Typical data:
Active ingredient: 100%
Water content: max. 1%
HLB-value: 16
pH (10%): 5-7
Density (50C): 1.04 g/cm3
Solubility: soluble in water, ethanol, acetone, petrol, toluene, xylene, chlorinated hydrocarbons
Application:
MULSIFAN RT 27 can be used as an universal emulsifier for many products such as e.g. lard, coconut fatty acid, stearic acid, natural and synthetic waxes as well as for the formulation of olein greasing agents and the stabilization of latex emulsions.

MULSIFAN RT 69:
Emulsifier for olein, fats and oils: solubilizer for perfumes and essential oils
Chemical composition: Ethoxylated triglyceride
CTFA adopted name: PEG-40 Castor oil
CAS-No.: 61791-12-6
Properties:
Aspect: yellow paste
Ionic character: nonionic
Typical data:
Active ingredient: 100%
Water content: max. 1%
HLB-value: 13
pH-value (10%): 6-8
Density (20C): 1.09 g/cm3
Solubility: soluble in water, ethanol, isopropanol, toluene, xylene, chlorinated hydrocarbons
Application:
MULSIFAN RT 69 can be used in the formulation of olein greasing agents, in the manufacture of emulsions of fats and oils; and it works as solubilizer for perfumes and essential oils.

ZSCHIMMER & SCHWARZ: Basic Material for Cosmetic Material (Continued):

Emulsifiers(Continued):

MULSIFAN RT 141:
Solubilizer for perfumes and essential oils
Chemical Composition: Ethoxylated sorbitan monolaurate
CTFA adopted name: Polysorbate 20
CAS-No.: 9005-64-5
Aspect: straw-coloured, viscous liquid
Ionic character: non-ionic
Typical data:
Active ingredient: 100%
Water content: max. 2%
HLB-value: 15-16
pH (10%): 5-7
Density (20C): 1.09 g/cm3
Solubility: soluble in water, ethanol, isopropanol, acetone, benzene, toluene, chlorinated hydrocarbions

MULSIFAN RT 146:
Solubilizer for aromatics and essential oils
Chemical composition: Ethoxylated sorbitan monooleate
CTFA adopted name: Polysorbate 80
CAS No.: 9005-65-6
Aspect: yellow, viscous liquid
Ionic character: non ionic
Typical data:
Active ingredient: 100%
Water content: max. 2%
HLB value: 15-16
Density (20C): 1.09 g/cm3
pH (10%): 5-7
Solubility: soluble in water, ethanol, isopropanol, acetone, benzene, toluene, chlorinated hydrocarbons
Application:
For dissolving aromatics and essential oils in water.

MULSIFAN RT 203/80:
Solubilizer for perfumes and essential oils
Chemical composition: Fatty alcohol polyglycol ether
CTFA adopted name: Pareth-25-12
CAS-No.: 68131-39-5
Aspect: colourless liquid
Ionic character: non-ionic
Typical data:
Active ingredient: 80%
Water content: 20%
Cloud point: 89-94C
pH (10%): 5-7
Density (20C): 1.04 g/cm3
HLB-value: 14
Solubility: soluble in water

ZSCHIMMER & SCHWARZ: Basic Material for Cosmetic Industry (Continued):

Amphoteric Surfactants:

AMPHOTENSID B 4:

Amphoteric surfactant for the formulation of hair shampoos and special cleaning agents

Chemical composition: Fatty acid amidoalkylbetaine

CTFA adopted name: Cocamidopropyl Betaine

CAS-No.: 61789-40-0

Aspect: yellow liquid

Ionic character: amphoteric

Typical data:

Solids: 35%

Active ingredient: 30%

Water content: 65%

Sodium chloride: 5%

pH (original product): 5

Density (20C): 1.04 g/cm3

Application:

AMPHOTENSID B 4 can be combined with anionic, non-ionic or cationic surfactants and has a sufficient lathering power. Due to its amphoteric character AMPHOTENSID B 4 has a good compatibility and is especially applicable in baby shampoos.

AMPHOTENSID GB 2009:

Amphoteric surfactant for the formulation of high quality hair shampoos and bath preparations

Chemical composition: Amphoteric coco imidazoline derivative

CTFA adopted name: Cocoamphocarboxyglycinate

CAS No.: 68650-39-5

Aspect: amber coloured liquid of high viscosity

Ionic character: amphoteric

Typical data:

Solids: 50%

Active Ingredient: 38%

Sodium chloride: 12%

Water content: 50%

pH (10%): 8-9

Density (20C): 1.14 g/cm3

Application:

Caused by its amphoteric character AMPHOTENSID GB 2009 is extremely mild to the skin. According to dermatological tests it was classified as non irritating to skin and eyes. Therefore it is especially suitable for use in baby shampoos.

AMPHOTENSID GB 2009 can be combined with non-ionic, anionic and cationic materials.

ZSCHIMMER & SCHWARZ: Basic Material for Cosmetic Industry (Continued):

Amphoteric Surfactants(Continued):

AMPHOTENSID 9 M:
Basic material for the formulation of bath preparations, hair shampoos and liquid body cleaning agents

Chemical compoosition: Coconut imidazoline derivative, modified with alkyl ether sulfate
CTFA adopted name: Cocoamphocarboxyglycinate (and) Sodium Laureth Sulfate
Aspect: yellow, viscous liquid
Ionic character: amphoteric/anionic

Typical data:
Active ingredient: 30%
Water content: 65%
Sodium chloride: 5%
pH (10%): 7.0-8.0
Density (20C): 1.09 g/cm3
Applications:
AMPHOTENSID 9 M is used in mild and high-foaming bath preparations, hair shampoos and body cleaning agents, either alone or in combination with other surfactants.

Other special products:

EXTRAKT ZS 8590:
Concentrated surfactant for the manufacture of bath preparations and hair shampoos
Chemical composition: Monoisopropanol ammonium fatty alcohol sulfate, modified
CTFA adopted name: MIPA-Lauryl Sulfate (and) Laureth-10
Aspect: yellowish paste
Ionic character: anionic
Typical data:
Active ingredient: 90%
Water content: 10%
pH-value (10%): 6.0-6.5
Density (20C): 1.03 g/cm3
Unsulfated amount: max. 3.5%
Inorganic salts: max. 3.5%
Application:
EXTRAKT ZS 8590 serves as basic material for the manufacture of exclusive bath preparations and shampoos. It has a good solubilizing efficiency for essential oils and other additives.

ZSCHIMMER & SCHWARZ: Basic Material for Cosmetic Industry (Continued):

Other special products(Continued):

EXTRAKT 52:

Basic material for the formulation of bath preparations, hair shampoos, liquid body cleaning agents

Chemical composition: Mixture of anionic, amphoteric and non-ionic surfactants
CTFA adopted name: MIPA-Lauryl Sulfate (and) Disodium Laureth-sulfosuccinate (and) Cocoamphocarboxyglycinate (and) Linoleamide DEA (and) Laureth-13
Aspect: clear, yellow, liquid
Ionic character: anionic
Typical data:
Active ingredient: 50%
Water content: 47%
pH (5%): 6.3-6.8
Density (20C): 1.07 g/cm3
Viscosity: 1700-2300 mPa-s
Application:

EXTRAKT 52 has been tested dermatologically. Caused by its good skin and eye tolerance EXTRAKT 52 is used for the formulation of bath preparations, hair shampoos and liquid body cleaning agents.

PERLGLANZMITTEL GM 4006:

Pearlescence additive to hair shampoos, bath preparations and liquid body cleaning agents
Chemical composition: Combination of non ionic materials with fatty alcohol ether sulfate
CTFA adopted name: Sodium Laureth Sulfate (and) Cocamide MEA (and) Glycol Stearate (and) Cocamidopropyl Lauryl Ether
Aspect: white, viscous product
Ionic character: anionic
Typical data:
Active ingredient: 30%
Water content: 70%
pH (10%): 7.0-8.0
Active detergent: 13%
Application:

PERLGLANZMITTEL GM 4006 gives a homogeneous, fine, nacrous lustre to cosmetic products as hair shampoos, bath preparations etc. PERLGLANZMITTEL GM 4006 is added at normal temperature before the product is diluted with water to its final concentration.

ZSCHIMMER & SCHWARZ: Basic Material for Cosmetic Industry (Continued):

Other special products(Continued):

PERLGLANZMITTEL GM 4055:

Pearlescence additive for cosmetic formulations
Chemical composition: combination of fatty acid glycol ester with fatty alcohol ether sulfate
CTFA adopted name: MIPA-Pareth-25 Sulfate (and) Glycol Stearate
Aspect: white to weakly yellowish, creamy

Typical data:
Active ingredient: 38%
Anionic active matter: 15%
Water content: 62%
pH (10%): 5.5-6.5

Application:

PERLGLANZMITTEL GM 4055 can be incorporated at normal temperature in formulations of hair shampoos, bath preparations etc. The application quantity amounts up to 10% according to the desired effect.

The stability of the glossing effect depends on a sufficient viscosity of the final product and can be achieved for example by addition of sodium chloride or fatty acid alkylolamide (PURTON).

PERLGLANZMITTEL GM 4175:

Pearlescent additive to hair shampoos, bath preparations and liquid body cleaning agents

Chemical Composition: Combination of fatty alcohol ether sulfate with a component giving pearly lustre
Aspect: white, viscous, pearly lustring

Typical data:
Active ingredient: 42%
Water content: 58%
pH (10%): 8.0-8.6
Active detergent: 22%
Sodium chloride: 1%

Application:

PERLGLANZMITTEL GM 4175 gives a homogeneous, fine, nacrous lustre to cosmetic products as hair shampoos, bath preparations etc.

PERLGLANZMITTEL GM 4175 is added at normal temperature before the product is diluted with water to its final concentration.

The application quantity amounts up to 10% according to the desired effect.

The stability of the pearly lustring depends on a sufficient viscosity of the final product and can be achieved for example by addition of sodium chloride or fatty acid alkylolamide (PURTON).

ZSCHIMMER & SCHWARZ: Basic Material for Cosmetic Industry (Continued):

PRODUKT GM 5019:
Basic material for the formulation of bath preparations, hair shampoos, liquid body cleaning agents
Chemical composition: Combination of surfactants, containing laurylsulfoacetate
CTFA adopted name: Disodium Laurethsulfosuccinate (and) Sodium Lauryl Sulfoacetate
Aspect: clear to lightly transparent, viscous yellowish liquid
Ionic character: anionic
Typical data:
Active ingredient: 40%
Water content: 60%
pH (5%): 6.0-6.5
Density (20C): 1.09 g/cm3
Viscosity (20C): approx. 10000 mPa-s
Application:
PRODUKT GM 5019 is used either alone or in combination with other surfactants for the formulation of mild and high-foaming bath preparations, hair shampoos, body cleaning agents.

TURKISCHROTOL 100%:
Solubilizer and refatting agent for cosmetics

Chemical composition: Sodium sulforicinoleate
CTFA adopted name: Sulfated Castor Oil
CAS-No.: 8002-33-3
Aspect: clear, yellow-brown liquid
Ionic character: anionic
Typical data:
Active ingredient: 85%
pH (10%): 7.0-7.5
Density (20C): 1.04 g/cm3
Total fatty acid content: 71%
Sulfonation degree: 31%
Solubility: clear soluble in dist. water; almost soluble in alcohol; dispersible in peanut oil and mineral oil
Application:
TURKISCHROTOL 100% is a solubilizing agent for essential oils and due to the contained oil portion as a refatting agent in shower baths, liquid soaps and handwashing agents.
For special applications Zschimmer & Schwarz delivers under the name TURKISCHROTOL 50% a product with approx. 45% active ingredient.

SUPPLIERS' ADDRESSES

Aceto Corp.
126-02 Northern Blvd.
Flushing, NY 11368
(718)-898-2300

Akzo Chemie America
300 South Riverside Plaza
Chicago, IL 60606
(312)-906-7500/(800)-227-7070

Akzo International Salt Co.
Abington Executive Park
Clarks Summit, PA 18411
(717)-587-5131

Albright & Wilson Americas
P.O. Box 26229
Richmond, VA 23260-6229
(804)-550-4300/(800)-446-3700

Alcolac
1099 Winterson Road
Linthicum, MD 21090
(301)-859-4900/(800)-ALCOLAC

Allied-Signal Inc.
Columbia Rd. and Park Ave.
P.O. Box 2332R
Morristown, NJ 07960
(800)-222-0094

Amerchol Corp.
P.O. Box 4051
Edison, NJ 08818-4051
(201)-287-1600

Aqualon Co.
Little Falls Centre One
2711 Centerville Road
P.O. Box 15417
Wilmington, DE 19850-5417
(302)-996-2000/(800)-537-7883

Avatar Corp.
7728 W. 99 St.
Hickory Hills, IL 60457
(312)-430-4200/(800)-255-3181

BASF Corp.
Chemicals Div.
P.O. Box 181
Parsippany, NJ 07054
(201)-316-3000/(800)-253-3904

Bio-Botanica, Inc.
75 Commerce Drive
Hauppauge, NY 11788
(516)-231-5522/(800)-645-5720

Brooks Industries Inc.
70 Tyler Place
South Plainfield, NJ 07080
(201)-561-5200

Capital City Products Co.
P.O. Box 569
Columbus, OH 43216-0569
(614)-299-3131/(800)-848-1340

CasChem, Inc.
40 Avenue A
Bayonne, NJ 07002
(201)-858-7900/(800)-526-1467

Costec Inc.
P.O. Box 693
Palatine, IL 60078
(312)-359-5713

Croda Inc.
183 Madison Ave.
New York, NY 10016
(212)-683-3089

Cyprus Industrial Minerals
8985 East Nichols, Suite 300
Englewood, CO 80112
(303)-643-5000/(800)-325-0299

Desert King Corp.
3802 Main St.
Chula Vista, CA 92011
(619)-427-7121

Dow Chemical U.S.A.
P.O. Box 1206
Midland, MI 48641-9940
(800)-447-4DOW

Dow Corning Corp.
Midland, MI 48686-0994
(517)-496-4000

Dragoco, Inc.
Gordon Drive
P.O. Box 261
Totowa, NJ 07511
(201)-256-3850

DuPont Co.
Chemicals and Pigments
Brandywine Bldg. 15305
Wilmington, DE 19898
(302)-774-2102/(800)-441-9442

Eastman Chemical Products Inc.
P.O. Box 431
Kingsport, TN 37662
(800)-EASTMAN

ECC America Inc.
5775 Peachtree-Dunwoody Road
Suite 200G
Atlanta, GA 30342
(404)-843-1551/(800)-843-3222

Finetex, Inc.
P.O. Box 216
Elmwood Park, NJ 07407
(201)-797-4686

Florida Food Products, Inc.
P.O. Box 1300
Eustis, FL 32727
(904)-357-4141/(800)-874-2331

Freeman Industries, Inc.
100 Marbledale Road
POB 415
Tuckahoe, NY 10707-0415
(914)-961-2100/(800)-558-8658

H. B. Fuller Co.
2400 Energy Park Drive
St. Paul, MN 55108
(612)-645-3401

GAF Chemicals Corp.
1361 Alps Road
Wayne, NJ 07470-3688
(201)-628-3000

GE Silicones
260 Hudson River Rd.
Waterford, NY 12188
(800)-255-8886

Givaudan Corp.
100 Delawanna Ave.
Clifton, NJ 07014
(201)-365-8000

Goldschmidt Chemical Corp.
Rt. 2 - Box 1299
Hopewell, VA 23860
(804)-541-8658/(800)-446-1809

B.F. Goodrich
6100 Oak Tree Blvd.
Cleveland, OH 44131
(800)-331-1144

W.R. Grace & Co.
Davison Chemical Division
P.O. Box 2117
Baltimore, MD 21203-2117
(301)-659-9010/9242

Haarman & Reimer Corp.
70 Diamond Road
P.O. Box 175
Springfield, NJ 07031
(201)-686-3132/(800)-422-1559

Hexcel Chemical Products
215 North Centennial St.
Zeeland, MI 49464
(616)-772-2193

Hoechst Celanese Corp.
Route 202-206 North
Somerville, NJ 08876
(201)-231-2000/(800)-526-4960

Geo. A. Hormel & Co.
Austin, MN 55912
(507)-437-5608

Huls America Inc.
Turner Place
Piscataway, NJ 08855-0365
(201)-981-5000/(800)-526-0339

Humko Chemical Division
Witco Corp.
P.O. Box 125
Memphis, TN 38101-0125
(901)-320-5800

ICI Americas, Inc.
Concord Pike & New Murphy Road
Wilmington, DE 19897
(800)-759-3500

Inolex Chemical Corp.
Jackson & Swanson Streets
Philadelphia, PA 19148-3497
(215)-271-0800/(800)-521-9891

Jojoba Growers & Processors Inc.
2267 South Coconino Dr.
Apache Junction, AZ 85220
(602)-982-1125

Lipo Chemicals Inc.
207 19th Ave.
Paterson, NJ 07504
(201)-345-8600

Lonza Inc.
Fair Lawn, NJ 07410-2692
(201)-794-2400/(800)-777-1875

Dr. Madis Laboratories, Inc.
375 Huyler St.
South Hackensack, NJ 07606
(201)-440-5000

McIntyre Group Ltd.
4851 S. St. Louis Ave.
Chicago, IL 60632
(312)-927-2401

Mearl Corp.
41 East 42nd St.
New York, NY 10017
(212)-573-8500

M. Michel and Co., Inc.
90 Broad St.
New York, NY 10004
(212)-344-3878

Miranol Inc.
P.O. Box 436
68 Culver Road
South Brunswick, NJ 08810
(201)-329-3900

Mona Industries, Inc.
P.O. Box 425
76 East 24th St.
Paterson, NJ 07544
(201)-345-8220

Napp Chemicals Inc.
199 Main St.
Lodi, NJ 07644
(201)-773-3900

Penreco
106 S. Main St.
Butler, PA 16001
(412)-283-5600/(800)-245-3952

Petrolite Specialty Polymers
6910 East 14th St.
Tulsa, OK 74112
(918)-836-1601

Pfizer Chemical Division
235 East 42nd St.
New York, NY 10017
(201)-420-7721/(800)-231-1590

Pilot Chemical Co.
11756 Burke St.
Santa Fe Springs, CA 90670
(213)-723-0036

Pokonobe Industries, Inc.
C.P. 814, Snowdon
Montreal, Quebec, Canada H3X 3X9
(514)-737-4099

Polyesther Corp.
P.O. Drawer BBBB
Southampton, NY 11969
(516)-283-4400

PPG/Mazer Chemicals
3938 Porett Drive
Gurnee, IL 60031-1281
(312)-244-3410

Proctor & Gamble
120 W. Fifth St.
Suite 502
Cincinnati, OH 45202
(513)-562-2655/(800)-543-1580

Protameen Chemicals Inc.
375 Minnisink Road
P.O. Box 166
Totowa, NJ 07511
(201)-256-4374

Quantum Chemical Corp.
Emery Division
11501 Northlake Drive
Cincinnati, OH 45249
(513)-530-7300

Quest International Fragrances USA
400 International Drive
Mount Olive, NJ 07828
(201)-691-7100

Rewo
Sherex Chemical Co.
5777 Frantz Rd.
P.O. Box 646
Dublin, OH 43017
(614)-764-6500/(800)-366-6500

Rheox, Inc.
Wyckoff Mills Road
P.O. Box 700
Hightstown, NJ 08520
(800)-866-6800

Rhone-Poulenc, Inc.
Princeton, NJ 08543-5266
(201)-297-0100

Roche Chemical Division
Hoffman-LaRoche Inc.
Nutley, NJ 07110
(201)-235-8075/8077

Rohm and Haas Co.
Independence Mall West
Philadelphia, PA 19105
(215)-592-3000

Sandoz Chemicals Corp.
4000 Monroe Rd.
Charlotte, NC 28205
(704)-331-7000/(800)-631-8077

Scher Chemicals, Inc.
Industrial West
P.O. Box 4317
Clifton, NJ 07012
(201)-471-1300

Schweizerhall, Inc.
3001 Hadley Road
P.O. Box 395
South Plainfield, NJ 07080
(201)-753-5000

Shell Chemical Co.
One Shell Plaza
P.O. Box 2463
Houston, TX 77252-9923
(713)-439-1000

Sherex Chemical Co., Inc.
5777 Frantz Road
P.O. Box 646
Dublin, OH 43017
(614)-764-6500/(800)-366-6500

Sonneborn Division
Witco Corp.
520 Madison Ave.
New York, NY 10022-4236
(212)-605-3912

Stepan Co.
22 Frontage Road
Northfield, IL 60093
(312)-446-7500/(800)-457-7673

Sutton Laboratories, Inc.
116 Summit Ave.
Chatham, NJ 07928
(201)-635-1551

Tic Gums, Inc.
4609 Richlynn Drive
Belcamp, MD 21017
(301)-273-7300/(800)-221-3953

TRI-K Industries, Inc.
466 Old Hook Road
Emerson, NJ 07630
(201)-261-2800/(800)-526-0372

Union Carbide Chemicals and
Plastics Co., Inc.
39 Old Ridgebury Rd.
Danbury, CT 06817-0001
(203)-794-2550

R.T. Vanderbilt Co., Inc.
P.O. Box 5150
30 Winfield St.
Norwalk, CT 06855
(203)-853-1400

Van Dyk
Main & William Streets
Belleville, NJ 07109
(201)-759-3225

Waitaki International Biosciences
55 Glen Scarlett Road
Toronto, Ontario
Canada M6N 1P5
(416)-761-4089

Witco Chemical Corp.
520 Madison Ave.
New York, NY 10022
(212)-605-3655/(800)-634-4010

Zschimmer & Schwarz
P.O. Box 2179
D-5420 Lahnstein/
West Germany

TRADE NAME INDEX

Trade Name	Supplier
ABIL	Goldschmidt Chemical Corp.
A-C	Allied-Signal Inc.
ACCOMEEN	Capital City Products Co.
ACCOMID	Capital City Products Co.
ACCONON	Capital City Products Co.
ACETULAN	Amerchol Corp.
ACT	Cyprus Industrial Minerals
ACUMIST	Allied-Signal Inc.
ACYLAN	Croda Inc.
ADOGEN	Sherex Chemical Co., Inc.
ADOL	Sherex Chemical Co., Inc.
AGRUDOR	Quest International Fragrances USA
ALBA	Sonneborn Division
ALBONE	DuPont Co.
ALCLOXA	TRI-K Industries, Inc.
ALCOLITE	United Guardian (Freeman Industries)
ALCONATE	Alcolac
ALDE-CON	Florida Food Products
ALDIOXA	TRI-K Industries, Inc.
ALIPAL	Rhone Poulenc
ALTALC	Cyprus Industrial Minerals
AMBERLYN	Quest International Fragrances USA
AMBER-SEC	Quest International Fragrances USA
AMBRASOL	Quest International Fragrances USA
AMBRETTILYN	Quest International Fragrances USA
AMERCHOL	Amerchol
AMERLATE	Amerchol
AMEROXOL	Amerchol
AMERSCREEN	Amerchol
AMIDOX	Stepan
AMINOFOAM	Croda
AMINO GLUTEN	Croda
AMINOXID	Goldschmidt Chemical Corp.
AMMONYX	Stepan
AMPHOSOL	Stepan
AMPHOTENSID	Zschimmer & Schwarz
AMPHOTERGE	Lonza
ANIONYX	Stepan
ANTHER	Quest International Fragrances USA
ANTIL	Goldschmidt Chemical Corp.
ANTI-STAT	Hexcel
APPLINAL	Quest International Fragrances USA
AQUALON	Aqualon
AQUANAL	Quest International Fragrances USA
AQUATHIX	United Guardian (Freeman Industries)
ARBOROMA	Quest International Fragrances USA
ARKOPOL	Capital City Products Co.
ARLACEL	ICI
ARLAMOL	ICI
ARLATONE	ICI
AROSURF	Sherex Chemical Co., Inc.
AROVA	Huls
ASCORBYL	TRI-K
ATLAS	ICI
AURA	Cyprus Industrial Minerals
AURANTOIN	Quest International Fragrances USA
AVALONE	Quest International Fragrances USA
AVAMID	Mona
AVANEL	PPG/Mazer
AVITEX	DuPont Co.

Trade Name	Supplier
BANGALOL	Quest
BBC MOISTURE TROL	Bio-Botanica
BEAUVERTATE	Quest
BENTOLITE	ECC America
BENTONE	Rheox
BE SQUARE	Petrolite
BIJU TX	Mearl
BIJU ULTRA TX	Mearl
BI-LITE	Van Dyk
BI-LITE ULTRALITE	Van Dyk
BI-LITE ULTRAPRESS	Van Dyk
BI-LITE ULTRAWHITE	Van Dyk
BIOCELL	Brooks Industries
BIO-CHELATED DERMA-PLEX	Bio-Botanica
BIO-CHELATED NEUTRAL HENNA PLUS	Bio-Botanica
BIO-CHELATED NUTRA PLANT	Bio-Botanica
BIO-CHELATED SAUNA DERM	Bio-Botanica
BIO-DANDRA PLEX	Bio-Botanica
BIODYNES	Brooks Industries
BIOMIN	Brooks Industries
BIOMIN ACQUACINQUE	Brooks Industries
BIOMIN CINQUE	Brooks Industries
BIOPLEX	Brooks Industries
BIOPOL	Brooks Industries
BIO-SOFT	Stepan
BIO-TERGE	Stepan
BOURGEONAL	Quest
BRETOL	Hexcel
BRIJ	ICI
BRILLIANTE	Cyprus
BRIPHOS	Albright & Wilson
BROMAT	Hexcel
BRONOPOL	Inolex
BROOKSWAX	Brooks Industries
BROPHOS	Brooks Industries
CACHALOT	M. Michel
CALFOAM	Pilot Chemical
CALYXOL	Quest
CAPITAL	Capital City Products
CAPMUL	Capital City Products
CAPROL	Capital City Products
CAPTEX	Capital City Products
CARBOPOL	B.F. Goodrich
CARBOWAX	Union Carbide Chemicals
CARTARETIN	Sandoz
CASTLE	Cyprus
CASTORWAX	CasChem
CEDEPAL	Miranol
CEDEPON	Miranol
CERAPHYL	Van Dyk
CERASYNT	Van Dyk
CERVOLIDE	Quest
CETATS	Hexcel
CETOL SUMQUAT	Hexcel
CETOMACROGOL	Croda
CHANDILYN	Quest
CHROMA-LITE	Van Dyk
CHRYSANTHAL	Quest
CISTULATE	Quest

Trade Name	Supplier
CITRATHAL	Quest
CITROFURAN	Quest
CIVET	Quest
CLAYTONE	ECC America
CLEARCOL	Croda
CLOISONNÉ	Mearl
COJOBA	Costec
COLLAMINO	Brooks Industries
COLLA-MOIST	Brooks Industries
COLLASOL	Croda
COLLNECTIN	TRI-K
COLLOID	Tic Gums
COSEPT	Costec
COSMERIN	H.B. Fuller
COSMETOL	CasChem
COSMOWAX	Croda
COVERA	Costec
CREMBA	Croda
CREMOPHOR	BASF
CRILL	Croda
CRODACEL	Croda
CRODACID	Croda
CRODACOL	Croda
CRODAFOS	Croda
CRODALAN	Croda
CRODAMOL	Croda
CRODAPEARL	Croda
CRODESTA	Croda
CRODYNE	Croda
CROLASTIN	Croda
CROMOIST	Croda
CRONECTIN	Croda
CROQUAT	Croda
CROSILK	Croda
CROTEIN	Croda
CRYOSEPT	Avatar
CRYSTAL	CasChem
CRYSTAL CROWN	CasChem
CUIRONAL	Quest
CUTICULIN	Paninkret Chemicals (Freeman Industries)
CYCLOCHEM	Alcolac
CYCLOMIDE	Alcolac
CYCLOMOX	Alcolac
CYCLORYL	Alcolac
CYCLOSHEEN	Alcolac
CYCLOTERIC	Alcolac
CYCLOTON	Alcolac
CYPRUS	Cyprus
DARVAN	R.T. Vanderbilt
DECUMAL	Quest
DERMAFORM	United Guardian (Freeman Industries)
DERMALCARE	Alcolac
DERMASOME	Brooks Industries
DETAINE	TRI-K
DEWBERRY	Quest
DIONIL	Huls
DIPSAL	Scher
DO CONTROL	Eastman Chemical
DOW CORNING	Dow Corning
DOWICIL	Dow Corning

Trade Name	Supplier
DRAGOSANTOL	Dragoco
DRAKEOL	Penreco
DRAKETEX	Penreco
DREWMULSE	Stepan
DRIVOSOL	Huls
DUOCHROME	Mearl
DUPICAL	Quest
DUPONOL	DuPont
EASTMAN	Eastman Chemical
EFA-PLEX	Brooks Industries
EFETAAL	Quest
EKTASOLVE	Eastman Chemical
ELASTEIN	Costec
ELASTOSOL	Croda
ELFACOS	Akzo Chemie America
ELINTAAL	Quest
ELINTAAL FORTE	Quest
EMCOL	Witco
EMERSOL	Quantum Chemical
EMERY	Quantum Chemical
EMPETAL	Quest
EMPHOS	Witco
EMPICOL	Albright & Wilson
EMPICRYL	Albright & Wilson
EMPIGEN	Albright & Wilson
EMPILAN	Albright & Wilson
EMPIPHOS	Albright & Wilson
EMPIWAX	Albright & Wilson
EMULGATOR	Goldschmidt Chemical
EMULPHOR	Rhone Poulenc
EMULSYNT	Van Dyk
EPITONE	Quest
EPOLENE	Eastman Chemical
EQUINOL	Quest
ESCALOL	Van Dyk
ESTRAN	Desert King
ESTRAN-LITE	Desert King
ESTRAN-PURE	Desert King
ESTRAN-ULTRA	Desert King
ETHA-COLL	Brooks Industries
ETHA-KERATIN	Brooks Industries
ETHA-SILK	Brooks Industries
ETHA-SOY	Brooks Industries
EVERNIA	Quest
EXTRAPONE	Dragoco
FELVINONE	Quest
FENYRANE	Quest
FIBRONEX	TRI-K
FIBRO-SILK	Brooks Industries
FINETEX	Finetex
FINQUAT	Finetex
FINSOLV	Finetex
FIORIVERT	Quest
FLAMENCO	Mearl
FLORANE	Quest
FLOROCYCLENE	Quest
FLOWTONE	ECC America
FLUILAN	Croda
FOAMATION	Desert King

Trade Name	Supplier
FOAM-COLL	Brooks Industries
FOAMOLE	Van Dyk
FOAM-SOY	Brooks Industries
FONOLINE	Sonneborn
FRESCILE	Quest
FRUITLISE	Quest
FRUTONILE	Quest
GAFAC	Rhone Poulenc
GAFAMIDE	Rhone Poulenc
GAFFIX	GAF Chemicals
GAFQUAT	GAF Chemicals
GANEX	GAF Chemicals
GANTREZ	GAF Chemicals
GARDAMIDE	Quest
GARDOCYCLENE	Quest
GELWHITE	ECC America
GEMTONE	Mearl
GENAMIN	Hoechst Celanese
GENAMINOX	Hoechst Celanese
GENAPOL	Hoechst Celanese
GERMABEN	Sutton Laboratories
GERMALL II	Sutton Laboratories
GLUCAM	Amerchol
GLUCAMATE	Amerchol
GLUCATE	Amerchol
GLYPROSOL	Brooks Industries
GYRANE	Quest
HARTOLAN	Croda
HEPTAVERT	Quest
HERBOXANE	Quest
HOE	Hoechst Celanese
HOSTACERIN	Hoechst Celanese
HOSTAPHAT	Hoechst Celanese
HOSTAPON	Hoechst Celanese
HOSTAPUR	Hoechst Celanese
HYDROBA	Jojoba Growers
HYDROCELL	Brooks Industries
HYDROCOLL	Brooks Industries
HYDROFOL	Sherex
HYDROKERATIN	Brooks Industries
HYDROKOTE	Capital City Products
HYDROMILK	Brooks Industries
HYDROSOY	Croda
HYSTRENE	Humko
IGEPAL	Rhone Poulenc
IGEPON	Rhone Poulenc
INCROCAS	Croda
INCROMATE	Croda
INCROMECTANT	Croda
INCROMIDE	Croda
INCROMINE	Croda
INCROMINE OXIDE	Croda
INCRONAM	Croda
INCROPOL	Croda
INCROQUAT	Croda
INCROSUL	Croda
INDUSTRENE	Humko
INONYL ACETATE	Quest

Trade Name	Supplier
INONYL ACETATE EXTRA	Quest
ISOJASMONE	Quest
ISOLONGIFOLANONE	Quest
IVARBASE	Brooks Industries
IVARLAN	Brooks Industries
JACINTHFLOR	Quest
JASILYN	Quest
JASMACYCLENE	Quest
JASMATONE	Quest
JASMIN	Quest
JASMOPYRANE	Quest
JASMOPYRANE FORTE	Quest
JESSATE	Quest
JOJOBEADS	Jojoba Growers
JOJOBUTTER	Jojoba Growers
JORDAPON	PPG/Mazer
KARAJEL	United Guardian (Freeman Industries)
KATEMUL	Scher
KATHON	Rohm and Haas
KATIORAN	BASF
KELATE	TRI-K
KEMAMINE	Humko
KEMESTER	Humko
KEMSTRENE	Humko
KERAMINO	Brooks Industries
KERAMOIS	Paninkret Chemicals (Freeman Industries)
KERASOL	Croda
KESSCO	Stepan
KESSCOLIN	Stepan
KESSCOWAX	Stepan
KLUCEL	Aqualon
KODAFLEX	Eastman Chemical
LACTIL	Goldschmidt Chemical
LANEXOL	Croda
LANPOLAMIDE	Croda
LATHANOL	Stepan
LAUREX	Albright & Wilson
LEATHER BASE	Quest
LEXAINE	Inolex
LEXAMINE	Inolex
LEXATE	Inolex
LEXEIN	Inolex
LEXEMUL	Inolex
LEXGARD	Inolex
LEXOL	Inolex
LEXQUAT	Inolex
LIGANTRAAL	Quest
LIGUSTRAL	Quest
LIGUVERT	Quest
LIMETTAL	Quest
LIPA	Mona
LIPACIDE	R.T. Vanderbilt
LIPAMIDE	Lipo Chemicals
LIPAMINE	Lipo Chemicals
LIPITEIN	Costec
LIPO	Lipo Chemicals
LIPOBEE	Lipo Chemicals
LIPOCOL	Lipo Chemicals

Trade Name	Supplier
LIPOLAN	Lipo Chemicals
LIPOMULSE	Lipo Chemicals
LIPONATE	Lipo Chemicals
LIPONIC	Lipo Chemicals
LIPOPEG	Lipo Chemicals
LIPO POLYOL	Lipo Chemicals
LIPOPROTEOL	R.C. Vanderbilt
LIPOQUAT	Lipo Chemicals
LIPOSORB	Lipo Chemicals
LIPOVOL	Lipo Chemicals
LIPOWAX	Lipo Chemicals
LIQUID LITE	Eastman Chemical
LIXETONE	Quest
LUBRAJEL	United Guardian (Freeman Industries)
LUBRAQUAT	United Guardian (Freeman Industries)
LUBRASLIDE	United Guardian (Freeman Industries)
LUNACERA	H.B. Fuller
LUNACERA ALBA	H.B. Fuller
LUNACERIN	H.B. Fuller
LUSANTAN	BASF
LUSTRA-PEARL	Van Dyk
LUSTRA-PEARL GLIMMER	Van Dyk
LUSTRA-PEARL SILK	Van Dyk
LUTROL	BASF
LUVIQUAT	BASF
LUVISET	BASF
LUVISKOL	BASF
LUVITOL	BASF
MACALOID	Rheox
MACEAL	Quest
MACKADET	McIntyre
MACKALENE	McIntyre
MACKAM	McIntyre
MACKAMIDE	McIntyre
MACKAMINE	McIntyre
MACKANATE	McIntyre
MACKERNIUM	McIntyre
MACKESTER	McIntyre
MACKINE	McIntyre
MACKOL	McIntyre
MACKPEARL	McIntyre
MACKPRO	McIntyre
MACKSTAT	McIntyre
MANZANATE	Quest
MAPROSYL	Stepan
MARLAMID	Huls
MARLINAT	Huls
MARLIPAL	Huls
MARLON	Huls
MARLOPHOR	Huls
MARLOPON	Huls
MARLOSOL	Huls
MARLOWET	Huls
MARLOX	Huls
MASIL	PPG/Mazer
MASILWAX	PPG/Mazer
MAYPON	Inolex
MEARLMAID	Mearl
MEDIALAN	Hoechst Celanese
MEFRANAL	Quest

Trade Name	Supplier
MELONIS	Quest
METHOCEL	Dow Chemical
MEVANTRAAL	Quest
MICHEL	M. Michel
MIGHTY SOFT	Eastman Chemical
MIKROKILL	Brooks Industries
MILKAMINO	Brooks Industries
MINERAL COLLOID	ECC America
MIRANATE	Miranol
MIRANOL	Miranol
MIRAPOL	Miranol
MIRATAINE	Miranol
MODULAN	Amerchol
MONAMATE	Mona
MONAMID	Mona
MONAMINE	Mona
MONAQUAT	Mona
MONATERIC	Mona
MONAWET	Mona
MONOSET	Eastman Chemical
MOSS BASE	Quest
MOUSSE DE MER	Quest
M-QUAT	PPG/Mazer
MULSIFAN	Zschimmer & Schwarz
MULTIWAX	Sonneborn
MUSK	Quest
MUSOL	Brooks Industries
MYACIDE	Inolex
MYRJ	ICI
MYTAB SUMQUAT	Hexcel
MYVACET	Eastman Chemical
MYVAPLEX	Eastman Chemical
MYVATEM	Eastman Chemical
MYVATEX	Eastman Chemical
MYVEROL	Eastman Chemical
NATROSOL	Aqualon
NATURECHEM	CasChem
NEOBEE	Stepan
NEOBERGAMATE	Quest
NEOBERGAMATE FORTE	Quest
NEOBERGAMATE SPECIAL	Quest
NEODOL	Shell Chemical
NEO HELIOPAN	Haarman & Reimer
NEOLAVANDATE	Quest
NEWSULFUR	TRI-K
NINOL	Stepan
NOVOL	Croda
NU-ANTIQUE	Mearl
NUOSEPT	Costec
OCTOPIROX	Hoechst Celanese
OHLAN	Amerchol
OLEO-COLL	Brooks Industries
OLEO-KERATIN	Brooks Industries
OLEO-SOY	Brooks Industries
ORGASOL	Lipo Chemicals
ORTHOLATE	Quest
OXETAL	Zschimmer & Schwarz
OXONE	DuPont
OXYPON	Zschimmer & Schwarz

Trade Name	Supplier
PALATINOL	BASF
PAROL	Penreco
PARSOL	Givaudan
PEARL-GLO	Van Dyk
PEARLWHITE	Mearl
PELARGENE	Quest
PENETECK	Penreco
PENRECO	Penreco
PENRECO REGENT	Penreco
PENRECO ROYAL	Penreco
PENRECO SNOW	Penreco
PENRECO SUPER	Penreco
PENRECO ULTIMA	Penreco
PEPTEIN	Geo. A. Hormel
PERFECTA	Sonneborn
PETIOLE	Quest
PETRONAUBA	Petrolite
PIVAROSE	Quest
PLASMANECTIN	TRI-K
PLURONIC	BASF
POLAWAX	Croda
POLECTRON	GAF Chemicals
POLYCHOL	Croda
POLYJEL	United Guardian (Freeman Industries)
POLYLAN	Amerchol
POLYMIST	Allied Signal
POLYPEPTIDE	Inolex
POLYPRO	Costec
PRENYL ACETATE	Quest
PROCETYL	Croda
PROCHEM	Protameen
PROCOL	Protameen
PROLAGEN	Costec
PROMYRISTYL	Croda
PRONECTIN	TRI-K
PROSTEARYL	Croda
PROTACHEM	Protameen
PROTALAN	Protameen
PROTAMATE	Protameen
PROTAMIDE	Protameen
PROTAN	Protameen
PROTAPHOS	Protameen
PROTAPON	Protameen
PROTAQUAT	Protameen
PROTASORB	Protameen
PROTEGIN	Goldschmidt Chemical
PROTERIC	Protameen
PROTOPET	Sonneborn
PROTOX	Protameen
PROVOL	Croda
PURECO	Capital City Products
PUR-OBA	Goldschmidt Chemical
PURTON	Zschimmer & Schwarz
QUADROL	BASF
QUAMECTANT	Brooks Industries
QUANTUM	Quantum Chemical
QUAT-COLL	Brooks Industries
REGAL	Cyprus
RETICUSOL	Croda

Trade Name	Supplier
REWOTERIC	Rewo
RHODIGEL	R.T. Vanderbilt
RHODOPOL	R.T. Vanderbilt
RHUBACITRIL	Quest
RHUBAFURAN	Quest
ROCHE	Roche
ROSANIA	Quest
SAFESTER	Lipo Chemicals
SAIB	Eastman Chemical
SANDOPAN	Sandoz
SAUTANE	Quest
SCHERCAMOX	Scher
SCHERCEMOL	Scher
SCHERCODINE	Scher
SCHERCOMID	Scher
SCHERCOPOL	Scher
SCHERCOQUAT	Scher
SCHERCOTAINE	Scher
SCHERCOTERIC	Scher
SCHEROBA	Scher
SERDET	Huls
SERUMPRO	Brooks Industries
SERVO AMFOLYT	Huls
SERVOXYL	Huls
SETACIN	Zschimmer & Schwarz
SHEERSKIN	United Guardian (Freeman Industries)
SICOMET	BASF
SICOMET INDIGOTINE	BASF
SILKALL	Paninkret Chemicals (Freeman Industries)
SILKPRO	Paninkret Chemicals (Freeman Industries)
SILKWHITE	Mearl
SILTEK	Petrolite
SILTEX	Petrolite
SINOCITRIL	Quest
SIPEX	Alcolac
SIPON	Alcolac
SIPONATE	Alcolac
SIPONIC	Alcolac
SIPOTHIX	Alcolac
SKIN LITE	United Guardian (Freeman Industries)
SKLIRO	Croda
SOLAN	Croda
SOLARSCREEN	CasChem
SOLLAGEN	Costec
SOLU-COLL	Brooks Industries
SOLULAN	Amerchol
SOLU-LASTIN	Brooks Industries
SOLU-MAR	Brooks Industries
SOLU-SILK	Brooks Industries
SOLU-SILK PROTEIN	Brooks Industries
SOLU-SOY	Brooks Industries
SOLU-VEG	Brooks Industries
SONNEBORN	Sonneborn
SONO JELL	Sonneborn
SONORA	Jojoba Growers
SORBISTAT	Pfizer
SORBO	ICI
SPAN	ICI
SPARKLE	Mearl
SPECTRA-PEARL	Van Dyk

Trade Name	Supplier
STABLETS	Pfizer
STAR	Procter & Gamble
STARFOL	Sherex
STEDBAC SUMQUAT	Hexcel
STEOL	Stepan
STEPANHOLD	Stepan
STEPANHOLD EXTRA	Stepan
STEPAN-MILD	Stepan
STEPANOL	Stepan
STEPANQUAT	Stepan
STEROTEX	Capital City Products
SULFETAL	Zschimmer & Schwarz
SUMQUAT	Hexcel
SUNVEIL	TRI-K
SUPER CORONA	Croda
SUPERFINE	Croda
SUPER HARTOLAN	Croda
SUPEROL	Procter & Gamble
SUPERPRO	Inolex
SUPER REFINED	Croda
SUPER SOLAN	Croda
SUPER SOLANGEL	Croda
SUPER STEROL	Croda
SUPERTI	United Guardian (Freeman Industries)
SUPRA	Cyprus
SUPRAFINO	Cyprus
SUPREME	Cyprus
SURFACTOL	CasChem
SURFINE	Finetex
SUTTOCIDE	Sutton Laboratories
SYLOID	W.R. Grace
SYLOX	W.R. Grace
SYNCROWAX	Croda
TAGAT	Goldschmidt Chemical
TAKANAL	TRI-K
TANGENIL	Quest
TAURANOL	Finetex
TAURATE	Finetex
TECH PET	Sonneborn
TECQUINOL	Eastman Chemical
TEGACID	Goldschmidt Chemical
TEGAMINE	Goldschmidt Chemical
TEGIN	Goldschmidt Chemical
TEGINACID	Goldschmidt Chemical
TEGO	Goldschmidt Chemical
TEGO-CARE	Goldschmidt Chemical
TENOX	Eastman Chemical
TETRAHYDROCONVALOL	Quest
TETRONIC	BASF
TEXTURE LITE	Eastman Chemical
THIXCIN	Rheox
TIC PRETESTED	Tic Gums, Inc.
TIMICA	Mearl
TOPNOTE	Cyprus
TOP ROSE	Quest
TRANSJOJOBA	Jojoba Growers
TRASEOLIDE	Quest
TRI-COL	TRI-K
TRI-DERM	TRI-K
TRI-K	TRI-K

Trade Name	Supplier
TRI-K SOYPRO	TRI-K
TRILANE	TRI-K
TRI-LASTIN	TRI-K
TRILON	BASF
TRI-QUAT	TRI-K
TRISEPT	TRI-K
TRISTAT	TRI-K
TRI-TEIN	TRI-K
TUBEROSE	Quest
TWEEN	ICI
TYLOSE	Hoechst Celanese
ULTRAFINO	Cyprus
ULTRAHOLD	BASF
ULTRATI	United Guardian (Freeman Industries)
UNICIDE	Lipo Chemicals
UNILIN	Petrolite
UNIPABOL	Lipo Chemicals
UNIPERTAN	Lipo Chemicals
UNIPHEN	Lipo Chemicals
UNITHOX	Petrolite
UNITRIENOL	Lipo Chemicals
UNITWIX	United Guardian (Freeman Industries)
UVATONE	Lipo Chemicals
UVINUL	BASF
VANCIDE	R.T. Vanderbilt
VANSEAL	R.T. Vanderbilt
VARAMIDE	Sherex
VARIFOAM	Sherex
VARION	Sherex
VARIQUAT	Sherex
VARISOFT	Sherex
VARONIC	Sherex
VAROX	Sherex
VARSULF	Sherex
VEEGUM	R.T. Vanderbilt
VELSAN	Sandoz
VELTOL	Pfizer
VELTOL PLUS	Pfizer
VELVET	Mearl
VERAGEL	Dr. Madis
VERDALIA	Quest
VERDILYN	Quest
VERDINAL	Quest
VERDORACINE	Quest
VERTAL	Cyprus
VERTELON	Quest
VICTORY	Petrolite
VIGOROSE	Quest
VIOTRIL	Quest
VISCASIL	GE Silicones
VITA-COS	CasChem
VOLPO	Croda
VYBAR	Petrolite
WAXENOL	CasChem
WECOBEE	Stepan
WICKENOL	CasChem
WITCAMIDE	Witco
WITCO	Sonneborn/Witco

Trade Name	Supplier
WITCOLATE	Witco
WITCONATE	Witco
WITCONOL	Witco
YLANG	Quest
ZETESOL	Zschimmer & Schwarz

Other Noyes Publications

EMULSIFYING AGENTS
An Industrial Guide

by

Ernest W. Flick

This book describes more than 1500 emulsifying agents which are currently available for industrial usage. It has been compiled from information received from numerous manufacturers and distributors of these products.

The data included represent selections from manufacturers' descriptions, made at no cost to, nor influence from, the makers or distributors of the materials. Only the most recent information has been included. It is believed that all of the products are currently available, which will be of utmost interest to readers concerned with product discontinuances.

Emulsifying agents find uses in industries ranging from food preparation and processing to drilling fluids, from cosmetics and pharmaceuticals to heavy duty cleaners and degreasers for metal finishing. Textile manufacturers, pulp and paper processors, and agricultural product manufacturers also account for large quantities of emulsifiers. Another important application is emulsion polymerization systems which yield coatings, polishes, sealants and adhesives. Obviously, the market for these products is broad and new applications are constantly being developed.

Types of raw materials used for emulsifying agents include alkoxylates, alkanolamides, fatty acid esters, imidazolines, lecithins, quaternaries, phosphate esters, natural gums, glycerides, and sulfonates.

The book lists the following product information:

(1) Company name and product category,
(2) Trade name and product number,
(3) Product description including properties and applications, as presented by the supplier.

The following companies are represented in the book:

Akzo Chemicals Inc.
Alcolac
Alkaril Chemicals Ltd.
Amerchol Corp.
American Cyanamid Co.
Angus Chemical Co.
BASF
Capital City Products Co.
CasChem Inc.
Central Soya
W.A. Cleary Products, Inc.
Climax Performance Materials Corp.
Cyclo Chemicals Corp.
DeSoto, Inc.
DuPont Co.
Eastman Chemical Products, Inc.
FMC Corp.
Goldschmidt Chemical Corp.
Grindsted Products, Inc.
Gumix International
Harcros Chemicals Inc.
Henkel Corp.
Hercules Inc.
Humko Chemical Division
ICI Specialty Chemicals
Jordan Chemical Co.
Lanaetex Products, Inc.
Lipo Chemicals Inc.
Lonza Inc.
Mayco Oil & Chemical Co.
Mazer Chemicals
Milliken Chemicals
Mobay Chemical Corp.
NL Chemicals
Patco Products
Quantum Chemical Corp.
Sandoz Chemicals
ScanRoad Inc.
Shell Chemical Co.
Werner G. Smith, Inc.
Sonneborn Division
Stepan Co.
Unichema Chemicals, Inc.
Van den Bergh Food Ingredients Group
Westvaco Corp.
Witco Corp

ISBN 0-8155-1225-2 (1990) **6" x 9"** **324 pages**

Other Noyes Publications

COSMETIC AND TOILETRY FORMULATIONS
Second Edition

by

Ernest W. Flick

More than 1800 cosmetic and toiletry formulations are described in the Second Edition of this well-received and useful book. The book is based on information obtained from industrial companies and other organizations. The data represent selections from manufacturers' descriptions, in their own words, made at no cost to, nor influence from, the makers or distributors of these materials.

Only the most recent formulas have been included. It is believed that all of the trade-named raw materials listed are currently available, which will be of utmost interest to readers concerned with raw material discontinuances.

Spurred by a strong economy, cosmetic and toiletry sales, a multibillion dollar market, have been increasing at 6 to 7% annually, thus making the information in the book particularly interesting to anyone considering new products or process variations.

Each formulation in the book is categorized by a description of its end use. The formulations include the following: a list of each suggested raw material; the percent by weight of each raw material; a recommended formulation procedure; and the formula source, which is the company or organization that supplied the formula.

The formulations in the book are divided into fourteen categories as shown below. In addition a valuable section on **Trade-Named Raw Materials** is included, which lists trade-names, a brief chemical description, and the supplier's name. The final section contains **Suppliers' Addresses** and will no doubt be a useful tool to the reader.

Section titles are listed below. Parenthetic numbers indicate the number of formulations per topic.

1. **Antiperspirants and Deodorants (44)**
2. **Baby Products (64)**
3. **Bath and Shower Products (138)**
4. **Beauty Aids (189)**
5. **Creams (230)**
6. **Fragrances and Perfumes (12)**
7. **Hair Care Products (381)**
8. **Insect Repellents (11)**
9. **Lotions (157)**
10. **Shampoos (206)**
11. **Shaving Products (52)**
12. **Soaps (91)**
13. **Sun Care Products (180)**
14. **Miscellaneous (82)**
15. **Trade-Named Raw Materials**
16. **Suppliers' Addresses**

ISBN 0-8155-1218-X (1989) **6"x9"** **964 pages**

Other Noyes Publications

INDUSTRIAL SURFACTANTS

by

Ernest W. Flick

This book describes almost 3500 surfactants which are currently available for industrial use. The book will be of value to technical and managerial personnel involved in the specification and use of these products.

Industrial surfactants find uses in almost every industry, from asphalt manufacturing to carpet fibers, from petroleum production to leather and fur processing. Examples of the types of chemicals used as surfactants are fatty alcohol sulfates, alkylolamides, alkoxylates, sulfosuccinates, amines, quaternaries, phosphate esters, acid esters, block copolymers, betaines, imidazolines, alkyl sulfonates, etc. The market for these products has, and is expected to continue to have, a steady growth rate, of the order of 4 to 5%, and in some cases up to 10%, per year.

The data included represent selections from manufacturers' descriptions made at no cost to, nor influence from, the makers or distributors of the materials. Only the most recent information has been included. It is believed that all of the products listed are currently available, which will be of utmost interest to readers concerned with product discontinuances.

Products are presented by company, and the companies are listed alphabetically. Also included are a **Trade Name Index** and a list of **Suppliers' Addresses.**

The book lists the following product information, as available, in the manufacturer's own words: company name and product category, trade name and product numbers, plus a description of the product, as presented by the supplier.

The following companies are represented in the book:

Air Products and Chemicals
Akzo America, Inc.
Akzo Chemie America
Albright & Wilson, Inc.
Alcolac, Inc.
American Cyanamid Co.
American Hoechst Corp.
Angus Chemical Co.
BASF Corp.
Capital City Products Co.
Central Soya/ Chemurgy Division
Continental Chemical Co.
Croda, Inc.
DeSoto, Inc.
Dow Chemical U.S.A.
DuPont Co.
Emery Industries, Inc.
Emulsion Systems, Inc.
Ethyl Corp
Exxon Chemical Co./ Tomah Products
GAF Corp.
Goldschmidt Chemical Corp.
W.R. Grace & Co.
Henkel Corp.
ICI Americas, Inc.
Inolex Chemical Co.
Jordan Chemical Co.
Lonza, Inc.
Mazer Chemicals, Inc.
McIntyre Chemical Co.
Miranol, Inc.
Mona Industries, Inc.
Monomer-Polymer and Dajac Laboratories, Inc.
Monsanto Co.
Niacet Corp.
Olin Chemicals
Pilot Chemical Co.
Rohm and Haas Co.
Ruetgers-Nease Chemical Co.
Samson Chemical Co., Inc.
Sandoz Chemicals Corp.
Scher Chemicals, Inc.
Shell Chemical Co.
Sherex Chemical Co.
Stepan Co.
3M Corp.
Texaco Corp.
Arthur C. Trask Corp.
Union Carbide Corp.
R.T. Vanderbilt Co., Inc.
Vista Chemical Co.
Westvaco Chemicals Division
Witco Corp.

ISBN 0-8155-1173-6 (1988) **6"x9"** **743 pages**